How the Mind Comes Into Being

How the Mind Comes Into Being

Introducing Cognitive Science from a Functional and Computational Perspective

Martin V. Butz and Esther F. Kutter

Department of Computer Science and Department of Psychology
Faculty of Science, Eberhard Karls University of Tübingen
Tübingen, Germany

OXFORD

UNIVERSITY PRESS

OXFORD
UNIVERSITY PRESS

Great Clarendon Street, Oxford, OX2 6DP,
United Kingdom

Oxford University Press is a department of the University of Oxford.
It furthers the University's objective of excellence in research, scholarship,
and education by publishing worldwide. Oxford is a registered trade mark of
Oxford University Press in the UK and in certain other countries

First Edition published in 2017
Impression: 1

Published in the United States of America by Oxford University Press
198 Madison Avenue, New York, NY 10016, United States of America

British Library Cataloguing in Publication Data
Data available

Library of Congress Control Number: 2016946807

ISBN 978–0–19–873969–2

To our families

Preface

More than 2000 years ago — maybe as the first humans in the world — Greek philosophers have thought about the puzzling introspectively assessed dichotomy between our physical bodies and our seemingly, non-physical, mental minds. How is it that we can think highly abstract thoughts, seemingly fully detached from the actual, physical reality? Despite the obvious interactions between mind and body (we get tired, we are hungry, we stay up late despite being tired, etc.), until today it remains puzzling how our mind controls our body, and vice versa, how our body shapes our mind.

This textbook focuses on the embodiment of the mind in all its facets. Many other books on cognitive science focus on sensory information processing, motor output generation, reasoning, symbol processing, and language in separate rather disconnected chapters. This book integrates these aspects of cognition sketching-out their interdependencies and their potential ontogenetic (over a life-time) and phylogenetic (over centuries, millennia, and beyond) development. As a consequence, after giving a multidisciplinary background on the development of cognitive science as a research discipline, the book starts from an evolutionary developmental perspective, shedding light on how behavior traits and genetic codes may have developed on our earth. Next, we proceed with cognitive development during a life-time, focusing on reinforcement learning and the development of flexible behavioral capabilities by learning about and selecting amongst redundant alternative interactions with the environment. An excursion into a functional perspective on the human brain then leads to principles of perception and action generation, including the relations between these principles, leading to multisensory and multi-information interactions. Processes of attention and the different aspects of attentional processes then lead to principles of decision making, language, and abstract thought. One goal of the book is thus to sketch-out a developmental pathway towards higher, symbolic, abstract cognitive capabilities starting bottom-up with principled processes of bodily representations and body control, offering a novel perspective on cognitive science by taking a sensorimotor, integrative, embodied stance.

Meanwhile, the textbook offers a principled mathematical, functional, and algorithmic background about the processes that apparently underly neurocognitive representations, processes, adaptations, and learning in our brains. In particular, when considering phylogenetic, evolutionary development of the human mind, we introduce genetic algorithms and their functionality, revealing potentially rather simple developmental pathways and contrasting these with situations of highly unlikely evolutionary developments. Moving to reward-oriented adaptations of behavior, reinforcement learning is introduced and analyzed once again in terms of its capabilities and challenges. Interestingly, evolutionary learning algorithms appear to have much in common with a particular type of reinforcement learning algorithm. We also show, nonetheless, that our behavior and behavioral decision making cannot be purely reward-driven. We are able to make context-dependent decisions and choose alternative environmental interactions given current circumstances. Thus, we are highly flexible in our interactions with the environment. Where does the flexibility of our mind come from? We show how it is possible to alter interactions on demand – so to say, effectively doing what seems right given the current context – by means of abstract, predictive models.

With cognitive flexibility in mind, we then have a look at the current neuroscientific knowledge from a functional perspective. Several of the rather dedicated modules in our brain are then further evaluated and analyzed in further computationally-oriented detail. In particular, we proceed by bottom-up visual information processing and the possibility to include top-down influences into this processing stream. Once again computationally oriented, we show how such interactive information processing can be accomplished by means of Bayesian probability and predictive coding principles. The same principle also applies when information from various sensors needs to be integrated into one fused percept of both the environment and the own body. To interact flexibly and selectively with mental concepts, attention for behavior needs to be enhanced to principles of internal attentional processes, which select those mental representations that are currently most appealing. Once again, the bottom-up processes are combined with top-down processes to guide information processing for behavioral- and mental control.

Equal processes expand to principles of planning and decision making, for which an embodied value system needs to be integrated, and finally to language and abstract thought. Even with respect to language, computational principles can explain aspects of language evolution, including benefits of communication, the need to individualize your conversation partners, and principles of basic grammatical structures. Moreover, structures that are highly suitable for accomplishing behavioral flexibility and enabling more complex planning and decision making may determine the principle underlying the universal grammar, offer solutions to the symbol grounding problem, and bootstrap the very flexible compositionality offered by human languages.

In closing, we hope this book provides a new, intriguing perspective on cognitive science. While being a textbook with educational contents, the book also aims at shedding light onto the recent movement towards embodied cognitive science, offering answers to the questions of what embodied cognitive science may mean and what it may imply.

Acknowledgments

This book would not have come into being without the help of many people, including our families, friends, and many colleagues. Thank you for all the support and understanding.

As the first author, I would like to particularly thank my colleagues at the University of Würzburg for supporting me during the time when I gave the lecture – upon which large parts of the book are based on – for the first time in the summer term of 2008. In particular, professor Frank Puppe has encouraged me to present my perspective on artificial intelligence and cognitive systems back then. Moreover, professor Joachim Hoffmann has always provided his complete trust and support throughout my time at the University of Würzburg and beyond that until now. I am immensely grateful for all the inspiring discussions, comments, and constructive criticism throughout my career. Only due to his thorough and farsighted work in cognitive psychology – and beyond that towards cognitive science – have I understood and internalized the importance of predictions and anticipations for cognition. I am equally in debt to professor David E. Goldberg for all his trust and support throughout my time at the University of Illinois at Urbana-Champaign, IL (UIUC) and beyond that. Due to the highly inspiring discussion and analytical sessions with him, I have learned – amongst many other things – to appreciate the importance of understanding a complex system's functionality by means of facet-wise analysis. I would also like to thank my colleagues at the University of Tübingen for supporting my work here in many ways. Finally, I would like to thank my wife for her understanding for all the late nights and her continuous support, and my kids for always welcoming me home and reminding me in their beautiful ways that a very important part of life has nothing to do with research.

As the second author, I would like to express my deepest gratitude to Martin for all the trust and support he put in me. In countless passionate and constructive discussions with him he helped me understand how to see and forge links between all the different aspects of cognitive science. During the past years in his team I have internalized the impact and importance of a holistic, unifying understanding of cognition in all its facets. I am immensely grateful that he offered me this unique opportunity to become coauthor of this book. Finally, I am so grateful to my family, who provided me with unfailing support and continuous encouragement throughout the project.

Comments and suggestions for the actual book came from numerous people. We found the enthusiasm that was returned when we detailed the planned book's content particularly encouraging. Several anonymous reviewers from OUP have provided very useful suggestions including to keep a balanced perspective, which is hopefully maintained successfully. Moreover, we would like to thank the Cognitive Modeling team for reading book chapters and providing very useful contents suggestions. We are particularly grateful also for PD Oliver Herbort, who managed to give detailed and highly valuable comments on the first chapters in these busy times. Moreover, we would like to thank Carolyn Kinney for proofreading the book in immense detail and very thoroughly and also Kenneth Kinney for his detailed proofreading effort.

In closing, we would like to thank professor Harold Bekkering, professor Wayne Gray, and professor Karl Friston for their wonderful book endorsements and all the support beyond that. We very much hope, of course, that the book will be well-received by many other researchers, students, and anybody with the longing to understand "how our minds come into being."

Contents

List of Tables

Chapter 1

Embodied Cognitive Science

1.1 Introduction

Over the last twenty years or so it has become increasingly clear that purely symbolic approaches to cognitive science, as well as to artificial intelligence are insufficient to grasp the intricate beauty of our minds. We are able to draw analogies and understand metaphors that go far beyond symbolic forms of representations and associations between them. For example, we immediately understand a phrase such as our "circle of friends", although our friends never really come together in a circle or form a circle. Nonetheless, the term "circle" suggests that our group of friends forms a close and interactive, circular relationship. Similarly, we understand the phrase "grasping an idea", although an idea is not a physical object that can be grasped with our hands. Nonetheless, "grasping" suggests in this context that we got hold of an idea and can thus handle it, manipulate it, and think about it.

Thousands of examples can be made in this respect (Bergen, 2012; Johnson, 1987; Lakoff, 1987; Lakoff & Johnson, 1999), suggesting that our brain cannot be a *symbol processing machine*. Rather, it seems that our brain may be characterized as a *highly distributed, parallel, neuro-dynamic system*. The problem with this latter characterization, however, is that the functionality of such a system is very hard to imagine. How are analogies understood or drawn by such a system? How can metaphors be understood by a highly distributed, parallel neuro-dynamic cognitive architecture?

Analogy making, but also many other examples of innovative and intricate, interactive thoughts, and behavior have led to the development of *embodied approaches* to cognition. This turn to embodiment, indeed, has established itself or has at least been discussed in various disciplines, including philosophy, neuroscience, psychology, artificial intelligence, and the overarching field of cognitive science. Although an *embodied mind* that is *grounded* in *sensorimotor experiences* may sound appealing, many open questions remain: What are the implications of an embodied mind? How may embodied, sensorimotor processes and representations lead to suitable abstractions and thus "higher-level" cognition? How embodied is the human mind really?

The aim of this cognitive science textbook is to shed light on these questions by introducing the basic concepts that constitute embodied cognitive science, and sketching-out relations of these concepts and their involvement in shaping our human minds.

Two particularly puzzling facts are discussed in this chapter to further motivate the need for embodied cognitive science approaches: the fact that the brain controls the body and the second fact that, vice-versa, the body controls the brain. In order to establish this symbiosis, embodied cognitive science suggests that the body and the surrounding world in which we grow up and live in shapes the brain such that effective, bidirectional body–brain interactions become possible. In order for us to understand such interactions and implications on a deeper level, we strongly stress that cognitive science needs also to be approached by means of computational techniques in the pursuit of a functional perspective.

We believe that observations of particular, intricate cognitive interactions — such as those revealed by analogies — alone are not sufficient to foster a deeper understanding of how the mind may come into being. Rather, only with computational models based on mathematical and information-theoretic principles, cognitive processes can be understood. Thus, while sketching-out how the mind may come into being, we detail computational principles that need to be involved to realize the different cognitive aspects necessary to reach our abstract reasoning, language, and thought capabilities.

1.2   Our brain controls our body

From an introspective, self-conscious perspective probably every one of us is convinced that we are responsible for our own bodily decisions and actions. This seemingly obvious fact, however, has been highly challenged by philosophers and neuroscientists for decades. One particular challenge was put forward by experiments conducted by the physiologist and psychologist Benjamin Libet (1916–2007). Using brain-imaging techniques (EEG in this case, cf. Section 7.5.3), Libet published the results of a series of experiments in 1979, which indicate that the decision, when to press a button, was made by the brain before the participant claimed that he or she had made the actual decision. This puzzling contradiction between conscious decision making and the actual brain activities that are involved in the decision-making process has led to many heated discussions and their implications are still being questioned, reinterpreted, and debated today.

Nonetheless, it is clear that the neural signals that control our body (especially our muscles, but also glands and other organs to certain extents) come from our brain. By means of our brain, we make decisions and accordingly activate behavior by moving our body in our world. Mostly fully subconsciously, our brain controls our heartbeat. Breathing is controlled consciously and subconsciously to a certain extent. Even our intestines are luckily not fully controlled by the organs alone, but they also communicate with the brain to coordinate certain needs, such as when the stomach indicates hunger. Even the circadian rhythm, that is, the night-time-day-time rhythm, which makes us sleepy at night and keeps us somewhat awake during the day (even when we are actually really tired) is controlled mainly subconsciously by our brain.

In addition, even those behaviors that we seem to choose consciously, such as grasping, walking, or writing, are controlled to a much lesser degree by conscious processes as we typically tend to think. Actually, consciously thinking about it typically somewhat slows down or even disrupts automatic behavior, indicating that conscious control is typically much slower than subconscious control processes. Meanwhile, it seems that well-automatized control processes have access to bodily details that are hardly accessible by conscious awareness. Try to walk a few steps while controlling every single muscle in the legs! It seems impossible because the control access to our muscles is very indirect. Is this a problem? Should we be scared of falling much more than we are because we cannot really control our muscles fully consciously? Obviously not!

Indeed, we have to be more scared of stumbling when we suddenly become aware of our walking or our posture, for example, when we sometimes become aware that we are being watched. We will see that it is highly beneficial for our cognitive capabilities that lots of our behavioral capabilities are automatized, thus actually relieving our higher-level cognitive capabilities from the burden of taking care of such low-level control demands. What is accessed by cognitive control are somewhat abstracted encodings of our behavioral capabilities. Shedding light on the nature of these encodings and mechanisms, and thus on how our brain controls our body, can give us insights about how abstracted forms of cognition may develop in our brain.

1.3 Our body controls our brain

Also the other direction seems obvious, but it implies rather different aspects about the functionality of our brain. Our stomach indicates that we are hungry and leads us to search for something to eat, to prepare food, and to eat. When our body indicates pain, we direct attention to this uncomfortable pain and attempt to relieve this uncomfortable state of mind. Pain thus causes discomfort, which in turn leads to behavior that we believe will help ease this discomfort. Even subtle bodily postures, such as when we smile – regardless if intentionally or unintentionally – induces a somewhat more positive perspective on our environment – leading, for example, to a more positive classification of cartoon figures (Strack, Martin, & Stepper, 1988). Thus, states of our body, as well as perceived states of the surrounding environment generally influence our brain activity.

Over the last two decades or so, however, a much more far-reaching perspective has started to establish itself in the cognitive science community and beyond. This "embodiment turn" in cognitive science suggests that our cognitive capabilities are highly dependent on bodily processes and on the environment with which our body continuously interacts (Ballard, Hayhoe, Pook, & Rao, 1997; Barsalou, 1999; Clark, 1999, 2013; Engel, Maye, Kurthen, & König, 2013). In particular, it suggests that the representations and processes found in the brain and their ontogenetic development over a lifetime are based on the body with its bodily morphology, including its sensory and motor capabilities. As a consequence, seemingly highly abstract, somewhat symbolic cognition is grounded in and has developed from sensorimotor, bodily interactions with and experiences about our environment. As a result, abstract cognition inherently reflects the structure of our world and cognition is inherently adaptive (Anderson & Schooler, 1991). This book thus focuses on how and which abstract encodings and processes may develop, based on experiences about our own body and about interactions with other objects, entities, animals, and humans.

1.4 Our body and our world shape our brain

The main implication for cognitive science itself then is essentially the need to not only consider abstract, symbolic thought as the field of study, but to also work on revealing how our brain controls our body and how our body controls our mind. In order to become a success story, and to ultimately explain how our mind comes into being, it is necessary to understand the development of language, abstract cognition, and conscious thought based on our sensory and behavioral capabilities, and the experienced interactions with our world.

Even though our personality traits are predisposed by our genes, they are shaped by the body and the experienced environment. Starting as a fetus, and continuing until and throughout adulthood, cognitive development of the brain is inevitably grounded in a body that experiences the outside world. Our mind grows and develops while interacting with and by means of interacting with the world, and in particular with the present, particular structures in the world. These worldly structures by no means only encompass physics and objects, but also include other humans, especially our friends and family, and the societal and cultural structures. Clearly, education, communication, scripture, photography, TV and other media, and generally all sorts of training of a certain expertise also shape our mind, and our capabilities to perceive and think about our world, including present, past, and future. Thus, myriad worldly influences, mediated by our body, influence the way we think, the way we perceive ourselves, including our perceived place in society, and our abstract thought capabilities. Intelligence and consciousness certainly do not exist from birth in their adult form.

Although already born with many clever mechanisms and capabilities, these capabilities develop further and are shaped by the available experiences. The ultimate goal of cognitive science is to unravel the development of our mind, and the concurrent development of brain structures and neural representations dependent on the body and the environment in which each of us grows up, which each of us experiences, and which each of us actively explores given individual behavioral capabilities, cognitive priorities, and environmental stimulations.

1.5   Our brain develops for a purpose

When considering such an interactive, developing mind from an ontogenetic (during a lifetime) and phylogenetic (over generations) developmental perspective (cf. Chapter 4), it has to be kept in mind that the brain does not develop for the purpose of being intelligent or being able to think symbolically per se. Rather, the massive, disproportional brain development in humans must inevitably have a purpose, that is, it must have been and most likely still is beneficial in terms of survival and reproduction. Seeing that most other species are not as capable as humans to have abstract thoughts, intelligence does not appear to be the ultimate goal that evolution strives-for, and also not the ultimate solution for ensuring survival and reproduction.

Besides this phylogenetic, evolutionary puzzle, the following ontogenetic puzzle may be considered the most challenging one in cognitive science: how can abstract thought develop so naturally in most human beings? How is it that all of us start smiling, crawling (or variations of it), and walking along our developmental path toward being a toddler? How is it, that we become language-ready and are capable of learning any language while being a toddler, seemingly without any extraordinary mental effort? How do we manage to integrate ourselves in our communities, typically without any major or even deadly conflicts? Embodied cognitive science emphasizes that these abilities develop naturally because of our particular genetic predispositions, which influence the way in which the brain builds progressively more abstract, higher-level structures given embodied, sensorimotor experiences.

Moreover, embodied cognitive science emphasizes that our cognitive development is driven by behavior, which in turn is driven by goals, which in turn are driven by bodily and mental needs. Curiosity, which is already visible in newborns, is a great example: newborns were shown to preferably look at biological motion stimuli, or at least stimuli that somewhat resemble biological motion, when compared with random motion patterns (Bidet-Ildei, Kitromilides, Orliaguet, Pavlova, & Gentaz, 2014; Pavlova, 2012). Also faces are particularly interesting from very early on. Meanwhile, sensorimotor contingencies are explored, such as how ones own hands look when moving them before ones eyes, or how an object can be made to behave in a predictable manner, or how an object sounds when manipulating it. This exploratory behavior starts very early in life—definitely at birth—but it is probably already present while the infant is still in the womb.

Fetuses have been shown to suck on their own thumb and to know quite a lot about their own body when born (Rochat, 2010). These and other results suggest that behavior is goal-directed from very early on (von Hofsten, 2004). Goals can have various forms, may be ill-specified by the brain, or may be highly abstract. Regardless how they are exactly encoded, though, they seem to be responsible for selecting and executing most of our behavior. Thus, cognition may be viewed as selecting goals, rather than behavior. Desirable and seemingly currently achievable goals are the ones that are selected, and that then cause the execution of the behavior that is believed to reach the selected goal.

Thus, the brain develops to achieve goals. Goals are selected and activated in turn for the purpose of satisfying bodily and mental needs. Mental needs in turn appear to be driven by genetic predispositions, such as a curiosity for face-like stimuli, which in turn is grounded in the body, development, and the available sensor and motor capabilities. Thus, from this perspective the brain develops to effectively select and control behavior – not to develop abstract thoughts or intelligence. Abstract thoughts and "higher-levels" of cognition develop in turn to be able to select goals in a maximally efficient and versatile manner, ultimately in order to be able to adapt to and survive in diverse environments.

1.6 Computational knowledge is necessary

While the purpose of this chapter is to motivate embodied cognitive science and sketch-out its implications, a secondary purpose is to motivate the functional, computational approach pursued in this book. For studying cognition, including the involved mental development and behavioral decision making and action control mechanisms, we believe it is absolutely necessary to understand the underlying computational principles. While we hope that the reader may find the previous paragraphs somewhat intuitive and possibly even obvious to a certain extent, they only "describe" the embodied perspective on cognitive science. Details about the implications of this perspective, such as exactly which structures can be expected to develop, and exactly how these structures may develop, have not been given.

In order to show that and how the sketched out development toward higher levels of cognition can actually take place, the involved learning processes and developing encodings need to be specified further by means of computational models and implementations thereof. By detailing and implementing computational mechanisms that can mimic particular aspects of cognitive development and particular cognitive abilities, we can ensure that our brain can implement similar mechanisms, at least in principle. Moreover, once we understand the involved computational mechanisms, we may even be able to build smart machines and intelligent agents in the real world, and possibly also in virtual worlds. In essence, we will gain a deeper and more precise understanding of what cognition really is about. Thus, this book introduces the necessary, fundamental computational mechanisms underlying cognition and embeds them into the embodied perspective of cognitive science.

1.7 Book overview

The book is partitioned into two main parts: chapters 2–6 sketch-out how embodied cognitive science has developed and then detail the key computational mechanisms that govern evolution, ontogenesis, learning, and the development of anticipatory, goal-directed behavior and cognition. Chapter 7 offers an introduction to the neuroscience of the brain in preparation of the second part of the book. Chapters 8–13 then go into more detail about how the mechanisms and challenges put forward in the first part appear to be realized and solved by our brain, moving from basic sensory, multisensory, and sensorimotor control processes, over decision making and behavioral coordination, to language and abstract thought. The concluding chapter (Chapter 14) wraps up *How the Mind Comes Into Being*, pointing-out many aspects that could not be touched upon in detail, but that are nonetheless typically closely related to the covered material and sketching-out an overall perspective on the put-forward material, its inherent interrelations, and its implications.

In particular, the first part initially focuses on the historic development of our knowledge about brain and mind (Chapter 2). The development of the philosophy of mind particularly is discussed in detail, considering the mind–body problem from various perspectives and emphasizing that the mind can only develop by means of a body, which interacts with and thus experiences itself, as well as the outside environment. Based on these assumptions, it is shown why the traditional approach to artificial intelligence, which had focused on logic and symbol manipulations, has contributed only so much toward an explanation of how the mind comes into being. To solve fundamental problems in cognitive science, such as the mind–body problem, but also the symbol-grounding problem, the frame problem, or the binding problem, the embodiment of the mind has to be acknowledged (Chapter 3). Embodied cognitive science essentially puts forward that abstract, symbolic cognition is grounded in sensorimotor experiences. The rest of the book essentially focuses on how abstract, symbolic cognition may actually be grounded in these experiences.

Acknowledging the embodiment perspective, Chapter 4 suggests that the development of the mind–body complex during ontogenesis, starting from conception and going beyond adulthood, needs to be put into focus. Cognition dominantly develops by learning from experiences gathered in interaction with the environment. Evolution has laid-out the right body and the right brain, with sufficient computational capacity and suitable structuring, to develop the cognitive capabilities of a healthy human. Moreover, it appears that evolution has laid out a suitable developmental pathway, along which body and mind co-develop. The chapter thus first provides information about cognitive psychological development. Next, it details evolutionary mechanisms and provides technical details on how evolution appears to work from a computational perspective, highlighting its capabilities, as well as its limits. In conclusion, the paper sketches-out how much evolution may have contributed to human cognition and how actual human cognition appears to develop, co-determined by evolutionary predispositions during ontogenesis.

Cognitive development goes hand-in-hand with behavioral development, such that Chapters 5 and 6 consider development in detail focusing on behavior. Chapter 5 considers reward-dependent learning, covering psychological learning theories as well as theories from machine learning. In particular, the Rescorla–Wagner model is related to reinforcement learning (RL). RL is then introduced in detail, contrasting model-based with model-free RL, as well as value-function learning with direct, policy gradient-oriented behavioral optimization. In the end, we ask the question how far we can get with RL and related techniques, and we acknowledge that in order to be able to act flexibly, goal-directed dependent on the current circumstances, and in order to develop a form of "understanding" of body and environment, knowledge about action effects and thus predictive forward models are needed.

As a consequence, Chapter 6 contrasts different types of forward models and sketches-out some typical learning techniques. More importantly, though, the chapter focuses on the realization of flexible, anticipatory, goal-directed behavior, which is only possible by forward model-based predictions and forward model inversions. Model-based predictions enable forward anticipations, including information filtering, the generation of surprise, and the identification of external entities and agents. The inverse application of forward models results in goal-oriented, *active inference* and can thus generate highly flexible, dexterous behavioral decision making and control processes.

With the help of the neuroscience overview provided in Chapter 7, sensory processing is then considered in detail. Chapter 8 points out that visual processing undergoes a hierarchical cascade of processing stages, extracting redundant and complementary aspects of information about one's own body and the environment from the visual stream. Different information aspects typically contributed in different manners to the extraction about what is going on in the outside environment, providing information about depth, surfaces, edges, colors, motion, and so forth.

Chapter 9 then focuses on the interaction of bottom-up, incoming sensory information with top-down expectations about this information. In particular, *Bayesian, generative, graphical* models are introduced. Along these lines, it is also emphasized that predictions can be forward in time, expecting changes due to motion and forces, but also top-down, expecting corresponding information, such as incoming sensory information. The internal estimations about the outside environment appear to dynamically change over time and they adapt to the experienced regularities in the sensory information. Current models of visual processing thus emphasize the interaction of top-down spatial and feature predictions with bottom-up, incoming sensory information. This interactive information processing principle is also the reason for various types of illusions, some-of-which are surveyed in the end of this chapter.

After having considered a single sensory stream in detail, Chapter 10 emphasizes multisensory interactions, where multiple sensory modalities provide once again redundant and complementary sources of information about body and environment. Thus, generally, similar information processing principles apply as for single sensory processing. However, it is emphasized that the learning of multisensory information first needs to focus on learning the structure of the own body, to be able to transform modal sensory information into other modalities posture-dependently. To enable flexible information transformations, the brain appears to learn multiple spatial encodings and spatial transformations. These enable not only the effective correlation of multimodal sensory information, but also the effective filtering of own bodily motion, as well as robust object and behavior recognition. Besides body-relative spatial encodings, also multisensory *cognitive maps* develop, which support behavioral planning and decision making in navigational tasks.

With the knowledge of all these types of multisensory information, their flexible correlations across frames of reference, the involved spatial encodings, and the Gestalt encodings in hand, the focus then falls on attention in Chapter 11. Attention is essentially necessary to enable the selection of and focus on those information aspects that are currently behaviorally relevant. Seeing that we are, so-to-say, continuously bombarded with a huge amount of sensory information, it appears that our brain needs to filter most of it in order to enable the execution of focused, goal-oriented planning, decision making, and control processes. Starting with psychological phenomena of visual attention, we point-out that *world is its own best model*, such that attention needs to direct cognitive resources only to those aspects of the environment that seem relevant. Dynamic neural fields and dynamic Bayesian information processing are good candidates to realize such attention-based, focused information processing.

One of the most important questions with respect to cognition, however, is how attention decides on what is actually relevant. Chapter 12 shows how this relevance determination is grounded in motor control. Motor control in our brain is, similar to visual information processing, structured in a cascade of interactive processing loops. As a result, higher cognitive processes do not need to fully control or even plan each individual muscle twitch; rather, they can focus on higher-level planning, action decision making, and action control. On this higher-level, *segmentations* of environmental interactions into *events* and *event transitions* become important. Events can be characterized as a short or extended period of time during which particular interactions take place. Event transitions are those cases where these particular interactions change significantly. With suitable segmentations in hand, hierarchical planning and abstract, action- and event-oriented decision making becomes possible. Moreover, relevance becomes determined by means of encoding events, event transitions, and conditional structures, which specify when an event or event transition can typically take place.

With such behavior-oriented abstractions in hand, finally, language comes into play. After providing a short overview over the most important language components, Chapter 13 focuses on how language may be grounded in behavior and the abstractions of behavior for enabling, goal-directed, hierarchical action decision making and control. It turns out that there are several indicators that Chomsky's generative *universal grammar*, may actually be grounded in an *action grammar*, from which progressively higher abstractions can develop. Following the path of language development, we then put forward how language enables a new means to think about the environment, including other humans, animals, and also the self, opening up a new level of cognition. Meanwhile, the symbol-grounding problem is solved by grounding word meanings in the sensorimotor experiences, as well as in the other, already gathered linguistic experiences. Language and cognitive control by means of language thus essentially enables to think on a linguistic level, which is closely tied to the experiences gathered, but which nonetheless enables the detachment of thought from actual, concrete experiences. Thus, the human mind has come into being.

In the concluding chapter, we highlight several important aspects that should have received more detailed coverage. We relate them to the mechanisms, processes, encodings, and their interactions put forward in this book, and thus hope to put them into perspective. Finally, we summarize the book's overall point on how the mind comes into being and dare a glimpse at highly relevant future research directions.

Chapter 2

Cognitive Science is Interdisciplinary

2.1 Introduction

Cognition may be characterized as a mechanism that processes information and knowledge in a meaningful way. Clearly, however, this and any other definition is somewhat unsatisfactory because cognition is a very broad term that can be comprehended and used in various contexts, addressing various aspects of the mind, its functionality, the involved processes, the brain, forms of representation, or types of computation. Thus, the term cognition comprises many aspects and involved concepts. A better, slightly more detailed description of cognition may be: "a mechanism that processes sensory information and ultimately translates it into *meaningful* behavior," where "behavior" includes motor behavior, but also the direction of attention, thought processes, and learning, and "meaningful" emphasizes that the behavior must be linked to some sort of purpose or goal, such as energy intake, communication, or information gain.

Note how these definitions did not exclude any animals, not even plants for that matter, explicitly. Nonetheless, depending on the complexity of a particular task that involves information and knowledge processing, only humans may show particular forms of cognition, such as the ability to play chess, for example, or soccer. Interestingly, both, playing chess and playing soccer have been posed as artificial systems challenges. Playing chess was solved in 1996/1997 by the IBM computer "Deep Blue", albeit arguably by an algorithm, which does not really resemble cognition at all. Playing soccer is pursued by various teams around the globe in different robotic and simulation leagues, which have by now reached a rather high level of sophistication. In both cases, however, regardless of how cognitively inspired the involved algorithms and methodologies may actually be, none of the programs gets anywhere close to understanding *why* they are actually playing the game. They cannot even think of or understand the question. An ultimate goal of cognitive science may thus be characterized as unraveling our ability to ask the following questions (without answering them satisfactorily): "Why am I here in this world? Why can I think – including thinking about myself? Why can I actually ask these questions?"

Typical "higher levels" of cognition, which cognitive science mainly focuses on, include forms of attention, learning and adaptation, problem solving, motivations and emotions, memory, speech and communication, intelligence, and consciousness. How do these processes work? What is their functionality? What are the basic underlying encodings on which they unfold and which they manipulate? What are the mechanisms that bring these individual aspects of "higher level" cognition about? How do they interact? When does a system become intelligent or even reportably conscious, such that it can come up with the famous statement of René Descartes: "*Cogito ergo sum*"?

Greek philosophers were the first who wrote down thoughts on this matter in a form that is accessible still today. Starting with an excursus into the philosophy of science, we survey some of the fundamental questions and metaphors that some of these philosophers advanced, and how these thoughts developed over the last two-and-a-half millennia. Clearly, the last two centuries have boosted this development more than any time before that, and possibly the last twenty years with the establishment of the Internet has sped-up this development even further. Inevitably, scientific developments go hand-in-hand with cognitive development, understanding, and forms of intelligence, at least to certain degrees.

Meanwhile, however, science has developed hundreds if not thousands of branches, disciplines and subdisciplines, sidetracks, and hypes, but also failures, false beliefs, false theories, and false methodologies. Seeing that the ability to gather and analyze data nowadays, which exceeds the ability of doing so twenty years ago possibly by a million times or more, it may be time to put things back together to generate a holistic, integral image of cognition as a whole. To do so, the interdisciplinary subject of cognitive science considers insights from anthropology, philosophy, biological evolution, neurobiology, psychology, cognitive neuroscience, linguistics, and related disciplines, and fuses the respective knowledge with that of computational neuroscience, computer science and informatics, robotics, mathematics, engineering, and artificial intelligence. This endeavor seems only possible – if at all – with the help of the Internet, near instant knowledge availability, and present computer technology. This book attempts to tie the disciplines together and focuses on one of the ultimate goals of the sciences, which is to understand how the mind comes into being.

To further an understanding of the premise of this book, we first take an historical stance starting from old Greek philosophy. We then make an admittedly large jump into the renaissance, where René Descartes' thoughts led to the appreciation of the *mind–body problem* with all its implications for later philosophy including empiricist David Hume and epistemologist Immanuel Kant. Moving on to the scientific developments in medicine and biology, we shed light on the basic hardware components of our brains and the development of our current knowledge about it from the biological perspective. Finally, with the establishment of psychology as its own field of empirical science, behavior came into focus. Unfortunately, this trend soon reduced the mind to its actual behavioral components exclusively, leaving out thoughts and reason. After the second World War, computer science and informatics were established, and people began to think seriously about the mind as being a computer after all. This led to the perspective of the mind as a symbol-processing machine, somewhat forgetting its behavioral purpose.

When research began to return to *embodiment* in the early 1990s, behavior was put back into play. However, until today it has been rather hotly debated and questioned (i) how important embodiment is for understanding the mind and (ii) what are the actual implications of an embodiment perspective for the mind and its development. Thus, the next Chapter focuses fully on the embodiment turn in cognitive science in order to shed detailed light on the implications of the embodied approach to cognitive science. Putting the disciplinary pieces together again without making the embodiment perspective overly explicit for now, we end this chapter by introducing David Marr's *three levels of understanding*, stressing the need to understand aspects of cognition not only along his three levels, but also to understand the interplay among these three levels.

2.2 Philosophy

The desire to understand our nature and the nature of our existence can be traced back to the oldest known human cultures. The fact that very early humans tended to bury their dead indicates that these humans had the capacity to grieve, and thus to empathize and realize the implications of death. Providing the dead with tools or ornaments indicates that they thought about the possibility that there was an afterlife, and the hope that mental life and the self do not cease to exist upon death. Explicit reports of grief due to the death of a friend can be found in the oldest literary texts, including the three millennia old Epic of Gilgamesh. Stories of gods, half-gods, and their interactions with humans clearly show that very early humans were able to imagine things, and that these imaginations are even somewhat comparable with the imaginations that we have nowadays.

Probably the old Greeks were the first, however, who made philosophical thoughts about life and cognition explicit. They were the ones who most fundamentally influenced scientific development of modern philosophy and cognitive science. Natural philosophers, including Socrates, Plato, and Aristotle, were the first to attempt to explain the nature of things and the underlying physical principles. Initially, however, most of these principles were embedded in strong beliefs about mythology and religion.

During the fifth century before Christ, theories about cognition progressively gained recognition. General, abstract questions were considered regarding science, knowledge, and cognition. These developed over the centuries until today:

- In *epistemology*, the science of knowledge and cognition, the insight emerged that humans will never be able to really understand the world and the universe as a whole. "How much knowledge can we actually gather?" is still an open question today.

- The *philosophy of science* emerged out of epistemology and considers how knowledge can actually be gained. Methodology, preconditions, and goal-oriented experimentation were recognized as fundamental pillars of proper scientific rigor. How can knowledge actually be gathered properly? With the emergence of ever new measurement techniques, technical knowledge, and new tools of analysis, these questions need to be asked over and over again.

- The *sciences of the mind* project epistemology and the philosophy of science onto humans, leading to questions such as how come we can use our cognitive capabilities to gain knowledge? What differentiates us from plants and other animals? Where does the mind actually come from? How can our seemingly non-physical mind control our physical bodies? The first humans partially attempted to answer these questions by developing god(s) and religion. Embodied cognitive science offers a fully integrated view on the matter. However, many puzzles and details remain to be solved.

2.2.1 Epistemology and its development

Starting with Greek philosophy, the development of epistemology over the last three millennia shows how the science of the mind is rooted in fundamental questions of scientific knowledge and approaches to science, both of which will reoccur when considering the evolution of biology, neurobiology, and psychology.

Old Greek philosophy

Probably Socrates (∼470–399BC) was the initial pioneer of old Greek Philosophy and many corollary scientific disciplines, which have their roots in Greek epistemology and the science of knowledge in general. Although we do not have any writings by Socrates himself, he must have been a highly influential and probably intriguing and somewhat weird personality, who bothered his fellow Athenians with questions about the meaning of life, death, knowledge, and cognition.

His most influential student Plato (∼428–∼348BC) developed the questions posed by Socrates more explicit, asking questions such as what "knowledge" actually means. While different from an opinion or a belief, it remains unclear if "knowledge" can be equated with our sensor-based experiences. What is knowledge exactly? An important idea in Plato's theory of knowledge is the concept of an idea itself and a dichotomy resulting from this concept: everything in this world, according to Plato, has two expressions. The one is the perceptible one, which is the one that humans can sense, explore, and experience. Due to our bodily, sensory, and physical restrictions, however, this perceptible dimension is inevitably limited and prone to errors. The other one is the perfect idea of the thing, which can only be recognized and appreciated by thought and reason.

An example by Plato is the concept of a circle: we all have probably the perfect idea of a circle in our heads. An illustration and actual realization of a circle, however, will never be an actual perfect circle in its ultimate precision. Additionally, the idea of a circle typically encompasses more than the mathematical circle: we can think of road circles, our circle of friends, or a circular process, all of which are conceptually circles applied in a different realm of thought.

The resulting dualism also led to Plato's perspective on the mind and soul, which we will revisit later: only the immaterial and immortal soul is capable of generating true ideas and true knowledge. As a consequence, it is this immaterial, immortal soul that exists beyond and can control the mortal body.

The central tenet of this perspective is nicely put forward in Plato's "Allegory of the Cave", in *The Republic,* in which Socrates is portrayed as having a conversation with Glaucon. The allegory characterizes the essence of the resulting dualism:

> SOCRATES: [...] Imagine human beings living in an underground, cavelike dwelling, [...]. They have been there since childhood, with their necks and legs fettered, so that they are fixed in the same place, able to see only in front of them, because their fetter prevents them from turning their heads around. Light is provided by a fire burning far above and behind them. Between the prisoners and the fire, there is an elevated road stretching. Imagine that along this road a low wall has been built-like the screen in front of people that is provided by puppeteers, and above which they show their puppets.
>
> [...] Also imagine, then, that there are people alongside the wall carrying multifarious artifacts that project above it statues of people and other animals, made of stone, wood, and every material. And as you would expect, some of the carriers are talking and some are silent.
> GLAUCON: It is a strange image you are describing, and strange prisoners.
> SOCRATES: They are like us. I mean, in the first place, do you think these prisoners have ever seen anything of themselves and one another besides the shadows that the fire casts on the wall of the cave in front of them?
> [...] What about the things carried along the wall? Isn't the same true where they are concerned?
> [...]All in all, then, what the prisoners would take for true reality is nothing other than the shadows of those artifacts. (Reeve, 2004, p. 209.)

The dualism of knowledge and beliefs is thus vividly illustrated: the prisoners perceive a reality that is not actually the "true" reality and, as the allegory continues, even if a prisoner would step out of the cave, understand more about the actual reality, and come back and report on it, the prisoners will have a hard time understanding what this person may try to tell them. Even worse, they may become angry and try to kill him due to their ignorance and the resulting fear of the deeper truth concerning their reality.

While being visionary, characterizing the fact of our inevitably limited knowledge (remember also Heisenberg's Uncertainty Principle, according to which it is basically theoretically impossible to measure the exact state of a particle, including its velocity and its mass), and even expressing the fact that some humans are still scared of deeper knowledge and truth, one important issue was neglected in the Cave Allegory: we can interact with and manipulate our environment! That is, we can probe it for particular matters of truth and experiment with it. We can test certain hypotheses, by probing cause-and-effect relations, properties of things, or simply the identity of things. In this way we can, literally speaking, "grasp" a new idea, probe it, verify it, and also disprove it. Ideas that were never falsified despite thousands of tests become knowledge, become facts, which are generally accepted.

Thus, despite the fact that we all live in a cave with restricted perceptions, we can intently direct and manipulate our perceptions by our actions. Furthermore, we can augment our perceptions by tools, such as measurement gadgets, and by manipulating the thing that is to be perceived. For example, we may measure the weight of an object by a scale or we may turn an object around to verify its identity. Embodied cognitive science essentially supposes that such sensorimotor interactions – where motor actions probe sensory perceptions and thus systematize sensory perceptions and, in retrospect, current and desired sensory perceptions lead to further motor actions – pave the development of the human mind including abstract cognition and the conceptualization of knowledge and truth itself.

Perhaps even more important than Plato to our understanding of our world and our mind, are the words of his student Aristotle (384–322BC). Aristotle has produced treatises on all the scientific and other knowledge disciplines imaginable at the time including politics, ethics, logic, rhetoric, linguistics, biology, physics, and metaphysics. In fact, his work has not only influenced most researchers until now, it has also led to the foundation of some of these scientific disciplines.

For cognitive science perhaps most significant is the "Organon" (Greek for tool or method) collection of Aristotle's six works on logic. Aristotle set forth any important basic concepts for developing scientific theories, which are still relevant today in mathematics, philosophy, linguistics, and many other disciplines. For example:

- The general principle of formulating definitions and hierarchical classifications.

- The definitions of propositions, where a proposition is a sentence with a unique truth value (true or false), as well as the first important rules of propositional calculus.

- Several types of proofs and their applicability, including *deduction*, that is, drawing concrete conclusions out of general premises, and the reverse way, *induction*, that is, drawing general conclusions out of concrete premises.

Besides aspects of definitions, truth, and logic, Aristotle thought deeply about the reasons for our existence and about the world in its existence as a whole. For example, he stated that:

> Knowledge is the object of our inquiry, and men do not think they know a thing till they have grasped the 'why' of [20] it (which is to grasp its primary cause). (Aristotle, 2014, KL 9434-9436.)

The "why" question essentially asks about the reason for the existence of a particular thing, a particular occurrence, or a particular behavior. In answering a why question, Aristotle argues that four different answers can be given about the causes of the why:

- *The material cause*: the cause may be the actual material that a certain thing consists of – thus offering a purely physical, particle-based explanation of a certain thing or behavior. For example, a door may be made out of wood, thus it may burn, burst, have a certain weight, and so forth. Another example may be a flying bird that is rather light in its biological material and has feathers, thus being able to fly, to tumble in the air, to be soft, and so on.

- *The formal cause:* the cause may be the actual form a certain thing has, that is, the shape with all its implications. Certain shapes may cause certain effects, movements, or behaviors. For example, a door may block a passage due to its shape; the bird's shape allows it to fly and glide through the air by means of its wings.

- *The efficient or moving cause:* the cause may be the efficiency, that is, the force that has generated a certain thing or sets a certain thing in motion. For example, when opening a door the force of our hands pushing or pulling the door results in the door swinging open; similarly, the bird uses its muscles to flap its wings.

- *The final cause:* maybe the most debated and open cause, the final cause specifies the actual end of a particular thing, event, force, or motion. For example, the door may be opened in order to move through it; the bird may flap its wings to fly from A to B.

The final cause especially has deeper implications in cognitive science and particularly embodied cognitive science: behavior typically appears to be goal-directed, that is, selected and to a large degree controlled by the goal that an animal intends to reach or achieve. From this perspective, goal-directed behavior is essentially generated by final causes, which cause the generation of efficient causes (that is behavior), which then manipulate the form, relations, and material of entities. Thus, from a cognitive science perspective Aristotle has set a framework for explaining behavior as being driven by final causes, that is, current motivations, intentions, and involved goals. Lifting this perspective to society as a whole, Aristotle's four causes can be related to cultural evolution and the involved, ever continuing generation and development of new human-made artifacts, where cultural evolution is driven by the continued human endeavor to achieve new goals.

However, to a large degree also natural, physical objects, such as a mountain, can be explained by these four causes. Only the ultimate, final cause seems to remain obscured, given that we are not satisfied with a final cause explanation of the type: "the mountain grew to give way to the motion of tectonic plates" or "the volcanic mountain grew because pressure distributions in the inner earth had to be released on the surface."

In the following sections, we will introduce several other levels of explanations, including Nikolaas Tinbergen's four fundamental questions in biological science and David Marr's *three levels of understanding* in psychological research. Aristotle's four causes may be applicable to each of these taxonomic distinctions, in each case systematizing the principles underlying the respective distinctions. Material and form can be manipulated by forces such as motion, heat, electricity, etc., and the development of things and even brains can only be guided by these principles. The relevance of the final cause may be underestimated in the literature in this respect. It has often been denied and even Aristotle himself may not have thought about its actual full implications. Nonetheless, the final cause may be one of the most important principles that led to the development of our conscious minds and even to our self-conscious, reflective souls.

Rationalism, empiricism, and materialism

Somewhat surprisingly and most unfortunately, the world took a different turn after these initial deep thoughts. During the Roman empire, which mainly focused its potential on conquering, reigning, engineering, and "bread and games" for the people, religion played a progressively more important role. Philosophy was often reduced to ideas about how the existence of god can be proven and other theological matters. It was not until the 15th century that things started moving again. Societies in Europe changed and the Renaissance led to the beginning of the modern western world. Meanwhile, philosophy of mind became to an increasing extent a religion-independent philosophical subdiscipline, reconsidering the thoughts of Socrates, Plato, and Aristotle. Besides questions about to what extent thought itself and reason may actually be possible, the question about the ontogenetic development of thought and reason also became a matter of consideration.

As we have seen, Plato was already distinguishing between a perceiving body and a cognitive soul. Only with the latter, he argued, are insight and knowledge possible. This differentiation was prominently discussed further by René Descartes (1596–1650). Similarly to Plato, Descartes contrasted sensory perceptions and cognition, and hypothesized that the objective structure of our environmental reality can be grasped *a priori* via thought and knowledge, fully independent of any sensations. Indeed, he even suggested that perceptions can be misleading. For example, dreams may appear almost like actual, physical experiences, but, according to Descartes, do not yield any novel knowledge. The bases for knowledge, then, are not actual physical perceptions, but rather *ideae innatae*, that is, innate ideas, which were supposed to exist *a priori* from birth and which are used to structure, develop, and derive further knowledge based on this *a priori* knowledge. Examples of such *a priori* knowledge were assumed to encompass the knowledge about the existence of god, an immaterial, cognitive soul, as well as the material body. Later on, Descartes derived the mind–body problem from these theories. The dichotomy between body and soul and the resulting problems are also reflected in Descartes' famous statement *"Dubito, ergo Cogito, ergo sum"* ("I doubt, so I think, so I am"): thought defines who we are, not experience.

The supposition that the mind defines who we are led to the development of *rationalism*, in which the insights about our reality are based purely on reason and thought. Protagonists of the concurrent philosophical movement of *empiricism* strongly denied this belief. Empiricism denied the immaterial soul or mind and hypothesized that learning and development may rather be based on experiences alone. The English philosopher John Locke (1632–1704), for example, wrote that the theories of Descartes are completely implausible and contradictory. Instead, he supposed that the human mind may be viewed as a *tabula rasa*, that is, as an empty sheet of paper that is filled by means of experiences:

> Let us then suppose the mind to be, as we say, white paper void of all characters, without any ideas. How comes it to be furnished? Whence comes it by that vast store which the busy and boundless fancy of man has painted on it with an almost endless variety? Whence has it all the materials of reason and knowledge? To this I answer, in one word, from experience. (Locke, 1690, p. 95.)

Thus, according to empiricism, only experiences shape the development of the mind, soul, and ideas by means of reflection, abstraction, generalization, and recombinations of ideas.

The Scottish philosopher David Hume (1711–1776) is well-known for his empiricist standpoint. Hume particularly stressed the importance of anticipation and prediction. According to Hume, the brain is not a passive observer, but rather an active inference system. Ideas are connected and developed by means of measures of similarity, contiguity, and causation. He states that:

> [...] We have said, that all Arguments concerning Existence are founded on the Relation of Cause and Effect; that our Knowledge of that Relation is deriv'd entirely from Experience; and that all our experimental Conclusions proceed upon the Supposition, that the future will be conformable to the past. [...] (Hume, 1748, p. 62–63.)

Thus, perhaps the brain makes or at least relies on the supposition that the future will resemble the experienced past and present. It does not, however, assume the existence of particular things such as god, the soul, or reason itself.

Besides the emphasis on experience, spiritualism was increasingly criticized by the *materialistic movement*. The French medic and philosopher Julien Offray de La Mettrie (1709–1751), who died rather early and was possibly murdered due to his radical points of view, put forward that even the soul may be the result of complex bodily and brain functions. In his most renown book *L'homme machine* (Man a machine) (de la Mettrie, 1748), which he published anonymously in 1748, he put forward the pure materialist point of view, for example, stating that:

> It is not enough for a wise man to study nature and truth; he should dare state truth for the benefit of the few who are willing and able to think. As for the rest, who are voluntarily slaves of prejudice, they can no more attain truth, than frogs can fly. (de la Mettrie, 1748, p. 1.)

Materialism most strongly emphasizes that once one explains something as spiritual, there is no more reason to investigate it any further. Imagine, for example, explaining a flood by a river spirit: the river spirit substitutes and thus makes obsolete any further explanations about what may have caused the flood. Thus, spiritual explanations are highly contra productive, precluding any further scientific progress.

Similar to Mettrie, but more careful about his statements, the French writer, philosopher, and academic Denis Diderot (1713–1784) further propagated the materialistic take on science and nature. During his lifetime, Diderot was most engaged in contributing to and publishing the "Encyclopédia, or a systematic dictionary of the sciences, the arts, and crafts" between

1751 and 1772 in 28 volumes. Meanwhile, though, Diderot wrote several other volumes that he hid away and asked a friend to publish after his death. In these volumes, Diderot fully focused on and emphasized the importance of purely materialistic explanations of phenomena. He put forward that also physical behavior, which was often believed to be caused by an immaterial soul, may simply be generated by suitable structural organizations and catalyst processes. Dead matter is simply organized in a non-living manner, such that alive and dead matter may be considered different forms that are based on the same organizational principles. Diderot also suggested that evolutionary hypotheses may be able to explain the world, if we give it enough time. Diderot even mentioned the possibility of cloning by taking one human cell – assuming that inevitably each cell of the human body must contain the whole blue print.

Diderot concludes that also thought itself needs to be understood purely by means of mechanistic, naturalistic explanations; and more so, that we need an ethic of truth where religion has no place when we really want to coexist with the rest of nature. Religion he considered as a movement of ignorance, which was created simply due to the fact that we do not know enough about ourselves. Essentially for Diderot, this is the ultimate humanism – the acknowledgment of ignorance, the search for truth, and an ethic of truth.

Diderot has foreseen not only many scientific developments, but he may be viewed as being a radical – pushing for an ultimate truth that is purely based on naturalistic explanations. Seeing that many scientists are still struggling to accept the possibility that the mind and consciousness may in the end be grounded fully in naturalistic, biological structures, and processes, Diderot's foresight was indeed groundbreaking.

Logical empiricism and critic of rationalism

All three of these philosophical conceptualizations, rationalism, empiricism, and materialism, significantly influenced the developing scientific methodologies. Logical empiricism combined empiricism with abstract reasoning and logic, developing an inductive scientific approach. Critical rationalism, on the other hand, starts from a theory, and verifies or falsifies it by means of empiric experiments. Both developments attempt to stick to the materialistic point of view, but they combine it with a means of abstraction for enabling rationalistic explanations.

A group of philosophers from Vienna including Moritz Schlick (1882–1936) and Paul Rudolf Carnap (1891–1970) developed the basic ideas in the 1920s. The basis for cognitive development is experience, from which an inductive scientific methodology can be derived as follows:

- The basis of developing a theory is provided by sensory data, which are gathered by identifying systematic variations across individual subjects.

- The data is then used to test, verify, or falsify developed hypotheses. If they are verified, then inductive logic allows the development of a theory.

This logical empiricist approach, however, has a very important weakness, which lies in the induction itself. The Austrian–British philosopher Karl R. Popper (1902–1994) was in close contact with the group from Vienna. He argued that the inductive approach is problematic because in can lead to incorrect beliefs (which Aristotle had already realized as well). Moreover, its correctness is not really provable in a mathematical sense. Only the re-application of the induction itself provides a proof, which, however, leads into a circular chain of arguments (hypothesis, induction, induction ... which never ends in a full proof).

A short discussion on the relationship to mathematics is essential at this point. Mathematical proofs by induction are possible because the underlying number system itself is generated by induction (for example, starting from zero and creating all natural numbers by simply adding 1 to any already existing number). Thus, while mathematical proofs by induction are valid, a philosophical or cognitive science proof attempt by induction suffers from problematic inductive steps. Whatever it is, the generalization to all cases is extremely difficult because it typically cannot be proven that all possible cases can be reached by the inductive step.

Due to this problem, Popper proposed deduction as the solution – essentially inverting the inductive process. This deductive approach was later termed *critical rationalism*:

- The starting point in this case is not the data, but a theory, which is based on considerations based on rationality and reason, and by which hypotheses are derived. In doing so, the inductive step is avoided.

- The empirical gathering of data then is used to verify the hypothesis: falsification is possible by finding examples that contradict the hypotheses, otherwise its believability increases.

Clearly, also this approach is problematic: the empirical data needed to falsify the hypothesis is not necessarily easy to gather. Any attempt to falsify a hypothesis may be difficult for two reasons. First, the hypothesis may not be falsifiable at all with the available measurement techniques, such as when developing a model of a cognitive capability that mimics the cognitive capability correctly, but the underlying mechanisms of the model can neither be verified nor falsified. Secondly, the hypothesis may be falsified, but the falsification, which inevitably must be based on empirically gathered data, may be incorrect, because the refuted model never considered (that is, modeled) the influences that led to the particular, model-contradictory data patterns. Thus, the model may very well be correct, despite its (incorrect) falsification.

In conclusion, hypotheses and cognitive models that are hard to falsify may still be valuable, but certainly falsifiable models are preferable. Meanwhile, hypotheses that are actually falsified should not necessarily be discarded immediately. On the one hand, they may be adapted, enhanced, or modified. On the other hand, exceptional cases may be identified that led to the falsification, so that the hypothesis may be true in the general case, but not in particular, exceptional cases. In this light, it may not come as a surprise that there is the saying: "The exception proves the rule" – seeing that rules and hypotheses may be applicable in the general case, but typically for any rule and hypothesis one can find exceptions.

Philosophy of science

In the further development of the two approaches to science, the two perspectives continued to blend into each other. The physicist, historian, and philosopher of science Thomas S. Kuhn (1922–1996) conceptualized the resulting continuous scientific developments from a paradigm-driven scientific perspective. The concept of a paradigm characterizes a scientific approach to inquire particular interactions scientifically. A paradigm may develop due to the availability of a new technique as a result of a technical advancement or simply due to a new methodology of testing certain interdependencies or causal relations.

According to Kuhn, science progresses by means of currently accepted theories and methodologies. The key point is the current acceptance, that is, even though some accepted theories and methodologies may be false, they may still be considered scientific at the time. Good examples are the science of *phrenology* – research that attempted to deduce functional capabilities and modular processes in the brain by analyzing the shape of the skull – as well as the works of the German physician Franz A. Mesmer (1734–1815), who propagated the belief of some form of *animal magnetism*, which was assumed to result in natural energetic transferences between entities. While eventually disproved, Mesmer is responsible for the verb *to mesmerize*, that is, to fascinate or to hypnotize in such a way that one believes false theories or claims. Scientific paradigms thus reflect the consensus of a particular scientific community, rather than scientific truth itself.

Moving even one step further, Kuhn developed a general theory of scientific development, which was set forth in his very influential book *The Structure of Scientific Revolutions* (Kuhn, 1962):

- Scientific branches generally commence without any prior knowledge about the branch – certainly though with general prior knowledge. Kuhn termed this stage of development the pre-paradigmatic phase. During this phase, scientists are able to conduct broad exploratory research, which typically is much less efficient than focused research.

- During this development, somewhat successful methodologies are identified as valid approaches to conduct inquiries on the subject matter. By passing these on to other researchers in a growing community, normative scientific approaches develop leading to focused research inquiries.

- When the dominant paradigm is questioned, however, then a crisis may develop leading to a sort of scientific revolution (small or large), which generates paradigm shifts and develops new dominating paradigms.

While the concept of a paradigm cannot be defined precisely (Kuhn himself modified it several times), the main point concerning scientific development is the fact that many scientific revolutions occurred over the last centuries (in seemingly ever increasing frequency). Most well-known may be the refutation of the Ptolemaic view that the earth is the center of the universe, in favor of the heliocentric model of Nicolaus Copernicus (1473–1543). Also Charles Darwin's (1809–1882) theory of natural evolution led to the development of a wholly new perspective on humans and their position in the world.

In psychology, behaviorism (discussed later) is often viewed as a scientific revolution, which has partially brought about the experimental approach to psychology. Behavioristic paradigms, such as conditioning behavior by means of positive and negative rewards, are still applied today. In later chapters, we will see how behaviorism can be understood computationally and how it is implemented by means of reinforcement learning (cf. Chapter 5). However, we will also see that there is more to cognition than behavioral conditioning by means of reinforcement learning.

2.2.2 Philosophy of mind

Epistemology and scientific development went hand in hand with the development of the philosophy of mind. Probably very early in the development of the human race, humans have started to think about the mind, the soul, where we are going after death, and so forth. This is evidenced by the discovery of graves, as well as signs of worship and the development of superstition and mythology. The question of what makes and allows us to think and feel, as well as the question of what makes us different from other animals and thus uniquely human, have probably been around from the very beginnings of the human race. Deeper and more concrete questions developed over the centuries, characterizing consciousness itself and facets of it, as well as the concept of *qualia*, that is, the qualitative, conscious, and subconscious experiences of our feelings in the here and now.

Once again the old Greeks

Plato was probably the first who made the conflict between body and soul explicit, as mentioned in Section 2.2.1. The important implications of the mind–body dualism was explored in his essays on the matter. Body and mind (or soul) are considered as being two mutually independent entities. The mortal body serves the immortal soul as its residence, or negatively expressed, as its prison. The immaterial soul unites emotions, reason, and free will. Despite this dichotomy, however, Plato considered the capability of movement as an important property of the soul, thus attributing souls also to animals and even celestial bodies. The problem of how the immaterial soul then may control the material body, however, was apparently not addressed by Plato.

Aristotle further differentiated the dichotomy by moving the soul closer to the material body in his book *De Anima*. He considered the soul, or mind, as inseparable from the body and as manifested in the body itself:

> Suppose that the eye were an animal— sight would have been its soul, for sight is the substance of the eye which corresponds to the account, the eye being merely the matter of seeing; when seeing [20] is removed the eye is no longer an eye, except in name – no more than the eye of a statue or of a painted figure.
>
> [...] as the pupil *plus* the power of sight constitutes the eye, so the soul *plus* the body constitutes the animal.
>
> From this it is clear that the soul is inseparable from its body, or at any rate that certain parts of it are (if it has parts)— for the [5] actuality of some of them is the actuality of the parts themselves. (Aristotle, 2014, KL 18274-18276;18281-18284, *On the soul*, book II.)

Thus, according to Aristotle, the soul – possibly further characterizable in this case as the essence of life itself – is thus a property of any living organism, and possibly even every moving entity that has the capability of consumption, growth, and replication. In animals and humans, Aristotle additionally considered perceptual and behavioral capabilities as essential. However, Aristotle also postulated that, in addition, humans have a soul independent of body and mind (Greek *nous*), which gains knowledge and insights by means of abstractions of sensations, perceptions, and emotions, resulting in the ability to reason and to think in abstract terms.

The homunculus problem

René Descartes, whom we already have mentioned, as one of the founders of rationalism, reignited the discussion on mind and body. According to his treatises, the body is nothing more than a mechanical machine that obeys the laws of nature, and thus does not require any soul or mind. In contrast to Plato and Aristotle, he denied that plants and animals may have souls in any form (which was probably also necessary to avoid problems with Christianity at the time). However, because humans are able to talk, think, and behave intelligently–which, according to Descartes, is impossible solely by the laws of nature–another immaterial and immortal substance must necessarily exist, which is only available to humans, and which is attributed to an immaterial soul (Latin *mens*).

Seeing that Descartes was not only a philosopher, but also a natural scientist, he himself was not quite satisfied with the resulting dualism between body and soul. Thus, he addressed the question: how can the immaterial soul control a material body? How are interactions between the two entities possible? An answer was found by identifying the epiphysis cerebri – or pineal gland – which is located centrally in the brain and thus was somewhat plausibly considered as the center of the soul. Still during Descartes' time, Galen's theory of ventricles, in which nerves were considered to be hollow canals through which fluids flow that stem from the ventricles, was considered general scientific truth (another example of a false belief in the history of science). Descartes additionally assumed that nerves contain fibers, which allow the coordination of very small valves in the hulls of the ventricles. Once stimulated, the valves will open, fluid will flow, and as a consequence an image of the perception will be generated. Even movements are possible by means of the coordinated opening and closing of the available valves. Albeit extremely far-fetched, Descartes thus imagined aspects of the actual functionality of neurons in a very imaginative and analogy-making fashion.

The theory about the epiphysis was soon refuted. However, the philosophical inquiry on the mind–body problem is still being discussed and fully satisfactory answers on where the mind comes from and on the origins of the "soul" in each human are still missing. Maybe the best caricature of the problem was proposed in a thought experiment by the American philosopher Daniel Dennett (*1942): Imagine there was a center for consciousness and the soul for that matter. If this center was located at some center in the brain, then a little human, that is, a little mind would necessarily need to be present at this position. With one further step then the whole concept is put *ad absurdum*: what about this little center then? Would there not necessarily be another even smaller little human necessary within this center? ... thus leading to an infinite regression without actual explanation. The little human is replaced by an even smaller human and so forth, essentially highlighting that is is impossible to explain mind and soul by a homunculus.

Qualia and consciousness

Mind and soul are often directly associated with the term "consciousness", which is equally hard to define or specify:

- Consciousness in the intransitive sense refers to the contrast of being awake versus being asleep or even unconscious. This contrast still seems to be objectively distinguishable and can thus also be investigated experimentally.

- The question of whether our consciousness is responsible for our action choices and intentions – or even if it gives us "free will" – is much harder to investigate because it is difficult to define it precisely. The intention to eat or sleep my be attributed to genetic predispositions and innate urges. However, where does the intention to be creative, play music, draw a picture, or recite or even write a poem come from?

- It becomes even harder when considering the phenomenology of consciousness, that is, the conscious experience of our body with all its perceptions and the current conscious state of mind. The Scottish philosopher David Hume (1711–1776) identified this problem, which was later termed the *Qualia problem*, as follows:

  > We cannot form to ourselves a just idea of the taste of a pineapple, without having actually tasted it. (Hume, 1789, p. 5.)

  In his empiricist perspective, he essentially highlights that it seems impossible to judge in an exact manner how something actually feels or tastes without ever having experienced it. For this reason, is seems also impossible to experience the exact way others experience a particular stimulus or situation. This qualitative, conscious experience seems to be a purely personal matter.

- Finally, the reflective self leading to self-consciousness remains an open question. What does it mean to be conscious about one's own life? What does it mean to be conscious about our own consciousness including all its facets?

Recently, with the continuously improving capability of brain imaging, it has become easier to monitor brain processes or brain activities in real time in particular situations or over short durations. Neuroscientists have searched for particular brain centers that bring about consciousness. Once again, it seems that the homunculus problem kicks in. As suggested by Daniel Dennett, it would come as a big and very mysterious surprise if we find one center in the brain that gives us consciousness. The explanatory power of such a discovery would be highly questionable. Nonetheless, neural correlates that are present while we are conscious have been identified, suggesting that a distributed, highly interactive neural network gives rise to consciousness.

The American philosopher Ned Joel Block (*1942) proposed a thought experiment that highlights the problem in an illustrative manner, called the *China–Brain*: imagine we give all Chinese people a cell phone and exact instructions when to press which number in reaction to which incoming call. This sounds generally similar to activations of neurons in the brain, thus, imagine (alternatively) that each Chinese person plays the role of a neuron in a highly distributed network. Could this resulting network (of interconnected cell phones or neurons) develop consciousness – or at least the imagination of a particular object?

A similar thought experiment was suggested by the American linguist John Searle (*1932): Imagine a *Chinese Room*, in which a human lives. This human cannot understand Chinese. However, he has a huge huge database of symbols and a large book of rules, which give instructions about how to handle incoming inquiries in Chinese. Now imagine further that scientists pose questions to the person in the room and the person seems to answer them in a very intelligent manner. In fact, the scientists may be incapable of determining with certainty if the person in the room understands Chinese. Is the person in combination with the room then conscious of Chinese? Does the person or the person in combination with the room actually understand Chinese?

Putting these thoughts into a computer system that works with rules and replacing the human with the central processing unit (CPU) of the computer, a similar question arises for a purely technical system: can an artificial system equipped with enough symbols and rules, which manipulate the symbols, become intelligent or even conscious? When is the system more than just a symbol manipulation machine? When does it reach some form of consciousness? Many parts of this book will address exactly these questions from different perspectives.

2.2.3 Philosophy of language and forms of representation

Our capability to talk, comprehend speech, write, and read is often considered the Holy Grail of human cognition and consciousness. Therefore, linguistics is also inseparably tied to cognitive science. Besides the complexity of phonology and syntax, which are part of any existing language, the interconnection with meaning and pragmatics is fascinating. Many questions arise in this respect:

- Which word is used when and in what context?

- How can symbols and words have particular meanings and how are they linked to their particular meanings?

- How does the meaning of a sentence often beyond the meaning of the individual words in the sentence?

- What is the exact role of grammar and how is it tied to semantics?

- How has the compositionality in language, that is, the ability to generate an infinite number of meaningful sentences out of a finite set of words, developed?

Embodied cognitive science essentially postulates that language builds upon the prior capability of interacting and perceiving interactions with the environment in a meaningful manner. When we speak, we essentially verbalize observations or interactions with the world that are already prestructured by our experiences gathered when interacting with the world. This perspective, however, is still considered hypothetical, and not even a sufficiently satisfactory proof-of-principle exists.

And again the old Greeks

Not surprisingly, Aristotle had some thoughts on this matter as well. In his essay *De Interpretatione* (Latin for About Interpretations) he posits an initial manifestation of the fact that words are generally arbitrarily linked to the object or meaning they are referring to (exceptions are onomatopoeias, which are words that sound like their meaning, such as cuckoo, roar, or buzz):

> Now spoken sounds are symbols of affections in the soul, and written marks symbols of spoken sounds. And just as written marks are not the same for all men, [5] neither are spoken sounds. But what these are in the first place signs of— affections of the soul— are the same for all; and what these affections are likenesses of— actual things— are also the same. (Aristotle, 2014, *De Interpretatione*, KL 808-812.)

Plato acknowledged that one may name identical things differently, but he assumed further that every entity in this world can be associated with one perfect "name", and only this perfect name can reflect the true nature and idea of the entity that it names. The challenge for a philosopher then is to find this true name. Plato explains this point as follows:

> I may illustrate my meaning by the names of letters, which you know are
> not the same as the letters themselves, with the exception of the four, ϵ, ν, o,
> ω; the names of the other letters, whether vowels or consonants, are made up of
> letters which we attach to them; but so long as we introduce the meaning of the
> letter, and there can be no mistake, the name which indicates the letter is quite
> correct. Take, for example, the letter beta – the addition of η, τ, a, gives no
> offense, and does not prevent the whole name from having the value which the
> legislator intended – so well did he know how to give the letters names. (Plato
> & Jowett, 1901, p. 632.)

Plato thus believed in the concept of ideas, as discussed, and words are mere referents for
the idea behind it. For example, the color term "red", according to Plato, could be used
for various things – but any red object is only an exemplar reflection of the actual meaning
of red – an example of the "redness" idea. Thus, only the idea behind a word can reflect
the true meaning of the word – examples will always be insufficient. In conclusion, Plato
asks the question where the words come from, seeing that there never is an uniquely true
example of their meaning. Do terms, such as red, exist per se – so to say *a priori* – or are
they actual human constructs?

Symbols in language

A pioneer of modern linguistics was the Swiss scientist Ferdinand de Saussure (1857–1913),
founder of the linguistic structuralism. According to him, languages are nothing but symbol
systems that have an underlying relation between particular elements of the system, which
need to be reconstructed to be understood. As Aristotle had already put it: "The whole is
more than the sum of its parts."

In this respect, a dyadic structure developed, contrasting the actual utterance of the
speaker from the actual perception of the listener. The *signifier*, which is the auditory
utterance of the speaker, influences what is *signified*, that is, how the listener interprets the
perceived signifier. The assignment of meaning by the listener is fully dependent on the
context and the interpretation of the heard sounds. Misunderstandings are certainly great
examples of this dyadic system.

Saussure's dyadic structure is also related to the *triadic symbol model* of the American
philosopher and logician Charles S. Peirce (1839–1914):

> A Sign [...] is constituted a sign merely or mainly by the fact that it is
> used and understood as such, whether the habit is natural or conventional, and
> without regard to the motives which originally governed its selection. [...].
> It is of the nature of a sign, and in particular of a sign which is rendered significant
> by a character which lies in the fact that it will be interpreted as a sign. Of course,
> nothing is a sign unless it is interpreted as a sign; but the character which causes
> it to be interpreted as referring to its object may be one which might belong to it
> irrespective of its object and though that object had never existed, [...] (Peirce,
> 1960, 2207–8.)

A sign thus represents something only because it is interpreted in this way, whether by
convention, intellectual inference, or simply by usage. The sign refers to an object, which
may be an *idea*, such as an entity or thought. However, the object the speaker refers to may
not be identical to how the listener interprets the perceived sign. Thus, a trichotomy arises
where the symbol refers to an entity via an interpretation.

Peirce's symbol definition not only encompasses the representation of things and formal
aspects of language, but it is also embedded in a complex speech and cognitive theory, which
we cannot discuss in detail here. Nonetheless, Peirce's trichotomy sets forth interesting
aspects of object relations: Pierce differentiates *icons*, that is, perceivable symbols that are
related directly to the referred item (for example, a skull that indicates death and thus
poison); *indices*, that is, indicators that refer to a different object or situation due to their

indicatory character (such as a knock on a door, which indicates a visitor); and *symbols*, that is, symbolic referents that gain their particular meaning only by convention in the actual culture, language, and context in which they are used (for example, an upright index finger may ask for ´attention' or indicate the number "one" depending on the context; similarly, nodding with the head may mean yes or no depending on the culture).

This three-fold differentiation is particularly interesting because Peirce allows that essentially anything can become a symbol as long as it is used in a meaningful context. The interpretation of the symbol in the particular context then leads to our actual interpretations and conclusions in respect to the actual context. In the light of this very general interpretation of a symbol then, the question arises how the interpretation of a symbol can actually lead to a general convention of comprehension. That is, how does a symbol get its conventional meaning? How is a symbol's meaning grounded in our world? This *symbol grounding problem* will be addressed in further detail in Chapter 3 (cf. Section 3.4.1).

Inspired by Peirce and influenced by the Vienna circle (cf. Section 2.2.1 on logical empiricism), the American philosopher Charles W. Morris (1901–1979) proposed the now generally accepted differentiation between syntax, semantics, and pragmatics. Semantics refers to the relation of a symbol to the meaning of the object, item, or thing that is referred.

Syntax addresses the grammatical relation between symbols. Finally, pragmatics refers to the association of symbols to each other and the associated correlations, imaginations, and interpretations of the symbols in interaction.

Pragmatics led to the *speech act theory*, which was put forward by the American philosophers John L. Austin (1911–1960) and John Searle (*1932), according to which any meaningful utterance can be viewed as an action in itself that realizes a change in the environment – and primarily a change in the state of mind of the listener with all the involved interpretations. Utterances, such as promises, orders, warnings, invitations, or statements of facts, viewed from this perspective are thus actions with their associated goals and intentions. For example, a statement such as "I am not going" is a piece of information that is given to the listener, with the goal to inform the listener that the speaker is not going to the place or event that is currently under consideration.

2.3 Biology

While philosophers attempt to understand cognition, the mind, the soul, and the brain from a rather abstract, symbolic, and theoretical perspective, biology starts from the other side, focusing on the organic mechanisms and structures in the human body, regardless if there is a mind or soul or not. How does the human mind thus function in the human organism?

As we had seen, Aristotle indeed acknowledged the bodily aspects of ourselves in his formulation of the four causes, which are necessary to produce a complete explanation of something: material, formal, efficient, and final cause. Moreover, he emphasized that:

> It is manifest that the soul is also the final cause. For nature, like thought, [15] always does whatever it does for the sake of something, which something is its end. To that something corresponds in the case of animals the soul and in this it follows the order of nature; all natural bodies are organs of the soul. (Aristotle, 2014, KL 18385-18387.)

This implies that the final cause, that is, the end or ultimate goal of an action or a thought is equivalent to the mind (or "soul"), which functions for the sake of its body following the order of nature.

More than 2000 years later, the ethologist and Nobel laureate Nikolaas Tinbergen (1907–1988) reconsidered the questions of the "how" and the "why". Along the lines of Aristotle's four causes, he proposed four basic questions in biological sciences:

- *Mechanisms and causations:* Which organic structures and mechanisms lead to which capabilities and properties?

- *Ontogeny:* How do these structures and mechanisms develop over a lifespan beginning with the DNA?

- *Adaptation:* Which behavioral strategies and characteristics are developed by an animal to increase the probabilities of survival and success in reproduction?

- *Phylogeny:* How did a whole species change over generations?

Embodied cognitive science typically attempts to integrate tentative answers to Tinbergen's four questions: Evolutionary bodily and neurocognitive adaptations over centuries and millennia are considered. Moreover, the *ontogenetic development*, which is ultimately driven by the phylogenetically evolved genetic biases, is investigated. Thereby, the main questions are how behavioral and cognitive flexibilities develop over a lifespan (ontogeny) and how they are adapted to current circumstances (adaptation). Along these lines, also the question how learning works is addressed. Finally, our behavioral and cognitive capabilities and flexibilities in the here-and-now are investigated, that is, how behavioral and cognitive decision making and control works. In this formulation, embodied cognitive science may be considered to be highly biologically-oriented, attempting to integrate all four questions of Tinbergen into one embodied cognitive theory.

To reach this integrated cognitive science perspective, however, cognitive science came a long cultural-phylogenetic way. We thus proceed with a short excursus into the history of neurobiology and the development of the evolutionary theory. Even now it seems that our minds struggle to consider the possibility that body and mind may have developed solely as a result of biological, and particularly evolutionary, processes, and mechanisms. Religion still plays an important role in making sense of our world, ourselves, and others – and this was certainly even stronger in the past, rooting neurobiology in mysticism and religion. Nonetheless, bits of truth can be found from the very beginning.

2.3.1   Neurobiology

Greek and Egyptian antiquity

In the case of neurobiology, the documented texts on the brain reach as far back as 4000 years. An Egyptian document written on papyrus, the *Papyrus Edwin Smith*, is one of the oldest medical documents found to date. It contains an extensive description of the brain with its gyri and sulci in the neocortex, the meninges, and the connection of the brain to the spinal cord. Surprisingly, differentiations between nerves, blood vessels, and tendons were not found, and the function of the brain was probably still unknown and apparently considered unimportant. While liver, stomach, and intestines were often preserved in mummies, the brain was apparently discarded.

Ancient Greek medicine was strongly influenced by their Egyptian neighbors. Aristotle also believed that the mind resides in the heart, whereas the brain was considered an organ that is responsible for cooling the body:

> For this reason it is, that every sanguineous animal has a brain; whereas no bloodless creature has such [25] an organ, [...]. For where there is no blood, there in consequence is but little heat. The brain, then, tempers the heat and seething of the heart. (Aristotle, 2014, *De Anima*; KL 28035-28037.)

About 500 years later, the Greek medic and anatomist Galen (ca. 129–200/216AD) published an extensive treatise (*Methodi medendi*) about the medical knowledge of the time, enhanced with his own theories. His theories particularly depended on observations from sheep, pigs, and other animals, as the dissection of humans was not allowed due to religious beliefs. In particular, his theory on ventricles was considered valid for the next several hundred years and throughout the Middle Ages. According to Galen's theory, nerve tracts are hollow canals, which connect muscles and sense organs with the brain. The brain substance was already believed to be involved in cognitive processes somehow, however, the communication with muscles and sense organs was believed to be mediated by the *pneuma psychikon* (Greek for *rational soul*). Again, we see how bits of truth developed and how some of the interpretations are plainly wrong due to the lack of deeper knowledge. While the brain was now correctly believed to mediate communication with sense organs and muscles, the mechanisms reflect the best imaginable knowledge of the time combined with some mysticism due to the lack of better tools for investigation.

Toward contemporary neuroscience

While in the Middle Ages medical knowledge stagnated or even fell back to more religious-driven beliefs, even losing knowledge that was acquired in past generations, the Renaissance enabled the resurrection and enhancement of medical knowledge. Examinations of dead human bodies became acceptable, enabling more detailed anatomical studies of the human body. The Italian universal genius and inventor Leonardo da Vinci (1452–1519) produced a first catalog of extensive sketches and drawings of anatomical structures. With the invention of electricity in the 18th century not only as a matter of scientific inquiry, but also as a scientific paradigm, new insights became possible. One of the most well-known experiments at the time is that of the medic and anatomist Luigi Galvani (1737–1798), who showed in 1780 that frog legs move under the influence of static electricity. This experiment clearly rejected the theory of fluids in the ventricles and progressively replaced it with a neural-electricity theory.

The parallel inventions of the telescope and the microscope, which both combine several optical lenses, enabled totally new insights and ways of understanding biological mechanisms, processes, and forms of life (around 1600 by Zacharias Janssen, Hans Lipperhey and Galileo Galilei, as well as a slightly different type of telescope by Johannes Kepler). With the ability of seeing structural details 100 times larger than recognizable with the naked eye, biology developed a new understanding of organic tissue and the subdiscipline of cell biology was born. In 1665, the English universal genius Robert Hooke (1635–1703) discovered and documented that cells are the elementary parts of any plant, naturally leading to the investigation of cells in animal tissue, bacteria, and the discovery of single-cell organisms. In 1839, the German physiologist Theodor Schwann (1810–1882) hypothesized in his work "Microscopic researches into the accordance in the structure and growth of animals and plants" (Schwann, 1839) one of the foundations of modern Biology: cells form the basic building blocks of all plants and animals, including their individual parts, such as organs, hair, skin, muscles, etc.

Around the same time, the Italian physiologist Camillo Golgi (1843–1926) developed a method (silver staining) with which nerve structures could be visualized. Golgi thus discovered that the brain consists of millions of interconnected neurons. Even more importantly, he discovered that cells consist of many organelles, that is, subunits, one of which is now called the *Golgi apparatus*. He also discovered the *Golgi tendon organ*, which is a somatosensory organ in the muscle that senses changes in muscle tension. By means of the silver staining technique, the Spanish medical doctor Santiago Felipe Ramón y Cajal (1852–1934) was able to show that the millions of neurons in the brain communicate with each other via highly specialized neural connections, the so-called *synapses*. As a result of the discovery that the brain does not consist of one blended substance but rather of highly compacted, structured, and systematically interconnected cells, referred to as the *neuron doctrine*, Golgi and Cajal received the Nobel price for medicine and physiology in 1906.

Despite their discovery, the means of communication between cells still had to be unraveled. The German physiologist Emil du Bois-Reymond (1818–1896) uncovered a first part of the puzzle: starting off with the electrophysiological works of Galvani with frogs, he could show that there is a difference in electric potential between the inner part of a neural cell and the surrounding, which is now referred to as the *resting potential*. Moreover, this resting potential could be manipulated by means of neural stimulations leading to a characteristic change in the potential – first at the stimulated location and then moving along the nerve tract. The *action potential* was uncovered! However, how do neurons change the resting potential by themselves without external electrical stimulation? The German pharmacologist Otto Loewi (1873–1961) answered this question by showing that the communication between neurons via the action potential can be realized by discharging chemical substances, the now called *neurotransmitters*, at the synapses.

Taking a theoretical standpoint, the British physiologists Alan L. Hodgkin (1914–1998) and Andrew F. Huxley (1917–2012) proposed the highly influential mathematical model of cell communication, now called the *Hodgkin–Huxley–Model*, which is capable of modeling and simulating action potentials, including the properties of all or nothing (fire or not) and the refractory period (when drifting back into the resting potential). They received the Nobel Prize in medicine and physiology in 1963. As cognitive scientists, who are naturally interested in how the brain works from a functional perspective, Hodgkin and Huxley laid the foundation for computational neuroscience, as well as for important functional considerations in cognitive science. For the first time one could imagine that a computer could simulate neural communication, suggesting that perhaps a computer could explain or predict the behavior of neural cells and potentially even whole biological organisms. Taking this thought one step further, one could speculate that a computer could replace the neurons themselves, thus simulating or even becoming a neural-cognitive organism, possibly exhibiting animal or even human behavior and intelligence. Conversely, it also became imaginable that our brain could be nothing other than a very complex machine.

Naturally, other scientists at the time attempted to comprehend the functionality of the brain also on a more macroscopic level. A highly influential but totally incorrect theory of phrenology was developed by the German anatomist Franz Joseph Gall (1758–1828). According to his theory, certain traits of personality could be determined by certain extensions of brain regions, which in turn are reflected in the form of the skull. Unfortunately, this belief was used by German Nazis to foster racism – even though the theory had been clearly refuted by that time.

A much more scientifically sound approach, which is still used today, is the examination of patients with particular brain lesions. Their observed physiological and psychiatric deficits are believed to be correlated with the brain regions that were affected by the lesion, which, however, initially could only be located postmortem. Lesion studies, for example, suggested that prefrontal cortical areas control or at least influence social competence. Also the two most important and well-known language-related areas were discovered in this manner: the French medical doctor Pierre Paul Broca (1824–1880) discovered the now termed *Broca area* in the left interior frontal cortex, which needs to be functional to allow speech production. Meanwhile, the German neurologist Carl Wernicke (1848–1905) identified the now termed *Wernicke area* in the left posterior temporal cortex, which needs to be function to allow auditory speech comprehension.

Another important development was pushed by neuro-anatomists, such as the German Korbinian Brodmann (1868–1918). In his highly influential book from 1909 with the title *Vergleichende Lokalisationslehre der Großhirnrinde in ihren Prinzipien dargestellt auf Grund ihres Zellenbaues* (*Comparative localization studies of the cerebral cortex based on its cellular architecture*), Brodmann separated the cerebellar cortex into 52 areas according to their histological structure. These areas are now referred to as *Brodmann areas* and will be presented in further detail in Section 7.3.

Besides anatomical analyses, also brain imaging and electrophysiological single-cell recordings slowly became available and are still being brought to perfection. These advances led to the discovery of columnar cell arrangements, which were originally discovered by the neuroscientist Vernon B. Mountcastle (1918–2015) in the 1950s in the somatosensory cortex of cats. Soon afterwards, in 1959, the neurophysiologists David H. Hubel (1926–2013) and Torsten N. Wiesel (*1924) discovered similar cortical cell arrangements in the visual cortex, where individual neurons selectively respond to particular visually presented edge orientations. Once again a discovery that was worth a Nobel Prize: in 1981 Hubel and Wiesel received the Nobel Prize in medicine and physiology for the discovery of the information processing principles underlying the visual cortical system of the brain.

Over the last decades, neuroscience divided into several subdisciplines, each of which uses particular methods and is interested in particular neuroscientific questions. While they are certainly not fully separable, one can contrast the following subdisciplines:

- *Cognitive neuroscience* investigates the general neural processes underlying complex behavior, language, and even consciousness and imagination.

- *Systemic neuroscience* investigates the organization and functionality of individual neural systems, such as a particular sensory or motor system, as well as memory.

- *Molecular and cellular neuroscience* goes down to the level of individual neurons and small neural assemblies, focusing on the properties of cell membranes and ion channels, how proteins interact with neurotransmitters, precisely how and when an action potential is invoked, and even how particular genes are involved in giving rise to synaptic plasticity, modifying, adapting, and building up new neural connections.

- *Computational neuroscience* focuses on modeling molecular cell behavior and cell assemblies, but also systemic organizations of neurons and neural assemblies. The focus in this case is on mathematical models of the neural structures to develop a functional understanding of brain processes and to be able to simulate brain development, adaptation, and actual communication. The Hodgkin and Huxley model introduced previously laid the foundation for this scientific approach.

In summary, from a neuroscientific perspective, the knowledge that neurons are the basic building blocks in the brain is still rather new. The insights gained over the last few decades are huge and have produced an image and an understanding of the human brain that goes far beyond anything that was imaginable 150 years ago. Greatly simplifying, the most fundamental discoveries include the fact that neurons communicate via action potentials, neural information is topologically organized, local neural ensembles form cortical columnar structures selectively encoding particular stimulus properties in a systematic fashion, and individual neurons typically selectively respond to one particular stimulus property.

These units of signal processing in the neocortex constitute parts of the fundamental principles underling brain organization, development, and functionality. Systematic organization in particular may imply topological organizations in brain areas where the actual topology is yet unknown. In addition, the interplay between different topologies, the encoded particular stimulus properties, and the information exchange across different topologies still needs to be understood in detail and will be discussed in subsequent chapters. Considering the different levels of neuroscientific inquiry, cognitive science typically focuses on the cognitive and systemic levels as well as on higher-level computational inquiries. Molecular and cellular processes, however, may also be considered at times, for example, when attempting to verify hypothesized mechanisms of learning, adaptation, and information processing or when considering processes of neural degeneration leading to the development of cognitive deficits (such as Parkinson's or Alzheimer's disease).

2.3.2 Evolution

The mythology of nearly any civilization produced a story of the origins of life, earth, and humans. Once again ancient Greek philosophers may have been the first who tried to provide rational explanations about the origins of life and humans. Anaximander of Miletus (610–546BC) assumed that plants and animals developed from warm mud and humans developed out of a fish-like species significantly later. Empedokles (495–435BC) believed that life developed from moist mud. But he assumed that individual parts (such as body parts or organs) developed individually and united later on. Moreover, he assumed that only those species survived that had useful body part combinations. Based on Empedokles, Aristotle postulated:

> For nature never makes anything superfluous or in vain. She gives, therefore, tusks to such animals as strike in fighting, and serrated teeth to [25] such as bite. [...] Nature allots each weapon, offensive and defensive alike, to those animals alone that can use it; [...] and she allots it in its most perfect state to those than can use it best; and this whether it be a sting, or a spur, or horns, or tusks, or what it may of a like kind. (Aristotle, 2014, KL 28421-28426.)

Despite this assessment, Aristotle also believed that all species, whether primitive or complex, remain the same and are thus inalterable. Of particular interest is Aristotle's *epigenetic theory*, according to which the structures and parts of an organism are shaped only during the individual, ontogenetic development.   Although to some extent replaced in the 17th century by preformationism, epigenetic theory is under reconsideration today, seeing that certain genetic expressions appear to be influenced by the life circumstances of the mother during pregnancy or even before that, as documented for example in relation to the Dutch famine in 1944. Several ideas of Greek philosophy can be found in modern evolutionary theory:

- Humans have developed from pre-human species.

- Only bodily and neural shapes, forms, and parts that are somewhat useful for the species will emerge.

- Each species is equipped with those bodily parts and capabilities that are particularly useful to do what the species does.

We will see that these considerations are manifested in evolutionary theory in terms of the evolution of new species by means of natural selection. Moreover, this natural selection depends on a fitness for survival and reproduction, which in turn depends on the ecological niche in which the species lives and which it shares with other species.

At the start of modern age, and the realization that the catholic church is not inerrant, progressively more doubts about the biblical account on the history of creation developed. The French zoologist Jean-Baptiste Lamarck (1744–1829) was the first to attempt to put forward a consistent and encompassing theory of human evolution.   He proposed that simple life forms developed over centuries and millennia into more and more complex species, modifying and adapting the body parts to the encountered environment as determined by successful usage. Giraffes, for example, develop long necks because they again and again attempted to reach the fresh leaves in treetops. Lamarck thus postulated a directed evolutionary process that strongly depends on the experiences of a species during its lifetime. This does not appear to be the case – at least in the direct manner Lamarck proposed.

Evolutionary theory in its present form was born with the publication of the British biologist Charles R. Darwin's (1809–1882) most famous book *The Origin of Species* in 1859 (Darwin, 1859). The main postulates of his book are:

- All live forms are in a continuous struggle and stand in competition with each other for life-essential resources, including water and nutrition.

- Small property differences within the individuals of a species develop due to mutation (random changes) or due to a recombination of the inherited, parental properties.

- Good properties and property changes are passed on to subsequent generations by means of the general principle of the *survival of the fittest* realized by means of *natural selection*. Fitter individuals will live longer and reproduce more often, thus having a higher chance of passing particular suitable traits on to the next generation.

- In this manner, fitter species and subspecies will occupy the respective ecological niches and will lead to the extinction of less fit species, whose organisms rely on the identical or similar natural resources for survival and reproduction.

Besides the actual implications of this evolutionary theory, it should also be noted that the evolutionary theory, founded on biological observations and facts about species, their traits, and interactions, manages to offer a ground-breaking theory *without* actually understanding the microscopic mechanisms that enable this theory to actually work – genes and DNA had not been discovered, yet. Nonetheless, the Darwinian theory of evolution holds true until today with few modifications and enhancements.

Darwin was very much aware of this lack of microscopic knowledge about how traits may be passed on to the next generation. The general belief at the time was still dictated by preformationism to large extents, which suggested that in each sperm and each ovule of the mother resides a miniature form of the whole organism. Although the discovery of individual cells and cell properties had refuted preformationism to a certain degree, how and which traits are passed on to the offspring via sperm and ovule remained unknown. Darwin suggested a mechanism called *pangenesis*, He assumed that each cell may have a gemmule, which stores experiences about bodily activities and later on moves to ovule or sperm to pass the stored trait onto the offspring. Similar to Lamarck, Darwin's pangenesis principle thus implies the inheritance of traits that developed and were utilized during a lifetime, although put forward on a cell level rather than on a cognitive level. Thus, Darwin's theory is often contrasted to Lamarck's theory of evolution because Darwin postulated natural selection as the main driving force of evolution, whereas Lamarck focused on directed evolution due to actual, direct, and fully explicit evaluations of the traits of a species during a lifetime.

At the same time that Darwin was developing his theory on evolution, the Austrian priest and natural scientist Gregor J. Mendel (1822–1884) came up with the answer to the question what is actually passed on to the offspring. Focusing on plants and species of peas in particular, Mendel formulated his famous now-called *laws of Mendelian inheritance*, which he derived from the evolutionary development of individual pea plants when selectively crossbreeding them. Initially overlooked, Mendel's laws actually mark the birth of modern genetics and the whole field of genetic biology, thus constituting a scientific revolution in the sense of Kuhn.

Further discoveries over the following century led to the building blocks of biological genetics, identifying that genetic encodings can be found on deoxyribonucleic acid (DNA), which constitute chromosomes. Chromosomes in turn separate in two parts and recombine forming new types of cells, which contain the recombined genetic information. The American molecular biologist James D. Watson (*1928) and the British bio-chemist Francis Harry Compton Crick (1916–2004) uncovered the double helix structure of the DNA, which is made of four basic nucleotides – or letters – forming the genetic alphabet (guanine (G), adenine (A), thymine (T), or cytosine (C)). Watson and Crick received the Nobel Prize in medicine and physiology in 1953 for this discovery.

On a more philosophical level, the British biologist Richard Dawkins (*1941) in this 1976 book *The Selfish Gene* (Dawkins, 1976) modified and enhanced the implications of Darwin's theory with his theorization of selfish genes. Dawkins argues that each individual life form inherits a randomly combined selection of genes from its parents. Thus, metaphorically speaking, natural selection actually leads to a competition between "selfish" genes, which compete for survival. The organisms themselves only indirectly fight for survival and repro- duction driven by their selfish genes. Dawkins theory not only provides many examples and metaphors about the principles of natural selection, it also provides an explanation for al- truism, which Darwin was always puzzled about. It makes sense to help a close relative even in life-threatening situations because the relative is likely to possess a significant number of the genes the altruistic individual possesses. In his further works, including "The blind watchmaker" and "Climbing mount improbable" (Dawkins, 1986, 1997), Dawkins positioned himself as a protagonist who fostered and further established Darwin's theory of evolution against the still present religiously-motivated belief in creationism. In particular, he pro- duced clear examples and explanations about how the diversity of life on earth can develop solely based on random mutations, recombination, and natural selection, leading to a bal- anced design of each species and their interplay in the natural niches on earth, such as the savanna, the rain forest, or a coral reef. Moreover, he plots a pathway for the development of humans with our cognitive and intellectual capabilities.

Seeing that humans are also the product of such continuous evolutionary processes based on natural selection, it can be assumed that generally more genetic material will be passed on to next generations that was particularly suitable in the given circumstances. These circumstances are nowadays certainly strongly influenced by many social and cultural factors. Nonetheless, due to these facts, we can assume that also the human genetic material is still in flux and is continuously evolving. Moreover, it also implies that human genetic material may not be the "best" material possible for any purpose it may encode for – be it intelligence, the ability to interact socially, language, tool usage, or similar – but it was certainly produced by natural evolution to foster some of these traits.

Question about the origins of life and particularly human life are not only exciting, but also full of implications about cognition and the human mind. Genetic predispositions and their phylogenetic development in species over centuries should also be considered in conjunction with ontogenetic development, that is, gene expression and consequent organismic growth, development, and state of life at any point in time. While genes determine general development, the environment, the nutrition, the social interactions, and other environmental factors ultimately determine cognitive, intellectual, and bodily development under the given gene-determined constraints. It may be best to phrase these implications in terms of questions instead of answers:

- Why did particular behavioral traits, as well as bodily and brain structures develop in the way they did?

- Why did only humans develop such a complex, auditive communication system as manifested in existing languages? How is human language so significantly distinct from the communication forms and proto-languages found in other animals?

- How did the different cognitive subsystems in our body and brain develop over the millennia separately and in interaction, as well as how do they develop ontogenetically in interdependence with the encountered environment, including other humans and animals?

- What is the nature of the niche that humans evolved into? Will this niche develop inevitably during any process of natural evolution, that is, will human-like intelligence and human forms of language inevitably be developed by natural evolution?

- With respect to other animals, which mechanisms, bodily traits, and capabilities have developed to optimize survival and reproduction in their respective ecological niche? Which are these bodily traits and capabilities that are decisive for survival and reproduction, and what are the determinant features of the species-respective ecological niche within which it evolves and develops?

In later chapters we will see how these considerations go beyond cognitive science and biology. When striving to understand the design of artificial cognitive systems, such as smart robots, bodily morphology, ecological constraints, and niche properties matter as much as they do in natural selection. In fact, one may speak of cultural and economic forms of evolution that are driven by human kind, where a market niche is comparable to a free ecological niche, such that evolution may cause a new product to evolve that covers this niche.

Meanwhile, it is possible to learn from biological systems by considering their morphology, their sensory and motor capabilities, and their cognitive complexity. One important lesson is that most animals do not understand the world in the detail humans do. Thus, understanding is not really necessary for survival. Moreover, another lesson is that some animals have bodily traits that exceed human capabilities in various respects. Thus, evolution does not necessarily tend toward maximal complexity. It appears that *balanced designs* are striven for by natural selection and evolution, optimizing only where necessary, while drifting where possible.

2.4 Psychology

While mathematics, physics, biology, or chemistry established themselves as natural sciences centuries if not millennia ago, psychology as a matter for scientific inquiry was a sub-discipline of philosophy for centuries. Only at the end of the 19th century was it established as its own discipline and as a separate subject in universities. As do philosophy and biology, psychology comprises such a large spectrum of scientific inquiries that we cannot cover them all satisfactorily in this book. Once again, we will focus on key aspects and particularly those aspects that are most relevant for cognitive science.

A pioneer of modern psychological research, the German Gustav Theodor Fechner (1801–1887), may be called the father of psychophysics, which he postulated as a matter of scientific inquiry in his treatise from 1860 on the *Elemente der Psychophysik* (*Elements of Psychophysics*). Fechner pleaded for a scientific approach to psychology that must be based on experimentally assessed data. Almost 20 years laster, in 1879 Wilhelm Wundt (1832–1920) founded the first institute of experimental psychology worldwide in Leipzig. It was officially recognized as an institute by the university in 1883. In the United States, William James (1842–1910) can be considered as the founder of psychology. In his influential work *Principles of Psychology* from 1890 (James, 1890), which is still cited today, James not only summarized the available knowledge on psychology (even considering neural correlates for psychological phenomena), he also formulated four basic psychological methodologies of scientific inquiry, that is, psychological paradigms: (behavioral) data analysis, introspection, psychological experiments, and statistical comparisons.

2.4.1 Behaviorism

At the beginning of the 20th century *behaviorism* established itself as the most influential scientific paradigm of psychological research. Behaviorism stressed the importance of psychology as a well-founded scientific discipline that ensures reproducibility and falsifiability of psychological theories. Thus, behaviorism focused on observables and particularly on observable and measurable behavioral adaptations. In its purest form, behaviorism makes the following assumptions:

- Behavior and behavioral adaptations are based on simple learning mechanisms, which are based on experiences of the organism about. Essentially, it is assumed that the organism forms *stimulus-response associations*.

- This behavior is observable, predictable, and explainable without the need to refer to internal, mental, cognitive processes – essentially avoiding speculations about unobservable mental states.

Behaviorism thus focused on learning and adaptation of behavior, and established two basic forms of reward-driven learning, which are commonly referred to as *classical conditioning* and *operant conditioning*.

Classical conditioning has its roots in the experiments of the Russian physiologist Ivan Pavlov (1849–1936), who examined dogs behavior before and during feeding. In particular, he observed how dogs start salivating before actually starting to eat. He then determined whether a stimulus, such as the sound of a bell, would lead to the dog salivating even without the presence of food. He was, in fact, able to make the dog salivate by pairing the sounds of a bell with the beginning of a feeding event in close temporal proximity – with the bell essentially enabling the anticipation of the feeding event. For his work, Pavlov received the Nobel Prize in medicine and physiology in 1904 in recognition of his work on the physiology of digestion, through which knowledge on vital aspects of the subject has been transformed and enlarged (Pavlov, 1904).

Abstractly speaking, classical conditioning is the learning of a pairing of a *conditioned stimulus* (CS; such as the sound of the bell) with a meaningful, *unconditioned stimulus* (US; such as the food), which leads to a typical *unconditioned reaction* (UR; such as salivating).

Once the CS and US are paired sufficiently often and consistently, CS–UR are associated by the organism, such that the CS alone is sufficient to invoke the UR reaction – essentially transforming the UR into a then called *conditioned reaction* (CR). Based on the work of Pavlov, the American psychologist and co-founder of behaviorism John B. Watson (1878–1958) transferred the studies of Pavlov to human, experimenting with, for example, fear-inducing stimulus pairings. In 1920 he published the "Little-Albert-Experiment", done with his collaborator Rosalie Rayner, in which a little boy was reported to develop strong fear toward a white rat, to which he previously showed affection, by pairing the appearance of the rat with scary sounds. In the further development of this experimentally induced phobia, the little boy was reported to generalize this fear even to fur in more general and other, similar animals.

Operant conditioning ties back to the American psychologist Edward L. Thorndike (1874–1949) and his work on how cats and dogs can learn to open a cage that is locked by a simple mechanism. Based on these experiments and resulting observations, he postulated the *law of effect*, according to which behavior that leads to a positive consequence, such as the release from a cage and the discovery of food outside the cage, will be reinforced and thus expressed more often in similar contexts. In contrast to Pavlov and Watson, operant conditioning stresses the modification and adaptation of behavior itself, whereas classical conditioning focuses on the pairing of already available and typical unconditioned behavior to novel conditioned stimuli.

In the 1930s, the American psychologist Burrhus Frederic Skinner (1904–1990) continued the work of Watson and Thorndike. He expanded the behaviorist stimulus-response learning theory further, by presenting positive and negative reinforcers not only at the end of a trial or interaction episode (as did Thorndike, Watson, or Pavlov), but also immediately after the individual presented particular, initially typical spontaneous behavior. In this way, he was able to teach animals complex behavioral sequences. This principle is still partially used when training animals for circus shows. Similarly, the principle manifests itself when training dogs by the clicker training method, where the click serves as the indicator of positive reinforcement before the reinforcement is actually presented, very similar to Pavlov's bell sound. The click essentially allows the trainer to indicate which behavior of the dog was the good one in much closer temporal proximity as would be possible with bits of food. As a consequence, much faster training of behavioral sequences is possible.

Behaviorism may thus be considered as a very important first step toward an emancipation of psychology as its own science. The pure methodological approach based on well-founded, replicable psychological behavior-based experiments was an important step toward scientific inquiries without false beliefs or overly theoretical approaches without the possibility for falsification. Additionally, the dependence on metaphysical or introspective states was no longer necessary. As seen by clicker training, basic principles of behaviorism, such as the concept of a reinforcer, are still matters of research.

Nonetheless, behaviorism sketched-out an overly simplified picture of the animal and human mind and behavior. The neglect and denial of cognitive processes and cognitive, internal states and encodings, oversimplified the matter to large extents. Many aspects of animal, and to an even larger extent human, cognition was fully neglected: how is innovative thinking and problem solving possible? How can goal-directed, anticipatory behavior be established by a purely, reinforcement-driven system? How are intricate things such as empathy, social interaction, or linguistic communication possible without more intricate learning and cognitive capabilities? A classic experiment by the German biologist and psychologist Wolfgang Köhler (1887–1967), co-founder of Gestalt psychology, illustrates the matter. Köhler observed that apes are able to solve a complex problem (such as reaching a banana that is out of reach) by an apparent sudden insight (realizing that they can stack boxes to reach the banana). Clearly, insights beyond pure reinforcer-dependent, conditioned behavioral adaptations appear possible.

2.4.2   Constructivism and developmental psychology

While behaviorism and reward-based learning dominated much of the field early in the 20th century, many questions remained or were raised in parallel with the idea of reward-based learning and adaptation. A key question was: how does the mind develop ontogenetically during a lifetime? Are the principles of conditioning sufficient to explain our versatile behavioral and cognitive capabilities? How can the human mind control and learn to control its body in a goal-directed and controlled fashion?

The German philosopher and psychologist Johann Friedrich Herbart (1776–1841) was one of the first to explore these questions. He asked: how is it possible to act voluntarily, that is, to strive for goals and thus to act goal-directedly? The *ideomotor principle* of psychology was born, which suggests that our brains start learning by associating bodily motor actions, which may be initially invoked by simple inborn reflexes or even simple muscle contractions, with the sensory effects that are registered after the motor action was executed. Later on, when the effect is desired again (regardless if consciously or sub-consciously by the presence of particular stimuli), the associated motor action can be invoked and can be further optimized over time. William James phrased the core of ideomotor theory as follows:

> An anticipatory image, then, of the sensorial consequences of a movement, plus (on certain occasions) the fiat that these consequences shall become actual, is the only psychic state which introspection lets us discern as the forerunner of our voluntary acts. (James, 1981, p. 501.)

Similar assessments came also from the Würzburg school of psychology as put forward by Narziß Ach (1871–1946) when talking about determinant tendencies:

> Determinant tendencies may be thought of as causes that start with a peculiar mental idea about a goal and that lead to a determination in the sense of or according to the meaning of the idea.
>
> [Unter den determinierenden Tendenzen sind Wirkungen zu verstehen, welche von einem eigenartigen Vorstellungsinhalte der Zielvorstellung ausgehen und eine Determinierung im Sinne oder gemäß der Bedeutung dieser Zielvorstellung nach sich ziehen.] (Ach, 1905, p. 187, own translation)

Albeit the ideomotor theory was strongly criticized by the head of the American Psychological Association (the APA) Edward L. Thorndike (1874–1949) in 1913, as a result of its relationship at the time to occult settings and mystic relations between effect and resultantly executed cause, the theory has experienced a revival over the last few decades. Essentially, the theory stresses the important idea that animal and human infants initially learn to associate self-generated sensory and motor stimulations with each other, thus starting to make sense of the world by first understanding the peculiarities and systematicities of their own body.

Most likely motivated by behaviorism, but also by the ideomotor principle, developmental psychological aspects were investigated further. The highly influential swiss psychologist Jean Piaget (1896–1980) postulated for the first time that cognitive development may be considered as an active, constructive process, which is driven (i) by the structure of the available knowledge, (ii) by the assimilation of novel, acquired knowledge into the available structures, and (iii) by the accommodation of novel knowledge by means of a restructuring process. Piaget theorized that typically four stages of cognitive development are passed. First, only simple reflex and perceptual schemata (sensorimotor schemata) are used. Later on, preoperational, cognitive processes are established. These processes are combined and transformed over time into concrete, operational processes, which can be flexibly employed. With additional refinements, these capabilities are developed further into formal-operational, abstract, and highly systematic capacities for generating thought.

Piaget's theories were derived from various behavioral experiments – many of them done with his own three children. His methodologies were highly questionable in retrospect –

most of his scientific inquiries were exploratory, based on reports and observations, without any possibility for replication. Standardized protocols or statistical evaluations were not conducted.

One of the most famous series of experiments done by Piaget is the demonstration of the *A-not-B search error*. Piaget reported to hide an object visible to a child in location A (for example, behind a pillow), which was easily recovered at the same location by the child. However, when he hid the object visible to the child at location A and then at location B (still visible to the child), the child tended to search for the object at position A (where it was previously successfully retrieved) and not at position B. He observed this error with babies aged between eight and twelve months, but not later than that.

How can such behavior be explained, which can, by the way, also be observed in studies with animals? Piaget assumed that the schema of object permanence had not fully developed, yet. Another possibility is that the child had an immature memory system. A third possibility is that the child may not yet be able to control her behavior sufficiently voluntarily, such that the child is incapable of disregarding the previously successful action at location A, even though location B was observed as well.

A similar approach was also pursued by the Russian developmental psychologist Lev S. Vygotsky (1896–1934), who also assumed that learning is an active and constructive process. In contrast to Piaget, however, he strongly emphasized the importance of social interactions, coining the term *Zone of Proximal Development* (ZPD), which

> [...] is the distance between the actual developmental level as determined by independent problem solving and the level of potential development as determined through problem solving under adult guidance or in collaboration with more capable peers. (Vygotsky, 1978, p. 33.)

Decisive for effective mental development and a precursor for independent problem solving of a particular task is scaffolding, that is, the guidance and encouragement by caretakers and peers. A good example may be learning to ride a bike: in the beginning it is hard to steer, pedal, and balance at the same time. So parents assist with the balance until the child manages to balance on her own. Similarly, balance bikes bootstrap the biking capability simply by disentangling pedaling from balancing and steering, thus enabling the child to learn to ride a bike by a simpler, two-stage process.

Vygotsky was furthermore convinced that learning and cognitive development are a life-long matter, which stands in strong contrast with Piaget's believe that development has matured once the formal-operational stage is reached. Life-long learning is a key-term in various current research directions and becomes ever more important in our so quickly changing society with its electronic gadgets. This life-long learning, according to Vygotsky, is strongly influenced by culture and the individual social network. As a consequence, mind, ideas, and values are psychological instruments that are adapted according to the propagated and assumed values in the experienced society.

To summarize, while psychology is still focusing most of its resources on the study of the adult human mind with its capabilities and peculiarities, constructivist psychology beginning with the ideomotor principle has fostered the importance of cognitive development right after birth – or even before that. The consequent developmental, constructive process is assumed to progressively make more sense of the encountered world, systematically structuring it according to the gathered experiences and the involved regularities in these experiences. In doing so, individual experiences from interactions with physical objects and social experiences from interactions with peers, caretakers, and other humans and animals strongly influence cognitive development. Finally, the constructivist psychology usually assumes that cognitive processes are goal-directed, seeking and processing information about the world actively, rather than observing and analyzing it passively.

2.4.3 The cognitive turn

While in the United States behaviorism dominated the field of psychology in the first half of the 20th century, in Europe behaviorism was less popular, and cognitive processes were considered in many of the published scientific works. However, the experimental protocols were admittedly often much less objective than they should have been. In Germany, the Würzburg school of psychology established by Oswald Külpe (1862–1915) investigated how humans are able to solve complex problems, using self-reporting during a problem solving session. The Würzburg school was strongly criticized by Wilhelm Wundt and the Leipzig School of Psychology, because of its questionable methodology. Nonetheless, even self-reporting is still used especially in conjunction with but typically only as an addendum to other measurable tests.

Meanwhile, *Gestalt psychology* was established by Max Wertheimer (1880–1943), Wolfgang Köhler (1887–1967), and Kurt Koffka (1886–1941), propagating the idea that the whole Gestalt, that is, the form or configuration of a thing, dominates the importance of the parts. Gestalt psychology essentially investigated the critical constitutive of perceiving a whole object given only minimalistic parts, such as how we are able to perceive a full human in motion when seeing only some systematically moving dots, or how we are able to perceive a whole glass, even if significant parts of it are hidden behind an occluder.

In the United States, the American psychologist Edward C. Tolman (1886–1959) was one of the key pioneers of the belief that there is more to psychology than behaviorism, and particularly, that learning takes place even in the absence of concrete reinforcers. Although Tolman called himself a behaviorist – most likely attempting to stress the fact that he solely focused on controlled behavioral experiments in psychological research — he moved forward and proposed modifications to the rather static stimulus-response learning principle put forward by Watson, Skinner, and others. He introduced inner states and intervening variables (between stimulus and response), which according to him may still be investigated and explored by behaviorist methodologies.

Tolman is particularly well-known for his thorough experiments with rats and their behavior in "T-mazes" (mazes with one or many T-junctions). In particular, he observed that rats learn the outline of a maze by simply allowing them to explore it without the provision of any obvious reinforcer. He concluded so, by observing that rats, who had the opportunity to explore a maze for several days without a reinforcer, were much faster in learning a path to a provided reinforcer afterwards when compared with rats that had not the opportunity to explore the maze beforehand. In various other experiments he confirmed this observation, suggesting that a form of *latent learning* takes place, that is, learning without immediate consequence and without the dependence on current reinforcement. The knowledge about the maze achieved by latent learning mechanisms can then be employed at will once a particular goal (as manifested by a positive reinforcer, such as food, at a certain location in a maze) is perceivable and desirable. The rat thus derives a shortest path estimate to the desired goal by employing the *cognitive map* of the maze it had previously learned latently.

Besides latent learning of cognitive maps, another key player in the cognitive turn was the American linguist Noam Chomsky (*1928), who propagated the idea of a generative transformation grammar and the principle of an *universal grammar* in the human mind. In his books *Syntactic Structures* (1957/2002) and *Aspects of the Theory of Syntax* (1965), Chomsky opposed the descriptive tradition of linguists (Chomsky, 2002, 1965). Instead, he took the developmental standpoint, and asked the question how we may learn the complex grammatical rules and structures of our languages. His work and thoughts were certainly influenced by various well-known publications. One classical work came from the French linguists and philosophers Antoine Arnauld (1612–1694) and Claude Lancelot (1615–1695), called the *Port Royal Grammar* published in 1660. This work essentially postulated that grammar is a universal mental process that encompasses inner, meaning constitutive aspects as well as outer, phonological aspects. Also the German linguist, philosopher, and politician

Wilhelm von Humboldt (1767–1835), who had assumed that language is generally a rule-based system, apparently influenced Chomsky's ideas.

Chomsky's formalization of a generative grammar, which formalizes sentences by a finite number of symbols and *production rules*, even entered the research realms of theoretical computer science and formal logic (cf. Chapter 13 for further details). Cognitively speaking, Chomsky's universal grammar offers an answer to the question how we humans are apparently able to produce and comprehend an infinite amount of sentences, including sentence that were never encountered before. Possibly, the resulting cognitive flexibility, which is supported by human languages with their universal grammatical structures, is one of the key factors that distinguishes humans from other animals.

From a developmental psychological standpoint, Chomsky fueled the nature-nurture debate, which asks the question: how much of our language competence is inborn, that is, determined by the genes (nature), and how much is acquired during a lifetime under the influence of peers, caretakers, and society (nurture)? Watson assumed that everything can be trained and learned by the principles of conditioning:

> Give me a dozen healthy infants, well-formed, and my own specified world to bring them up in and I'll guarantee to take any one at random and train him to become any type of specialist I might select – doctor, lawyer, artist, merchant-chief and, yes, even beggar-man and thief, regardless of his talents, penchants, tendencies, abilities, vocations, and race of his ancestors. I am going beyond my facts and I admit it, but so have the advocates of the contrary and they have been doing it for many thousands of years. (Watson, 1930, p. 82.)

In contrast, Chomsky opposed this assumption by his theory of a *universal grammar*. According to Chomsky, this universal grammar encodes general, basic grammatical principles, which are universally available to or acquired during infancy by any human child. The available *Language Acquisition Device* (LAD) is used to learn the particular language the child is exposed to – by embedding the principles of the universal grammar in the particular language structure.

Over subsequent decades various studies have been conducted on the development of language in children, including mother tongue studies, second language studies, and bilingual studies. One particularly appealing example in favor of the universal grammar is the fact that sign languages used by deaf-mutes has a complexity that is very much comparable with the grammatical structures of spoken languages. Even more intriguing is the example of the *Nicaraguan Sign Language*, which was developed by deaf-mute children in the 1980s in Nicaragua with hardly any supervision by adults and which was further refined when it was passed on from elder to younger children. In this case, a complexity emerged that is covered by the universal grammar principles. Regardless of whether it is inborn or developed in interaction with the environment during the first months or first few years of life, the capability of developing a language that reflects universal grammatical structures, but at the same time is unique, remains as one of the Holy Grails of scientific research in cognitive science.

The works of Chomsky influenced, among others, the Canadian psychologist Albert Bandura (*1925) and led him to question behaviorist theories. In 1963, the *Bobo doll study* laid the foundation for a theory of learning by observation. Bandura had small children (between four and five years of age) watch a movie in which an adult beat-up, kicked, and scolded a plastic doll called Bobo. After that, the children watched one of three endings of the movie: the aggressive behavior of the adult was either positively evaluated by another adult, or it was negatively evaluated, or it was not further commented upon. Then the children were taken into another room where among other toys also the Bobo doll could be found. Bandura observed the expectable: the children showed similar aggressive behavior toward Bobo when the previously observed behavior was positively evaluated or not commented upon, but they did not show this behavior when it had been negatively evaluated. Bandura thus showed that the reinforcer does not need to affect the child itself (as behaviorists would assume);

behavior can also be modified solely by learning from observation. When reinforcement is involved, it suffices to observe the reinforcement, which, nonetheless, implies significant cognitive capabilities: the observed needs to pay attention to the interaction, interpret the observed behavior in the context correctly, mirror this behavior onto her own behavior accordingly, and remember the observed interactions accurately.

2.4.4 Memory

While psychologists acknowledged that mental processes are inevitable to be able to explain all observable behavior, the big question about how these mental processes work, function, and are structured, is still a matter of inquiry. It remained particularly unresolved how learned information – be it via observations or active interactions – is ordered, stored, and accessed on demand. Memory models thus became a matter of scientific inquiry.

Once again, these questions were not driven purely by behaviorism, but inquiries on the nature of memory had already been raised in the 19th century. The German psychologist Hermann Ebbinghaus (1850–1909) focused his experimental psychological research on learning and memory performance. As early as the 1870s he conducted systematic learning experiments, as for example on the learning of a sequence of meaningless syllables. Albeit his main subject was himself, perhaps bringing into question the validity of the test, he made highly important observations, which are considered valid until today:

- Over time learned items are forgotten. The resulting curve of forgetting can be described by a negative exponential function, suggesting that we forget most right after learning and exponentially less over time.

- He also observed that the order of the presented syllables plays a crucial role: syllables that are presented in the beginning (*primacy effect*), as well as in the end (*recency effect*) of a sequence, are remembered best.

The methods put forward by Ebbinghaus are still being applied in studies on memory.

William James also contributed to the study of memory. He distinguished between knowledge that is currently directly and consciously accessible (primary memory) from knowledge that first needs to be actively remembered (secondary memory). Today, the terms *short-term* or *working memory*, and *long-term memory* are used to refer to these two contrasting types of memory.

In the 1960s, the perception of memory was strongly influenced by the development of computers, so that human information processing and memory were directly compared to the memory system in a computer.

> Computers take symbolic input, recode it, make decisions about the recoded input, make new expressions from it, store some or all of the input, and give back symbolic output. By analogy, that is most of what cognitive psychology is about. It is about how people take in information, how they recode and remember it, how they make decisions, how they transform their internal knowledge states, and how they transform these states into behavioral outputs. [...] The terms are pointers to a conceptual infrastructure that defines an approach to a subject of matter. Calling a behavior a *response* implies something very different from calling it an *output*. It implies different beliefs about the behavior's origin, its history, and its explanation. Similarly, the terms *stimulus* and *input* carry very different implications about how people process them. (Lachman, Lachman, & Butterfield, 1979, p. 99.)

Cognitive psychology at the time was ready and willing to propagate the computer metaphor of the brain.

Still very influential is the multi store, modal model of memory proposed by the American psychologists Richard C. Atkinson (*1929) and Richard Shiffrin (*1942), which was strongly influenced by the computer metaphor. Their model distinguishes three crucial components, which have different capacities, memory sustainability properties, and information encoding structures:

- *Sensory registers* encode physical properties as registered by the sensors, such as visual, auditory, or haptic information. Principally, Atkinson and Shiffrin assumed that sensory registers have infinite capacity; the information, however, is assumed to be stored only for maximally a few hundred milliseconds. Everything that is not further processed, which is mediated by attention, is forgotten after this short period of time.

- *Short-term memory* is the active memory part where individual units of "thought" can be temporarily stored and maintained over an extended period of time spanning several seconds to a few minutes. The short-term memory capacity was assigned the infamous number of 7 ± 2 units, which has been downscaled to rather 4 ± 2 units. Short-term memory is assumed to actively process and maintain information gathered from the sensory registers and combine with units of long-term memory. To date, it remains unclear where or how the 4 ± 2 units are stored, which types of units can be actively maintained, and how do maintained units interact with other units.

- *Long-term memory* is assumed to be manifested in a semantic network of meaningful representations that is assumed to have a theoretically infinite capacity. Stored bits of information can be accessed automatically by particular processes. However, it is assumed to be possible that the ability to access particular memory items fades over time, leading to their inaccessibility.

Many experiments have consistently confirmed the general correctness of these distinctions. Several studies by the British psychologist Alan Baddeley (*1934) and Graham Hitch, however, raised doubts about the nature of short-term memory. Observing that parallel tasks led to selective interferences, they developed their *working memory* perspective, propagating a further modularization of short-term memory into:

- a *phonological loop*, in which phonological, language-based information (in the form of sequences of phonemes) is maintained by active repetitions;

- a *visual-spatial sketch pad*, in which dominantly visual information is stored, further separated into spatial aspects, such as object positions and object motion, and visual aspects, such as form, color, and object identity.

- a *central executive*, which coordinates the transfer of information between long-term memory and the other two components.

Baddeley and Hitch enhanced their memory model further by an *episodic buffer*, which is assumed to be a multimodal working memory component that is capable of storing relevant information about encountered interaction episodes.

Even though nobody now would doubt the existence of a mechanism that brings about working memory, the strict compartmentalization of working memory has been questioned over the last decades (cf., for example, Ericsson & Kintsch, 1995; Rubin, 2006). Not only are sensory information other than visual and auditory information often not considered (such as proprioceptive information, taste, or smell), but even more importantly interactions between the assumed memory components are often neglected and the purpose of working memory itself remains barely addressed. In addition, the nature of the selectivity by which items are chosen to enter working memory – presumably coordinated by the central executive – has hardly been addressed.

The *Levels-of-Processing Theory* somewhat addresses this question, postulating that the depth of processing of an item is crucial to how well it will be remembered (Craik & Lockhart, 1972). However, the depth of processing remains somewhat ill-defined. Moreover, it remains unknown how the selectivity of the depth of processing comes about – how is it that we remember particular aspects of an episode selectively in much more detail than other aspects? How does our brain decide which aspects of a scene are important and are thus processed in further depth?

In this book we put forward the embodied approach to cognitive science to give partial answers to these questions. This embodied approach essentially suggests that those aspects will be processed in further detail that are assumed by the brain to possibly be "behaviorally" relevant, or rather, relevant for maintaining internal *homeostasis*.

2.5 Bringing the pieces together

While we had given a short discussion on embodied cognition in the first chapter, this chapter has focused on how the different strands of scientific history led to where we are today in cognitive science. Influences can be identified from philosophy, particularly epistemology and the philosophy of mind, biology, particularly neurobiology, and evolutionary theory, as well as from psychology, particularly learning, behavioral, and developmental psychology. All strands of scientific development were recently heavily influenced by the development of computers and other electronic devices. Philosophical questions, such as could a computer in principle think, neuroscientific questions, such as is a neuron nothing more than a processing unit in a biological computer, as well as psychological questions, such as is the brain a symbol processing machine, is language based on an inborn universal grammar, or is memory nothing more than a physical storage device, all imply that the computer metaphor is omni present. In fact, it seems that cognitive science would always get stuck on qualitative explanations about how the mind works if the computer, as well as principles of mathematics, statistics, and information processing, were not available. The rise of the computer paradigm as a method for cognitive modeling may lead to a deeper understanding of the brain, its capabilities, and functional properties.

However, as seen in our short history of science, a new paradigm comes with huge potentials, but also with many pitfalls. Scientific care is necessary to be able to validate computer models and computer analogies – and better yet to falsify such models and analogies in order to reveal the true beauty and diverse capabilities of our brains.

In this vein, the British mathematician, computer scientist, and psychologist David Marr (1945–1980) – now considered one of the founding fathers of computational neuroscience – compared the functionality of the brain with that of computers. Inspired by this comparison and aware of potential pitfalls, he suggested that human information processing or even more generally human thought, or the human mind, needs to be understood on *three levels* of understanding:

- the computational level,

- the algorithm level, and

- the hardware level.

He described the computational level as follows:

> The most abstract is the level of *what* the device does and *why*. [...] The whole argument is what I call the *computational theory* [...]. Its important features are (1) that it contains separate arguments about what is computed and why and (2) that the resulting operation is defined uniquely by the constraints it has to satisfy. (Marr, 1982, p. 22f.)

Marr illustrated this level using a cash register. The machine adds up any combination of numbers (what) in order to produce a final value, the sum that the customer has to pay (why). Constraints are that the machine does summations properly, false values are not acceptable, and that the machine does not charge for nothing, that is, no item yields zero as the sum. Note how this level of analysis is already found in Aristotle's concept of the *efficient cause* and the *final cause*, where the former specifies the *what* and the latter the *why*.

> The second level of the analysis of a process [...] involves choosing two things:
> (1) a *representation* for the input and for the output of the process and (2) an
> *algorithm* by which the transformation may actually be accomplished. [...] this
> second level specifies the *how*. (Marr, 1982, p. 23.)

With respect to the example of the cash machine, the question is raised of how numbers may be represented and how additions may be accomplished. When choosing a binary form of representation or even the Roman numerals, for example, the rules for addition differ from those necessary to work with the Arabic decimal system. Aristotle's concepts of the *formal cause* are most present in this case, but the *efficient cause* is also relevant in that the form determines the available meaningful manipulations.

Finally, Aristotle's material cause – that is, the actual implementation on a physical device – emphasizes that the material and its properties ultimately determine the computational progress and its manifestation. In Marr's words:

> This brings us to the third level, that of the device in which the process is to
> be realized physically. The important point here is that [...] the same algorithm
> may be implemented in quite different technologies. (Marr, 1982, p. 24.)

Clearly, various forms of representation and physical implementations are possible. An analog cash register may consists of cylinders on which the numbers zero to nine are placed and for which one full rotation leads to one further rotation of the next cylinder. In an electronic machine, on the other hand, numbers will be represented in a binary manner and translations between binary representations and decimal visualizations need to be available, as well as the ability of the machine to add numbers in the binary format. Thus, Marr essentially stresses the fact that the same goals can be accomplished by different means and with different control algorithms. He furthermore stresses the importance of understanding cognitive processes on all three levels, tying together neurobiology and neuroscience with psychology, and even the philosophy of mind. Proper analyses on each of these levels and even more importantly, interactions between these levels, are a matter of ongoing research in cognitive science.

For example, in respect to the memory models, qualitative, partially philosophical explanations focus on Marr's first level of explanations, albeit often not satisfactorily. An explanatory attempt to form an embodied cognitive science perspective may be that memory is the maintenance of acquired knowledge over time (the what) in order to interact with the environment in smarter ways in the future (the why). The memory models by Atkinson and Shiffrin, as well as the modifications by Baddeley and Hitch are on the second, that is, the algorithmic level, addressing the how question. Representations in short-term memory are characterized by items in these models, differentiated by modalities in Baddeley and Hitch's perspective, which are manipulated by a "central executive". The exact algorithm of the central executive and the forms of representation of particular items, however, are a matter of current research.

Neurobiology and neuroscience in general address mainly the third level, that is, the hardware question: how does the physical device, that is, the brain, accomplishes the required computational processes and brings about the involved forms of representations? Due to this somewhat disciplinary research, however, bridges between the different levels need to be built, focusing on the question how the brain with its neural structures and mechanisms can generate a short-term or working memory that typically exhibits a somewhat itemized nature of the current memory contents. Once again, computer models seem inevitable to bridge the emergent explanatory gaps between Marr's levels of understanding.

To provide another analogy, Marr's three levels may be compared with the three scientific disciplines physics, chemistry, and biology. Physics focuses on the hardware, with its functionality and properties. Chemistry builds on top of physics putting physical components together in such complex ways that it required a whole new discipline focusing on the subject of more complex composite forms of material and matter. Finally, biology builds on chemistry, but investigates the nature of life, cells, and whole organisms. As a cognitive scientist one should be aware that the three levels of understanding proposed by Marr should always be taken into consideration and one should also be aware at which level the current analysis is conducted. Furthermore, one should be aware that once an understanding of a cognitive process, mechanism, or form of representation on all three levels is believed to be accomplished, a full understanding is still likely to be illusive as long as the interactions between the different levels were not considered or fully understood, yet.

In a somewhat comparable manner, psychological research has developed several subdisciplines, that may be characterized as follows:

- *General psychology* investigates the nature and functionality of diverse cognitive aspects, including perception, learning, memory, problem solving, reasoning, attention, motivation, emotion, and speech. The level of analysis is mainly qualitative residing on the first, most abstract level of Marr's classification, although sometimes reaching into the second level when proposing an actual model, such as the introduced memory models. Neural correlates are typically hardly considered.

- *Biological psychology* is the one that focuses on the physiological correlates of behavior and experience. Matters of analysis are not only the hardware of the human body, including brain and muscle activities, but also blood pressure, heart rate, and other somatic markers.

- *Developmental psychology* focuses on how cognitive and behavioral capabilities are acquired during development and in adulthood. How do the individual physical and cognitive systems develop over time? How do they interact? Which ones are inborn – genetically determined – and which ones develop in interaction with the environment?

- *Evolutionary psychology* asks questions about the evolutionary roots of biological, developmental, behavioral, and cognitive traits in humans, including, for example, language evolution, tool usage, or social cooperation.

- *Social psychology* is concerned with the importance of interacting with other humans, such as understanding other humans, showing empathy, or being able to communicate. It also addresses implications for the development of the individual mind in social interaction. Moreover, individual differences between humans and groups of humans – focusing, for example, on the development of personality traits and intelligence – are considered.

An analogy to Tinbergen's four basic questions of biological research does not seem to be far fetched: Biological psychology addresses the mechanisms and causations that lead to particular capabilities and properties of mental processes. General psychology addresses these mechanisms and causations from Marr's higher levels of understanding. Moreover, the adaptation question is addressed when addressing the question how behavior can be manipulated and adapted in certain situations. Developmental psychology focuses on the ontogeny, that is, the cognitive development of a lifespan. Phylogeny may often be considered only insufficiently, which is also due to the lack of knowledge about our ancestors beyond 5000–10,000 years.

Cognitive science is designated to bring all these subjects and levels of understanding together and build bridges between them in order to enable the development of a satisfying answer to the question "how the mind comes into being". To illustrate the difficulty of the task, we end this chapter with an east-Asian allegory: imagine a group of blind people who attempt to learn what an elephant is like by tactile inspection. One of them may grab the trunk and experience an elephant as a type of snake. The other may touch a leg and associate the term elephant with a somewhat soft, but leathery tree trunk. Another may only get hold of one of the Elephant's tusks, interpreting it as a sharp and dangerous spear. Thus, all three men have a picture of an elephant that is very different from each other. None of them, however, has perceived the animal as a whole. Can the different experiences be put together and linked into a whole?

Cognitive science often focuses on the task of understanding the whole. However, inevitably, due to the complexity of the human mind, parts also need to be understood in detail before the whole can be addressed. The implication is that good communication between the involved disciplines and subdisciplines is inevitable in order to develop an overall understanding of cognition. Moreover, a functional understanding seems necessary, which allows both the transfer of information between the different levels of understanding, as well as the transition of one mechanism into macroscopic others. An example may be the nature of working memory, which inevitably needs to emerge due to the properties of neurons, the nature of their activations, and their interactions in the brain. Various other bridges between levels of understandings and functional explanations will be put forward throughout the rest of the book.

2.6   Exercises

1. Characterize the mind–body problem in your own words. How is Descartes famous statement "Cogito Ergo Sum" related to that problem?

2. The view that consciousness is a "homunculus", which observes sensory information and makes motor decisions, can be put *ad absurdum*. Why and how? Give a possible alternative computational characterization of consciousness.

3. Even if a computer was generated that claims to be fully conscious and seems to be highly intelligent, why would the qualia problem still stand?

4. John Locke and David Hume were the protagonists of the empiricist movement in philosophy. Shorty explain their standpoint. In which way does the empiricist movement avoid the homunculus problem. Give an example of how we may learn about cause–effect relations from experience during early cognitive development.

5. Denis Diderot and Julien de La Mettrie stressed the materialistic point of view on the world including cognition. From their point of view, where does prejudice arise from and what is the reason for the existence of religion?

6. Contrast the inductive from the deductive scientific methodology. In which manner do these logical empiricist approaches to knowledge go beyond the ideas of empiricism – or do they actually contradict empiricism?

7. Contrast a word from the idea that the word refers to.

8. Given a word in one language, it is often not possible to find an exactly corresponding word in another language. Why might this be the case? Why is it nonetheless the case, that ideas can be communicated and shared across languages, even if one person dominantly thinks in one language and the other person dominantly in another one?

9. Summarize and contrast the main contributions to neuroscience of Broca and Wernicke with those of Mountcastle, Hubel, and Wiesel.

10. Contrast phylogenetic from ontogenetic development. In which computational manner do the two developmental processes interact?

11. In which fundamental manners does behaviorism differ from constructivism? What is the role of the "mind" in behaviorism and what is its role in constructivism?

12. Why does constructivism go hand-in-hand with developmental psychology?

13. Relate the Noam Chomsky's idea of a universal grammar to the nature–nurture debate. Taking an empiricist, constructivist perspective, where may the language acquisition device in children come from?

14. Relate the idea of a central executive, similar to a computer's central processing unit, to the homunculus problem. What is, generally speaking, the alternative to such a central executive in the brain?

15. Imagine an electrical device in the kitchen, such as a toaster, a coffee machine, a microwave, or a stove.

    - David Marr has introduced three levels of understanding cognition. Explain the imagined device on all three levels to reach a full understanding of it.

    - More than 2000 years earlier, Aristotle has introduced four causes to answer the question "why" something exists. Explain the existence of the kitchen devices according to Aristotle's four causes.

16. Relate Tinbergen's four basic questions in biological science to Aristotle's four causes.

Chapter 3

Cognition is Embodied

3.1 Computers and intelligence

While we have so far focused on the sciences relating to the mind, the brain, and biological systems, over the last century a new driving force has developed – computer science in general and artificial intelligence (AI) in particular have established themselves as further fundamental foundations of cognitive science. The possibility of imagining that machines may be able to think on their own one day – and even do so already to a certain extent – leads to thoughts and ideas that were hardly imaginable before the development of the first machines.

The development of complex machines and computers in the 19th, and much more so in the 20th and 21st, centuries has fostered the computational perspective on cognition immensely. Nonetheless, the old Greeks and Egyptians had already constructed the first machine-like systems. Some statues were equipped with intricate control mechanisms – lifting their arms or blinking their eyes – effectively making them appear even more god-like for the average believer. Thus, the first seemingly intelligent artifacts already existed in antiquity.

In the 12th century AD, the first written reports about automatons with human appearance can be found. These automatons were able to pour a drink by a complex mechanical mechanism or they were able to "play" an instrument, even though only to a rather creepy extent. Three centuries later, Leonardo da Vinci imagined many other such mechanisms and proposed a myriad of inventions, which also included intelligent automatons. In all the reported cases, however, these automata were "pre-programmed", in that they were mechanically structured such that they would exhibit some form of seemingly intelligent, human-like behavior. The simulation of thought itself, however, was hardly imagined. This was also probably due to the philosophical and religious take on nature at the time – it was basically unimaginable, but also implicitly forbidden to imagine that thought, and thus the soul, may be mimicked by a machine, let alone that a machine may have its own thoughts and ideas.

Only in the late 18th century did science and philosophy slowly begin to think that animal – and possibly even human – behavior was not driven by an internal spirit or soul, but by processes and material mechanisms that may ultimately be explained by natural ways of reasoning. This movement of materialism and naturalism, as introduced in the previous chapter with respect to the protagonists de La Mettrie and Diderot (cf. Section 2.2.1), led to the first explicit considerations that one day a machine might actually exhibit human-like cognitive capabilities.

A first seemingly "intelligent, thinking" machine was indeed introduced shortly after such ideas spread across Europe. The inventor, Wolfgang von Kempelen (1734–1804), presented the Chess-Turk, a human-sized puppet that was able to move chess figures on its own and corrected mistakes by shaking its head. Seeing that this machine was, in fact, capable of winning against human opponents, for a while it was considered a phenomenal invention.

However, although the mechanics of the puppet were quite refined and intricate, actual decision making and control was done by a human, who was hidden inside the apparatus. Indeed, the German saying : "Das ist bestimmt getürkt" (literally: certainly this is "turked"; that is, certainly this is a cheat) developed from Kempelen's invention.

At that time, much more pressing than trying to mimic the human mind, however, was the development of machinery that could support humans at work. Not surprisingly the term robot comes from the Slavic word *robota*, which may be translated as "enforced labor" or even "slavery". First successes came from mechanical calculators, as developed by the astronomer Wilhelm Schickard (1592–1635) and the philosopher and polymath Gottfried Wilhelm Leibniz (1646–1716) in the 17th century. The machines could solve basic calculus tasks with much higher numbers and much faster than an average human could. Leibniz also demonstrated the advantages of using a binary system in machines, rather than the decimal system.

The mathematician and inventor Charles Babbage (1791–1871) may be called the "father" of modern computers. In the 1820s he developed his Analytic Engine – a mechanical, steam-powered machine that was equipped with control, storage, and processing units. The Analytic Engine was supposed to be capable of accomplishing various kinds of computational tasks, although it never reached full functionality. Concurrently with Babbage's inventions, the first computer algorithms were developed by his coworker Ada Lovelace (1815–1852), who also suggested that machines may one day think. She, however, disagreed with the possibility of reaching human-like innovative thought, stating that:

> The Analytical Engine has no pretensions to *originate* anything. It can do *whatever we know how to order it* to perform. (Lovelace, 1842, her italics, as quoted by Turing, 1950, p. 450.)

Note, however, how Lovelace contrasted the idea of original thoughts and ideas with systematic algorithms that simply process instructions, which were programmed by humans. It took another couple of decades until scientists started to seriously wonder: what if the machine starts to write its own instructions ...?

A decisive contribution to this development was made by the English logician and philosopher George Boole (1815–1864), who developed the Boolean Algebra in the 1850s. It contains only the binary states *true* and *false*, as well as the logical operators *and, or,* and *not.* In fact, even today the Boolean Algebra provides the basis for all computer architectures, as in the end all computations are based on huge concatenations of true and false states, which are signaled by differences in the flow of electricity.

The first fully functional, albeit mechanical, computer was built by the German engineer Konrad Zuse (1910–1995) in 1941. His Z3 machine was a binary, fully automatic, freely programmable machine, which was very slow compared with current computers and also broke down after only two years. The first electronic, fully functional computer was developed in the United States in 1946, called ENIAC. Even more significant may be the work by the American electrical engineer Nathaniel Rochester (1919–2001), who developed the computer IBM-701, which was IBM's first commercial scientific computer and on which, amongst other things, the very first artificial intelligence algorithms were tested. Soon the industrial production of computers and, thus, the unstoppable computerization of our world began. The first storage media came in the form of simple cards with holes in them, which had originally been used to control automatized weaving looms. In the 1960s, these storage devices were replaced by magnetic tapes and discs, eventually leading to the development of the extremely fast digital storage media available today.

Besides these mere computational advancements, most significant for the further development of cognitive science and artificial intelligence may have been the ideas, thoughts, and solutions put forward by Alan Turing (1912–1954) in the first half of the 20th century. As a British logician and cryptanalyst (also known as the genius who helped to decipher the German Enigma, which led to a very important turn in the intelligence capabilities of the Allies in the Second World War), Turing was a visionary computer scientist before the actual development of fully functional (and sufficiently fast) computers. His contributions to theoretical computer science on the *decision problem* (Is there an algorithm that can decide in finite time if an input satisfies a certain property or not?) and on *computability theory* (Can an algorithm exist that is able to determine for any input of a potential set of inputs a correct answer?) are invaluable. His most important contribution is certainly the Universal Turing Machine, which boiled-down the capabilities of a computer to very few highly simple symbol manipulation and storage mechanisms. In fact, with the help of the Universal Turing Machine, he was able to show that the decision problem is not solvable in the general sense. Moreover, he showed that anything that is *computable* – which can be calculated by an algorithm – is also computable by the Turing Machine.

3.2 What is intelligence anyway?

Due to the development and availability of computers, and the rise of a general understanding that the computability principle is very powerful, scientists began to think about the meaning of "intelligence". What if the brain was nothing more than a huge computer that can solve any type of computable problem? What does this imply about the uniqueness of human beings and philosophically about the question of free will? Can we then still be made responsible for our actions? Vice-versa, what if a computer is built that matches or even surpasses the cognitive capabilities of humans? Is this computer then as intelligent as we are? Is it as alive as we are? As self-responsible? As conscious?

How does such a computer need to function? Which capabilities does it need to become "intelligent"? Before we address these questions and the historic development of answers to them, we first need to ask the question, "What is intelligence?" We will see that intelligence can be defined in a variety of ways, so that the task of developing intelligent computers or robots is certainly not as straight-forward as partially believed in the 1960s and 70s.

3.2.1 Early conceptualizations of intelligence

Before the rise of computers, early definitions of "intelligence" were developed with the goal of assessing the mental capacities of school children. The French psychologist Alfred Binet (1857–1911) and his student Théodore Simon (1872–1961) developed first assessment mechanisms. They also attempted to clarify the term "intelligence":

> Nearly all the phenomena with which psychology concerns itself are phenomena of intelligence; sensation, perception, are intellectual manifestations as much as reasoning. [...] in intelligence there is a fundamental faculty, the alteration or the lack of which, is of the utmost importance for practical life. This faculty is judgment, otherwise called good sense, practical sense, initiative, the faculty of adapting one's self to circumstances. (Binet & Simon, 1916, p. 42.)

This very general characterization of intelligence indeed contains certain aspects of intelligence, which we will later use as a general definition – the capability to adapt (mental and physical behavior) to current circumstances. How can we determine or assess this capability?

In the following years, Binet and Simon developed approximately 30 different kinds of exercises, which focused on assessing particular competencies. Driven by their belief that the intellectual development of a child follows a particular path – each child with its own pace – the "Binet–Simon tests" were designed to identify the intellectual, developmental age of a child. Note that the famous child psychologist Jean Piaget (1896–1980), who was a student of Simon, was very much influenced by these ideas in the advancement of his stage-wise developmental theory. The Binet–Simon tests were modified and developed over the decades in various ways, now being manifested in the generally well-known assessment of an intelligence quotient (IQ).

At about the same time, the English psychologist Charles Spearman (1863–1945) published his two-factor theory of intelligence. He observed that the results produced by one person in different exercises, which were generally different from those developed by Binet and Simon, typically correlated with each other. His conclusion was that:

> [...] all branches of intellectual activity have in common one fundamental function (or group of functions) [the g-factor], whereas the remaining or specific elements of the activity [the s-factors] seem in every case to be wholly different from that in all the others. (Spearman, 1904, p. 284.)

Instead of analyzing the different capabilities of an individual in a differential psychological manner, Spearman attempted to grasp the general nature of intelligence.

Once again, in contrast to Spearman, the American Psychologist Louis Leon Thurstone (1887–1955) proposed seven independent primary factors of intelligence, termed primary mental abilities, which included calculus, language, spatial cognition, memory, deductive capabilities, word fluency, and the capability to conceptualize. Empirical studies led to contradictory results, however. While in some experiments a differentiation (no correlation) between these factors was possible, correlations were detected in others. Thus, intelligence was understood either as consisting of facets of "intelligence", which may be separated to certain extents, or as being characterized by one general intelligence factor.

3.2.2   Further differentiations of intelligence

As a student of Spearman, the American psychologist Raymond B. Cattell (1905–1998) developed, together with his student John L. Horn (1928–2006), the concept of a fluid and crystalline intelligence:

> [...] there are those influences which directly affect the physiological structure upon which intellectual processes must be constructed – influences operating through the agencies of heredity and injury: in adulthood development these are most accurately reflected in measures of fluid intelligence. And on the other hand there are those influences which affect physiological structure only indirectly through agencies of learnings, acculturations, etc.: crystallized intelligence is the most direct resultant of individual differences in these influences. (Horn & Cattell, 1967, p. 109.)

Thus, according to Cattell, intelligence is not a universal, static property, but it is shaped and developed by means of learning and individual experiences. This learning and development is coupled with the culture within which a person grows up, which ties back to Vygotsky's developmental psychological considerations. Acknowledging that a culture and individuals within that culture are interactively co-existing and co-developing, a reconciliation was offered to the intense debate of *nature versus nurture*, that is, the debate if intelligence is determined by the genes or by experience: both factors mutually influence and shape each other!

Besides the recognition that intelligence develops, is fluid, and is influenced by the environment in which each individual grows up in, others pursued the idea that intelligence has different facets to it. In 1984 the American psychologist Robert J. Sternberg (*1949) published his "Triarchic Theory of Intelligence", which is very functional- and process-oriented. Essentially, he suggested that the fundamental factor that determines intelligence is how each individual applies her or his individual information processing capabilities to the experiences gathered while interacting with the environment. According to Sternberg, intelligence can be split into three major aspects: *analytical aspects*, which focus on the capability to recognize regularities and structure in the environment; *creative aspects*, which focus on how well novel situations are handled, and how well and flexible automatized processes can be performed; and *contextual practical aspects*, which focus on how one applies ones knowledge, and ones own practical and behavioral abilities in the given circumstances. Thus, while focusing on general intelligence, Sternberg emphasized three functional, information processing aspects, which critically contributed to intelligence.

In contrast to the rather unitary process-oriented perspective of Sternberg, the American psychologist Howard E. Gardner (*1943) developed his theory of multiple intelligences. He postulated seven fundamental intellectual capabilities: linguistic, logical-mathematical, visual-spatial, musical-harmonic, body-perception, interpersonal, and intrapersonal. In 1999, he added a form of naturalistic or possibly biological intelligence, which refers to the ability to understand other species and signs in nature. A form of spiritual intelligence was also mentioned. Gardner's theory is still widely discussed, but often is considered not concrete enough to be of actual value to cognitive science. Nonetheless, Gardner's theory emphasizes that humans may be proficient in various aspects of "intelligence", and these aspects are not restricted to abstract forms of reasoning.

To summarize, as originally characterized by Binet and Simon (Binet & Simon, 1905), intelligence may be characterized by processes of short- and long-term adaptations, which determine how knowledge and experiences are applied in novel situations in an adaptive manner. Most researchers in cognitive science now agree that "intelligence" should not be restricted to symbol manipulation or logical forms of reasoning, but rather intelligence needs to be understood as a flexible, fluid, adaptive process, which manifests itself in different contexts and situations in different manners. Due to these different manifestations, different forms of intelligence can be distinguished to a certain extent and each individual is typically not equally skilled in all of these forms. In essence, several individual functional, information-processing skills determine the particular intellectual capabilities of an individual.

3.3 Symbolic artificial intelligence and its limitations

With a general grasp of intelligence in the 1950s – well before the current sophisticated understanding of intelligence and adaptivity – and with the general idea in mind that computers may, indeed, theoretically be as intelligent as humans, the first rise of research on artificial intelligence began. Is it possible to generate human-like thought machines, which can help us solve pressing engineering and mathematical problems? The first enthusiasm led to exciting speculations of how quickly human thought may be simulated.

The birth event of artificial intelligence is often considered the Dartmouth Conference, which took place in 1956 at Dartmouth College in Hanover, New Hampshire (USA). Renowned researchers from different fields, including John McCarthy, Marvin L. Minsky, Claude E. Shannon, and Nathaniel Rochester, formulated a clear aim to study "artificial intelligence" – a term that was coined in this proposal. The goal of the meeting was clearly formulated and was expected to lead to significant progress in one (!) summer project:

> We propose that a 2 month, 10 man study of artificial intelligence can be carried out during the summer of 1956 at Dartmouth College in Hanover, New Hampshire. The study is to proceed on the basis of the conjecture that every aspect of learning or any other feature of intelligence can in principle be so precisely described that a machine can be made to simulate it. An attempt will be made to find how to make machines use language, form abstractions and concepts, solve kinds of problems now reserved for humans, and improve themselves. We think that a significant advance can be made in one or more of these problems if a carefully selected group of scientists work on it together for a summer. (McCarthy, Minsky, Rochester, & Shannon, 2006, p. 12.)

Although no solutions could be presented at the meeting or shortly after, key protagonists in the further development of this young field attended the conference, including, in addition to McCarthy et al., also Arthur Samuel, Herbert A. Simon, and Allen Newell. As a consequence, symbolic artificial intelligence received a primary research focus, although biological, neural network-mimicking structures were also considered.

3.3.1 Symbolic problem solving

An influential first AI system was the General Problem Solver, developed by Herbert A. Simon (1916–2001) and Allen Newell (1927–1992) in the late 1950s after the Dartmouth event (Newell, Shaw, & Simon, 1959; Newell & Simon, 1961). As suggested at Dartmouth, the aim was to solve it all, that is, to create an algorithm that was able to solve any type of problem that could be formulated. The software developed generally follows the principle of problem reduction: step-by-step, the overall problem is reduced into simpler subproblems, which are then individually solved and put back together to yield an overall solution. The procedure is now often termed *means–end analysis*, which is certainly a very human-like problem approach: the final goal is approached step-by-step by the available means. While the General Problem Solver was successful in that it could solve some limited logical theorems and geometrical problems, less well-defined problems stayed out of reach.

As a result of the failure to really create a general "general problem solver", the research focus in AI shifted to individual facets of intelligence, thus addressing smaller, typically well-defined problem domains. In the following years, many *expert systems* were developed, which were specialized to solve or provide support for individual problem domains.

In particular, the game of chess gained a research focus very early. Already in the 18th century the aforementioned Mechanical Turk gained significant interest. The computer pioneers Konrad Zuse and Alan Turing actually developed first sketches of potential chess programs, the realization of which was yet impossible, due to the limited computer power. In the 1950s, Claude Shannon (1916–2001), who is now considered the founder of *information theory*, published his ideas on the creation of a chess program (Shannon, 1950). Indeed, he laid out the basic principles of most of the chess programs that are currently on the market: forward tree search, combined with minimax-based pruning and a proper evaluation of board constellations.

In the end, though, it was not until 1997 that the computer "Deep Blue" from IBM successfully beat the chess world champion at the time, Garry Kasparov (May 11, 1997, result: 3.5-2.5 Deep Blue vs. Garry Kasparov). As originally proposed by Shannon, Deep Blue's main approach was based on good evaluation functions of board constellations and on tree search. The evaluation function, however, was much more complex than originally thought (more than 8000 features), and the number of moves in the sequences considered reached over 30 million positions and included up to 20 moves into the future.

Analyses of how humans actually play chess have shown that the anticipation of potential sequences of moves goes much less deep and is much more selective, and the "evaluation function" in humans – and especially in chess experts – is much more elaborate. The hard part, however, is to implement this "human evaluation function". Over the last decade or so, evaluation functions have significantly improved and brute-force forward search is often replaced by more selective forward search processes. As a result, standard chess programs reach very high performance levels without too much computational effort.

In many other board games, computers have reached human-superior performance (such as Scrabble, Checkers, Othello, or Backgammon). Even for the game of Go, which was considered a remaining challenge until very recently, a computer program is now available that has successfully beat one of the best Go players of the world (Silver et al., 2016). Go has a huge number of possible game states and the branching factor – when planning ahead – is equally huge. This is why brute-force simulations of all possible moves have no chance of being effective. Planning heuristics, more intelligent board evaluations, and sub-strategies are absolutely necessary to succeed. Indeed, a partially randomly exploring, so-called *Monte Carlo tree search* led to the first breakthrough in Go in 2007 (Coulom, 2007; Gelly & Silver, 2011). Essentially, particular positions are evaluated not by considering all possible continuations, but by considering only a few partially heuristically, partially probabilistically selected continuations, averaging across them. The addition of learning the evaluation function by means of a clever combination of deep-learning techniques with reinforcement learning techniques from games, which the algorithm partially played in simulation against itself, has led to the recent success (Silver et al., 2016).

Board games with discrete states thus remain marginally interesting with respect to cognitive science. Due to their discrete states and symbolic forms of representation, computers can use brute-force computational techniques (such as efficient search), which are easy to realize with a computer but much harder with our brain. Even the implementation of brain-like solutions to board games has been accomplished to certain extents – particularly the mentioned Go program, which partially uses deep learning neural network techniques, is revealing in this respect. However, even this program does not really understand the purpose of the game, let alone the implemented training routines. In this sense, the further development of computational models that reveal the neurocognitive mechanisms in our brain that enable us to play, and become proficient in, particular board games – or also in video games for that matter – still promises to generate important insights for cognitive science.

To summarize, the overly optimistic goal of developing intelligent systems in one summer – or at least setting the foundations to do so – generated much enthusiasm over the following years. It also resulted in the insight that there is more to intelligence and cognition than mere symbol processing and logic. Prognoses about success were delayed: in 1958 Simon and Newell expected the development of a competitive chess playing computer over the next 10 years. In 1965, it was estimated that in 20 years a human-like machine would have been developed, but we are still rather far from the latter. Expert systems became the research focus in artificial intelligence for a while and gained some success, in particular in well-defined niches and problem domains. Due to too much optimism and the failure to live up to it as fast as prognosticated, a first "AI-winter" developed in the late 1970s, resulting in reduced funding and limited research progress.

3.3.2 Symbolic linguistic processing

In addition to logic-oriented problem solving, the problem of human language capabilities was of early interest. In 1966 the German-American computer scientist Joseph Weizenbaum (1923–2008) proposed the "ELIZA" program, which may be considered the mother of all chat bots available today. The most famous version was the DOCTOR version, a simulation of a psychotherapist, which could indeed generate a reasonable natural conversation with a user:

> Human: My boyfriend made me come here.
> *Computer: YOUR BOYFRIEND MADE YOU COME HERE?*
> He says I'm depressed much of the time.
> *I AM SORRY TO HEAR YOU ARE DEPRESSED.*
> I need some help, that much seems certain.
> *WHAT WOULD IT MEAN TO YOU IF YOU GOT SOME HELP?*
> Perhaps I could learn to get along with my mother.
> *TELL ME MORE ABOUT YOUR FAMILY.*
> My mother takes care of me.
> *WHO ELSE IN YOUR FAMILY TAKES CARE OF YOU?*
> My father.
> *YOUR FATHER?*

Despite an overwhelmingly positive resonance at the time, ELIZA is very limited. ELIZA works by processing simple scripts in which rules are combined with pattern matching techniques. Key words and phrases are extracted – such as "mother", "need help", etc. – and put together in an appropriate response, typically faking an empathetic reaction or generating a targeted question. In the case of no proper match, the last sentence is simply rephrased as a question, or a standard question is asked. Weizenbaum himself was rather negatively surprised by how many users were fooled for significant amounts of time by ELIZA.

An interesting next step toward a more proper, general language capability was taken by the American computer scientist Terry A. Winograd (*1946), who developed as his dissertation project at MIT, under the supervision of Marvin Minsky, from 1968 to 1970 the program SHRDLU. The program could give information about and interact with a blocks world simulation, in which various cubes, pyramids, and other objects could be placed. Due to the contextual restriction, a rather impressive program could be built that made SHRDLU much more "intelligent" than ELIZA: the program could not only follow instructions and manipulate its world accordingly or give information about the current state of the world, it could also learn new terms and actions, plan to generate particular object constellations, report about past actions and constellations, and try out new actions. Here, is an excerpt from the SHRDLU demo:

> Human: Pick up a big red block.
> *Computer: OK. (does it)*
> ...
> Find a block which is taller than the one you are holding and put it into the box.
> *BY "IT", I ASSUME YOU MEAN THE BLOCK WHICH IS TALLER THAN THE ONE I AM HOLDING.*
> *OK. (does it)*
> ...
> Can a pyramid be supported by a block?
> *YES.*
> Can a pyramid be supported by a pyramid?
> *I DON'T KNOW.*
> ...
> How many things are on top of green cubes?
> *I'M NOT SURE WHAT YOU MEAN BY "ON TOP OF GREEN CUBES". DO YOU MEAN:*
> *1 - DIRECTLY ON THE SURFACE*
> *2 - ANYWHERE ON TOP OF ?*
> ...
> Had you touched any pyramid before you put the green one on the little cube?
> *YES, THE GREEN ONE.*
> When did you pick it up?
> *WHILE I WAS STACKING UP THE RED CUBE, A LARGE RED BLOCK AND A LARGE GREEN CUBE.*
> Why?
> *TO GET RID OF IT.*
> `(http://hci.stanford.edu/~winograd/shrdlu/index.html)`

In terms of linguistics, SHRDLU was much more versatile than ELIZA, while ELIZA worked with simple pattern matching, SHRDLU implemented syntactic and semantic tools. However, in the end SHRDLU failed to be expanded. It was not directly applicable in other contexts, and it was never applied to more complex worlds. Nonetheless, SHRDLU had many properties of a computer chat bot that seemed to understand its little world consisting of simple objects.

There are several reasons for the failure to expand such programs and to develop general symbolic- and logic-based problem-solving mechanisms, which are discussed in further detail in the remainder of this chapter. In the following chapters, mechanisms and techniques are then introduced that shed light on more human-like, cognitive aspects of thought and reason, as well as on their foundations. The implications from traditional logical and purely symbolic artificial intelligence approaches thus are that there is more to human intelligence than logic and symbolic forms of reasoning. In the book, we put forward the view that in order to succeed in creating truly intelligent and helpful artificial systems, lessons from cognitive development and embodied cognition need to be considered from a functional and computational perspective.

3.4 Hard challenges for symbolic processing systems

Despite the successes with logic and symbol-based AI systems, these techniques have encountered important limitations over the last decades. Expert systems have been applied successfully in various domains, but even in the particular domains in which they have been applied, human experts still often surpass their computer-based competitors.

How can domains be characterized in which traditional AI systems largely fail? Is it the mere complexity of our environment? Board game domains and related tasks are usually mainly deterministic with few stochastic components, such as dice or when shuffling cards. The states are discrete and mostly accessible. Thus, in each state it is clear what can be done next and the uncertainty can be expressed mathematically by means of probability theory. Moreover, usually all players know which game interactions have happened so far (again possibly with some uncertainty).

These properties stand in rather sharp contrast to the real world. States are not well-defined but are continuous. States are also not fully observable – for example, our eyes provide only certain information about the environment. Some environmental properties (such as radioactivity) cannot be sensed at all by our bodies. The number of possible hidden states is so huge that it seems impossible to account for all of them. Even the existence of particular entities, forces, and causes in the environment, which determine the hidden environmental state, needs to be learned from scratch, given sensory information, whose precision can also only be deduced from experience.

In relation to these hard challenges of learning about the complexity and diversity of states in the world, our brains need to solve fundamental problems for being able to develop abstract cognitive abilities. The *symbol grounding problem* asks what do symbols actually stand for? The *frame problem* asks what is relevant for succeeding in the world? Finally, the *binding problem* asks which things, properties, and information sources belong together, and should thus be fused into distinct entities?

3.4.1 Symbol grounding problem

The symbol grounding problem may be characterized by the following three questions:

- What do symbols actually stand for?

- How are symbols related to worldly entities and ideas?

- Where does the semantics, that is, the meaning of symbols come from?

The main problem is that in our real world things, items, objects, properties, events, behaviors, etc. come in a seemingly infinite number of forms and types. When learning a symbol, it seems impossible to identify exactly those properties to which the symbol refers to. Take the example of a "dog". Dogs come in various forms and shapes, colors, fur lengths, strengths, speeds, noisiness, smelliness, etc. What do dogs have in common? They typically have four legs, they bark, have fur, have a particular body shape and head form, and also exhibit particular behavioral manners and motion dynamics. These may be common properties, but already the identification of these properties, their exact characterization, and estimations of importance of each property seem far from straight-forward. Despite these challenges, even at the age of two, children are typically fully capable of identifying dogs and of generalizing their knowledge to uncommon examples of that species. Even cartoon drawings of a dog are typically easily recognized at that age.

Thus, *object categories* form in our brains very early on – probably even before actual language capabilities develop. These categories are then ready to be mapped onto symbols, that is words, given ongoing and concurrent language development.

Note also how these object categories seem to be very robust recognizers, showing invariance against distortions, occlusions, and the point of view from which we observe a particular object. We recognize a dog from the side and from the front, from the top and even from the bottom if necessary, although this last perspective would certainly be more challenging. We recognize a dog looking out of a car window or out of its kennel, when it sits, sleeps, or begs. We recognize a dog under different lighting conditions, under different fur conditions, and even when mostly covered under a blanket. Thus, our perceptual recognition system is very invariant with respect to size, orientation, lighting, and partial occlusion. It is essentially able to recognize particular categories – and particular identities for that matter – under many different circumstances and despite this without overgeneralizing.

One may think that animals pose a particularly strong challenge to such an invariant category recognition and identification system, but artifacts and machines pose similar problems. Consider, for example, the category "car": it becomes quickly obvious that cars come in very different forms, colors, and shapes, but a young child has no problem understanding (realizing) that a (motor) trike is neither a motorcycle nor a car, and a bobby-car is not really a car, either, although it is certainly more similar to a car than to a bicycle. Consider another example, the general category "container", that is, anything that may contain something else. Visual information alone is not sufficient to identify all possible containers – such as bags, mugs, bottles, buckets, trailers, but also even ship containers, airplanes, bodies, houses, and so forth. Thus, a non-visual, conceptual representation seems necessary to characterize the term "container".

These examples show that symbols do not only describe visual properties and they thus cannot exist only as visual-based encodings in our brains. Symbols of our language are inherently *conceptual*, describing a conglomerate of particular properties, binding them into a meaningful, but very abstracted, symbolic form.

Meanwhile, this conglomerate of properties that are bound into a symbolic category can also help us to link associations and to generate ideas, which may go beyond our imaginative abilities that are available without the support of symbols. Consider the task of drinking water. We can think of various forms in which we can transport water to our mouth: a glass, a mug, or a bottle are rather obvious forms. But our own hands may also serve well enough for the task, or even a sheet of paper could be used. In all these cases, the particular objects (including our hands) would all be used as (temporary) containers, which are sufficiently suitable for the task at hand. Another alternative would be to side step the transportation problem altogether and simply drink directly out of the faucet. In all cases, we assign the symbol "drink" to the behavior.

Thus, symbols are grounded in various particular and selective invariant aspects that characterize entities in our world or interactions with our world. These aspects are typically not merely visual properties – apart from a few exceptions such as color names (although even those are context-dependently assigned as we will see in later chapters) – but are grounded in various perceptual modalities and characteristic environmental properties. As these characteristics are not equally meaningful for us, we tend to select those that are of a particular relevance or interest. For example, young children love to point out cars and dogs but much less so trees, houses, or chairs. Things have a particular attractiveness to us from an early age and this attractiveness develops with our knowledge over time.

The American psychologist James J. Gibson (1904–1979) proposed a *theory of affordance*: objects in our world have particular behavior-grounded meanings. By exploring our environment actively with our body, we can realize what we can manipulate and in which way. To act upon our environment in an ecologically effective manner, we need to know what is out there and what can be manipulated how, and what affects us in what way. Things in our world thus afford particular interactions, which are indirectly primed by the mere observation of those things. These affordances change during bodily and cognitive development, and are particularly dependent on the current manipulatory abilities. A bobby-car, for example, becomes particularly interesting when a child is able to ride it.

Thus, symbols not only describe entities in the environment in a sensory manner, but they also describe the behavioral semantics of things. What can I do with a particular object? What happens when I interact with a particular object? What is the consequence of a particular action? Which properties of an object are particularly relevant for executing a particular action successfully? Answers to these questions seem to be categorized and symbolized in our minds. One grand challenge for cognitive science is to develop systems that reliably find or even learn to find answers to these questions – ultimately with computational models. In later chapter, we will greatly expand on these issues and also show that similar questions can be asked, and similar mechanisms and encodings are involved when conceptualizing and symbolizing entities, and whole systems that are hardly accessible by means of sensorimotor interactions, such as physics, mathematics, biology, economics, or even politics.

3.4.2 Frame problem

Let us assume for a moment that we have created an agent that has solved the symbol-grounding problem. It has identified the objects in its environment and knows about their properties, their relevance for the agent itself, as well as their relationships to each other – however it may have done so. Even with such an agent, we face a second major challenge for cognitive systems: How can the agent meaningfully represent changes in its environment? In which frame of reference and frame of relevance should the agent consider executing a particular environmental interaction, especially when considering a formal description? John McCarthy, together with Patrick Hayes, identified this problem in 1968 in his work on *Philosophical Problems from the Standpoint of Artificial Intelligence* (McCarthy & Hayes, 1968).

At first glance the questions may appear to be simple, but the complexity arises when considering the details. Assume we have a room in which in one corner ($Loc1$) two boxes ($B1, B2$) are stacked on top of each other. By means of logical expressions, we could formalize the situation as follows:

$$\texttt{at}(B1, Loc1) \ \wedge \ \texttt{at}(B2, B1) \tag{3.1}$$

($B2$ thus is on top of $B1$). A robot shall now move $B2$ from its current position to another corner of the room ($Loc2$), which results in the following:

$$\texttt{result}(\texttt{move}(B2, B1, Loc2)) \ = \ \texttt{at}(B2, Loc2) \ \wedge \ \neg\texttt{at}(B2, B1) \tag{3.2}$$

The truth value of $\texttt{at}(params)$ depends on the current situation s, which is fluent, that is, it changes with each interaction. Therefore, we need to include the situation as an additional parameter to be able to maintain consistency in the logical system. The result is *situational calculus*, which includes the current state, s, in its logical expressions, such as $\texttt{at}(params, s)$.

However, this method is still too simple. Before we can actually execute an interaction with the environment, we need to assure that the action is actually executable in the current situation. In particular, box $B2$ needs to be the top box to be graspable. That is, in the current situation s, the state of box $B2$ needs to be clear, where the concept `clear` can logically be defined by:

$$(\forall objs : \ \neg\texttt{at}(objs, B2, s)) \ \Leftrightarrow \ \texttt{clear}(B2, s), \tag{3.3}$$

where $\forall objs$ indicates that the equivalence "$\Leftrightarrow$" is true for all objects $objs$ present in the scenario. Moreover, the other location, $Loc2$, needs to be clear to be able to position a box onto it, which makes the logical expression even more involved, yielding:

$$\begin{aligned} \texttt{applicable}(\texttt{move}(B2, B1, Loc2), s) \ \Leftrightarrow \\ \texttt{at}(B2, B1, s) \ \wedge \ \texttt{clear}(B2, s) \ \wedge \ \texttt{clear}(Loc2, s) \ \wedge \ B2 \neq Loc2 \end{aligned} \tag{3.4}$$

From a logical perspective, however, we cannot know the state of other objects in the next state s'. Particularly objects that did not change due to the result of the movement pose an additional challenge. The state of these other objects has not been directly carried over into the next state s'. We thus do not really know, for example, if $at(B1, Loc1, s')$ still holds.

To be precise and logical, for any action it is necessary to introduce *frame axioms*, by means of which it is defined if and for which objects non-affected states stay the same. In our example we have to specify that any object $o1$ that is not replaced by the action will still be located at the same position. Moreover, we can specify that it remains clear when it was clear before. Formally, this becomes increasingly tricky to express:

$$\forall o1, o2, l1, l2, l3, s :$$
$$\mathtt{at}(o1, l1, s) \wedge o1 \neq o2 \Rightarrow \mathtt{result}(\mathtt{move}(o2, l2, l3), s) = \mathtt{at}(o1, l1, s') \wedge \qquad (3.5)$$
$$\mathtt{clear}(o1, s) \wedge o1 \neq l3 \Rightarrow \mathtt{result}(\mathtt{move}(o2, l2, l3), s) = \mathtt{clear}(o1, s)$$

Note, we logically do not need to write the second equation explicitly in this case, because `clear` is defined given the current *at* situation. However, it needs to be logically re-computed for all objects after each manipulation. With the additional axiom we are capable of deducing all consequences of a `move` action, being thus able to transfer the current state s directly into the next state s'.

Clearly, our real world does not only consist of boxes, locations, and transport actions. A simple calculation makes the fast blow-up in the number of necessary expressions very explicit: assume that our environment is defined by M properties and we can execute N possible actions, then we have to define MN additional frame axioms. This may still be doable, but then consider that for all objects in the world we need to process in each situation all frame axioms to process their current situational properties into the next property. Soon, this process becomes very cumbersome and logic-based computations become slow.

While artificial intelligence has by now developed more efficient techniques to represent action-effects logically – for example, by assuming that unaffected aspects of the environment stay the same – the frame problem has an even greater significance when we realize that different objects typically have different properties. What about object properties such as weight and size, or the bodily capabilities of a human or robot, such as the force that can be exerted or the arm length and flexibility available? To ensure the success of an interaction, any property may have particular behavior-determining influences.

When reconsidering the real world properties discussed already, it soon becomes clear that it seems nearly impossible to account for all possibly relevant properties. It is a wonderful feature of the brain that we are typically able to continuously maintain a good balance between detail and generalization, that is, between precision and noise. We consider those frames that seem relevant to ensure successful behavioral executions but ignore behaviorally-irrelevant aspects. Similarly, we typically consider relevant behavioral consequences successfully, while ignoring irrelevant ones.

The following famous story of the American philosopher and cognitive scientist Daniel Dennett illustrates this frame problem in a vivid manner:

> Once upon a time there was a robot, named R1 by its creators. [...] its designers arranged for it to learn that its spare battery [...] was locked in a room with a time bomb set to go off soon. [...] There was a wagon in the room, and the battery was on the wagon, and R1 hypothesized that a certain action which it called PULLOUT(WAGON,ROOM) would result in the battery being removed from the room. Straightaway it acted, and did succeed [...] Unfortunately, however, the bomb was also on the wagon. R1 [...] didn't realize that pulling the wagon would bring the bomb out along with the battery. Poor R1 had missed that obvious implication of its planned act.
>
> [...] "Our next robot must be made to recognize not just the intended implications of its acts, but also the implications about their side-effects, by deducing these implications from the descriptions it uses in formulating its plans." They called their next model the robot-deducer R1D1. [...] it too hit upon the idea of PULLOUT(WAGON,ROOM)[...] It had just finished deducing that pulling the wagon out of the room would not change the color of the room's walls, and

was embarking on a proof of the further implication that pulling the wagon out would cause its wheels to turn more revolutions than there were wheels on the wagon - when the bomb went off.

"We must teach it the difference between relevant implications and irrelevant implications," said the designers. "And teach it to ignore the irrelevant ones." So they developed [...] the robot-relevant-deducer, R2D1. When they subjected R2D1 to the test [...] they were surprised to find it sitting, Hamlet-like, outside the room [...] "DO something!" its creators yelled. "I am," it replied. "I'm busily ignoring some thousands of implications I have determined to be irrelevant. Just as soon as I find an irrelevant implication, I put it on the list of those I must ignore, and..." the bomb went off. (Dennet, 1984, p. 128)

The frame problem thus does not pose the challenge to differentiate between behaviorally or task-irrelevant, and relevant aspects of the environment, but rather it poses the challenge of being able to decide from moment to moment quickly and effectively which aspects in the environment to consider and which ones to ignore.

Humans do this kind of decision making all the time – most of the time without being aware of it. Recent research has shown that our eyes act in anticipation of the current task, scanning a scene in the continuous search for the next relevant information sources. When reading, our eyes are already on the next word while we still read the current one (in fact, they are often even further ahead than that). Before our hands start to grasp an object, our eyes already ascertain the perfect position to grasp the object to accomplish the successive task. This selective, very anticipatory and behavior-oriented, selective information processing appears to take place in all kinds of decision making and behavioral control tasks, including processes of fully abstract thoughts.

3.4.3　Binding problem

Related to the *symbol grounding* and the *frame problem*, although much less discussed in the philosophical domain of cognitive science, is the *binding problem*. Here, the main issue is how we manage to bind things together into one overall percept, although a "thing" is composed of – and thus also decomposable into – several aspects and components.

Take, for example, the Necker cube shown in Figure 3.1(a). We could very easily say that we see a couple of lines, or that we see two rectangles that are connected with oblique lines. However, what most of us do actually see is a cube. Moreover, typically our three-dimensional interpretation of seeing this cube is switching between either side being further to the imaginary front. As the Necker cube nicely illustrates, our brain binds items together into an object percept (a Gestalt in terms of *Gestalt psychology*, cf. Section 2.4.3), and this binding seems to work without any conscious effort. Similar observations can be made when viewing an image of a *Rubin vase* (Figure 3.1(b)), which were first examined by the Danish psychologist Edgar John Rubin (1886–1951) in relation to figure-ground segregation and which were often referred to by Gestalt psychologists.

The binding problem not only addresses the challenge of binding individual percepts into a Gestalt, though. It poses the more general question of how we perceive objects as actual whole objects. Why is a tree a tree and not a stem with often something above that stem? Why is a car a car and not wheels that are carrying a coach? ... or a blue entity of a certain size? Why, when we see a bottle, do we see a bottle first and only possibly next, the lid, especially when it is currently of relevance? Or take a closet: why do we not first perceive the doors of it, although they are typically the biggest part?

Visual explanations are not enough in these cases. We tend to name the global Gestalt first – of an object or a thing in general. Then we look at its parts. We attempt to integrate parts into a global object. Interestingly, there are patients with a cognitive defect termed *simultanagnosia*, who literally cannot see the wood for all the trees. They perceive individual items, but not the context. In a supermarket, they may be able to identify butter, milk,

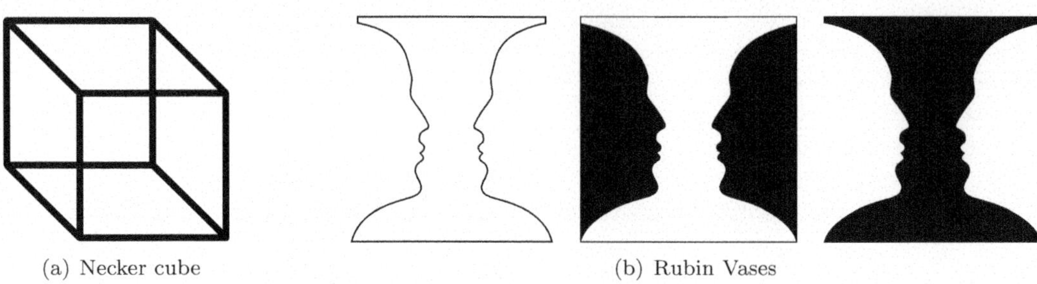

(a) Necker cube   (b) Rubin Vases

Figure 3.1: The Necker cube nicely illustrates how lines are bound into a perceptual whole, that is, a cube. Moreover, it illustrates how the brain actively processes its cube interpretation, switching between two equally plausible three dimensional interpretations. Similarly, in the Rubin vase depictions either the face or the vase is perceived. To see both interpretations concurrently is virtually impossible. Depending on the coloring, either the face or the vase tend to dominate the perceptual interpretation.

yogurt, etc., but they are typically not able to perceive the cooling compartment. When shown a cartoon story, they may identify individual entities but are unable to integrate them into an interactive scene ... such as that "Mickey Mouse is aiming at a kangaroo with a water gun", which remains "Mickey Mouse" / "kangaroo" / "water gun".

David Navon developed the challenge of identifying a global letter that consists of many local letters, as shown in Figure 3.2. While a typical human participant tends to see the global letter faster than the local letters, simultanagnosia patients can typically identify the local letters, but not the global one. They somewhat get stuck in local processing and cannot identify the global Gestalt, presumably being overruled or overwhelmed by local cues.

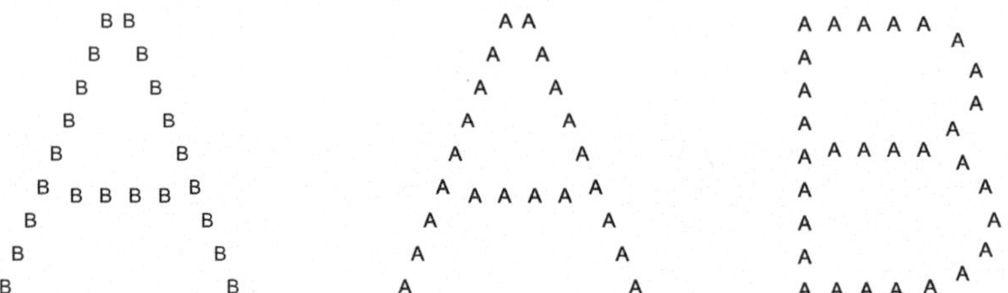

Figure 3.2: Navon figures show a global letter that consists of many local ones, where the local ones may be different from the global one.

Healthy human brains manage to bind local cues into one consistent and complete form or Gestalt. For example, an "elephant" may be identified and learned first, and later the "trunk" of an elephant may become nameable. Subconscious visual and higher-order processing integrates individual visual cues into one global Gestalt, binding the parts together and thus making way for naming proper and relevant entities, rather than individual parts. When further differentiating the global Gestalt, also individual parts become individualized, identifiable, and nameable. Nonetheless, typically global precedence persists.

Such subconscious processing predispositions and their functional foundations also help solve the grounding problem. Moreover, they help to focus on the relevant aspects of a scene, solving the frame problem. In later chapters we will introduce several mechanisms that set the stage for these perceptual processing and interpretation capabilities.

3.5 Neural networks

In parallel with symbolic processing mechanisms and "traditional artificial intelligence", neural network approaches to intelligence have been pursued. Early pioneers of artificial neural networks were inspired by the discovery of neurons as the apparent fundamental processing units. In the 1980s, artificial neural networks were subsumed under the term *connectionism* as an alternative to symbolic cognitive science approaches and symbolic AI.

The discovery of biological neurons started in the 19th century and was detailed in Section 2.3.1. The first artificial neural networks (ANNs) were proposed in 1943 – before Hodgkin and Huxley's biological neural cell model – when the cyberneticist Warren McCulloch (1898–1969) and the logician Walter Pitts (1923–1969) proposed a mathematical neural model in which a neuron was characterized as a binary state system (states "1" or "0"). Each neuron was thus a logical processing unit, returning either true of false. In accordance with the biological model, the axon hillock was mimicked by a threshold mechanism, where the sum of input activities led to a *true* output (that is "1") when it was larger than a particular threshold value and *false* (that is "0") otherwise. In addition, they were able to show that a finite network of these simple neurons is theoretically capable of finding a solution to any computable problem (McCulloch & Pitts, 1943).

The Canadian psychologist Donald O. Hebb (1904–1985) also significantly contributed to the development of connectionism. In 1949 he proposed a bridge between neural activities and cognitive association learning (remember the law of effect by Thorndike, Section 2.4.1):

> When an axon of cell A is near enough to excite a cell B and repeatedly or persistently takes part in firing it, some growth process or metabolic change takes place in one or both cells such that A's efficiency, as one of the cells firing B, is increased. (Hebb, 2002, p. 62.)

This *Hebbian Learning Rule* can be easily formulated mathematically:

$$\Delta w_{AB} = \eta \cdot a_A \cdot a_B \tag{3.6}$$

and is often simply characterized as "what fires together, wires together." The activation of the connection between neuron A and neuron B is determined by the respective activities a_A and a_B. The stronger both neurons fire at the same time, the more the connection strength w_{AB} is increased, where the increase is controlled by the learning rate η.

The neural model of McCulloch–Pitts and the Hebbian learning rule constituted the main ingredients to the now well-known *perceptron model*, which was developed by the American psychologist and computer scientist Frank Rosenblatt (1928–1971). In its original form, the perceptron mathematically described the behavior of one neuron. Input signals into the neuron were weighted differently and these weights were learned by increasing the connection weight w_{AB} when the output of neuron B, that is, a_B is smaller than the target signal t_B and the neuron A was active, while it is decreased when $a_B > t_B$ with neuron A being active:

$$\Delta w_{AB} = \eta \cdot (t_B - a_B) \cdot a_A \tag{3.7}$$

A trained perceptron is thus essentially capable of differentiating input activities into two separate groups (that is, states true and false). This learning rule is now known as the *delta-rule* and is still being used – typically in enhanced forms – even in most of the current *deep learning neural network* architectures.

Despite this pioneering work, Marvin Minsky (1927–2016) pointed out that the perceptron itself is much less powerful than initially thought. Each perceptron is essentially capable of linearly separating data into two classes. With one layer of perceptrons it is thus impossible to solve the XOR problem, which is the problem of assigning input states 10 and 01 to class 1 and states 11 and 00 to class 0. In his 1969 book *Perceptron,* Minsky pointed out this problem, questioning the capability of perceptrons to model intelligent or cognitive mechanisms (Minsky & Papert, 1969). This rather harsh critique also contributed to the development of the aforementioned first AI-winter, ceasing further research on ANNs for some time.

With some important enhancements, however, ANNs have now proven to be as powerful as originally imagined with almost unimaginable potential. Several protagonists led the way to this development of which we can name only a very few here. Paul Werbos, David Everett Rumelhart, James McClelland, and several others developed, more or less in parallel, the *backpropagation algorithm* in the 1970s (Werbos, 1974; Rumelhart, McClelland, & the PDP Research Group, 1986; Rumelhart, Hinton, & Williams, 1988). This algorithm generalizes the delta rule to *multilayer perceptron* ANNs. Moreover, the perceptron model was generalized, such that a neuron was equipped with a differentiable threshold function – such as a *sigmoid function*. With these generalizations, ANNs regained popularity in the 1980s, promising the development of general intelligence.

Somewhat concurrently and as an alternative to backpropagation learning, which is believed to be neurally rather implausible, self-organizing networks were developed, for example, by the Finnish engineer Teuvo Kohonen (*1934). These networks are considered comparable to biological networks, supposing that similar stimuli should be encoded in the brain in closely connected areas. The innovative aspect about self-organizing networks is the importance of the network topology: neural connections close to the center of stimulation are modified more strongly than those further away. In this way, neural lattices can be learned that can reflect the topological structure of the incoming stimulations. In later chapters, particular applications of neural networks for learning behavioral control mechanisms and spatial, topological representations will be discussed (cf., for example, Section 10.2.3).

In recent decades several further advancements have been made, yielding the most capable machine learning systems available in many domains, including image classification, as well as speech processing to a certain extent. These ANNs are currently referred to as *deep-learning ANNs*, because they contain many layers, backpropagating the supervised error signals deeply backwards along these layers. Deep ANNs often include recurrent, neural connections, such as connections to neighboring neurons in the same layer (often termed lateral connections) or also connections that project information backwards into a previous layer. Moreover, the incoming neural activities are sometimes combined multiplicatively instead of additively and the activation functions are systematically selected among, for example, linear activations, rectified linear activations, and nonlinear activation functions. Finally, particularly when facing visual processing tasks, deep ANNs contain *convolution layers*, where a local neural kernel – such as a local edge detection matrix – is applied on any position in the image, essentially revealing the presence of the encoded kernel anywhere in the image (Krizhevsky, Sutskever, & Hinton, 2012; LeCun, Bottou, Bengio, & Haffner, 1998). In particular this latter structure seems to be implemented – albeit in a different manner – also in our visual cortex.

Concurrently, *generative ANNs* – today often termed *deep-believe networks* – were developed by the British cognitive psychologist and computer scientist Geoffrey Hinton (*1947) and others, deriving them via free energy formulations and Boltzmann Machines (Ackley, Hinton, & Sejnowski, 1985). The now rather popular *Restricted Boltzmann Machine* (RBM) was proposed by Paul Smolensky (*1955) in 1986, originally termed "Harmonium" (Hinton, 2002; Smolensky, 1986). RBMs are believed to approximately model fundamental information processing mechanisms similar to the brain. In fact, the British neuroscientist Karl Friston (*1959), who has also co-developed many fundamental brain imaging analysis tools, has proposed that *free energy-based inference* models, similar to the ones that allow the derivation of learning algorithms for RBMs, may constitute the fundamental learning concept in our brains (Friston, 2009).

3.6   Embodied intelligence

Despite the enthusiasm about artificial intelligence in the early years, we have seen that "Good Old-Fashioned Artificial Intelligence" (GOFAI) has not succeeded in establishing itself universally. On the other hand, the great hope for ANNs is currently being confirmed to a certain extent in various fields, including image recognition tasks and speech processing. However, these ANNs are nowhere close to actually "understanding" what they are doing or why they are doing something. That is, the current successful ANNs convert input into output to optimize something, to satisfy certain error criteria, or to create some provided target output. They do typically not attempt to model how the problem actually works, that is, they do not attempt to "understand" the mechanisms underlying the problem.

Due to the limited successes of GOFAI and ANNs in the 1980s and early 1990s, which was partially due also to the limited computational power, as well as due to the recognition of the frame problem, the symbol grounding problem, and the binding problem, a second AI winter spread throughout the globe, questioning the value of AI once again. Hardware problems also contributed to the questions surrounding AI. The Lisp-machines, on which nearly all AI-programs had been developed until then, became inefficient and were no longer economically feasible. More modern computers developed by Apple, IBM, and others, offered more powerful alternatives, but often prevented the transfer of the previous programs. Thus, previously developed code was lost because the effort to re-implement it on the new machines was not invested.

In the search for alternatives, one important realization was that our brains do not function without a body. In fact, the neural development of the brain goes hand-in-hand with the development of the body. Only a few weeks, if not days, after conception the fetus in the womb of a pregnant woman already actively explores its body and develops neural structures in accordance with the gathered bodily experiences. The open question to all of those approaches – and even to the currently very successful deep learning ANNs – is the question of how cognition and intelligence can develop "from scratch" during ontogenesis.

When considering neurocognitive development, it is obvious that the brain can learn only from the incoming sensory signals and the interactions of these signals with its own motor behavior. Thus, learning is largely determined by the sensorimotor experiences gathered while interacting with the body and the outside environment, besides neurally encoded learning predispositions. Moreover, during this neurocognitive development, learning is strongly influenced by the developing sensory and motor capabilities of ones own body.

Insights from biological research have shown that many animals – even some with rather small brains – exhibit complex behavioral capabilities. Questions such as how an ant manages to find its way back to its nest, as well as how an insect manages to avoid obstacles during flight and how it manages to maintain a certain height above the ground, have been answered to some extent. These insights suggest that some seemingly challenging problems for cognitive systems can be solved by means of cleverly arranged sensory-to-motor couplings, with hardly any involvement of learning, adaptation, or active motor control. Moreover, these insights suggest that some seemingly challenging problems may best be solved by a combination of simple subproblem solvers, which do not really know anything about each other, but are activated when necessary just in time in a coordinated manner. Inspired by these findings, researchers in cybernetics and synthetic biology designed the first seemingly intelligent machines, which showed emergent behavioral patterns simply by combining a suitable bodily architecture with simple sensory-to-motor wirings. From a computational cognitive science standpoint, it needs to be acknowledged that behavior is realized by a body, in which genetically predetermined wirings and morphological arrangements can yield complex behavioral capabilities, with hardly any neural control.

The concept of *embodiment* thus puts forward the idea that sensory and motor processes, as well as physical and neural processes are in continuous interaction with each other. Each process accomplishes particular tasks, thus providing abilities. These abilities ease the work that is to be accomplished by other processes. Moreover, each process induces structural constraints. For example, a complex behavior may not need to be controlled by a high level cognitive process. Rather, it may only need to be selected and instantiated appropriately, obeying the structural constraints induced by lower-level control processes and considering the current circumstances. Given an appropriate instantiation, the actual control then is accomplished by the activated lower-level, embodied control processes. To understand cognition as a whole, it is thus mandatory to develop an understanding of the interactions between these processes, of the computations done by each process, and of the encodings that govern the computations within and between the processes.

3.6.1　Embodied biological processing

A large part of the motivation for studying embodied cognitive systems and emphasizing the embodiment of the mind comes from research in behavioral biology. Many animals can exhibit astoundingly clever patterns of behavior, despite small brains and very little time to learn much at all.

In the following, we crudely survey four such behavioral observations and relate them to embodiment, emergence, as well as to the processes that cause the observed behavior:

- Frogs can detect prey and danger, and act accordingly.

- Insects are very skilled in avoiding obstacles and in maintaining a certain height while flying.

- Four-legged animals exhibit few particular types of locomotion.

- Ants quickly build trails to suitable food locations.

These examples will illustrate that our interpretation of behavior often differs from the mechanisms that are causing it. Moreover, they illustrate that simple sensorimotor couplings can result in *dynamic morphological attractors*, which yield stable behavioral patterns on the one hand, but which also offer themselves as codes, which can be associated with symbols.

How frogs distinguish prey from predators

For many frogs the distinction between prey, that is, food for themselves such as a fly, and predators is obviously highly relevant for survival. In the former case, a suitable expulsion of the tongue in the right direction at the right moment can lead to food consumption and internal motivational satisfaction. In the latter case, however, this behavior would be fatal. Instead, a well-timed jump into the pond or a freezing behavior may be preferable. How do frogs do it?

As an external observer, we might tend to interpret the behavior as human-like, that is, the frog monitors its environment, classifies objects into preys, predators, and irrelevant other objects, and acts accordingly. It turns out, however, that such an interpretation is far from the truth. In fact, the neural realization of such explicit classification and decision-making mechanisms is rather complicated, neurally challenging, and difficult physically given the visual capabilities of the typical frog.

Nature has apparently helped frogs to sidestep these neurocomputational challenges. Instead, neurons have been detected in the deeper visual cortex that signal prey versus predator, by focusing on simple visual features. Neural detectors for prey react to quickly moving, dark stimuli in front of a hardly moving, lighter background. Neural detectors for predators, on the other hand, react to large dark areas that are expanding. The brain of the frog wires these detector neurons to suitable behavioral responses. These connections are tuned and adapted during the lifetime of the frog. However, the main detectors seem to be genetically pre-wired. Thus, the frog is equipped with a survival-relevant, highly suitable detector and motor primitive system (such as jumping, freezing, tongue expulsion), which are suitably pre-wired, enhancing the survival chances of the frog significantly without spending too much neurocomputational energy.

How flies know how to maintain a safe distance

Another example of coupling sensory signals with motor commands comes from the world of flying insects. Their compound eyes actually turn out to be more useful than previously thought – the individual light detector in each compound is oriented relative to the neighboring ones in such a way that the visual flow, that is, how fast the visual input flows from one compound to the neighboring one gives direct information about the distance to the visual stimulus. Fast visual flow indicates closer proximity. Thus, the faster the visual flow the more the insect should avoid the direction toward that visual flow when avoiding obstacles. Given appropriate visual flow detectors, the insect can control its flight direction with respect to these visual flow signals to maintain a safe distance from objects it is currently not interested in.

To confirm this, flight tunnels have been built in which the ground or the sides can be moved backward or forward, directly influencing the way a bee is flying. Given that the ground is moving forward or backward, the bee will fly lower or higher, respectively. Similarly, a side of the tunnel is avoided less or more strongly, depending on whether the side is moving in the direction of flight or in the opposite direction, respectively. Landing behavior can be understood in a similar manner: the legs prepare for touchdown when the visual flow expansion is maximal (cf. Section 8.5.4).

In conclusion, the behavioral strategies of flying insects show that it is not always necessary to know exactly what is out there, that is, to do explicit computations to make behavioral decisions to, for example, maintain a certain distance. Rather, loose sensorimotor couplings often suffice. In flying insects, this is possible due to a suitable sensor arrangement (radial detectors with particular arrangements of a compound eye) and due to the signal detectors (visual flow), which have developed to allow a direct transfer to motor encodings, such as directional signals that influence flight direction. Seeing that the morphology of insect eyes with simple motion detectors computes distances, such an embodied form of computation is called *morphological computation.*

Morphological attractors in four-legged locomotion

The third example highlights another important aspect of embodied cognition, where the body's morphology interacts with motor control processes. Consider a baby gazelle in the Serengeti: after it is born it needs to be able to run with its mother as soon as possible and it takes less than 30 minutes to accomplish this. Is the brain really learning all the necessary computations for stabilizing the body and properly using the legs in such a short period of time? It turns out that it is actually not as hard as originally thought. The four legs are arranged in such a way and have such self-stabilizing properties that the actual control commands issued by the brain are merely issuing coordinated oscillating walking signals to the four legs. The rest is accomplished by the legs themselves. Each muscle has its own *Golgi-tendon organ,* which helps to generate basic reflexes and self-stabilization mechanisms. The muscle arrangements result in self-stabilization mechanisms as well. Not surprisingly, muscles are usually arranged in pairs with agonist and antagonist, resulting in a mutual stabilization to avoid over-stretching. Together this arrangement enables walking as fast as possible when necessary.

When considering the typical different types of walking that, for example, a horse can execute, it turns out that there are only a few particular ones. In fact, the different types have even received their own terms, including walk, trot, canter, gallop, and pace, and are even referred to as a group as *horse gaits.* All the particular horse gaits can be characterized by the coordinated motion of the legs (how many are on the ground at a certain point in time, which leg is moved concurrently with which other leg). On the other hand, *ambling gaits* are intermediate types of horse movements, which rarely can be stably maintained over time by the average horse. Thus, horse gaits form motion attractors, which are mostly generated by a particular type of coordinate leg movement at a particular speed. Most of the dynamics and self-stabilization involved are actually handled by the individual leg muscles, the overall coordination of motion direction, and various other physical dynamics, which unfold while the horse is moving. Thus, dynamic morphological motion attractors develop due to the interaction of the body with the environment, driven by dynamically oscillating control mechanisms. Note that besides the much easier coordination of leg movement during locomotion when contrasted with alternative, full-blown neural control architectures, the resulting dynamic attractor states can also be symbolized by giving a name to each possible attractor state. Thus, also the symbol grounding problem is tackled in this manner, grounding symbols by associating them with embodied, morphological attractors.

Ants and swarm intelligence

Besides intriguingly processed and morphologically arranged sensor to motor couplings, the intelligence of swarms plays an important role in forms of embodied intelligence. Bees and ants have been intensively studied by biologists over the last decades and it has become clear that the intelligence of the individual bee or ant is not very deep. However, there is a collection of clever behavioral routines and means of communication that yield intelligent social behavior.

As first suggested by the biologist Edward O. Wilson (*1929) in 1962, ants leave pheromones indicating the path to a valuable food source for their tribe. However, ants do not do this purposefully, nor are they aware of the consequences of doing so. They do not deposit the pheromones only when they have found a very effective path to the food source; ant trails develop emergently because shorter paths to a food source will inevitably be discovered more often than longer paths when initially randomly searching for food. Moreover, a heuristic algorithm to find the way back to the nest helps in this emergent process. Finding the way back to the nest is, depending on the particular ant species, typically accomplished by maintaining a general idea of the direction back to the nest via path integration, as well as by memorizing a crude snapshot view of the nest. When approaching the nest, the memorized snapshot view is compared with the current view and the insect moves in that direction in which the difference between the two views decreases. In this way, pheromones get progressively more densely deposited on the shortest routes to a valuable food source (and back to the nest). Thus, an ant trail has formed without any single ant being aware of it.

In conclusion, social animals are capable of creating emergent structures that may be stable and valuable for the society without any of the individuals actually knowing their purpose. Interestingly, such ideas have also been turned into effective, distributed, social search and optimization algorithms, called *swarm intelligence* and, even more particularly, *ant colony optimization* algorithms. In fact, taking a far-fetched analogy, one could say that similar things happen in human societies. Take, for example, a large economy. Nobody seems capable of fully understanding its overall functionality – let alone proving that it will be stable for a certain period of time. Too many interacting factors are influencing the overall system. Nonetheless, given that these factors are mutually influencing each other in a positive manner – such as leaving pheromone trails that indicate energy sources (whatever this may exactly be in an economic system with all its niches) – and are thus maintaining a certain kind of equilibrium, there is hope that the system will be beneficial for the whole society.

Summary and conclusion

These four examples suggest that seemingly intelligent behavior may not always be based on very complex, neural, cognitive, or even symbolic coordination mechanisms. Rather, clever couplings of sensors with simple processing routines and with simple motor activities often suffice. The interaction of the horse, its legs, and its muscles, which causes the emergence of the very typical horse gaits, suggests that at least in similar cases, sensorimotor dynamics can result in morphological attractors, which are suitable, stable, but dynamic states, and which can be associated with discrete symbols, thus alleviating the symbol grounding problem. Finally, subtle interactions with the environment can lead to emergent group behavior, which yields efficient, seemingly intelligent environmental interactions without actually understanding or having computed such interactions in an explicit, goal-oriented manner.

3.6.2 Embodied artificial intelligence

The insights from biology about morphologically intelligent, self-stabilizing, control processes (frogs, flies, four-legged motion) and emergent optimization processes – such as the shortest path optimization in ants – were also noticed by researchers working in artificial intelligence. Seeing that the great hopes for purely symbol processing systems had mostly failed, alternatives were considered. Biological insights were inspiring in this respect. The general insight, which received close consideration, is that the mind cannot exist without a body, which implies that the body with its morphology, including its sensory and motor capabilities, must shape the mind. Moreover, the mind–body interactions and the involved sensorimotor control loops may often work independently of each other, somewhat like a loose collection of clever behavioral capabilities, such as the suction reflex and grasp reflex in infants.

In his inspiring philosophical, but also information-based, technical book *The society of mind* (1988), Marvin Minsky points out important implications:

> What magical trick makes us intelligent? *The trick is that there is no trick.*
> The power of intelligence stems from our vast diversity, not from any single,
> perfect principle. (Minsky, 1988, p. 308, author's emphasis.)

Minsky takes this concept of vast diversity to the extreme in his book, proposing that a simple collection of unintelligent agents leads to the emergence of intelligent, complex behavior, and even consciousness. While his agents may have been oversimplified and their interactions are kept very sketchy, truth certainly lies in the fact that our neurons do not know why they interact with other neurons in the way they do. Nonetheless, they enable the development of human intelligence. Inspired by this work and the insights from biology, research on cognitive systems has created embodied control systems and systems that are equipped with a number of more-or-less independent little cognitive or control experts. Intelligent behavior then emerges from the loosely coordinated interactions of these experts with the environment.

Sensorimotor interactions and behavior control loops

Behavioral intelligence inevitably starts with sensorimotor control loops, because only behavior directed toward some form of goal, such as a food source, is after all a helpful behavior. So it does not come as a big surprise that even the simplest bacteria are able to swim – in a somewhat biased random walk pattern – toward sugar or other sources of energy. Already in these simple species, sensors, which can, for example, detect sugar concentrations, are coupled with motor behavior – in this case with very simple flagellar propulsion – that directs a bacterium toward a food source.

In various more advanced forms, reflexes make use of the same principle, coupling particular sensory stimulations to particular motor primitives. Many common examples are well-known, such as the grasp and suction reflexes in infants previously-mentioned, the blink-reaction due to a puff of air, or the patellar reflex leading to the sudden jerk of the knee joint.

The neuroscientist and cyberneticist Valentino Braitenberg (1926–2011) published a very intriguing book on this subject (Braitenberg, 1984), in which he developed the *Braitenberg Vehicles*. These vehicles are little robots with two wheels, which are controlled directly by their sensors. That is, the strength of the sensor readings are directly wired to the motor activities. Thus, a wheel may turn faster the brighter the surrounding. With proper sensory arrangements and sensor-to-motor mappings, robots can be built that approach or avoid light sources, without ever computing the actual direction toward the light source (cf. Figure 3.3).

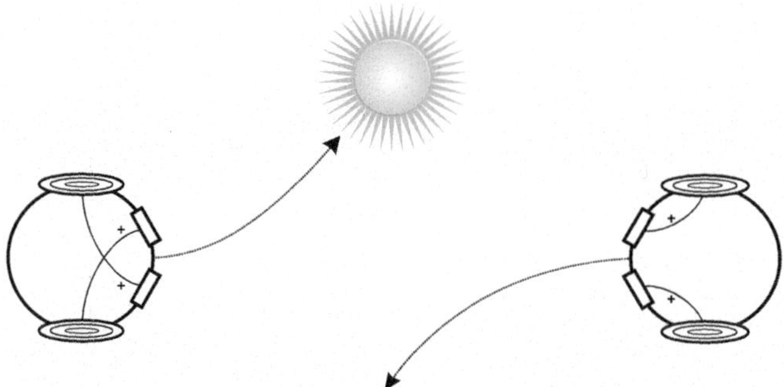

Figure 3.3: Dependent on the wiring, the sensor arrangement, and the sensor properties, a simple Braitenberg vehicle will turn and drive toward or away from a light source.

Similarly, insect-like robots have been built that exhibit wall following behavior by mapping visual flow detector information directly onto wheel speeds. Cleverly designed ornithopters mimic insect flight, self-stabilizing the body during flight by a simple low-hanging, self-stabilizing balance point mechanism. Jumping robots of various kinds have also been built. In most of these cases, the control mechanisms themselves are very simple and most of the necessary self-stabilization is realized by the bodily morphology, rather than by computational means.

Another interesting experiment shows that swarm robots may accomplish interesting, behavioral tasks without being explicitly programmed for them. The Swiss Robots or *Didabots* are a team of simple wheeled robots that react to nearby infrared sensor signals in a Braitenberg manner. Only two sensors are attached facing the right front and left front of the robot (cf. Figure 3.4). They are placed at the front of the robot about 10cm apart from each other. A signal on either side causes the wheel on the opposite side to turn backwards, thus avoiding obstacles and walls. If there is no signal, the robot moves forward. These robots are then put into a walled arena with boxes. The boxes are small enough such that when a box is positioned directly in front of the robot, it is not detected by the infrared sensors and the robot continues to move forward pushing the box along. As a result, it appears as if the robots are working together, pushing the boxes into one pile (cf. Figure 3.4). Clearly, the robots simply act on their sensory signals; they do not know anything about boxes. While this experiment works with a single robot, the swarm particularly helps to avoid robots from getting stuck, simply by receiving a suitable push from another robot.

The most extreme example of a cleverly engineered morphological intelligence, however, comes from the challenge of creating a human-like walking robot. In contrast to many robots that are fully controlled during locomotion, *passive walkers* can go as far as their weight pushes them down a long ramp. That is, without sensors or motors, but only with a suitable body design, *passive walkers* put one foot in front of the other, and so forth, as a result of a well-engineered combination of a low balance point, well-positioned and properly constructed joints, sufficiently large feet, and properly applied springs or rubber bands. These bodily features together cause the walker to swing from side to side, thus moving the legs and walking forward. Various other robots have been designed to mimic the walking of animals, such as six-legged walking robots, jumping-based means of moving forward, as well as dog-like motions (Pfeifer & Bongard, 2006).

All of these show that a properly designed morphology can lead to very stable behavioral patterns, just like the horse gaits, and the behavior of frogs and insects. While the computational costs are minimized, behavioral effectiveness is maximized. In all these cases, the particular sensorimotor loop that dominates behavioral control at a certain point in time leads to a behavioral attractor, such as a stable forward motion, obstacle-avoidance flight, or directed tongue expulsion. That is, the morphologically based behaviors, which are only

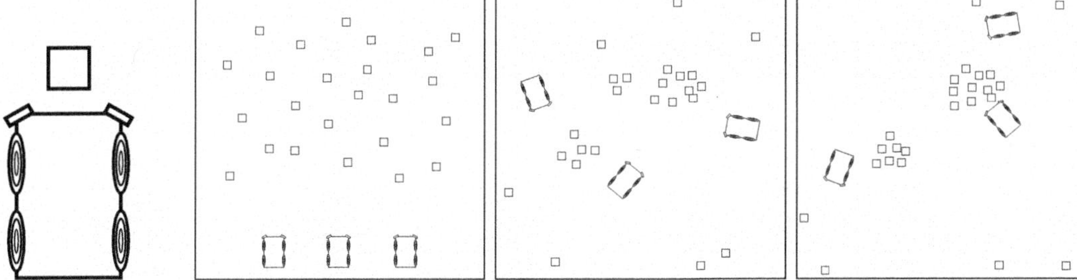

Figure 3.4: Didabots nicely illustrate how a swarm of robots can generate seemingly intelligent, goal-directed behavior, which emerges simply due to the robots interactions, the sensor arrangements, and the senory-motor couplings. The simple sensory arrangement and sensory-to-motor couplings (left-hand side) in each robot result in the observable behavior. As a result, the small robot swarm seems to intentionally push the boxes together into piles – a progression of which is schematically illustrated by the three displays of robot-boxes arrangements (right-hand side).

partially (if at all) controlled by computational means, offer stable behavioral attractors. Interestingly, the attractors may very well be associated with symbols, and thus can help to solve the symbol grounding problem. For example, words denote many stable, partially dynamic activities – such as lying, sitting, walking, running, jumping, etc. – all of which are somewhat dynamic behaviors that unfold stably over time while being executed.

Subsumption architecture

The frog example of tongue expulsion versus hiding behavior points out that multiple behavioral control routines may be ready in parallel, but may be activated only given the proper circumstances. The selective activation of currently appropriate behavior essentially enables behavioral flexibility and thus selective goal-directedness.

How this selection can be accomplished in robots, however, is not as yet fully clear. The first GOFAI attempts with pure logic and symbolic approaches did not get far. As an alternative, the Australian computer scientist and head of MIT's Computer Science and Artificial Intelligence Laboratory between 1997 and 2007, Rodney A. Brooks (*1954) developed many robots that followed the *subsumption architecture* principle. This kind of architecture combines loosely coupled, modularized, parallel processes, each with a particular functionality. More complex processes can be generated and selectively activated as desired, given the proper circumstances.

A subsumption architecture can easily mimic the selective behavior of a frog described earlier, but more complex simulations are also possible. One illustrative example is given by the robot "Herbert", which was designed to search and collect (or "steal") soda cans:

> The laser-based soda-can object finder drove the robot so that its arm was lined up in front of the soda can. But it did not tell the arm controller that there was now a soda can ready to be picked up. Rather, the arm behaviors monitored the shaft encoders on the wheels, and when they noticed that there was no body motion, initiated motions of the arm, which in turn triggered other behaviors, so that eventually the robot would pick up the soda can.
>
> The hand had a grasp reflex that operated whenever something broke an infrared beam between the fingers. When the arm located a soda can with its local sensors, it simply drove the hand so that the two fingers lined up on either side of the can. The hand then independently grasped the can.
>
> Given this arrangement, it was possible for a human to hand a soda can to the robot. As soon as it was grasped, the arm retracted.

> The same opportunism among behaviors let the arm adapt automatically to
> a wide variety of cluttered desktops, and still successfully find the soda can.
> (Brooks, 1990, p. 8.)

The robot Herbert does not really know about soda-cans or grasping routines. Rather, a
well-designed laser scanner detects soda-like objects and couples the directional signal to
the motors of the robot, moving it toward the object. It then positions the robot in such a
way that the detected soda-can-like object is positioned suitably for initiating the grasping
mechanism. Once the wheels stop, the arm extends and the simple grasp-reflex assures
that anything is grasped that comes between the gripper's fingers. As Brooks says, these
opportunistic behaviors smoothly complement each other, yielding rather complex object
interaction routines.

The subsumption architecture thus combines several ideas in a modular manner, yielding
emergent, seemingly intelligent behavior. Without explicit representations of what is out
there or how things may behave, and thus following the principle that *the world is its
own best model*, modules in the subsumption architecture are typical sensorimotor control
routines. These routines usually consist of:

- An *onset*, that is, a conditional encoding that reacts to sensory signals or the activities
  in other modules.

- A *sensorimotor control routine*, which maps sensory or processed sensory information
  onto motor control outputs.

- A *stopping or inhibitory mechanism*, that is, a conditional encoding that reacts to
  sensory signals or the activities in other modules and causes the sensorimotor control
  process to stop.

As frogs react appropriately to food or enemies and as flies maintain a good distance above
ground without actually representing the food, the enemy, or the distance to the ground,
robots controlled by subsumption architectures can develop rather clever and complex, seemingly goal-directed environmental interactions. Despite this big potential, neither the controlled robots nor the subsumption architectures controlling them know what they are doing.
Particularly, they do not know anything about the consequences of their actions. As a result, despite their ability to induce clever environmental interactions, the systems' behavioral
flexibility is rather limited. Without further processes, the coordination of the opportunistic
control routines is emergent, but hard-coded. For example, the robot Herbert will grasp
anything – regardless if it is actually graspable or if the presented object should actually be
grasped. Versatile, selective behavior can only be initiated with more elaborate control and
decision-making processes.

Behavioral coordination and executive control

All embodied AI agents introduced thus far are *reactive*. None of them in any way consider
the effects of their own actions. Thus, it is essentially impossible for them to inhibit an action
in order to avoid particular action consequences. To increase behavioral flexibility in this
direction, the AI needs to have a way of choosing between different behavioral alternatives
on the fly.

To enable the AI to choose between different behavioral alternatives, two options are
possible. First, behavioral choices can be optimized by, for example, reinforcement learning or other related behavioral optimization algorithms, essentially tuning the conditional
structures of each behavioral primitive in accordance with the other behavioral primitives.
Second, the AI may be enhanced by predictive and anticipatory capabilities, thus enabling
the AI to simulate action consequences and thus to choose those actions that are anticipated
to lead to the most desirable action consequences.

When endowing the AI with the latter (that is, predictive, anticipatory capabilities),
behavior can become explicitly goal-oriented. That is, *before* actual behavior is initiated, the

next goal is decided upon and made concrete, so that behavior can be maximally optimized to achieve the chosen goal. Note, however, that the goal choice itself needs to consider the possible benefits of achieving alternative goals and the likely respective efforts necessary for achieving them.

Interestingly, there are many indications that human and animal brains make use of both behavioral choice mechanisms, reinforcement-based and anticipatory-based. Very simple creatures, such as amoebae, are able to learn to avoid the dark by means of properly applied rewards and punishments. The more advanced, and especially socially-oriented, a species is, the more it tends to anticipate its behavioral consequences and makes behavioral decisions dependent on these consequences. Clearly, humans have the most advanced system in this respect, with a high-level, far-reaching anticipatory control system, which is able to take various social and cultural considerations into account.

Besides the action choice, which may be more-or-less explicitly goal-oriented, behavioral execution should also be monitored to increase behavioral flexibility. It indeed appears to be the case that many animals have such sensorimotor forward models in place to monitor current behavior-caused sensory effects. These forward models enable us to filter the sensory consequences on the one hand to optimize information processing by fusing predictions with the action-consequent sensory feedback, and on the other hand, they enable us to detect unexpected sensory events more quickly and reliably. This mechanism enables the detection of environmental changes that were caused by others by means of contrasting them with the anticipated effects caused by our own motor activities.

In this respect an important distinction with respect to motor control needs to be made: Behavior can be controlled in an *open-* or *closed-loop* manner. Open-loop control is essentially the execution of a behavioral program, such as a dynamic sequence of motor commands, without considering sensory feedback. Closed-loop control is a control process that makes each motor decision based on current sensory feedback. As sensory information is often delayed, open-loop control is typically faster. Direct sensorimotor couplings are closed-loop control processes with the advantage of having very little information processing delay. Thus, it becomes obvious that a subsumption-like architecture can be very effective: simple behavioral primitives may be executed open-loop – such as a reflex. Slightly higher-level behavioral primitives may need to become closed-loop. However, by coupling the involved sensorimotor coordination processes very tightly, hardly any processing delay may result. Even higher-level behavioral control loops may become progressively anticipatory, invoking and coordinating the activations of lower-level control processes.

Versatile and highly adaptive intelligent systems likely need to consist of various modularized lower-level and higher-level modules. Each module may be optimized in the involved sensorimotor couplings, associated forward models, and goal encodings. Moreover, the coordination of the modules may be optimized. The following chapters introduce the basic techniques for developing such behavioral versatility and adaptability in much further detail.

3.6.3 Embodied cognitive agents

As seen from the examples of embodied biological agents and embodied AI systems, in contrast to traditional symbolic and logic-based AI approaches, embodied systems face environmental and, thus, particular real-world challenges:

- Interactions with the environment are *non-episodic*, that is, they are continuous without obvious boundaries. Current and future actions often depend on previous actions and previous environmental states, so that episodes of particular environmental interactions cannot be easily segmented.

- The environment is *dynamic and continuous*, and thus it is difficult to partition it into discrete, symbolized states. Thus, the symbol grounding problem is an even harder challenge in the real world. Questions need to be answered, including which state in the environment, which properties, which types of interactions, and which aspects of these should be symbolized?

- Moreover, the environment is typically *non-deterministic* and *not fully accessible*. Regardless how much information we have available about our environment, uncertainties will remain. This is elucidated by Heisenberg's uncertainty principle, which essentially states that it is impossible to know the exact state of a particle at any given point in time (because measuring its location precisely will obscure its current motion and vice versa). However, also in the "normal" world with its entities and objects, plants and animals, the exact state of the world is hardly fully accessible. Even worse, these uncertainties will be partially very hard to quantify.

- Timing is also essential in the real world. Time is continuously moving forward and at certain points in time it is essential to act immediately without any time for further thought or elaboration. Thus, behavioral control processes need to be fast enough and behavioral decision making needs to be timely.

- Finally, when considering social interactions, other agents in the world have their own minds, which are largely inaccessible. Seeing that their actions can affect us in significant ways, taking them into account poses another serious challenge.

Embodied systems, however, do not consider these challenges from a traditional AI perspective: they do not attempt to fully discretize their environmental interactions into separate episodes, but rather they employ flexible sensorimotor control modules that are activated when necessary or desirable. Seeing the dynamics of the world, embodied behavioral systems do not primarily attempt to symbolize the world, but first explore it to identify stable sensorimotor interactions that lead to predictable effects. The resulting attractor states then offer themselves for symbolization. Since the world is non-deterministic and not fully accessible, approximations need to suffice. Here, the principle of the world as its own best model applies most significantly: it is simply not necessary to re-represent the environment with all its details by means of symbols or logic. The focus needs to lie on those accessible aspects of the world that are behaviorally relevant. In effect, evolution has developed suitable sensory processes and detectors that consider primarily those aspects of the environment that appear to be most important for behavioral success. With respect to the timing problem, enhanced subsumption architectures with very fast, reactive processes, and higher-level coordination processes seem most suitable. Such an architecture can enable fast reactions when necessary, but also allows for further deliberation when possible.

Therefore, one can speak of *fully embodied cognitive agents* when considering agents that face the challenges of the real world by interacting with it by means of their bodies. To characterize these agents, which include animals and humans, further, one can state that fully embodied cognitive agents are:

- *Continuously interacting* with their environment by means of sensorimotor control processes, which are supported by morphologically intelligent mechanisms.

- *Continuously exposed* to the laws of physics in the real world, experiencing highly complex dynamic interactions. Once again, however, the body morphology can alleviate the resulting control challenges.

- *Computationally limited* because of the fast environmental interaction dynamics, as well as the concurrent computational speed and capacity limitations imposed by the possible neural information processing mechanisms, brain size, and computational complexity.

- *Strongly dependent* on their bodies with their individual bodily morphologies. The morphology largely determines which particular sensorimotor control processes, involved perceptual capabilities, learning, adaptation, and versatile behavioral selection mechanisms can be developed.

In conclusion, embodied cognitive agents face challenges that differ from those of traditional GOFAI agents and they appear to solve these challenges by means of mechanisms that differ from GOFAI approaches. With the help of their bodily capabilities, they often circumvent the full GOFAI challenge preferring to consider only those aspects relevant for the successful execution of behavior, leading to survival and reproductive success. The rest of this book sheds further light on the involved mechanisms, which may be essential for developing embodied cognitive agents and also human cognitive capabilities.

3.7 When have we reached artificial, human cognition?

The final question in this chapter addresses the extremely challenging question of when an artificial, possibly embodied artificial system may be considered to have reached human-like cognition. Can fully embodied cognitive agents develop their own mind? Could they even be able to think and feel like a human?

Even when we can imagine the question, an answer is hard to identify because the question is hard to define. What do we mean by "developing their own minds"? How can we test that an artificial system is actually able to "think" or "feel" like a human?

Can machines think?

Alan Turing acknowledged this dilemma in his article *"Computing machinery and intelligence"* (1950), which set forth the *Turing Test*:

> I propose to consider the question, "Can machines think?" This should begin with definitions of the meaning of the terms "machine" and "think". [...] Instead of attempting such a definition [such as a Gallup poll] I shall replace the question by another, which is closely related to it and is expressed in relatively unambiguous words. The new form of the problem can be described in terms of a game which we call the 'imitation game'. (Turing, 1950, p. 433.)

The beauty in the imitation game lies in its simplicity: three players were originally proposed to be involved in the game: a man (A), a woman (B), and an interrogator (C). The interrogator cannot see nor hear the voices or any other sounds from A and B, but he can communicate with them (for example, via typewritten messages). Moreover, the interrogator does not know if A or B is the man, and the task is thus to identify who is the man and who is the woman. The game is called the imitation game, because both, A and B, are supposed to pretend to be male or female, thus attempting to fool the interrogator so he gets the sexes wrong.

Inspired by this game, Turing then asked the following question:

> "What will happen when a machine takes the part of A in this game?" Will the interrogator decide wrongly as often when the game is played like this as he does when the game is played between a man and a woman? These questions replace our original, "Can machines think?" (Turing, 1950, p. 434.)

Turing thus proposes playing the game with a human and a computer, both attempting to appear as human as possible. What would be the implication then when the computer is perceived as human as often as not? Would this machine then indeed "think"? Would it then be actually "human" with all the involved cognitive capabilities and even feelings?

Frames of reference and the Chinese Room problem

Even now the imitation game (known as the *Turing Test*) is being played in various forms, including simulations of artificial agents in games that are supposed to play as humans do. It has been claimed several times that the Turing Test was solved by chat bots in restricted scenarios and limited time frames. In particular, a limited number of questions were allowed, the users were not actually aware that a machine might be answering the questions, or the users were not instructed to interrogate if a machine may actually be answering the question, but were rather just confronted with the system. As we have seen, the chat bot ELIZA reached quite high performance levels and current applications certainly excel in these capabilities. However, if the Turing Test is made explicit and is not limited to a few questions, AI programs can still be unmasked rather easily. Especially when the interrogator is informed about the computational mechanisms and information sources that have been incorporated in an AI program, non-human like answers can be triggered rather quickly.

However, let us assume for the moment that these programs become so complex, are possibly neurally implemented, and are so successful in producing answers to these interrogations that even after hours of interrogation the program cannot be distinguished from a human. What would the actual implications be? The American philosopher and linguist John Searle (*1932) has attempted to illustrate the implications with the allegory of a "Chinese room":

> Suppose that I'm locked in a room and given a large batch of Chinese writing. Suppose furthermore [...] that I know no Chinese, either written or spoken [...]
>
> Now suppose further that after this first batch of Chinese writing I am given a second batch of Chinese script together with a set of rules for correlating the second batch with the first batch. The rules are in English, and I understand these rules [...]
>
> Now suppose also that I am given a third batch of Chinese symbols together with some instructions, again in English, that enable me to correlate elements of this third batch with the first two batches, and these rules instruct me how to give back certain Chinese symbols [...] in response to certain sorts of shapes given me in the third batch.
>
> [...] after a while I get so good at following the instructions for manipulating the Chinese symbols [...] that from the external point of view – that is, from the point of view of somebody outside the room in which I am locked – my answers to the questions are absolutely indistinguishable from those of native Chinese speakers. Nobody just looking at my answers can tell that I don't speak a word of Chinese. [...]
>
> As far as the Chinese is concerned, I simply behave like a computer; I perform computational operations on formally specified elements. For the purposes of the Chinese, I am simply an instantiation of the computer program. (Searle, 1980, p. 417f.)

From Searle's perspective, the computer would still be a symbol-processing machine incapable of having a mind although it perfectly answers all questions in a human-like manner. In the article, Searle contrasts advocates of *Strong AI*, who would concede a program that solves the Turing Test (at least a very hard version of it) as having a mind, with advocates of *Weak AI*, who would not.

If we tend toward the *Weak AI* stance, which is also taken by Searle, what is it, that humans have that at least GOFAI machines do not have? Intuitively, there seems to be more than symbol manipulation. We have intuition and we have qualitative feelings. Robots – in our typical understanding of the term – cannot feel, which is often referred to as the *Qualia* problem. Where do our qualitative feelings, such as joy or pain, actually come from? How exactly is a feeling invoked? Why does a feeling feel as it does?

As a result of our feelings and social minds, we tend to attribute meaning to particular items, objects, situations, other humans, etc. We even tend to attribute meaning to simple animated figures, as well illustrated by the Austrian social psychologist Fritz Heider (1896–1988) and his collaborator and psychologist Marianne L. Simmel (1923–2010), who generated videos in the 1940s of simple circular and rectangular figures that are interacting with each other in a systematic manner (Heider & Simmel, 1944). Simply by the way in which the relative object motions unfold, we typically immediately perceive a little social story played out by the involved geometric entities and can identify friend and enemies, aggressive behavior, or affection.

As the Heider and Simmel videos illustrate, but also as many other emergent phenomena show, we tend to interpret observations in a biased human manner. Ant highways are interpreted at first glance as intelligently planned and constructed streets. Robot behavior is interpreted as goal-oriented, such as the Didabots that "intend" to pile-up blocks. Objects in Heider–Simmel-like videos are interpreted as intentional, socially interactive agents. This problem that the observer is interpreting observations to the best of its knowledge, from its own perspective, is sometimes termed the *frame-of-reference problem.* This problem also needs to be considered when addressing the question "can machines think?" – and when considering the closely-related Turing Test. If a machine solves the test satisfactorily, we might be willing to attribute to it very human features; we may even fall in love with it as put forward in the 2013 Hollywood movie "Her". However, this willingness may be influenced by our frame-of-reference and, indeed, may not be an objective truth.

Perspectives on behavior and cognition

When attempting to explain the behavior of other entities including humans, Daniel Dennett put forward three levels of explanation, or perspectives, that should be considered:

1. *Physical laws* need to be considered when monitoring how things behave under the laws of physics.

2. *Design principles* need to be taken into account, essentially considering that things behave in a certain way because they were designed to do so.

3. *Intentional aspects* need to be considered when physical laws and design principles cannot account for the observed behavior. Then one may consider that the entity acts like a rational agent according to its internal goals and intentions.

Dennett's considerations about physical laws and design principles are closely related to what Fritz Heider had termed the *external attribution* (Heider, 1958). For example, when we watch a "Heider–Simmel video", we may see how a ball that just hit another ball changes direction in a certain way. Our interpretation then may be physical, that is, we attribute the ball's behavior to the impact and resulting changes in energy caused by the hit. We may also include knowledge about the designer, who, for example, has created snooker balls with certain properties such that the balls physically behave in a particular manner. Dennett's third stance is comparable with Heider's *internal attribution*: when, for example, the ball suddenly behaves in a way that cannot be predicted by physical and designer-based explanations alone, the causes of particular behaviors must lie within the agent itself, such that the observed agent's intentionality must be responsible for the behavior.

As implied by Heider and Simmel's videos, but also by many other examples, such as our tendency to over-interpret emails from an emotional stance, it appears that our brain quickly attributes intentions to objects, and particularly to those objects whose behavior cannot be explained directly from a physical or a designer-based perspective. Therefore, we should keep in mind that when observing particular agents, items, animals, and humans acting in certain ways, we should attempt to take different perspectives when trying to explain or to understand how the observed behavior came about. These perspectives should consider alternative intentional, designer, and physical explanations, as well as interactions between these levels of explanation.

Taking a step back and considering the overall implications of the fact that *cognition is embodied*, we can see that AI and cognitive science researchers have often erroneously thought about intelligence in a totally symbolic, human-educated, abstract fashion. While doing so, they have often overlooked the role of cognitive development.

Maybe this is the most important message for cognitive scientists: do not overlook the developmental, embodied perspective! Programs that may be programmed to answer questions in certain ways, that even may have been trained to learn to answer these questions given huge text corpora, the knowledge on the web, etc., have been programmed. They have not experienced actual consequences of their actions in their environment. Unlike humans, they have not developed their own embodied concepts and symbol interpretations, which are grounded in all their experiences and mediated by their bodies. Rather, they have made detailed statistical analyses, extracting systematic relations between words and sentences. They have thus identified commonalities between words, sentences, pictures, and possibly even videos, which may allow them to pass the Turing Test.

For a cognitive scientist, the challenge is to identify how the human mind manages to solve the symbol grounding problem, the frame problem, and the binding problem to develop human-like understandings of the world from the bottom up – shaping the mind during development. It is this challenge that is pursued in the remainder of this book.

3.8　Exercises

1. Why is it still nowadays hard for many of us to accept that a machine may be able to really think on its own one day? Relate your arguments to Ada Lovelace's beliefs.

2. List various types of "intelligence" that may be attributed to a particular person?

3. An intelligence test measuring the IQ of a person focuses on a particular type of intelligence. Characterize this type in your own words.

4. What may be the difference between interpersonal and intrapersonal intelligence according to Howard Gardner?

5. In which way is the ability of adaptation related to intelligence?

6. Relate the limitations of symbolic AI to the symbol grounding problem.

7. Come up with an everyday example that can illustrate the frame problem as good as Daniel Dennett's R1-R1D1-R2D1 robot example does.

8. The three Rubin face-vase illustrations in Figure 3.1(b) differ in which figure is dominantly perceived. Why are the faces in the center depiction more prominent than the other two?

9. Discuss and contrast the binding mechanisms that must be at work in the Necker cube, the Rubin Vase, and in Navon figures. When attempting to focus on one interpretation/entity in each figure, on which aspect(s) should our mind's attention attempt to focus on?

10. Implement simple neural network learning mechanisms. Learn to associate number patterns of a simple digital seven-segment display with individual number nodes. That is, learn to associate a seven dimensional vector a with a ten dimensional indicator vector b.

    - Implement the Hebbian learning rule and observe the changes in weight. Vary the sampling of the numbers and observe the effects in the weight changes over time. Given uniform number sampling, observe in which way the "imagination" of the digital display of a number can be generated by activating the corresponding value in b.

    - Implement the perceptron model and analyze the resulting classification accuracies. Again analyze performance changes due to data sampling variations.

    - Add random noise to a and b and analyze the respective degradations in performance.

11. Come up with some examples from nature of particular traits of animals or plants and loosely attempt to explain them in terms of the principles of embodied biological processing and morphological intelligence.

12. Design Braitenberg vehicles!

    - Design a Braitenberg vehicle that is increasingly active the brighter its surrounding and that seeks such surroundings by equipping it with three sensors and an appropriate wiring.

    - Design a Braitenberg vehicle that avoids light sources, but seeks heat sources.

13. Consider with respect to a "balanced design" the usefulness of equipping a snail with human-like eyes.

14. Consider the consequences of a "balanced design" for artificial cognitive robots. Imagine a particular scenario, for which a robot may be useful and discuss the usefulness of particular sensors, motors, and its general morphology. Propose a robot design (specifying its morphology, sensors, and motors) that may be particularly suitable given the imagined scenario.

15. Imagine a robot that is supposed to clean your kitchen without giving any further instructions or signals. Make the real-world challenges concrete that this embodied cognitive agent will have to face.

16. Discuss if a machine that passes a complex Turing test may have human thought or not. Contrast the standpoints of weak AI and strong AI in this respect.

17. Why do humans continuously face the frame-of-reference problem (often without realizing) when attempting to understand a particular process or phenomenon?

Chapter 4

Cognitive Development and Evolution

4.1   Introduction

We saw in the last chapter that higher levels of cognition must be inevitably grounded in lower-level, sub-symbolic, sensory-, and motor-processing routines. Cognition and intelligence thus develop in embodied brains, which learn and shape their knowledge, based on the encountered experiences with the environment, as well as evolutionarily predetermined developmental constraints on neural and bodily levels. For example, the body develops in such a way that crawling becomes physically possible only after a few months of development and walking even later than that. Another example is the eyes, which develop in the first year of the infant – initially providing a very fuzzy, noisy image of the environment, which then progressively becomes crisper. On the neural level, a very obvious example is the neurological changes encountered during puberty, including the extensive myelination of nerve fibers. Thus, cognition develops *ontogenetically*, that is, from conception onwards throughout a lifetime; and this development is influenced by genetic predispositions, the environment, and interactions between these factors on bodily, neural, and cognitive levels.

When applying this developmental stance to cognitive science, it is necessary to understand that development has many facets. Think about your own life. Your first memories may reach back to kindergarten or even slightly earlier, but hardly anybody claims to remember anything in particular under the age of 2 years or so. However, development certainly started earlier than that. When studying newborn infants, developmental psychologists have discovered remarkable capabilities, one of which is the following: infants are equipped with several types of reflexes, including the palmar grasp reflex and the rooting reflex – both of which are extremely useful during development. For now, let's focus on the latter. The rooting reflex works as follows: when touching the cheek of an infant, she orients herself toward the touch and attempts to suck on the thing that caused the touch – so to say in evolutionary-determined anticipation of the mother's breast and thus of receiving milk. The most important point, however, is that when the infant touches her cheek with her own hand or finger, she does not show the rooting reflex. How is she able to suppress the reflex in this case? The only explanation seems to be that the infant "knows" that she has just touched her own cheek and thus it is not worthwhile to orient herself toward the touch. The only way she can know this is that she has a sufficiently accurate, postural image of her own body. The conclusion is that this knowledge – as the behavioral effects show up right after birth – must have been acquired *before* birth, while developing inside the mother's womb or via genetic knowledge encodings.

This example illustrates how important it is to acknowledge that cognitive development does not start only after birth, but rather it starts shortly after the actual conception of the embryo. In fact, the first neurons are already developing in the embryo few weeks after conception. Thus, the brain is not a *tabula rasa* when the infant is born – as, for example, William James believed about 120 years ago – but it is already full of knowledge and behavioral capabilities. Clearly, within the womb the experiences are limited, but these limitations may actually be advantageous: the developing knowledge concentrates on knowledge about the fetus's own body, and basic sensory and motor capabilities.

Purely genetic knowledge encodings also contribute to the developmental progression. The rooting reflex is most likely present due to genetic encodings – as are many other reflexes, including the grasping reflex. It remains an open question as to how our genes generate such reflexes and precisely which details are genetically encoded. Seeing the variety of species and their extremely intricate developmental patterns – just think of the metamorphosis of a caterpillar into a butterfly – many intricate developmental, and possibly even cognitive, aspects may be encoded in our genes. The common ancestors of humans and apes lived on our planet not that far back in the past, in terms of evolution, so that particular human traits are likely to have much in common with ancestral pre-human species. Thus, there is some hesitation when attributing to a genetic code the development of a neural code that encodes a postural map of the whole body; growing that body, and then sensing and activating it by motor activities seems much easier. Nonetheless, when considering cognitive development, evolution must not be forgotten.

This chapter thus addresses *ontogenetic development*, that is, cognitive development starting from conception and continuing through adulthood, as well as *phylogenetic development*, that is, the evolutionary development of the human species. In the former case, several examples and capabilities will be discussed, which shape the way we think and which allow us to understand progressively more about our lives. In the latter case, the focus will lie on genetic algorithms and evolutionary computation techniques, which allow not only a glimpse of the probable power of evolution, but also of several very significant limitations. Thus, while the former will give us an understanding of how our mind develops during our lifetime, the latter will allow us to make crude estimates of which encodings developed by means of evolution and which encodings probably developed mainly during ontogenesis.

4.2 Ontogenetic development

Cognitive development during ontogenesis, that is, from conception through to death, has many aspects to it and is shaped in various ways. Here, we focus on the development from the first cells until the first years of life. Before doing so though, we must define and differentiate ontogenetic development.

Generally, ontogenetic development may be defined as permanent, developmental changes of some properties of an individual, including bodily as well as mental changes. Typically, particular developmental changes systematically correlate with age, that is, they typically take place within a certain age range. In contrast to Piaget's rather strict theory of developmental stages (cf. Section 2.4.2), it is now known that ontogenetic development does not necessarily progress linearly and the age correlations vary greatly. Nonetheless, ontogenetic development is a fact and exhibits particular systematicities. Developmental changes themselves can also be of different kinds, including quantitative and qualitative changes, acquiring and also forgetting, differentiations, replacements, integrations, abstractions, and generalizations. The following types of changes may be differentiated:

- The simplest and most obvious form of development is *bodily development*, including bodily growth as well as the genetically predetermined and controlled development of particular bodily structures and capabilities.

- *Behavioral development* – such as the above-mentioned reflexes – are to a large extent genetically predetermined and typically are exhibited at certain times during development.

- *Cognitive development* typically co-occurs with bodily and behavioral development. Learning progresses throughout our lives, but the first years are crucial in shaping us – structuring, evaluating, and integrating our experiences.

- *Social development* can be contrasted with cognitive development. It is particularly important in humans, but it is also important in many other animals. We develop social competencies and interaction patterns while interacting in social contexts – with parents, other family members, caretakers, friends, peers, other humans, and even animals and artifacts (such as dolls, stuffed animals, or toys). The integration of the self in a social world, indeed, seems to be highly relevant for our cognitive development.

- *Language development* is related to social development, but certainly goes beyond it. Our mind seems to be *language-ready* at a certain point in time, at which sufficiently structured encodings of the world are present and language sounds, that is words and progressively complex sentences, are associated with particularly structured encodings.

Taking a computational view of development, it should be clear that cognition can only develop if suitable hardware and software is available. Hardware components include the body with its sensor and motor capabilities, as well as neural hardware, that is, the brain, which needs to be capable of processing particular information before this information can even be taken into consideration for co-determining ontogenetic development. This corresponds to Marr's third level of understanding – the hardware that implements cognition. The hardware essentially determines which algorithms can be implemented – on Marr's second level of understanding. These algorithms, akin to the the software of a computational device, determine which structures develop and how they interact with each other. They develop in a social world, and the developing system must be equipped with goals or needs, which must be grounded in hardware and processed by the software. Goals and needs – Marr's first level of understanding cognition – determine intentions, attention, and behavior because we are not passive, purely reactive machines, but active explorers, who exhibit preferences and predilections for particular aspects of our world from early in development.

It is the interplay of these features – and probably many more not mentioned here – that determine ontogenetic cognitive development with all its particular and intricate facets. In the following we provide further details on particular developmental stages, distinguishing prenatal development from the further development during the first years of our lives.

4.2.1 Prenatal development

Albeit largely ignored by cognitive science books and theories, the cognitive development prior to the birth of an infant should neither be ignored nor underestimated. While it was assumed for a long time that only bodily development occurred before birth, various indicators from developmental psychology research suggest that many fundamental cognitive capabilities actually develop before birth. In the approximately 9 months after conception, one sperm and one egg cell unite and shape a human being including the cognitive apparatus. Indeed, the neural tract develops from shortly after conception and in close interaction with the development of the fetus's body, gathering experiences even before birth.

From one cell to an organism

After the uniting of a sperm with an egg cell, a *zygote* is created, which determines development over the subsequent days and weeks. Starting from one zygote cell, an explosion of changes takes place, developing out of this one cell the fetus and ultimately a newborn baby with over 35 trillion cells ($> 3.5 \cdot 10^{13}$!) of diverse shapes and functions, including muscles, skin, hair, other forms of bodily tissue, and various organs including the brain.

Within a few hours of conception the zygote cell undergoes the first cell division. After 3–4 days, a cluster of cells of about 0.2mm has developed consisting of about 16 identical cells, called the *morula*. This morula develops into a *blastocyst* after another 1 or 2 days, undergoing several additional cell divisions, and developing into a spherical shape. If everything goes well, the blastocyst implants itself into the uterus, ensuring further development. Development then progresses – the outer part of the spherical blastocyst develops into the placenta and into other structures responsible for supplying the developing embryo, while the inner cell mass, called the *embryoblast* develops into the actual embryo.

In the third week of pregnancy, the embryonic phase begins to develop the embryoblast, which is still less than 1mm in diameter. First, three separate specialized cell-clusters are formed, which are referred to as the *germ layers*:

- The *endoderm* develops into the digestive tract and the respiratory system, as well as into liver, thyroid, and other internal organs and glands.

- The *mesoderm* develops into the blood circulation system, including the heart, kidneys, muscles, skeleton, and sexual organs.

- The *ectoderm* develops into the skin, nails, and teeth, but also the whole nervous system, including the brain and all sensory organs.

Shortly after the development of these germ layers, *gastrulation* begins to develop the inner organs, while *neurulation* begins to develop the central nervous system.   Parts of the ectoderm develop into the neural tube, through which neural crest cells migrate and develop into pigment cells, neurons, and others cell types. This formation process takes about 1 week, during which the basis for heart, ears, eyes, and the digestive tract also develop.

After about 4 weeks, the embryo has developed into its typical "C"-shape with a size of about 4mm. The heart, and particularly the heartbeat, can be recognized and even the buds of the extremities are visible. The neural crest has formed three bubbles on the head side, which develop into the basic parts of the brain. These subdivisions of the embryonic brain then form the basis for all further brain development, including the formation of the *neocortex* and the *cerebellum*, as well as all other subcortical structures and nuclei.

At the end of the eighth week of pregnancy, the embryo has reached about 1.5cm in size and weighs less than 1g. All the important organs have formed, the sensory system is functional, and the basic shapes of the extremities are present. The brain and medulla oblongata, forming the spinal cord, are in shape.   Thus, while the embryo has now developed all the major internal organs, its bodily structure, and the main brain components, clearly there is still a long way to go. Nonetheless, the basic structures are there and are already interacting.

Fetal cognitive capabilities

From about the tenth week of pregnancy, the embryo enters the fetal stage of development. It has reached a size of about 3.5cm in length from crown to rump and weighs about 9g. The head makes up about half of the fetus's size at this point – a further indication that brain development starts very early. During this phase, growth and cognitive development are the main focus, rather than structural differentiation. In addition to the body, the brain also undergoes further fundamental growth. For example, the folding of the neocortex starts in the fifth month of pregnancy, developing the sulci and gyri – most likely to expand the cortical surface to make space for more neurons.

Beginning with the fetal phase, the fetus undergoes elementary cognitive development, exhibited by various observable behavioral patterns. For example, the fetus shows a rudimentary circadian rhythm with phases of activity and rest. During active phases, the fetus moves around in the womb. It is able to explore its extremities, its mouth cavity, and other body parts using its extremities. Indeed, the fetus has been shown to suck its thumb, to scratch its body, and to show facial expressions. Basic reflexes such as the grasp reflex are also present before birth.

Especially the prenatal presence of reflexes indicates that sensory and motor systems are not passively registering sensory signals or randomly generating bodily movements, but are already interacting with each other. The sensory system is registering its body and the environment within the womb. The fetus registers brightness and darkness, acceleration, pressure, taste, and it even processes first sound impressions, such as what must be the very loud heartbeat of the mother, and external sounds, such as voice patterns or music. From about 6 months, the fetus reacts to the heartbeat and the breathing of the mother. After birth, a newborn shows a clear preference for the mother's voice and can be calmed by hearing her heartbeat.

Further behavioral indicators confirm that newborn babies have quite a good knowledge of their own body. For example, babies show particular reflexes, such as the rooting reflex described earlier, only when the activating stimulus is not self-generated. Moreover, the mouth opens in anticipation of the own finger when inserting it for sucking. Thus, 3 months of embryonic and 6 months of fetal development not only yield a newborn baby, which has then reached about 50cm in size and a weight of about 3500g, but also an embodied brain that is somewhat accustomed to its own bodily signals and ready to explore the outside world. These details show that cognition does not start only with birth, but the brain processes sensory, motor, and bodily signals from very early on; especially during the fetal phase, but even a few weeks before that, neurons develop and interact with each other and with the body. Thus, fetal development does not consist only of bodily growth, but also of mental, cognitive growth.

4.2.2 Cognitive development after birth: the first few years

After birth, the newborn infant is confronted with a whole new world. It knows already much about its own body, as well as some things about general sensory impressions, such as sound impressions, daily routines, and the heartbeat, but the new openness and air in the outside world must result in very different experiences, when compared with the rather confined space before birth. Suddenly there are no immediate barriers surrounding the infant. Breathing air for the first time must be a shocking and amazing experience. Voices are now much clearer and visual signals will soon become much clearer.

All these new impressions want to be explored, registered, and understood, and the child will need several years until a deeper understanding of the world can be verbalized. However, probably from the very first minutes after birth these new impressions are integrated into the knowledge that was accumulated before birth, leading to diverse further expansions, differentiations, and modularizations.

Physiologically speaking, about 85 billion neurons in the human brain, most of them already present at birth, want to be properly situated and connected. At birth, the brain weights about 350g, which is one-quarter of the adult brain. The weight increase during the further ontogenesis is mainly due to the growth of further neural connections and myelination of connections, although some new neurons can and do develop after birth. Myelination is the formation of a myelin sheath that surrounds parts of a neuron (mainly the axon) improving its signal transfer speed. Brain development is very costly: 50% of the energy needs of an infant are consumed by the brain, whereas this need reduces to about 20% in adults.

In the following subsections we consider some important aspects of bodily and concurrent cognitive development, understanding that only with the appropriate bodily capabilities can particular cognitive capabilities develop. Vice versa, only with sufficient bodily strength and motor control capabilities can further cognitive capabilities develop. For example, as long as it is impossible for an infant to sit up properly, it is likewise impossible for the infant to explore objects on a surface such as the floor or a table top.

Motor system

When first looking at the motor system, it soon becomes apparent that various motor capabilities are in place. Several reflexes are present at birth, which support breathing, successful swallowing, and thus milk and later also other nutrition intake, including the sucking reflex and the previously mentioned rooting reflex. Reflexes help to shape cognition in that they provide particular small, but highly suitable motor programs to successfully interact and to further explore the world.

Very soon after birth it can be seen that the oculomotor system selectively looks, for example, preferentially at biological motion when contrasted with random motion patterns (Pavlova, 2012). The eyes attempt to follow proximal stimuli, such as a bouquet of flowers or a friendly face. Even room edges are of particular interest, presumably to extract general structural information that is found ubiquitously in the experienced outside world.

With respect to manual interactions, soon after birth infants attempt to reach for objects that are in reach more than for objects that are beyond reach. Also, they do so more when they expect to be able to maintain bodily balance while executing the reach (Rochat, 2010). Obviously, the arm extends the hand in the appropriate direction, even though initially in a rather clumsy manner, requiring many corrective movements, as is also the case when executing eye saccades (von Hofsten, 2003).

During cognitive development, the reflexes are differentiated and progressively more controlled and suppressed, where necessary. Eventually, the reflex itself ceases to apply at all and is fully subsumed by goal-directed, intentional control processes. During this transition, however, the reflexes help to explore the world. For example, the grasp reflex yields object interaction experiences, which seem to be crucial for cognitive development, from very early on – facilitating the differentiation of object concepts and the development of the ability to use tools.

Important additional stages during development manifest themselves by the infant's ability to lift her head after about 3–4 months of age. This ability enables her to follow moving stimuli over more extended periods of time, as well as to explore objects and other things in the environment from additional visual angles. The ability to roll over develops at about the same time, similarly enabling the infant to follow stimuli over extended periods of time – apart from being able to change into a potentially more comfortable position on her own. After that, the ability to sit without additional support – typically achieved after about 5–6 months – develops, enabling the infant to see the world around in an upright, steadier fashion, as well as to explore the surrounding world more intently. Recent research in developmental psychology has shown that seeing, feeling, and interacting with objects, other materials, and fluids, for example in the high chair, can boost further cognitive development (Byrge, Sporns, & Smith, 2014; Libertus & Needham, 2010; Smith & Gasser, 2005).

Finally, crawling, cruising (furniture walking), and actual walking open up whole new fields of experiences, and usually develop between the age of 6 and 18 months. The first time a baby manages to reach an object that is out of reach by crawling, slithering, or scooting, must be highly rewarding. Suddenly, the spatial radius that can be manipulated, or considered for manipulation, significantly expands. Navigational skills and even path planning slowly become possible, and can be progressively differentiated. When starting to cruise alongside furniture, the baby learns to maintain an upright posture, supported by her holding hand. The reachability concept is further differentiated when the baby must estimate which items are in reach when cruising along suitable furniture, which ones require somewhat more strenuous crawling, and which ones remain out of reach, despite both means of locomotion. Finally, balance maintenance needs to be further differentiated when taking the first steps and progressing toward actual walking.

How all this actually develops functionally will be addressed in later chapters. For now it suffices to acknowledge that behavioral capabilities, which are initially dominated by selective reflexes, soon are co-activated, and progressively controlled by the infant's motivations, intentions, and goals. These reflexes, however, are helpful in shaping further cognitive development, differentiating the behavioral capabilities, and exploring the outside world meanwhile. Similarly, the morphological development of the body goes hand in hand with cognitive development. Novel behavioral capabilities expand the horizon and set the stage for new cognitive insights, and also vice versa, new cognitive insights make the brain ready to explore and learn new behavioral capabilities (Byrge et al., 2014).

Sensorimotor system

Motor behavior does not develop independent of sensory behavior, but in close interaction with it. While the motor system develops, the visual system and visual experiences are also structured and differentiated. Moreover, other sensory systems provide diverse and often complementary, redundant information about our world, our body, and our motor activities. As already discussed, touch sensations are correlated with bodily postures, seeing that, for example, the rooting reflex is not triggered by self-touching. The internal sense of proprioception, that is, sensory feedback about the state of the body's muscles and joints, not only enables the determination of ones bodily posture, it also gives information about the outside environment, such as barriers and the weight of objects.

Auditory information provides information about interaction events and is used to differentiate particular interactions. Moreover, sound gives information about other individuals, including their identity. For example, we know that newborn infants are able to identify the voice of their mother. The auditory system gives crucial information about the outside world and is processed from early on, leading to selective orientations toward the auditory stimulus. Taste and smell are also fully functional and are being differentiated from birth onwards.

The visual information available to an infant's brain, though, is qualitatively speaking not as good as the other sensory sources of information. The lens of the eye as well as the retina further develop during the first year of an infant's life and yield adult-like sensory information only at about 12 months of age, although a pretty clear image is available after 3–6 months. At birth, though, the physical properties of the eye and retina only allow qualitatively high visual acuity at very close proximities of under 30cm. To an infant, further distant visual cues are very blurry.

Recent cognitive models indicate that this visual inacuity may actually be advantageous for cognitive development in various respects, and thus may be considered a morphological form of computing that shapes cognitive development. For example, distance information about an object is mainly provided by the parallax between the two eyes when focusing on the same object. To determine the parallax though, the two eyes need to learn to focus on the same point in space – such as an object – quickly and accurately. Cognitive modeling work indicates that initial coarse image resolutions can facilitate this learning process (Lonini, Forestier, Teuliere, Zhao, Shi, & Triesch, 2013).

The development of hand-eye coordination seems to be supported by morphological development as well. As hands are typically closer than 30cm, hands can be explored in detail without the distraction of the items in the background, which are very blurry anyways. This inevitably focuses the cognitive development over the first months to near space, which is the space that is reachable for the infant. Similarly, seeing that further distances are blurry, details cannot be differentiated, but general contour and outline patterns can be learned, such as walls, corners, ceilings, or forests and mountains, leading to the accommodation of the infant to particular surroundings and scenes.

When able to sit up, the hand–eye coordination is further differentiated. Objects and other items, such as food and fluids, can now be explored in detail. This object-oriented learning process is also supported by the fact that the hands are already well-controlled and well-known, enabling the brain to filter out visual signals about ones hands. Attention thus focuses in even more detail on near space and objects in near space, actively exploring with the eyes, hands, and mouth.

During this visual and sensorimotor development, particular stimuli have particular statuses. For example, faces are particularly interesting from birth, particularly when they are sufficiently close. Among these faces, the faces of mother and father soon reach high significance. Meanwhile, the first genuine smiles pop up at about 3 months of age, and become progressively selectively targeted toward mom and dad, as well as any other close caretakers and family members.

Sensorimotor development does not focus only on hand–eye coordination. Even earlier, visual stabilization is necessary when inducing own head movements. Thus, beginning very early the infant develops visual self-stabilization mechanisms, which enable the maintenance of a stable image across saccadic eye and head movements. Once again, the rather blurry eyesight capabilities may help in this respect – matches across images are much easier to determine when the images are not very detailed. Progressively further differentiated stabilizations are necessary when the whole body starts moving, such as when rolling over, and even more so when starting to crawl and eventually to walk.

Meanwhile, the self-image develops and knowledge about ones body's motion manifests itself in behavior. Infants only 2 months old show signs of distinguishing their own motions from the motions of others. They become progressively aware of themselves as acting agents in the environment, noticing that they produce multimodal feedback by means of own actions. It thus appears that the sense about the own body develops from intermodal bodily perceptions and actions, which cause these perceptions (Rochat & Striano, 2000) It is this sense that then becomes "a public affair" between 2 and 3 years, leading to the development of self-consciousness, and the embedding of the self in the experienced social reality (Rochat, 2010).

After about 6 months, the baby learns to differentiate further depth cues in addition to the cues from disparity and the developing parallax, including clues from occlusions, textures, and size variances. With progressively better visual acuity, interest in these cues naturally increases and further differentiations of the cues come naturally, especially with redundant other depth cues, such as parallax and disparity, now readily available. Moreover, the interest in distant items increases, thus fostering the drive to crawl and walk.

The object manipulation capabilities also develop further, differentiating the grasp reflex in manifold ways and developing handedness. The initial inborn grasp is modulated and adapted to the object so that the hand starts to open in anticipation of object contact and the opening is adapted to the size of the object. Later on, the dominant fist-like grasp is differentiated into a radial-palmar grasp, an immature rake grasp, and a scissor grasp shortly thereafter. These are followed by a radial-digital grasp and finally, at about 10 months, the pincer grasp using thumb and index finger. After about 1 year, babies are able to grasp even small pellets with a fine pincer grasp (Johnson & Blasco, 1997).

These intricate hand–eye abilities develop and are further refined throughout our lives. Toddlers slowly learn to use tools such as a fork, properly in the third to fourth year. Walking and running are further refined and jumping becomes interesting. Moreover, other types of locomotion and climbing capabilities are explored (Broderick & Blewitt, 2006). Behavioral skill development thus goes hand-in-hand with the sensory processing capabilities, and particular developments foster other developments. Cognitive conceptualizations also go hand-in-hand with these sensorimotor developments, some of which are detailed in "Conceptualizations".

Conceptualizations

While considering bodily and behavioral sensorimotor development, it should be kept in mind that perception and behavior develop not for their own sake, but rather to be able to interact with the world progressively more successfully to accomplish ones current goals. To do so, two fundamental prerequisites need to be accomplished. First, it is necessary to know what type of object is actually currently present. Objects need to be categorized into, for example, food objects, sound-producing objects, and caretakers. Meanwhile, the manipulability of the individual objects needs to be differentiated, such as objects that are graspable, interactable, or throwable. Thus, meaningful object categories need to be formed that indicate object affordances (Gibson, 1979), that is, what is an object good for and how may one interacted with it.

An interesting additional question arises when considering these conceptualizations: which conceptualizations are inborn and which ones are acquired during development? Currently, the discussion about this question is still open. General agreement in cognitive science has not yet been achieved. Nonetheless, several core concepts can be distinguished, which are apparently present at birth or soon afterwards, suggesting that these are inborn or at least strongly pre-shaped during prenatal development.

Many studies of conceptualization in early infants are based on an *habituation paradigm*, where the longer focus on an event or object is interpreted as being more interesting for the infant or baby. Unexpected interactions appear to be viewed longer by infants than expected interactions. One illustrative example comes from the "Ernie & Elmo" experiments conducted by Simon, Hespos, and Rochat (1995): in one scenario 3–5-month-old infants were shown two Ernie puppets, which are then occluded. Next, visible to the infant, one of the Ernies was removed. When the occluder was then removed, the infants looked significantly longer at the scene when two Ernies were still present in contrast to when only one was present (cf. Figure 4.1). This also worked with addition, having initially only one Ernie, adding another one visible to the child behind the occluder, and then removing the occluder with only one Ernie behind it – in which case one Ernie was looked at longer than when there were two Ernies. These results confirmed a previous study conducted by Wynn (1992), but also enhanced it in that the infants appeared to generalize over the identity of the puppet. If Ernie changed to Elmo, the surprise was not as big as when two Ernies were expected but only one was shown after the occluder was removed. Thus, object identity seems to be detached from object numbers – an effect that may be explained by the brain's differentiation into the dorsal "where" pathway and the ventral "what" pathway (Goodale & Milner, 1992; Ungerleider & Haxby, 1994).

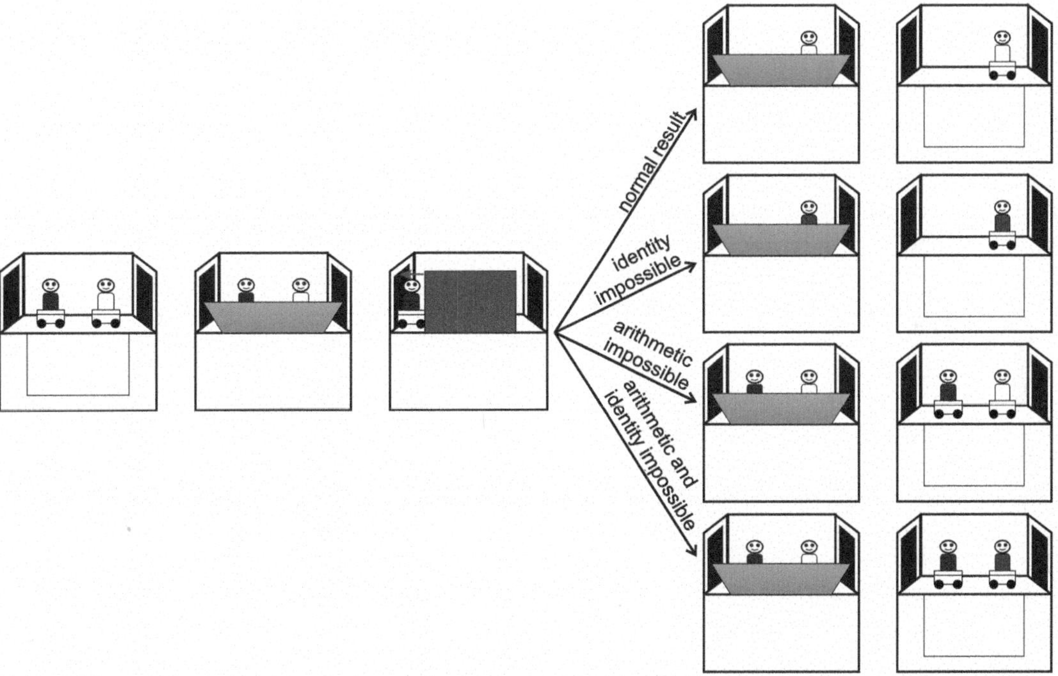

Figure 4.1: Five-month-old infants show signs of surprise when an object unexpectedly disappeared or appeared after occluder removal. [Modified and reprinted from *Cognitive Development, 10*, Simon, T. J., Hespos, S. J. & Rochat, P., Do infants understand simple arithmetic? A replication of Wynn (1992), 253–269. Copyright (1995), with permission from Elsevier.]

Similar experiments have investigated other core concepts and have further differentiated them. Focusing on objects and spatial concepts, Jean Mandler (Mandler, 2012), for example, has proposed the following conceptual primitives, which she assumes to be innate. She differentiates spatial concepts into a "path" concept with "start" and "end", as well as a "path to" concept, which leads to intentionality. A "link" concept specifies the knowledge that objects can be connected, and thus move together, or not. A "container" concept characterizes that a thing can be "in" other things and also can be moved "into" other things as well as "out of" other things. "Motion" of things is conceptualized and contrasted with "blocked motion" as well as with "things" being in "contact" or not. Finally, things can be moved "into" something else, "behind" something else, "out of sight", "into sight", or to a certain "location". Mandler proposes further that more complex concepts are formed out of these conceptual primitives. Other researchers argue even more strongly for development and embodiment, suggesting how these conceptual primitives may actually form based on experiences of sensorimotor interactions and sensory observations of particular object interactions (Butz, 2016).

A conceptualization of object permanence, and thus some intuitive physical knowledge, seems to be present very early in an infant's life. Baillargeon (1987) showed a solid screen that was rotated upwards like a drawbridge occluding anything behind it to 3.5–4.5 month-old infants. After a 180° rotation, the drawbridge was lying flat on the floor behind it. When a box was placed on the floor and the drawbridge moved through the box, the infants looked significantly longer at the event than when no object was present. Later in development, 6.5–7.5 month-old infants were shown to consider also the size and the consistency of the box (i.e. if the box could be flattened) in their judgments. Thus, all infants expected the box to still be present behind the drawbridge and that the box should block the path of the drawbridge, leading to a surprised, longer fixation when the drawbridge apparently moved through the box.

These results led to the questioning of Piaget's earlier conclusions that object permanence develops only much later in life, based on his observation of A-not-B errors in toddlers of slightly more than 1 year of age (cf. Chapter 2). The observation that children search for an object at the original location, needs to be explained in a different manner – most likely due to their inability to inhibit to search at the previously successful location despite the evidence that the object is no longer located there.

While a very basic animate versus inanimate object distinction appears to be present upon birth, other object categories are clearly learned only after birth. During the first year, an infant puts virtually everything into her mouth for exploring taste and edibility. Object categories, such as edible food objects, toys, and tools, develop during this time. Interestingly, systematicities can be identified during this object categorization and individualization development. For example, particular basic categories, such as cats, dogs, and cars, are learned before particular subcategories and also before more general categories, such as four-legged animals, Siamese cats, or Porsches. Many studies have indicated that categories are developed by distinguishing behavioral relevancies, forming equivalence classes over those relevancies (Hoffmann, 1996). Objects are considered similar that behave similarly and that have similar functional and motivational properties. As a consequence, generalizations are observable, such as food and drink items, glasses and mugs, or bicycles and motor cycles. In later chapters, we will introduce mechanisms and factors that can lead to the formation of these conceptualizations, and even to their linkage with linguistic, word-determined symbolic representations. Many researchers now agree, though, that thought and conceptualizations come before language and make the developing toddler *language-ready*. Language then further shapes and differentiates the conceptualizations. The initial conceptualizations, however, are present before language and significantly influence further cognitive development.

Development of item- and event-specific memory

All the aspects of cognitive development that have been discussed so far essentially manifest themselves in memory. Improvements of behavior, focus of attention, object category generalizations, and face and object recognition are nothing more than learned and thus memorized cognitive and behavioral capabilities. Item-specific and event-specific memory capabilities develop hand-in-hand with these other capabilities.

Due to the lack of language, however, other experimental paradigms are needed to determine the working memory capabilities of infants. *Habituation* is often used as the paradigm of choice (Gleitman, Gross, & Reisberg, 2011). Habituation essentially focuses on boredom: when the same object is presented multiple times in a row, the infant gets bored, or habituated, to it and thus pays less attention to it. When the same object is presented later on, the question is whether the infant is still habituated or not – if it has completely forgotten the interaction with or the observation of the object, behavior should be similar to a new object. If it remembers it, the interest should be lower. With such experiments, it was possible to show that 3-month-old infants can remember a visual stimulus for up to 1 day, while 1-year-old infants show a memory span of several days. Of course, these results need to be considered with some caution, seeing that the presented stimuli typically have no actual behavioral or motivational significance aside from their inherent novelty. Memory for exciting events, objects, or items may be better than that. Nonetheless, memory development was confirmed by means of the previously described paradigm.

One more behavior-oriented task has shown that behavior-grounded, procedural memory is stable much earlier in development. For example, Rovee-Collier (1997) connected a kids mobile with a string to the leg of an infant, so that leg movements resulted in controllable movements of the mobile. Infants connected to the mobile soon learned to move their legs more intently while focusing on the mobile. In this case, 2-month-old infants showed signs of remembrance when confronted again with the same set-up the next day, but not after 3 days. 3-month-old infants, however, showed signs of remembering after 1 week, and 6-month-old infants even after 2 weeks. Interestingly, the memory performance was also influenced by the similarity of the mobile during training and testing, and even by the pattern of the bed, indicating that the infant learned the interaction for a concrete situation and not in a more generalized fashion.

Social cognition, imitation, and the self

While the infant learns to explore her environment and her own body, the last important factors that determine cognitive development are social interactions and imitations. Various disciplines have shown that a very important trait that makes humans unique and distinguishes us from other animals are our abilities to cooperate, to share, and to develop a sense of fairness far beyond what animals are capable of (Tomasello, 2014). To be able to do so, it seems necessary on the one hand to perceive others as separate individuals – each with certain traits – and, on the other hand, to integrate the self into the experienced social and cultural groups.

The basis for social cognition develops right after birth and perhaps prenatally with the recognition of the mother as a separate, acting individual. After birth, *dyadic interactions* between infant and caretaker, and other people form the basis for further interactions. During these interactions the child learns about typical interaction patterns, behavioral responses, mimicking, eye contact, and language, including sound and tone. Some of these responses can be observed during social interactions, starting in the second month after birth. Imitations of facial expressions have been shown to be present even earlier than that. At that time or slightly earlier, the infant becomes able to differentiate self- and other-induced events in the environment – such as the motion of an object or the production of a sound. The infant knows that she can influence the environment in certain ways, such as when throwing a puppet to the floor or when shaking a rattle.

At about 3 months of age, infants typically have learned to direct their attention on a particular item over an extended period of time and are able to infer the current attention of their caretaker, leading to moments of *joint attention* directed toward the same item. At about 6 months, infants can be shown to focus on the action goal and less on the actual details of an action. For example, when observing a person grasping an object infants habituated to grasps to the same object more than to more similar grasps that were, however, directed to a different object (Woodward, 1998). In this way, *triadic interactions* develop between infant, caretaker, and object, where either the infant or the caretaker can initiate such interactions.

Beginning at about 9 months, infants start to understand pointing gestures as a request to direct their attention to the item pointed at and they also start to use these gestures themselves as a request to their caretaker. Understanding these gestures is then progressively differentiated leading to supportive behavior upon a request for a particular item, which can be observed after about 14 months and which develops into cooperative behavioral abilities shortly thereafter (Tomasello, Carpenter, Call, Behne, & Moll, 2005). For example, infants bring an object to their caretaker upon request issued by means of a pointing gesture, or they hold onto something temporarily for their caretaker. Soon, infants also show a certain form of rationality in their behavior, apparently considering the current capabilities or restrictions of the person issuing a request.

Importantly, infants show progressively more complex *perspective taking* abilities starting already in the first year (Moll & Meltzoff, 2011). That is, infants typically do not execute the exact kinematic bodily behavior observed, but rather assimilate it into their own frame of reference and contextual situation. In fact, infants typically focus on the goal of the observed action, rather than on the detailed means by which the goal was achieved (Bekkering, Wohlschlager, & Gattis, 2000). Moreover, infants consider the current abilities of the observed interaction partner, thus exhibiting clear rudimentary forms of abstracted behavioral simulations, which can also be related to empathy (Meltzoff, 1995; Tomasello, 2014).

These insights suggest that the infants use their own understanding of the world, including their knowledge about the interaction capabilities of others with the world, to infer behavioral motives and intentions (Gergely, Bekkering, & Kiraly, 2002). The infants appear to place themselves with their own behavioral abilities into the social context. As already mentioned, differentiations between self and other can already be noticed in infants at 2–3 months of age (Rochat, 2010). However, the actual exploration of ones body – such as a strange mark on the face – by means of the mirror only develops starting at about 18 months.

How are these self-perspectives projected into and differentiated with respect to other people? From the described social cognitive capabilities a theory of mind develops that progressively refines and differentiates other individuals, including their knowledge and behavioral capabilities (Rochat, 2010; Tomasello et al., 2005). From neuroscience it is known that *mirror neurons* in the brain not only encode ones own particular behavioral interactions with the environment but they are also active when another person executes a similar behavior (Rizzolatti & Craighero, 2004). These neurons have thus somehow solved the perspective taking problem, focusing on the actual observed environmental manipulation, rather than on the precise means by and perspective under which the interaction is accomplished. These perceptions of the self and others, by apparently similar neural means, are then put into a social context, leading to the assignment of particular roles to the self and to others in the social realm. After about 24 months, toddlers clearly re-recognize themselves in the mirror and also begin to show signs of perceiving themselves as an objectified, public entity (Rochat, 2010). It is at this point that toddlers begin to refer to themselves as "I" and also start to show behavioral patterns that apparently consider the consequences of being watched, indicating social feelings of, for example, shame, embarrassment, and guilt.

How all these developmental steps, including social development, are accomplished by our brains remains an open question. Nonetheless, the following chapters attempt to provide some hints that may lead to satisfactory answers. Before looking further into ontogenetic development and the involved mechanisms, however, we first turn to phylogenetic development, looking for suggestions of how the human species with all its particular developmental and cognitive traits has evolved phylogenetically.

4.3   Phylogenetic development and evolution

As cognitive development develops concurrently with bodily and brain development, it is worthwhile to consider the question of which aspects of these co-developmental processes are imprinted into our genes. To do so, we must first define what evolution actually is and give an overview of the basic mechanisms behind genes and their determination of bodily growth and development. We must also consider some principles underlying phylogenetic, evolutionary developments of species, and will introduce the concept of a *balanced evolutionary design*. This concept puts forward that evolution does not necessarily generate biological sensory organs that are maximally sensitive, or motor capabilities that allow extremely dexterous behavioral interactions with the environment. Rather, evolution balances the sensory and motor capabilities with the energy sources available. Furthermore, the concept of *ecological niches* emphasizes that our environment is full of particular subareas where particular energy resources, climatic conditions, substances, terrain properties, and other species can be found. The particular distribution in a niche determines in which way the present species compete with each other for reaping the available resources to survive and reproduce.

After that, we introduce *genetic algorithms*, which are implementations of the principles of evolution in the computer. Genetic algorithms reveal some basic computational principles that must have been implemented in one way or another by natural evolution. For example, *fitness* is implemented by principles of survival and reproduction. *Diversity* is accomplished by niche-dependent developments and coevolution. The concept of a *building block* will help to illustrate those particular components that appear to be encoded in our genes and that need to be available for proper recombination, restricted mutation, and possibly gene duplication.

4.3.1 A brief history of evolution science

After centuries of religious beliefs about the origins of species and particularly humans, many of which still persist today in various forms and countries, Charles R. Darwin (1809–1882) was innovative enough to propose a rather refreshing alternative scientific explanation (Darwin, 1859): The origin and diversity of the species on earth is the result of millions of years of gradual development over generations of species. Starting with the evolution of single cells, more complex species and ultimately humans developed due to the mutual interaction of two fundamental mechanisms or principles:

- *Variation:* individuals of a species are never completely identical but can be differentiated in their bodily, behavioral, and even cognitive capabilities, properties of which are passed on to the offspring of the respective individual.

- *Selection:* natural selection indirectly chooses those individuals for reproduction that are more effective in reaping the available resources, competing within and between species.

While the details and exact mechanisms underlying the variation and selection processes are still not fully understood, no current serious scientist questions these two general principles.

Origins of the theory of natural selection

Despite Darwin's groundbreaking accomplishment of establishing his theory of natural selection and thus his theory on the origins of species (Darwin, 1859), numerous natural scientists and philosophers had previously put forward theories on this matter. We know from written evidence that all cultures developed theories about the origins of the earth and life. Even Greek mythology has put forward explanations for the creation of life. Often these thoughts and ideas appear far-fetched, and somewhat amusing given our current knowledge.

For example, Anaximander of Miletus (610–546BC) proposed that plants and animals may have developed out of warm mud and humans developed later as descendants of fish-like creatures. Empedokles (495–~435BC) also believed that life had its origins in wet mud. He additionally suggested that initially only individual body parts may have developed, which when recombined randomly, sometimes yielded effective combinations and thus individual species. Aristotle was inspired by these ideas and postulated that nature allots only those body parts to a particular animal that are useful for that animal (cf. also Section 2.3.2). Despite its potential for explaining such allotment to natural selection, Aristotle believed that it was permanent and did not develop any further. He also suggested a first theory on epigenetics, according to which structures and organs of organisms develop dependent on environmental circumstances.

While these ideas are original, but far fetched, we can recognize grains of truths:

- Humans have evolved from more primitive forms of species, however they may have looked.

- Only useful structures and traits establish themselves and persist over time, which is closely related to the principle of natural selection, although the principle itself was not made explicit.

- Each species has those traits and structures that are maximally useful for its own survival, which can be considered as a precursor of the principle of *ecological niches*.

Birth of the modern theory of evolution

One of the first to propose a complete and rational theory of evolution was the French zoologist Jean-Baptiste Lamarck (1744–1829). He postulated that simple life forms had evolved over time into more and more complex life forms, by consistently and effectively using individual limbs and organs in interaction with the environment or conversely not using them. This led to the further development or the disappearance of the respective body parts. Giraffes, for example, have evolved their long legs because they used and stretched them to reach all the leaves in a tree. Lamarck thus essentially postulated an informed, directed evolutionary process. He also stressed an inevitable rise in complexity.

Both of these proposals were called into question by Charles Darwin's theory. In contrast to Lamarck's directed evolutionary process, Darwin' book on *The Origin of Species*, published in 1859, proposed and eventually established the *Theory of Natural Evolution*:

- All living beings continuously and interactively compete for the life-ensuring resources in the world, such as water and food.

- Individual differences within a species are due to small, random changes, that is, *mutations* and random recombinations, that is, *crossover*, of the parental properties.

- Some of these differences are more suitable than others, leading to a higher chance of survival, that is, *survival of the fittest*, and thus also to a higher chance of producing offspring.

- In this manner species and individuals of a species evolve that are fitter for survival and reproduction, causing mediocre species and individuals of their species to die out – and with them their mediocre properties.

Darwin's theory of evolution is a very good example of a theory that illustrates Marr's second level of understanding, also touching also on the first, most abstract level (cf. Section 2.5). On the most abstract, computational level, the theory is about species and their individual properties (the *what*) and, arguably, also about the reason for the identified process, which is natural selection. On the second level, the algorithm is described in detail, making intra- and inter-species competitions, survival of the fittest, mutation, and crossover explicit. Varied are the properties of individuals without making explicit how the variation may actually work. Thus, the realization in hardware, that is, the encoding of the individuals' properties by their genes, was not discussed because it was still unknown.

However, nearly in parallel Gregor J. Mendel (1822–1884) – an Austrian priest and natural scientist – made experiments with pea plants and formulated his rules of genetic inheritance. Mendel can thus be viewed as the father of all modern research on genetics and genetic inheritance. It took a while for Mendel's treatises to gain recognition, but eventually his theories were confirmed. With the establishment of Mendel's paradigm and more detailed research on the subject, progressively more insights were gained, including the existence of genes, their implementation by means of *deoxyribonucleic acid* (DNA), and their organization into chromosomes. Moreover, it became progressively clear that these chromosomes are recombined in embryonic cells, which then begin to replicate by means of cell divisions – thus initiating the development of the embryo. The American molecular biologist James D. Watson (*1928) and the British biochemist Francis H.C. Crick (1916–2004) discovered the double-helix-like structure of the DNA, within which nucleotide-types are opposing each other in pairs. In 1953, they received the Nobel Prize in medicine and physiology for this ground-breaking discovery.

Based on this knowledge, the British biologist Richard Dawkins (*1941) further modified and adapted Darwin's Theory proposing the principle of the selfish gene (Dawkins, 1976). He argues that living beings inherit a rather arbitrary subset of genes, so that the survival of the fittest focuses more on the inheritance of the fittest genes rather than on the survival of the species or the individuals of a species. Importantly, Dawkins' theory also facilitated explanations for altruistic behavior, such as the one developed in bees, termites, and ants, but also in other social species, including humans. Darwin did not have an explanation for such behavior. Dawkins wrote several additional books with the primary objected of explaining unresolved phenomena, which creationists put forward when arguing against theories of evolution (Dawkins, 1986, 1997). Dawkins explains many evolutionary phenomena by means of progressive random mutations and recombinations that interact with natural selection, leading to the balanced properties and capabilities of living beings.

4.3.2   Genetics in a nutshell

While the details of genetics are now being studied in various research fields, such as *transcriptomics*, we are going to focus on the very basic principles. DNA consists of four types of *nucleobases*: *adenine, cytosine, guanine*, and *thymine* (A,C,G,T), which constitute the alphabet, or *DNA-bases*, of the genes. The DNA's double helix structures consist of pairs of these letters – combining A with T and C with G. The result is a huge double-helix string of relatively stable genetic code. Each strand in the double-helix is held by a chemical backbone structure and other chemical bonding mechanisms, which we are not going to further elaborate upon here. We focus on the encoding capacity and the implications for evolution.

As the alphabet encodes four letters and two bits are necessary to encode four distinct symbols, each letter pair encodes two bits. With an approximate number of about $3.2 \cdot 10^9$ pairs in our genome, the encoded information is in the order of maximally 0.8 Gigabytes – which is, surprisingly, only slightly more than what fits onto a normal CD-ROM. Thus, although the genetic information held in our genome is large, it is not unimaginably big. Nonetheless, it certainly implements some very intricate encoding structures, which are still not fully understood.

As in a language, DNA is structured into words, which consist of sequences of letters. The basic set of words is encoded by three-letter pairs, the *codons*, which encode one of 20 basic amino acids as well as one "start-codon", which also generates one of the 20 amino acids, and three "stop codons". As codons can come in $4^3 = 64$ forms, there is room for redundancy in this encoding, with several codons encoding the same amino acid. The "start-codon" initializes the generation of a protein by converting the subsequent codon sequence into a protein until one of the "stop-codons" is reached, which ends this translation process and releases the produced protein. Since this process appears to be irreversible – the central dogma of molecular biology proposed by the British molecular biologist Francis H.C. Crick (1916–2004) – it seems impossible to encode information or experiences, which the protein might make, back into the genes. Thus, it is impossible that genes learn from the protein structures they reproduce – at least not through a direct inversion process.

Combinations of words can have lengths of up to more than one million bases pairs, and are in this form referred to as *genes*. Genes are not yet fully understood, but besides encoding proteins and combinations of proteins, switch-like structures are apparently present, enabling the expression of a certain gene under certain circumstances. Other mechanisms also appear to be encoded – such as transfer and messenger mechanisms – which can control which genes are actually currently expressed given the environmental, mainly chemical, circumstances. Genes can thus be expressed in various circumstances and often contribute to the development of a variety of structures and functions (called *pleiotropy*). On the other hand, a particular bodily structure or functionality is typically generated by a collection of genes (referred to as *polygeny*). Variations in the genes can lead to variations in the gene expressions (*alleles*), such as blue, brown, or green eye color.

Chromosomes encode chapters of words and are the most macroscopic building blocks of the genetic code. The set of chromosomes determines the characteristic *genome* of a particular species. Humans have 22 pairs of similar chromosomes, and one pair of differing chromosomes, which determine, among other things, the sex of the offspring. The 23 pairs constitute the human genome, which encodes about 25,000 genes in the above-mentioned approximately $3.2 \cdot 10^9$ basis pairs. These pairs, when unfolded, result in an approximately 2m long string of DNA basis pairs, and are found in every cell of our body. While this number seems high, and is among the highest in mammals, the record is much higher than this: the genome of the marbled lungfish (*Protopterus aethiopicus*) consists of $1.3 \cdot 10^{11}$ basis pairs and is believed to be the largest genome of all vertebrates. Yet, other research suggests that some plants and even particular amoeboid species may have even larger genomes.

During the reproductive process, 50% of the mother's and 50% of the father's genes are transferred to the offspring. This transfer is accomplished by means of two strands of RNA (*ribonucleic acid*), one of each of which is found in the father's sperm cell and the mother's egg cell. Mutations in genes can occur during the generation of RNA, during the fusion process of two strands of RNA into the offspring's DNA, as well as during replications of cells during embryonic, fetal, and other developmental growth processes, and continuously throughout life during cell replacement and repair. For evolutionary purposes the former are more relevant, leading to the variations Darwin discussed. These variations span from very small variations in single letter pairs, having hardly any effect, up to genetic mutations and even chromosomal mutations. Each RNA strand consists of one of each of the 23 chromosome pairs, where the reproduction process approximately randomly chooses one or the other of each chromosome pair. Given two parental DNAs, the result is an offspring DNA that recombines the parental DNA in $2^{23} \times 2^{23} = 2^{46} \approx 70 \times 10^{12}$ combinations.

To a certain extent, such variations change the *genotype*, where recombination radically recombines the genetic information from the mother's and the father's side. Crucial for evolution, however, is the effect of mutation and recombination on the *phenotype* of the offspring, that is, the actual resulting individual including its development, bodily and mental maintenance capabilities, and its reproduction capabilities. The genotype refers to the DNA, which encodes ontogenetic biological and neural development (which, of course, unfolds in interaction with the environment) and the involved bodily mechanisms, which keep the individual with all its bodily and cognitive functions going. The phenotype is the result of all of these processes, essentially defining the actual individual. The phenotype determines the fitness of the individual in its environment, because it determines if and how often the genotype of the specific individual is reproduced and inherited by subsequent generations. The variations of the genes that are transferred to the offspring occur upon reproduction and during the involved processes described previously. Thus, while genetic variations, including mutation and recombination occur on the genotypic level, the principle of natural selection is played out on the phenotypic level.

4.3.3 Evolutionary mechanisms

In addition to the biological mechanisms for developing an understanding of how the mind came about, it is even more important to understand the mechanisms that are a consequence of sexual reproduction, genetic encoding, and phenotypic expression of each individual's DNA. The resulting evolutionary mechanisms are not a consequence of only gene expressions, but rather of gene expressions combined with environmental interactions on the microscopic and macroscopic level, producing a fully embodied evolutionary mechanism. Although which genetic material is reproduced largely depends on the ontogenetically developing phenotypes, that is, the actual living individuals, natural selection also depends on the ecological niche, in which the phenotype lives. The niches is characterized by actual resource distributions, weather and climate properties, and changes over time, land-mass distributions and proximities, ocean currents and temperatures, and the distribution of species within the niche.

Recombination not only enables a huge number of possible mixtures of chromosomes, it also plays an important role in the determination of the distribution of the genetic pool of a particular species. Only as a result of sexual reproduction is the genetic material mixed within a species over and over again. This mixing, however, by no means unfolds completely randomly. Which genetic material gets a chance to recombine strongly depends on the genetic and allele pool available within a species. A species with a large group of individuals has a larger gene pool diversity. This diversity is enhanced, when the interaction between the individuals of a species is restricted, for example, due to a separating river, lake, or even an ocean. When genes are mixed again across these separated populations of a species, *gene flow* is observable. For example, a recent gene flow could be shown flowing from Africans to Native Americans. Africans typically have a group of genetic mutations in their genome that makes them rather resistant to some types of malaria. Before Columbus, Native Americans probably had none of these mutations in their genome – leading to a very high susceptibility to malaria, which spread through the populations of Native Americans as a result of their contact with the first settlers coming from Europe. As a result of this detrimental effect on Native American populations and the progressive mixture of Native Americans with the African slaves, most Native Americans now also have the mutation.

On the other hand, when a whole group of a species is permanently separated from another one (called the *founder effect*), an independent further evolution of the one species into two separate ones takes place. If a species is extinguished due to, for example, a natural catastrophe, one speaks of a *bottleneck effect*. In both cases, *genetic drift* occurs. The finches that Darwin studied on the Galapagos Islands are possibly the most well-known example of the founder effect. A few finches must have at some point – possibly due to a large storm – reached the remote Galapagos Islands, founding a new population. Due to the rather distinct environmental circumstances on the islands, totally new types of finches soon evolved, specialized for reaping the available food resources in the most effective way.

Another important factor that determines non-random mixtures of genetic material is the fact that individuals of a species typically do not mate randomly. Environmental circumstances must allow that a male and a female individual of a species can meet. Typically, the females decide with which potential partner they are willing to mate. While this choice often depends on the strength and power of the male individual, *assortative mating* can also be observed, which is the tendency to prefer partners that have like or unlike characteristics. Like characteristics suggest somewhat similar genetic material, leading to somewhat similar offspring, which may strengthen family and social bonds within a group. Altruistic behavior within a family clan makes more sense from this perspective because one helps other individuals with similar genes in accordance with Dawkins' "Selfish Gene" principle. On the other hand, preferences for unlike characteristics prevent inbreeding and fosters diversification, which can also be very important in strengthening the robustness of the species against diseases for example.

In addition to the mixing principles as a result of recombination and mate selection, it is the principle of the "survival of the fittest" – the inheritance of genes from the fittest – that determines the evolutionary process. Only species and individuals of a species that are well adapted to the environmental circumstances can survive long enough to succeed in reproduction – because ill-adapted individuals will die before they can reproduce. However, what does "well adapted" mean?

Adaptation always occurs within the *ecological niche* in which an individual develops and lives. A well adapted species is able to reap the resources of its ecological niche effectively.

Birds, insects, and bats, for example, have discovered the air space above the ground as their important ecological niche – enabling them to fly away from danger, to bridge large distances much faster, or to search for and hunt within and from the air for food.

Properties of an ecological niche are, however, never stable, but continuously in flux. This leads to the *co-evolution* of species, which Darwin referred to as "coadaptation". Given that a particular species evolves in a certain manner, another species may need to coevolve accordingly. For example, given that a new species of predators has evolved that can run faster than before, the most favored prey of this predator will also need to coevolve a faster running speed to prevent becoming extinct. Such developments are sometimes called *evolutionary arms races*.

Resulting evolutionary progressions also depend on other factors considering the available ecological niches. Insect eating bats, for example, can have evolved only once a sufficient number of insects flew through the air. In a more extreme example, plant-eating animals can have evolved only after sufficiently many plants existed. As a result, over millennia the genetic code has evolved into bodily plans, which distinguish the main classes of species. For example, the class of vertebrates includes fish, mammals, reptiles, birds, and amphibians. Even looking only at mammals it soon becomes obvious that while the basic bodily plan is the same in all mammals – exhibiting thus a *homologous structure* – the variations can be immense.

In addition, across classes of species similar bodily parts, that is, *analogous structures*, have developed. Winged animals span a large variety of classes of species, indicating the immense potential and resources available in the ecological niche "air space". Similarly, even mammals and birds have evolved fins again – essentially rediscovering the ecological niche water – at a point in time, though, where the niche was already heavily populated not only with competitors but also with potential prey. Thus, even though the biological mixing of the genetic material is mostly undirected, actual evolution is influenced by many factors that determine natural selection, including phenotypic influences and influences due to the available structure in the genotypic material.

With all these considerations in mind, let us finally consider the implications for cognition and the evolution of human intelligence. From the very first bacteria, it appears that intelligent capabilities are encoded in the genes. At first, intelligence focused on behavior, such as very simple reactive processes that link the sensors with the motors of a biological system in a suitable manner. For example, simple *Escherichia coli* bacteria tend to swim toward higher sugar concentrations by executing a biased random walk, moving forward faster while climbing the gradient, while randomly changing direction more often when descending the gradient (this is even simpler than a Braitenberg vehicle, cf. Section 3.6.2). Over more than 2 billion years, intelligence diversified in terms of bodies and their development, as well as neural, brain structures, and cognitive development. Birds and mammals arguably have the most intricate brain capacities and cognitive capabilities, although invertebrates, such as particular species of octopuses, have also shown highly intelligent behavior and learning capabilities.

Higher forms of intelligence almost always include forms of social intelligence. Indeed, there are many indications that social group interactions greatly foster the evolution of intelligent capabilities. The evolutionary niche into which humans have most evolved is probably social cooperation – offering the benefits gained from collaborative interactions on a group (or tribe) level (Tomasello, 2014). Group cooperation can also lead to the development of communication signals, starting from simple warning signals through to human language, which evolved most likely due to the immense benefits of coordinating social interaction, and distributing work load and responsibilities. In the end, the social niche has most likely co-determined cultural development and continues to do so. In later chapters, we will come back to these important components of human cognition.

4.4 Evolutionary computation

Once computers became available, scientists soon attempted to transfer the principles behind natural evolution into the machine. The resulting research community on evolutionary computation has developed into its own research area, which focuses not only on structural evolutionary optimization, but also on machine learning and artificial intelligence in general. In this section we focus on giving an overview of the basic principles behind the approach with the aim to develop an understanding of how the mind and cognition may have evolved. In particular, we focus on how particular structures may have evolved and highlight the caveats that natural evolution must have mastered.

Generally speaking, evolutionary computation involves the study of computer algorithms that implement aspects of the Darwinian principles of biological evolution, including natural selection, mutation, and recombination. An evolutionary algorithm evolves a population of genotypic encodings, which come in various forms of representations including binary and real-valued vectors, as well as tree-, graph-, or rule-structures, by simulating an artificial evolutionary process. In one form or the other, this evolutionary process implements a selection process – for deciding which individuals in the population get to reproduce their genes – and a reproduction process, which typically introduces some variations in the genetic material by means of mutation, crossover, and possibly further variation operators.

More or less independently, several evolutionary computation approaches have been proposed starting in the 1960s. In *evolutionary programming* a genotype encodes a computer program in the form of a finite state machine (Fogel, Owens, & Walsh, 1966). In the related field of *genetic programming*, computer programs are evolved in the form of trees (Koza, 1992). Meanwhile, the German research community proposed *evolution strategies*, in which the genotype is encoded by a vector of real numbers (Rechenberg, 1973). Also nearly concurrently with the development of evolution strategies, in the late 1960s and early 1970s John H. Holland (1929–2015) developed the concept of *genetic algorithms*, which encode the involved genotypes by binary vectors (Holland, 1975). Table 4.1 contrasts these approaches with each other.

Table 4.1: Major approaches of evolutionary computation

	Evolutionary programming	Genetic programming	Evolution strategies	Genetic algorithms
Proposed in	(Fogel et al., 1966)	(Koza, 1992)	(Rechenberg, 1973)	(Holland, 1975)
Evolution of	Finite automatons	Algorithms	Real-valued problem solutions	Binary problem solutions
Typical genotype	Graph encoding	Trees	Real-valued vectors	binary vectors
Main operators	Mutation	Mutation and recombination	Adaptive mutation	Recombination and mutation

In his 1975 book, Holland put forward a first relation of evolutionary algorithms with cognitive systems (Holland, 1975). He proposed a genetic algorithm that evolved condition-action-effect rules, showing that an evolutionary rule optimization is able to learn an effective, reward-oriented behavioral strategy. The implementation of his cognitive system was published in 1978 (Holland & Reitman, 1978), and later developed into another subfield of evolutionary computation, which is now typically referred to as the *learning classifier systems* research field.

In the following subsections, we first provide further details on how an evolutionary algorithm works in general and which operators and processing mechanisms are involved. Next, we explore the question of *when* an evolutionary algorithm works, that is, when it can be expected that an optimal problem solution can be found. Finally, we relate these insights to cognition and the development of human cognitive capabilities.

4.4.1 Basic components of evolutionary computation algorithms

Evolutionary algorithms are stochastic search algorithms, which ideally search for optimal problem solutions or sub-solutions. To do so, they make use of the principles behind natural selection. However, they do not encode genes by biological means (that is, DNA), but by means of abstract digital encodings, such as binary and real-valued vectors, or tree or graph structures. The genotypically encoded problem solution itself is the corresponding phenotype.

Given an encoding, the task of the evolutionary algorithm is to search through the space of genotypic encodings for an optimal solution to the problem at hand, that is, for an optimal phenotype. As in natural evolution, a *population* of *individuals* is evolved, where each individual encodes one particular genotype. It is this population of individuals that then undergoes a simulated evolutionary process, which iteratively decides which individuals undergo reproduction and which ones are deleted. This process typically depends on the quality of the phenotype of each individual, quantified by means of a *fitness* measure. The genotypes of the *offspring* are generated by taking those of the selected parents and varying them by means of a partially random process. Finally, the resulting next generation of individuals undergoes the same procedure.

In the form of an algorithm, a general evolutionary computation mechanism can be defined as follows:
Evolutionary algorithm:

1. *Initialization*: given a problem and a corresponding genotypic encoding of potential solutions to the problem, initialize a population of individuals by randomly generating genotypic encodings.

2. *Evaluation*: given the current population of individuals, evaluate them by means of the given fitness measure.

3. *Selection*: choose the better individuals for reproduction dependent on the determined fitnesses.

4. *Reproduction*: generate offspring by taking the genotypes of the selected parents, re-combining and possibly further varying the genotypic encodings.

5. *Integration*: the generated offspring form the new population. Possibly, some very good parents are kept.

6. *Iteration*: if not done, continue with this new generation of individuals in step 2.

In the following, let us focus on the first four processing steps. We assume that reproduction completely fills the new population with the generated offspring, so that step 5 is trivial. It should be mentioned, though, that sometimes some of the best parental individuals are kept in the population of the new generation, which is referred to as an *elitist strategy*. Some elitist strategies even compare the parental fitness with the offspring fitness, keeping, for example, the better half of all individuals. Elitist strategies most importantly prevent the forgetting of the best solution found thus far. However, for an understanding of the crucial mechanisms behind evolutionary computation, elitist strategies play a minor role.

Encoding and initialization of population

A crucial difference between biological evolution and evolutionary algorithms is certainly the chosen genotypic encoding. While DNAs have evolved into a hierarchical structure, which typically consist of chromosomes, which consist of genes, which consist of a basic alphabet of four nucleobases, in evolutionary computation a large variety of genotypic encodings have been used. Later we will show that particular hierarchical structures may be very well-suited for evolutionary development, while others may actually lead to disadvantageous evolutionary development. For now it suffices to be aware that the choice of encoding can strongly influence the artificial evolutionary process. For example, under the assumption that mutation operators should yield slight variations in the genetic encoding, it is disadvantageous when similar genetic encodings may encode very dissimilar phenotypic problem solutions – because in that case mutation would be similar to an approximately random search process. Thus, neighborhoods in genotypic space should also yield similar phenotypic problem solutions, albeit these two topologies will usually never map one-to-one.

Given a problem, and given the chosen genotypic encoding of problem solutions to that problem, actual genotypic codes need to be generated to initialize the population. If there is no further knowledge about the problem, a maximally diverse population of initial problem solutions is typically most advantageous, ensuring that the solution space is covered. However, when prior knowledge about the problem is available, which may suggest that particular problem solution subspaces do not need to be explored, biases or constraints can be included while generating genotypes. Without prior knowledge, though, the initialization should cover the plausible problem solution subspace in a maximally diverse manner, that is, approximately uniformly distributed, to minimize the probability of overlooking potentially superior problem solutions.

Evaluation

Given a population of individuals, evaluation is not as obvious as it is in natural evolution. Which individuals should be assigned which fitness value? In natural evolution, fitness is essentially indirectly defined by playing out the phenotype, that is, actual biological organisms in the real world, where the fitness of an individual is higher the more offspring it reproduces (irrespective of the actual quality of the offspring). In some approaches to artificial evolution, a simulation of a world with simulated organisms is used to determine the number and type of offspring an individual may produce. In most evolutionary algorithms, however, this process is replaced by a *fitness function*, which determines the phenotypic fitness of an individual, that is, it quantifies the encoded problem solution quality.

The fitness function may, for example, compute the quality of the product, such as an engine or the wings of an airplane, which may be encoded by a genotypic individual, or it may simulate an encoded control process and evaluate its quality. Similarly, it may run the encoded program and evaluate the quality of the result, such as the accuracy of the classification the program generated given some input samples. These few still rather abstract examples show that a large variety of fitness functions is imaginable. For an evolutionary algorithm to be successful, the fitness function is crucial because it needs to provide information that guides the evolutionary process toward better problem solutions. However, misleading fitness functions may also be encountered, which then typically hinder the evolution of an optimal problem solution. Essentially, a fitness function results in a *fitness landscape* in genotypic space, where the input space is given by the genotype and the landscape over this input is shaped by the fitness.

Let us look at a simple example, which illustrates aspects of an advantageous versus a disadvantageous fitness function. To do so, we define a genotypic space simply by a binary vector of a particular length L. Thus, the genotypic space is of dimension L and consists of 2^L different genotypic encodings. Three fundamental fitness functions can now be defined in this encoding space, which differ in their expected effect on evolutionary progress. We relate the fitness function in this case to the optimal problem solution, which may be a vector of all ones for simplicity. Note, however, that the following concepts hold for all binary optimal problem encodings (all combinations of zeros and ones).

First, a *one-max* function may be defined by $f(\mathbf{x}) = \sum_i x_i$. This function simply counts the number of ones in the particular problem solution. Thus, the more ones are in a particular solution and the closer the solution is to the optimal solution, the higher the fitness. When the fitness function is of this kind for a particular problem, then evolution will typically progress toward the optimal solution.

Second, a *needle-in-the-haystack* function may be defined by $f(\mathbf{x}) = \prod_i x_i$, which essentially only yields one when the vector consists of all ones, that is, when the vector encodes the optimal problem solution. In this case, the optimal problem solution can be viewed as a needle-in-the-haystack – all other solutions are of equally bad quality, such that there is no clue from these other solutions about where the optimal solution may lie. Such a fitness function is thus not very favorable for fostering evolutionary progress.

Third, a *trap* function can make things even worse:

$$f_{\text{trap}}(\mathbf{x}) = \begin{cases} L - 1 - \sum_i x_i, & \text{if } \sum_i x_i < L \\ L, & \text{otherwise.} \end{cases} \quad (4.1)$$

This function assigns the highest fitness to all ones but the second highest to the opposite, that is, all zeros. The more zeros in the evaluated genotype, the higher the fitness, except for when encountering only ones. This fitness gradient toward a bad solution essentially traps the evolutionary process into bad problem solutions, typically preventing the detection of the optimal problem solution.

Of course, similar fitness functions can be defined for real-valued genotypes. Table 4.2 illustrates the fitness functions for the binary case with $L = 4$. When combining several of the shown small binary problems into larger problems, where the fitness is defined simply as the sum of the fitnesses in the small problems, a building block processing challenge arises when facing needle in the haystack or trap problems. Similar three types of fitness functions for the real-valued case with $L = 1$ for a parameter range $x_i \in [0, 1]$ are shown in Figure 4.2. While these examples are very stereotypic, they nonetheless illustrate that an evolutionary process strongly depends on the fitness function. To find an optimal solution, a fitness that guides to an optimal solution is very helpful. Note that this fitness conceptualization suggests that in natural selection it is also very probable that evolution will sometimes optimize species toward local optima (such as toward '0000' in the trap function example). Acknowledging that fitness may indeed be misleading, This observation offers a partial explanation for the fact that our ecosystems are not necessarily very stable and can be easily and drastically infected by introducing a new species, for example, from another continent.

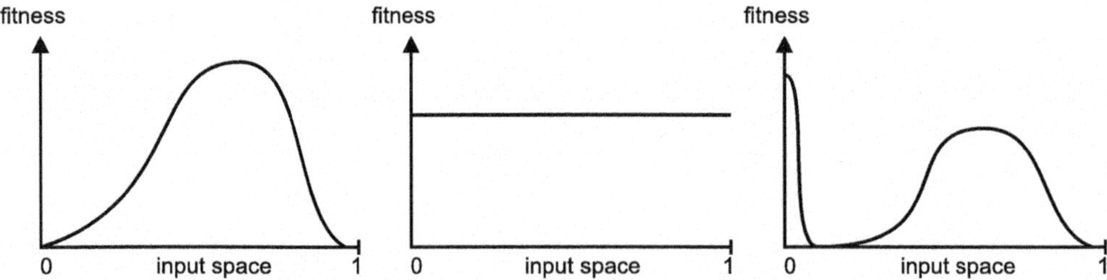

Figure 4.2: Fitness functions may lead to evolutionary progress or may require to take the right steps toward optimal solutions. This is the case for binary, as well as real-valued genotypic encodings. The three real-valued exemplary fitness functions show good fitness guidance (left), no fitness guidance (center), and misleading fitness guidance (right).

Selection

Fitness without selection and reproduction is certainly meaningless. Given a fitness for each individual in the current population, the actual selection process determines which individuals are allowed to reproduce. Once again, the selection process can be implemented in various manners and can lead to distinct influences on evolutionary progress.

Generally, selection will lie between two extremes. On the one hand, the very simple *max-select* selection process will always select the individual with the best fitness. This process

Table 4.2: When genotypes are encoded binary, three main types of fitness functions can be defined: simple, one max functions have good fitness guidance; the tough needle in the haystack problem yield not fitness guidance; the even harder trap function has a misleading fitness guidance, away from the optimum. Often such small problems are concatenated into bigger problems, resulting in challenging building block identification and recombination problems.

	One max	Needle	Trap
0000	0	0	3
0001	1	0	2
0010	1	0	2
0100	1	0	2
1000	1	0	2
0011	2	0	1
0101	2	0	1
0110	2	0	1
1100	2	0	1
1001	2	0	1
1010	2	0	1
0111	3	0	0
1011	3	0	0
1101	3	0	0
1110	3	0	0
1111	4	1	4

thus instantly destroys the current diversity in the population and focuses all further search power onto the neighborhood of the best individual. On the other hand, the very simple *random-select* selection process may ignore fitness altogether, simply selecting randomly from the current population. In this case, the search remains maximally broad, essentially yielding an (inefficient) random search process when being combined with variation operators during reproduction.

Between these extremes, various types of typically-applied selection processes can be distinguished. *Roulette wheel selection* illustratively characterizes the process of selecting individuals for reproduction by means of a fitness-weighted probability – akin to a roulette-wheel where the width of each slot on the roulette wheel covers a proportion of $f(\mathbf{x}_i)/\sum_j f(\mathbf{x}_j)$ for a particular individual i. The wheel is spun and the probability of selecting individual i thus corresponds exactly to the covered area on the imagined wheel. As a result, roulette wheel selection depends on fitness scaling. For example, when defining a new fitness function by $f'(\mathbf{x}_i) = e^{f(\mathbf{x}_i)}$ the fitness proportions change, focusing the selection process on the best individuals because of the exponential fitness scaling.

In contrast to roulette wheel selection, *tournament selection* does not depend on fitness scaling. This selection process simulates little tournaments between randomly selected individuals in the current population. The chosen tournament size may be denoted by $t_{\text{to}} \in [1, \infty)$. Given tournament selection with $t_{\text{to}} = 2$, for example, tournaments with two randomly selected individuals from the population are held and the better of the two is selected for reproduction. A non-integer tournament size essentially means that the tournament sizes are probabilistically chosen between the two neighboring integers. For example, a size of $t_{\text{to}} = 1.8$ would denote that the tournament is held in 80% of the cases with two individuals and in 20% of the cases with only one individual – choosing essentially a random individual for reproduction in the latter case.

Truncation selection also does not depend on fitness scaling. This selection procedure essentially chooses a particular proportion $t_{tr} \in (0, 1]$ of the better individuals for reproduction. For example, a truncation selection with $t_{tr} = 0.2$ will choose the 20% best individuals and reproduce only from that selected pool of individuals. In contrast to tournament selection, truncation selection completely bans mediocre individuals whose fitness lies below the t_{tr} best proportion to reproduce.

Selection thus typically focuses the reproduction process on the individuals with higher fitness. How strong this focus is and how much variability is typically found in the selected subpopulation strongly depends on the selection process, the parameters, and the fitness function. Despite these interdependencies, a *take-over-time* (TOT) can typically be determined, which quantifies the speed of this focusing mechanism. TOT estimates the number of generations necessary to fully converge to the best individual in the population, without considering variation operators. In the case of tournament selection with $t_{to} = 2$, for example, the best individual can be expected to be part of two tournaments and, because it is the best one, it will be chosen both times for reproduction. In effect, the proportion of best individuals in a population doubles in each generation and the take-over-time $\text{TOT}_{to}(2) = \log_2 N$, given a population size of N individuals in total and starting with one individual. The same holds true for truncation selection with $t_{tr} = 0.5$. For roulette wheel selection, however, it is impossible to determine the take-over-time without further information as it depends on the fitness scaling of the best individual with respect to the other individuals in the population. For example, when the best individual is only slightly larger in fitness value than the others, the take-over-time will be significantly higher than when the best individual's fitness is much larger.

Genotype variations

Given selected parents, evolutionary algorithms typically introduce variations in the reproduced offspring, As in natural evolution, these variations are on the genotype level. Variation operators in evolutionary computation come in many forms and strongly depend on the genotype encoding. Here, we focus on the very basic operators and discuss at how complex it can get. The two most basic and ubiquitous variation operators are genotype *mutation* and *recombination*.

Mutation   A simple variation of the genotype of an individual is called a *mutation*. Mutation is implemented in nearly every evolutionary algorithm and has been shown to be essential to ensure optimization success in various optimization problems. A fundamental distinction needs to be made when considering mutation in binary and real-valued genotypes.

In the binary case, a mutation flips a bit from zero to one or vice versa. The probability of flipping each bit in the genotype of an offspring is denoted by p_m. It is typically rather small. For example, by setting $p_m = 1/L$ it is ensured that on average one bit is flipped in every offspring's genotype. When a genotype is mutated in this way, the algorithm typically considers each bit for mutation, deciding randomly with the probability p_m if it is mutated or not.

Despite this rather simple mechanism, it is important to understand the evolutionary biases that are introduced by it. Bit mutations essentially result in a local search in binary space around the parent's genotype location. However, given that the parental genotype is very close to the optimum solution, it is rather unlikely that the last incorrect bits are mutated and no others. For example, given a parental genotype that is only one bit off the optimal one, the probability of generating the optimal one out of the given parental genotype is $p_m(1-p_m)^{L-1}$, because the crucial bit needs to be mutated and the others must not be mutated. Thus, mutation alone often takes a long time to find the optimal solution in the binary domain – especially in larger problem spaces L.

It is interesting to consider what happens in the evolutionary process when mutation is paired with random selection. Given, for example, a population of individuals that has a large number of ones (zeros) – say 80% – and undergoes random selection and mutation – it is highly likely that there will be a lower number of ones (zeros) in the subsequent generation.

Thus, mutation tends toward diversification, but also toward a uniform distribution of zeros and ones throughout the population.

In the real-valued case, mutation is typically defined by a possibly adaptive probability density. One of the most common ones is a normal distribution that is centered at zero and has a standard deviation of σ. During mutation, values are sampled from the distribution and added to the individuals' real-values in their real-valued genotype. In this case, the σ is crucial for the strength of the variations introduced. Small σs result in very little variation, while large σs cause stronger variations. Thus, σ is often adaptive typically changing from large values to progressively smaller values.

How fast should the σ tend toward smaller values? Should σ even increase in particular cases? Theoreticians in evolution strategies have shown that the *one-fifth rule* yields the optimum adaptation of σ – at least when the fitness landscape can be represented by a normal distribution (Beyer & Schwefel, 2002). The rule considers the evolutionary progress: when currently more than $1/5$ of the offspring is better than the parents, the evolutionary progress is considered good. As a consequence, σ is increased to speed-up the encountered progress. On the other hand, when less than $1/5$ of the offspring is better than the parents, then mutation seems to introduce too much variation, consequently lowering σ. In this way, it is ensured that the evolutionary progress does not stall when there is still room for optimization: when the process is close to an optimum, local optimization is the key to finding the exact optimum, thus decreasing σ to search locally. On the other hand, when the system is experiencing a clear gradient in one direction, about 50% of the offspring should be better than the parents. Thus, in this case σ should be increased to speed-up gradient ascent toward higher fitness regions.

The *covariance matrix adaptation evolution strategy* (CMA-ES) (Hansen & Ostermeier, 2001) has enhanced this idea to the case of oblique fitness landscapes, yielding an optimization algorithm that is independent of affine transformations of the genotypic encoding. That is, the algorithm yields similar optimization performance, given adapted initializations of individuals, even when the genotypic space is tweaked or stretched in any direction, when it is rotated, or when it is mirrored. The much simpler *Differential Evolution* technique, which combines mutation with crossover capabilities by mutating the real-valued genome of a reproduced individual considering the difference between its and another individuals genes, has also generated great performance in real-valued evolutionary algorithms (Storn & Price, 1997). In this latter case, the gradients toward better solutions is essentially locally estimated by considering the differences between genetic codes of selected individuals.

Recombination Besides mutation, recombination, which is also often called *crossover*, introduces another form of variation in the offspring genotypes. Similar to sexual recombination in natural evolution, recombination mixes parental genetic material.

Given two parents, three typically recombination operators can be distinguished. *One-point crossover* chooses one crossover point and swaps the encodings after this crossover-point in the genotypes of the two offspring individuals. When the encoding of the genotype is circular, in that the last bit is related to the first bit, *two-point crossover* is typically preferred, which chooses two crossover points and exchanges the genetic material between these two points. Finally, *uniform crossover* simply swaps each bit with a probability of 0.5 in the two genotypes of the offspring. While one-point and two-point crossover tend to separate genes that are further apart in the genome, uniform crossover does not consider distance in the genotype at all.

Recombination is particularly important in binary encoded individuals. Seeing that real-valued problems are typically lower-dimensional, the optimization process benefits much more from mutation. In the binary case, though, recombination is considered very important. It enables the generation of highly effective individuals from less effective parents. As we have discussed, mutation alone often fails to generate an optimal offspring from a sub-optimal parent that is only one or a few bits off the optimal solution. With crossover, given two solutions that are close to optimality, it is much more likely to generate a fully optimal solution.

Recombination is thus responsible for enabling the exchange of substructures in the parental genetic material. Such substructures are often referred to as *building blocks*. Given that the parental material each contains different fully optimized building blocks, recombination can help to combine these building blocks (by chance), potentially generating an individual that contains both locally optimized substructures. Selection plus recombination thus essentially strives to recombine building blocks.

However, depending on the crossover operator, building blocks may also be destroyed. For example, when choosing a crossover point within a building block in a one-point crossover, potentially ill-optimized building block material may be introduced into the material of the optimized parental building block from the other parent. If the structure of the building blocks, but not their actual code, is known, crossover can be tuned in such a way that building blocks, that is, a subsection of genetic locations, are never partitioned but are exchanged only as blocks. In this way, building block destruction by recombination can be prevented, fostering effective building block exchange.

Most of the time, though, the building block structures are not known. In this case, *estimation of distribution algorithms* (EDAs) can come to the rescue (Pelikan, 2005). EDAs analyze the genotypes in the parental population after selection and thus statistically estimate building block distributions. The recombination mechanisms in EDAs then take these estimates into account while producing offspring. With the help of EDAs, many binary optimization problems that were previously believed unsolvable were indeed solved. However, natural selection has most likely no EDA-like mechanism that ensures effective building block exchange. Rather, the chromosomal encodings and the local encoding of genes and protein "factories" probably help to increase the likelihood of effective building block exchanges.

4.4.2   When do evolutionary algorithms work?

Only the combination of the individual steps in an evolutionary algorithm can yield an effective evolutionary process. Selection without variations yields convergence to the present best individual. Mutation combined with random selection leads to a diversification of the population, converging toward a uniform genotype distribution. Crossover combined with random selection randomly shuffles the building blocks or the individual codes in a population without changing the proportion of values (that is, zeros and ones or real-values) present at each position in a genotype. In this subsection, we address the questions (i) when can good structures be expected to grow in a binary-encoded evolutionary algorithm and (ii) when can the interaction between the processes can be expected to generate effective genetic recombinations.

Schema theory

John H. Holland developed a *schema-theory* for genetic algorithms, which quantifies when good building block structures can be expected to grow in a population of individuals. Growth is quantified by considering the strength of selection, as well as the detrimental strength of unfortunate, destructive recombination events. Mutation effects can also be quantified by the theory.

To formalize the schema theory, we focus on the binary domain for which it was developed and define a building block in this domain. A building block, which Holland also referred to as a *schema*, can be specified by a particular subset of a binary genetic code with particular values for the individual bits in the particular subset. For example, the schema *10*1 encodes a subset of three bits on the second, third, and fifth location in the genome, which need to have values 1, 0, and 1, respectively. The representative individuals of this schema are: 01001, 01011, 11001, and 11011.

Any schema can be mathematically characterized by two properties, which are crucial to determine the likelihood of a schema's destruction or successful recombination. While we had defined the length of a genome by the letter L, the *defining length* d of a schema is defined as the distance from the first specified bit in the schema to the last specified bit. For example, the schema *10*1 from above has a defining length $d(\text{*10*1}) = 3$, while the schema ***1* has a defining length of $d(\text{***1*}) = 0$. In addition to the defining length, the *order* o of a schema is also important. The order quantifies the number of relevant bits, which are specified by a schema. Thus, for our exemplar schemata $o(\text{*10*1}) = 3$ and $o(\text{***1*}) = 1$.

Given these characterizations, the schema theory estimates a lower bound on the expected number of schema representatives in the next generation:

$$\underbrace{\langle m(H, t+1)\rangle}_{(1)} \geq \underbrace{m(H,t)}_{(2)} \cdot \underbrace{\frac{f(H,t)}{\overline{f}(t)}}_{(3)} \cdot \underbrace{\left[1 - p_c \cdot \frac{\delta(H)}{l-1}\right]}_{(4)} \cdot \underbrace{(1-p_m)^{o(H)}}_{(5)}, \tag{4.2}$$

where the individual parts of this inequality can be interpreted as follows:

- The expected number of individuals $\langle m \rangle$, whose genome will be representatives of a particular schema H in the next generation, that is, the generation at time $t+1$.

- The current number of individuals that represent schema H at the current time point t.

- The expected proportional change of individuals that represent the schema H assuming roulette wheel selection. The proportion is computed by relating the current average fitness of schema representatives $f(H,t)$ relative to the average total fitness in the population.

- The probability that the schema is not destroyed by one-point crossover, where the fraction encodes the schema-specific probability of selecting a crossing point within the schema and p_c specifies the cross-over probability.

- The probability that mutation flips a crucial bit, where p_m denotes the probability of mutation.

The schema theory thus quantifies the likelihood of the growth of a particular building block structure. It is an inequality because constructive events caused by suitable mutations or recombinations are not considered by the theory. For example, mutation may accidentally create a new schema representative or crossover may recombine two sub-building block structures yielding a full schema representative. While the schema theory is thus suited to quantify the likelihood of growth of particular existing building block representatives, it does not quantify the likelihood of evolving new building block structures.

Evolutionary discovery of new building blocks

While the schema theory quantifies when existing structures grow, the idea of a *control map* quantifies at least in an approximate manner under which circumstances it is likely that an evolutionary algorithm discovers novel building block structures (Goldberg, 1999). Clearly, to maximize the likelihood of discovery events, a large diversity in the current population is beneficial. However, assuming current sufficient diversity, the following interplay, which is dominated by selection and recombination, can be formalized.

The likelihood of evolutionary discoveries is essentially maximized by ensuring effective mixing of building block substructures by combining a suitable selection pressure with suitable recombinations. Overly strong selection results in an overly fast convergence to the current best individual, preventing the discovery of new structures. On the other hand, overly weak selection pressure (selecting nearly randomly) results in genetic drift – possibly in the direction determined by mutation – also preventing the fitness-oriented discovery of new structures. Thus, a medium selection pressure needs to be chosen.

Meanwhile, recombination must not be overly strong to prevent shuffling genetic encodings purely randomly. Recombination essentially needs to expect the existence of certain small sub-structures – somewhat like sub-building blocks – to generate building blocks of a larger order. Recombination, however, may not be overly weak either, because in that case mixing events may not occur frequently enough. The crossover probability p_c is the simplest way to change the strength of recombination.

When thinking of the convergence speed generated by selection pressure in relation to recombination, it becomes clear that the strength of recombination must increase when the selection pressure increases. Thus, a good balance between selection pressure and reproduction strength needs to be maintained. These thoughts have led not only to the development of balanced evolutionary computation approaches, but also to the development of structure-sensitive crossover operators, such as the previously mentioned EDAs. Also niching techniques have been explored to restrict the speed of convergence and maintain higher diversity in the evolving population – also fostering effective mixing. The control map in Figure 4.3 illustrates the sketched-out interactions between selection and recombination (Goldberg, 1999): overly strong selection results into immediate takeover of the best individual, preventing effective evolutionary recombination; in contrast, overly weak selection strength yields drift and random mixing. Similarly for recombination, overly strong recombination results in improper structural mixing and, in fact, too random shuffling; overly weak recombination, on the other hand, yields no or insufficiently fast structural recombination, preventing the discovery of new structures with high likelihood. Thus, the strength of both processes needs to be properly balanced to generate an effective, innovative evolutionary process.

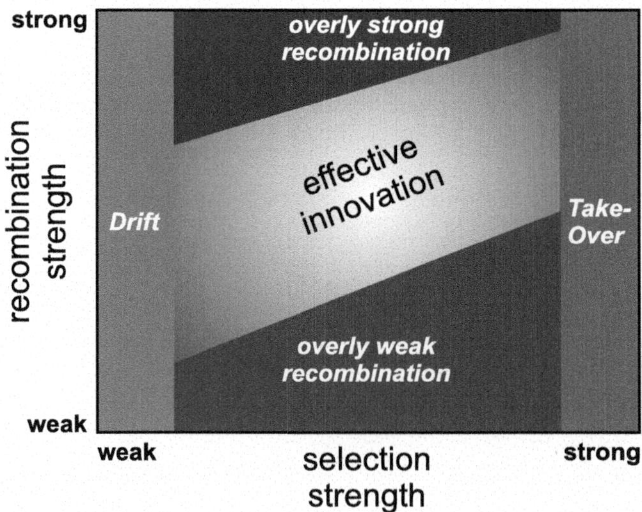

Figure 4.3: The theoretical control map for evolutionary algorithms shows that a good balance between structural recombination and selection strength needs to be maintained. [Re-sketched from an image published in *Evolutionary Design by Computers*, ed Peter J Bentley, Ch. 4. The Race, the Hurdle, and the Sweet Spot: Lessons from Genetic Algorithms for the Automation of Design Innovation and Creativity, pp. 105–118. Copyright Elsevier (1999).]

Overall considerations

Evolutionary computation approaches – despite their appeal with respect to natural selection and the original hope that optimization becomes much easier since natural evolution has worked rather well – have their caveats and need to be carefully implemented to ensure the unfolding of a successful optimization process. Thus, the success of evolutionary computation approaches depends on a good design of the individual components – leading to a *designer bias* in each specific evolutionary computation approach.

As is the case for many other machine learning algorithms, representation is the first key to a successful evolutionary algorithm. Genotype neighborhoods should yield similar phenotypes, so that mutations and recombinations in the genotype do not yield a random search process. Moreover, building block structures, that is, genes in the genotype, which encode particular substructures of the phenotype, should be situated close together in the genotype. Alternatively, recombination operators can be informed about or can attempt to automatically detect building block structures to prevent disruptive crossover events.

The initial population should be kept maximally diverse, possibly with a bias toward genotypic encodings that are believed to yield optimal solutions with higher likelihoods. The fitness function should not only identify the optimal solution, but it should also be designed in such a way that sub-solutions, which may lead to the optimal solution, generate higher fitness values. Mutation should not be overly strong in order to prevent the occurrence of too many detrimental mutation events in the binary domain. In the real-valued domain, adaptive mutation operators, such as the CMA-ES algorithm, are available that greatly increase the evolutionary success rate. Selection needs to be balanced, preventing drift and pre-mature convergence. Finally, selection should be balanced with recombination, especially when innovative recombination events are needed to find optimal solutions.

4.5 What can we learn from evolution?

With a basic understanding of natural evolution and evolutionary computation, possible implications for the evolution of human cognition can be considered. To do so, it is worthwhile to look at the progression of natural evolution.

Earth is believed to be about 4.6 billion years old. The first signs of life, *prokaryotes*, which are very simply cells, date back to about 4 billion years. These organisms already had a cell membrane, had glucose as their main energy source, and replicated by cell divisions during which the DNA was replicated. It then took about 1.5 to 2 billion years to evolve these cells further, resulting in first primitive multicellular organisms, *eukaryotes*, about 2.1 billion years ago. Vertebrates took about another 1.5 billion years, appearing on earth about 630 million years ago. Then evolution sped up. In relation to his modularized subsumption architectures (cf. Section 3.6.2), Rodney Brooks, the former MIT head of the Computer Science and Artificial Intelligence Laboratory, pointed out that:

> [...] mammals [arrived] at 250 million years ago. The first primates appeared 120 million years ago and the immediate predecessors to the great apes a mere 18 million years ago. Man arrived in roughly his present form 2.5 million years ago. He invented agriculture a mere 19,000 years ago, writings less than 5,000 years ago and "expert" knowledge only over the last few hundred years. (Brooks, 1990, p. 5.)

Evolution thus took a very long time to develop effective single cell structures, effective multicellular organisms, and the blueprint of most more intelligent lifeforms, that is, vertebrates. After that, the further differentiation into mammals, birds, reptiles, amphibians, and diverse types of fish only took another few 100 million years. The further evolution toward humans, starting with the common ancestors between man and the great apes, took only about 20 million years. Thus, the evolution of human intelligence, which enables reasoning, abstract thought, and sophisticated language usage took comparably little time to evolve.

To make these time-spans more understandable, let us project the evolution on earth over 4.6 billion years onto a 12-hour clock. At 0:00 time earth emerged. About 600 million years later, that is, at about 1:31 the first prokaryotes developed. About 2.1 billion years ago, at about 6:26, eukaryotes came about. 620 million years ago, that is, at about 10:23, vertebrates came about. Thus, it took about 80% of earth's history to develop effectively interacting organisms, with diverse, specialized individual cells. The first primates then appeared about 120 million years ago, that is, at about 11:41. *Homo sapiens* in its current form is believed to have evolved about 0.2 million years ago, which corresponds to about 11:59:58. The basis for developing higher forms of cognition thus lies in the very versatile design of individual cells, their versatile diversification and their integration into bodily blue prints, such as in that of vertebrates or mammals more particularly.

The insights gained from our examination of evolutionary computation have shown that effective genetic encodings are at least as important for evolutionary progress as are a good fitness function. The proposed time line for natural evolution suggests that single cells may have taken as long as they did to evolve multicellular organisms because complex additional mechanisms were necessary to succeed. Moreover, probably also the genetic encoding had to be optimized to prevent the destruction of fundamental genetic building blocks due to unfavorable mutation or crossover events. The same most likely also holds true for the transition from multicellular organisms to vertebrates.

As we know from prenatal, ontological development, the structure of the human body forms in the embryonic period over the first 10 to 12 weeks after the egg is fertilized. After that, differentiations take place that make the developing fetus more and more human-like, including brain development. Thus, the blueprint for embryonic and fetal development was most likely already present long before humans evolved, starting with the appearance of the first vertebrates, if not even earlier. This blueprint then evolved into a mammalian blueprint and then into ape like-creatures and finally into humans. In this latter progression, due to the short time periods, genetic building blocks can have changed only slightly, possibly duplicating or enhancing existing brain structures and shaping the developmental progression to the cognitive needs of the individual species. Unfortunately, the knowledge about the human genome and its evolution is still too limited to provide solid support for these theories.

An important consideration is the question of the original fitness benefit that led to the intelligent capabilities of humans. Many researchers now believe that it is the capacity to co-operate effectively in a group, going beyond a clear hierarchical group structure (Tomasello, 2014). The assumption that another member of the own tribe will be supportive seems to be a distinct feature that is only fully present in humans. Even preschoolers show the ability of judging fairness – independent of their cultural background – so that rewards are shared approximately equally when they were secured by a cooperative effort. If the reward was not secured cooperatively, though, sharing does not necessarily need to be fair. Trusted relations may also enable an effective division of labor, with particular duties for particular members of the clan, such as the traditional view of the men going hunting and the women focusing on gathering and child raising – although this view is subject to questioning. Regardless which divisions of labor took place, however, such effective, cooperative divisions of labor have clearly been shown to increase the chance of survival of the involved clan members and the reproduction of their genetic material.

Cultural evolution then, which is studied by anthropologists, reaches a whole new dimension of evolutionary progress. Especially once writing was developed, written laws including consequences for behavior against such laws, probably enabled the formation of larger clans and the development of the first advanced civilizations (Harari, 2011). Developments such as the invention of letterpress printing enabled an even faster distribution of knowledge and prevented detrimental loss of knowledge. Thus, cultural evolution enabled by language and writing progresses so fast because knowledge or new insights a particular human individual gains is not lost once it is written down. Now, as a result of the development of the Internet and the useful knowledge sources on the Internet, such as Wikipedia and online articles, knowledge availability is tremendously facilitated, fostering an even faster knowledge progression. Another thing that has sped up cultural evolution is the fact that the evolution of knowledge is directed, which stands in contrast to natural evolution, which is undirected. Cultural evolution is guided by our goals and by problems that we want to solve to make our life easier. Once a solution is found and written down, it is very difficult to be lost again. Thus, by means of language and writing, humans have opened up an evolutionary niche that enables cultural evolution and technical evolution in its current unprecedented form.

4.6 Exercises

1. It is known that fetuses tend to grasp and hold onto their umbilical cord, as well as onto their other arm. Discuss the differences between the two experiences and consider how particular aspects in these differences may help to start distinguishing their own body from other objects.

2. In which way does the absence of the rooting reflex, when an infant touches the own cheek, indicate that infants have knowledge about their own postural body schema.

3. Consider well-known developmental stages in humans throughout our lifetime and discuss their potential evolutionary usefulness.

4. Why may infants initially dominantly show spatial conceptualizations and only slightly later exhibit increasingly complex object conceptualizations and differentiations?

5. Why may the need to differentiate the self from others and to take on the perspective of others be important components that lead to the development of self-consciousness?

6. Define mathematically the one-max and trap fitness functions in general for any possible binary optimal solution vector $\mathbf{x}^*$.

7. Consider the following schemata according to John Holland's Schema theory: `10*1**1`, `1011***`, `***0***`, and `1011111`. Determine their order and their defining length.

8. Gray-codes encode binary numbers in a way that neighboring magnitudes can always be reached by exactly one bit flip. Why may such codes be more suitable when working with a binary mutation that flips bits randomly?

9. The traveling salesman problem is the problem of finding the shortest route through a number of cities, visiting all of them and ending up back in the start city. What is a good genotypic representation for a TSP solution? What could be the fitness? Which mutation operations may be useful? How could crossover be implemented?

10. Proof that in the schema theory $m(H)$ can also be interpreted as the expected proportion of individuals that represent schema H.

11. Reflect on the fact that no sensory organ is perfect and that different species are equipped with different sensory organs, which vary in their accuracy (for example, a snail with human eyes). Relate your thoughts to evolutionary niches.

12. Discuss the likely reason for the typical morphological position of eyes in predators (close together) versus prey (farther apart).

13. Why may the sclera, that is, the outer layer of the human eye, be white and typically clearly visible to others? Argue from an evolutionary perspective. Relate your answer to the typical human traits of social interaction and social cooperation, and to niche-based evolution.

14. Explain the term "Designer bias" with respect to evolutionary algorithms, but also with respect to cultural evolution.

Chapter 5

Behavior is Reward-oriented

5.1 Introduction and overview

In the last chapter, we saw that beginning with the fetus – if not in the embryonic stage – ontogenetic, neurocognitive development commences. That is, the fetus learns about its own body, as well as about sensory perceptions that do not come directly from its own body. In this chapter and Chapter 6, we focus on the fundamental learning principles that appear to guide this cognitive development. Here, we address *reward-oriented learning* based on positive reward and negative reward, that is, punishment, signals. In Chapter 6, we focus on *sensorimotor-oriented* learning, that is, learning mechanisms that extract regularities from sensorimotor contingencies and that can generate inference-based, anticipatory, goal-oriented behavior, given the learned sensorimotor knowledge.

To be able to learn from reward, the brain needs to be able to generate reward signals. These signals come from the bodily reward system, which monitors the bodily state, such as its current supply of fluids, fat, and oxygen, but also its state of health. Based on these internal states, the brain generates reward-oriented behavior, such as finding food, while avoiding punishments, such as getting hurt. In this chapter, we abstract these internal states, introducing a *motivational system*, which induces the self-motivated maintenance of homeostasis. Before focusing on such an artificial, self-motivated system, however, we take a step back and look at the principles of behaviorism and the development of the field of experimental psychology, which initially focused on reward-oriented behavior. Next, we introduce fundamental principles of reinforcement learning (RL) and the related policy gradient technique, which optimizes behavioral control parameters. Finally, we introduce the motivational system, showing how reward-oriented behavior can be grounded in bodily signals.

At the conclusion of this chapter, we address the fact that RL-based approaches are able to generate adaptive, reward-oriented behavior, but no "deeper" or "higher level", reflective cognitive processes. To generate such deeper cognitive processes, a system needs to develop internal, predictive models about the body and the outside environment. The involved sensorimotor, predictive model learning techniques then need to be combined with RL techniques to enable the development of a cognitive, living being that is able to generate anticipatory, goal-directed behavior and higher level, reflective cognition.

5.2 Reinforcement learning in psychology

As we discussed in Chapter 2, experimental psychology grew into its own scientific discipline in the beginning of the 20th century. With publications of Ivan Pavlov (1849–1936) and others on *classical conditioning*, it was shown that animals are able to anticipate future rewards. In particular, the animals were shown to associate a *conditioned stimulus*, such as the sound of a bell, with the subsequently occurring *unconditioned stimulus*, such as food.

The animals exhibited the typical reaction to the unconditioned stimulus, such as salivating, upon the presentation of the conditioned stimulus.

In contrast, *operant conditioning* techniques, developed by Edward L. Thorndike (1874–1949) and others, showed that animals are able to learn new behavioral patterns. To do so, dogs and cats were put into "puzzle boxes", which could be opened by various particular mechanisms. Thorndike showed that, over time, the animals learned to open the boxes more effectively, thus having learned the new behavior of opening a particular puzzle box.

Possibly the most important conclusion that Thorndike drew from his experiments, is the *law of effect*: learning from the effects of actions. Thorndike associated such effects, however, mainly with the *valence*, or motivational significance, the effect has. The opening of the uncomfortable box, for example, leads to a positive effect. In general, Thorndike thus postulated that when behavior is reinforced, it will occur more often in the future, while when behavior is punished, it will occur less often. Burrhus F. Skinner (1904–1990) later refined the mechanism behind it focusing on the effects of punishments, besides positive rewards.

With these investigations, it soon became clear that reward and punishment need to occur in close temporal and, if applicable, spatial *contiguity* to the action that is to be adapted. Even more important than contiguity, however, is the *contingency* of the behavior-dependent resulting effect. If the effect occurs only at a certain chance level, or if the effect occurs also randomly from time to time, then the observed behavioral adaptation, that is, learning, progresses much more slowly. Thus, the higher the contingency of the behavioral effect as well as the stronger its contiguity, the faster the behavioral adaptation. With these observations, it has been possible to train animals to perform stunts and even sequences of stunts using the additional technique of *shaping* (asking for progressively more complex behavior to get the reward) and *chaining* (requiring a progressively long sequence of behaviors). Clicker training, which is well-known for training particular behaviors in dogs, is an example of operant conditioning in combination with shaping and chaining techniques.

Knowing of such observations and qualitative explanations, Robert A. Rescorla and Allan R. Wagner developed first quantitative model of the involved learning processes., called the *Rescorla–Wagner–Model*:

$$\Delta V_t^A = K^A \cdot (\lambda^{US} - V_{t-1}^A). \tag{5.1}$$

The equation quantifies learning by the change of an assumed associative strength ΔV_t^A between a behavior or a stimulus A, and an unconditioned stimulus US, which results in a typical behavior (such as producing saliva). Since learning proceeds over time, index t specifies the current reinforcement event. The more events of that type were experienced, the closer the association is to the maximal association strength λ^{US} of an unconditioned stimulus US. In addition, the *saliency* of the encountered association K^A modifies the adaptation of the association strength.

Originally, Rescorla and Wagner put forward the equation as a quantitative, normative model for classical conditioning. However, its applicability to operant conditioning is also warranted, as in the latter case not reward, but behavior is associated with a given situation. Most importantly at the time, the Rescorla–Wagner model was the first one that could explain all observations that had been made by various researchers focusing on behaviorism. The model even predicted several additional behavioral patterns, including *extinction* of previously learned behavioral patterns, the *blocking* of new associations when a new stimulus is paired with the already conditioned one, and *conditioned inhibition* where an unconditioned stimulus becomes negatively conditioned when it blocks the occurrence of the expected positive reward.

Despite the focus on reward- and punishment-based experiments, Rescorla himself later stressed that the basic learning mechanism should not only focus on such scenarios. Rather, he hypothesized that learning by the proposed means can also explain discriminative behavior, that is, the learning of discriminative encodings, such as different types of food. Thus, while the original Rescorla–Wagner model was a great success for cognitive psychology and cognitive modeling, its implications go beyond reinforcement learning. Essentially, it stresses that learning always takes place when expectations (formalized by an association strength V_{t-1}^A) are violated, which was originally formalized by λ^{US}.

With this more general view, however, various other problems need to be solved. The *frame problem* is particularly challenging in this respect because, in order to anticipate a certain effect, the animal needs to identify those conditions and that behavior that were actually relevant for causing the effect – an endeavor that is very challenging. Even humans often have false beliefs, make up explanations for inexplicable phenomena, and tend toward superstition as a result of learning false associations, the rather simple Rescorla–Wagner rule needs to be differentiated and made more concrete. In the following, we focus on further developments with respect to reward-oriented learning. In Chapter 6, we then focus on learning predictive models, detached from actual reward, and face the challenge of learning relevancies.

5.3 Reinforcement learning

The Rescorla–Wagner model may generally be viewed as a model to learn associations between co-occurring stimuli. Similar learning mechanisms have been developed over the last century for the computer, many of which are closely related to the Rescorla–Wagner model. The whole discipline of *machine learning* has developed out of considerations of how a machine may be able to think and particularly learn to think. Especially as it seems impossible to prepare a machine for all imaginable situations beforehand (the *frame problem* once again), learning is as inevitable for machines as it is for humans when facing an open-ended environment.

Learning in artificial but also in biological systems can be separated into three types (Bishop, 2006): *supervised learning, unsupervised learning,* and *reinforcement learning.*

In *supervised learning* a distal teacher or supervisor is assumed to be available. This supervisor teaches the learner by providing correct answers or suggesting correct behavior. Classification learning is the most obvious type of supervised learning, where a learner is fed with exemplar data instances and corresponding classes – such as when learning to classify objects in images with deep learning artificial neural networks (Krizhevsky et al., 2012). After successful learning, the classification system – such as an artificial neural network – is expected to be able to classify images of objects, including novel images of learned objects, accurately. In addition to accurate classification, the importance lies in the ability to classify novel images – only if similar accuracy rates are achieved with novel images can one deduce that the system did not undergo an *exemplar learning process*, but a *generalizing learning process.*

Unsupervised learning is applied when no teaching signal is available. It is mainly useful when searching for general structures and patterns in data. Unsupervised learning is designed to compress the data into descriptive clusters, identifying data regularities. Over the last few decades, generative models have gained considerable interest, where the data is compressed in such a way that the learned generative model is maximally compact while still being able to re-construct individual data instances with sufficient accuracy. *Hidden Markov models* (HMMs, cf. Bishop, 2006; Rabiner, 1990), self-organizing maps (Kohonen, 2001), and *restricted Boltzmann machines* (RBMs, cf. Hinton, Dayan, Frey, & Neal, 1995; Hinton, Osindero, & Teh, 2006) are rather well known types of such generative models that can be trained by means of unsupervised learning.

Finally, RL stands somewhat in-between unsupervised and supervised learning, because no correct answers or behaviors are provided as feedback, only qualitative feedback in the form of rewards and punishments (Sutton & Barto, 1998). In obvious contrast to unsupervised learning, RL provides rewards. However, these rewards do not directly correct the behavior of the system, but just give feedback about the general quality of particular behavior. In this chapter, we focus on RL as the main form of reward-oriented learning of behavior. Later chapters, however, will also consider how supervised and unsupervised learning mechanisms can be combined with RL to improve behavior and learning even further.

5.3.1   RL problem

To formalize RL in a general manner, it is necessary to first formalize the environment in which RL is supposed to take place. Such an environment typically consists of two parts. First, the RL agent, which can roam around and manipulate the outside environment, as well as potentially its own body, needs to be specified. Second, the actual outside environment, which contains all information about the world, except for the one about the agent, needs to be defined. In simple RL problems, the agent is often defined by its current state with respect to the outside environment, or it is not made explicit at all but exists only in that it can execute interactions with the environment. The formalism typically used to characterize such agent-environment interactions is the Markov Decision Process (MDP), named after the Russian mathematician Andrei A. Markov (1856–1922), and enhancements there to.

Markov decision process

An MDP is essentially a process in which an agent iteratively interacts with a simulated environment. Similar to the rats in Tolman's mazes, for example, the agent may be simulated to move through a maze, getting a reward when finding food. Similar to the cats and dogs in Thorndike's experiments, the agent may be simulated as having various manipulative actions available and will be rewarded when it manages to execute the sequence of actions that leads to opening the box.

Formally we can describe the set of possible states in which an environment may be by $\mathcal{S}$, where each potential state $s \in \mathcal{S}$ specifies an exact, discrete state of the environment, typically including the state of the agent. At a particular point in time, t, the current state of the environment may be denoted by s_t. At each point in time t, the agent may then execute a certain action $a_t \in \mathcal{A}(s_t)$, which is possible in the given current situation s_t. The effect of an action is specified in state-transition probabilities, $\mathcal{P}(s_{t+1}|s_t, a_t) \in [0, 1]$, which specify the probability of reaching s_{t+1} from s_t, given a_t is executed. Besides this environmental agent model, the MDP also specifies the reward encountered as a consequence of an executed action. This *reward function* $\mathcal{R}(s_t, a_t, s_{t+1}) \in \mathbb{R}$ essentially specifies the reward resulting from the state transition caused by the executed action, where negative values can be interpreted as punishments.

Thus, an MDP is defined by the following four-tuple:

$$(\mathcal{S}, \mathcal{A}, \mathcal{P}, \mathcal{R}) \tag{5.2}$$

It is called a *Markov* decision process because it obeys the *Markov Property*, which refers to the memoryless property of a stochastic process, such as the decision process we are interested in. A memoryless stochastic process, which unfolds given a certain state $s_t \in \mathcal{S}$ and action choices $a_t, a_{t+1}, ...$, does not depend on previously encountered states before s_t. This can be formalized as follows:

$$Pr\left(s_{t+1}, r_{t+1}|s_t, a_t, r_t, s_{t-1}, a_{t-1}, \dots, r_1, s_0, a_0\right) = Pr\left(s_{t+1}, r_{t+1}|s_t, a_t\right) \tag{5.3}$$

and essentially implies that all information about the situations the world is in is contained in any state of the world s_t. Besides this first-order Markov process, where the dependency reaches back to the last state, *higher-order Markov processes* have also been described and conceptualized. An n-order Markov process is essentially a stochastic process whose stochastic progression depends on the last n states.

Note that the Markov property thus defines the world as *fully observable* given the last n states. It guarantees that there are no hidden states in the world that need to be assessed, for example, by epistemic, that is, information-seeking, actions. However, this does not imply that the agent knows what is best to do in the world. To learn what is best, it needs to interact with the world and essentially learn something about the (unobservable) state transition function $\mathcal{P}$ and reward function $\mathcal{R}$, which determine the world. In RL, this learning typically manifests itself in the learning of a *behavioral policy*.

Behavioral policy

In this formalized world, the agent is characterized by its behavioral strategy or *behavioral policy* $\pi : \mathcal{S} \rightarrow \mathcal{A}$, which specifies for all possible states of the environment $s \in \mathcal{S}$ an action $a \in \mathcal{A}(s)$, which is the action the agent will take in this state. The learning goal of this agent is to adapt this policy toward a particular optimality criterion.

To define such an optimality criterion, it is helpful to first define a *value function*, which specifies the value of a certain state or state-action tuple. The values, however, not only depend on the next reward that is possibly encountered, but on all the rewards that are expected to be encountered in the future given the agent executed its behavioral policy π.

The *state-value* function $V^\pi(s)$ specifies the expected reward when in state s and following the policy π from then on. Mathematically, this conceptualization can be written as follows:

$$V^\pi(s) = E^\pi \left(R_t + \gamma R_{t+1} + \gamma^2 R_{t+2} + ... | s_t = s, \pi \right), \tag{5.4}$$

where $\gamma \in [0, 1]$ is a fixed discounting factor that emphasizes the importance of a more immediate reward in contrast to reward in the more distant future and R_t denotes the reward encountered at time t. $V^\pi(s)$ specifies one value for each possible state $s \in \mathcal{S}$ for a specific policy π. As the policy π returns one action for each state, the value $V^\pi(s)$ essentially specifies the long-term expected reward when executing action $\pi(s)$ in state s and further following policy π. Reward knowledge about the other potential actions $\mathcal{A}(s) \neq \pi(s)$ in state s is not available in a state-value function.

In contrast to this state-value function, the state-action-value function $Q^\pi(s, a)$ specifies values for each possible action $a \in \mathcal{A}(s)$ for each possible state $s \in \mathcal{S}$ in the environment. Thus, mathematically the state-action-value function, which is often referred to simply as the *Q-function*, can be formalized by:

$$Q^\pi(s, a) = E^\pi \left(R_t + \gamma R_{t+1} + \gamma^2 R_{t+2} + ... | s_t = s, a_t = a, \pi \right). \tag{5.5}$$

While both value functions determine the expected future reward when following a behavioral policy π, the Q-function assigns Q-values to all possible actions $a \in \mathcal{A}(s)$ in all states s, the state-value function does not. As a result, the number of Q-values is by a factor $|\mathcal{A}|$ larger than the number of state-values, and it enables the direct consideration of alternative action outcomes in each environmental state.

Later we will consider several behavioral policies and illustrate their dependence on the state-value or the Q-value function. For now, however, we focus on behavioral optimality, and a nearly equivalent formalization of an optimal value function.

Optimal behavioral policy

Given the behavioral policy definition, it is now possible to determine an optimal policy by simply stating that the optimal policy is the one that maximizes the expected future reward under all circumstances. Indeed, given the Q-value function, this can be formalized mathematically by defining the optimal policy $\pi^\star$ as follows:

$$\pi^\star(s) := \arg\max_a Q^\star(s, a) \tag{5.6}$$

However, we do not know the optimal values $Q^\star(s, a)$. These values can be determined by the *Bellman equation*, named after the American mathematician Richard E. Bellman (1920–1984):

$$V^\star(s) = \max_a E \left(r_{t+1} + \gamma \cdot V^\star(s_{t+1}) | s_t = s, a_t = a \right) \tag{5.7}$$

$$Q^\star(s, a) = E \left(r_{t+1} + \gamma \cdot \max_{a'} Q^\star(s_{t+1}, a') | s_t = s, a_t = a \right) \tag{5.8}$$

The idea behind these two equations is essentially based on Bellman's principle of optimality and dynamic programming and the involved recursion:

> PRINCIPLE OF OPTIMALITY: An optimal policy has the property that whatever the initial state and initial decision are, the remaining decisions must constitute an optimal policy with regard to the state resulting from the first decision. (Bellman, 1957, p. 83.)

Bellman realized that the Markov decision process can be partitioned into the immediately encountered reward and the discounted future reward and that optimal behavior can be determined by considering the best behavior as the sum of immediate reward plus discounted, estimated future reward. Because the future reward cannot be explicitly acquired, it is estimated recursively.

To learn the optimal behavioral policy in such a way it is necessary to learn the optimal value function and derive the optimal policy from it. When the optimal Q-value function is learned, then the derivation of the policy is straight-forward as defined in Eq.(5.6). When the state-value function is learned, the policy cannot be derived directly, as will be further discussed later.

Learning the value function, however, is typically the hard part in such formalizations. As mentioned previously, Bellman developed a whole class of algorithms, which are able to solve problems that can be formalized in the way described. He called this algorithmic technique *dynamic programming*, which iteratively approximates all state-values $V^\star(s)$ or $Q^\star(s, a)$ by storing their values and iteratively updating the values by sampling state transitions randomly. However, to do so, the MDP needs to be fully accessible, that is, the learner needs to know about the possible consequences of any state action combination. In other words, the learner needs to have a fully accurate model of the environment available to learn by means of dynamic programming. Since we assume here that the learner has no such model available, we now focus on *model-free RL*, which relies on *temporal difference learning* principles.

5.3.2　Temporal difference learning

How then can an optimal value function be learned when no model about the environment is available? Two approaches are possible: (i) first learn a model of the environment and then derive the value function from the model; (ii) attempt to learn the value function directly while interacting with the environment, without ever learning the model. Of course, a hybrid approach is also imaginable, which we will introduce later. Here, we focus on learning the value function directly without learning an environmental model – called *model-free RL* or also *direct RL* – by iteratively interacting with the environment (Sutton & Barto, 1998).

Learning in model-free RL is based on the temporal difference between currently expected reward and actual reward, leading to an update of the respective value function estimate. Moreover, as Bellman's equation suggests, learning can be applied in every iteration, considering the currently encountered reward in combination with the currently expected discounted future reward. This approach directly leads to the temporal difference equation for state-value and Q-value functions as follows.

Given the agent is currently in state s_t, executes action a_t, leading to state s_{t+1} and the encounter of reward R_{t+1}, it is able to update the according value estimation based on the encountered temporal difference. When working with value functions and assuming that $a_t = \pi(s_t)$, then the following temporal difference update can be computed:

$$V^\pi(s_t) \leftarrow V^\pi(s_t) + \alpha \cdot [R_{t+1} + \gamma \cdot V^\pi(s_{t+1}) - V^\pi(s_t)], \qquad (5.9)$$

where $\alpha \in [0, 1]$ is a learning rate and the term in the brackets is the *temporal difference error*: the currently expected, to be encountered future reward when executing policy π in state s_t, that is, $V^\pi(s_t)$ is subtracted from the sum of currently encountered reward R_{t+1} plus the expected, discounted future reward $V^\pi(s_{t+1})$ in the just reached state s_{t+1}. A positive temporal difference error essentially indicates that more reward has been encountered than expected, while a negative value indicates the opposite. With respect to the currently executed action, a positive temporal difference error also suggests that the just executed action was, indeed, better than expected.

Analogous to the value function update, the Q-value function can be updated in a similar manner, yielding the following *state-action-reward-state-action* (SARSA) update rule:

$$Q^\pi(s_t, a_t) \leftarrow Q^\pi(s_t, a_t) + \alpha \cdot [r_{t+1} + \gamma \cdot Q^\pi(s_{t+1}, a_{t+1}) - Q^\pi(s_t, a_t)], \tag{5.10}$$

where the policy π determines both the current action a_t as well as the next action a_{t+1}. With the help of the Bellman equation, it can be shown that the estimates $Q^\pi(s_t, a_t)$ are guaranteed to converge to the exact values of $Q^{\pi*}(s_t, a_t)$, that is, to the exact Q-values with respect to a policy π under few additional assumptions.

Q-learning

An even more powerful approach than the SARSA-style update of the V- or Q-function, is the more direct approximation of the $Q^\star$ function by means of *Q-learning* (Watkins, 1989). The main idea of Q-learning is to estimate the Q-value function iteratively using temporal difference learning updated, but implementing the Bellman equation even more explicitly by separating the policy π from the values that are used for updating the Q-value estimates. Thus, Q-learning does not require that the action currently executed adheres to the current behavioral policy π. Rather, Q-learning updates its Q-value function given state s_t, the execution of action a_t – which may not necessarily be equal to $\pi(s_t)$ – and the resulting reward R_{t+1} and state s_{t+1}. Q-learning is also termed an *off-policy* RL technique, which does not require action execution according to policy π:

$$Q(s_t, a_t) \leftarrow Q(s_t, a_t) + \alpha \cdot \left[\left(R_{t+1} + \gamma \max_{a_{t+1}} Q(s_{t+1}, a_{t+1}) \right) - Q(s_t, a_t) \right]. \tag{5.11}$$

The crucial difference to the SARSA update is the maximum operator in the equation, which essentially assures that the best future discounted reward is used for updating the current Q-value $Q(s_t, a_t)$.

Watkins showed that the Q-value update converges, under few additional assumptions, to the optimal Q-value function $Q^\star$, as long as all actions $\mathcal{A}(s)$ are executed in all possible situations $s \in \mathcal{S}$ infinitely often in the long run, that is:

$$\lim_{t \to \infty} Q(s, a) \to Q^\star(s, a) \quad \text{and} \quad \lim_{t \to \infty} \pi \to \pi^\star. \tag{5.12}$$

Interestingly, SARSA and Q-learning are closely related to the Rescorla–Wagner equation introduced earlier. In all these update equations, learning depends on the difference between encountered and currently expected reward, where the estimate of the actually encountered reward equals λ^{US} in the Rescorla–Wagner model, and is estimated by the sum or currently encountered reward plus the discounted expected future reward in SARSA and Q-learning.

Q-learning example

An example of Q-learning in a simple maze environment should clarify things. Let us have a look at the tiny "maze" in Figure 5.1, which consists of only three states. In accordance with this maze, the following MDP can be defined:

- The maze consists of three states and a goal or **exit** state:

$$\mathcal{S} = \{A, B, C, \texttt{exit}\} \tag{5.13}$$

- Generally, the algorithm does not know about walls, so in any state motions in all four directions are possible:

$$\mathcal{A}(s) \;=\; \{N, E, S, W\} \; (\forall s \in S) \tag{5.14}$$

- The agent in the maze can either move into a wall and thus stay in the state it had been in, or it can move to an adjacent state. Expressed by a state-transition function, this may be formalized as follows, assuming the movement toward the goal state as a special case:

$$\mathcal{P}(s'|s, a) = \begin{cases} 1, & \text{if } s' \text{ is next to } s \text{ in direction } a \\ 1, & \text{if } s = C, a = S, s' = \texttt{exit} \\ 0, & \text{otherwise} \end{cases} \tag{5.15}$$

- The reward function may be defined in various ways. For example, a move into a wall may hurt and could thus yield a negative reward. Additionally, movement itself may cost energy and thus may also yield a small negative reward. Positive reward should be encountered when moving into the goal state. One of the simplest reward functions may be defined as encountering positive reward when the exit, that is, the food location is reached and zero reward otherwise:

$$\mathcal{R}(s, a, s') = \begin{cases} 100, & \text{if } s = C, a = S, s' = \texttt{exit} \\ 0, & \text{otherwise} \end{cases} \tag{5.16}$$

- Finally, to simplify things, we turn the MDP problem into an *episodic MDP*, where a current trial ends when the goal state $\texttt{exit}$ is reached. In an episodic MDP it is assumed that other things happen after the end of an episode, such that future reward is not considered, that is, $V^*(\texttt{exit}) = 0$.

Figure 5.1: Q-learning example in a simple maze.

While these specifications define the MDP problem, to learn a value function it is still necessary to specify a discount factor, which we set to $\gamma = 0.9$. Now it is possible to derive the optimal state-value function $V^\star$, as well as the optimal Q-value function $Q^\star$.

Let us first take the dynamic programming approach of Bellman to derive the state-value function $V^\star$ and the Q-value function $Q^\star$. We know that $V^\star(\texttt{exit}) = 0$ and that $Q^\star(\texttt{exit}, a) = 0 \; \forall a \in \mathcal{A}$, so we can initialize the learning process directly by starting with all values set to zero. Tables 5.1 and 5.2 show how the optimal value functions are approximated over time by means of dynamic programming.

Learning by dynamic programming assumes full access to the environment. From each row in the tables to the next, each state is probed and the state-values or state-action-values are updated by simulating the execution of all possible actions in each state. That is, equations (5.7) and (5.8) are applied as the update algorithms for each value estimation. As a result, we see that the estimates quickly converge to the optimal estimates – particularly when in such a small, discrete, and episodic MDP.

However, when we assume a more realistic scenario in which a living creature has to learn where it can find rewards, this creature has no access to a complete model of its world.

Table 5.1: Dynamic programming computation of the optimal state-value function $V^\star$.

Iteration	A	B	C
0	0	0	0
1	0	0	100
2	0	90	100
3	81	90	100

Table 5.2: Dynamic programming computation of the optimal Q-value function $Q^\star$.

Iteration	A,E	A,S	A,W	A,N	B,E	B,S	B,W	B,N	C,E	C,S	C,W	C,N
0	0	0	0	0	0	0	0	0	0	0	0	0
1	0	0	0	0	0	0	0	0	0	100	0	0
3	0	0	0	0	90	0	0	0	90	100	0	90
4	81	0	0	0	90	81	0	81	90	100	81	90
5	81	73	73	73	90	81	73	81	90	100	81	90

Thus, it can only learn iteratively by reward encounters and estimates thereof. To illustrate the progress while actively interacting with the environment, that is, the exemplary maze represented as an MDP, we consider the following, exemplar Q-learning updates. We set the learning rate to $\alpha = 0.5$ to illustrate the progress.

Let us assume that our agent starts in state C and – seeing that it currently has no knowledge (all Q-values are initialized with zeros) – chooses an action randomly. First, it may attempt to move north and find itself still in state C. Therefore, the following Q-learning update will be computed:

$$Q(C, N) \leftarrow Q(C, N) + 0.5(r(C, N, C) + 0.9 \cdot \max_{a \in \mathcal{A}(C)} Q(C, a) - Q(C, N))$$

$$= Q(C, N) + 0.5(0 + 0.9 \cdot \{0, 0, 0, 0\} - 0) = 0$$

Thus, essentially no change took place. Next, let us assume that the agent chooses to move south, encountering a reward and reaching the exit. In this case, the update is:

$$Q(C, S) \leftarrow Q(C, S) + 0.5(r(C, S, \texttt{exit}) + 0.9 \cdot \max_{a \in \mathcal{A}(\texttt{exit})} Q(\texttt{exit}, a) - Q(C, S))$$

$$= Q(C, S) + 0.5(100 + 0.9 \cdot 0 - 0) = 50$$

Because the goal is reached, the system is reset – say, for example, to state B.

To make some progress, we assume that in state B the agent chooses to move to the east, thus reaching state C. Note that any other action in states A and B will yield no change in the Q-value function estimates. The move from B to C via action O, however, yields a solid update:

$$Q(B, E) \leftarrow Q(B, E) + 0.5(r(B, E, C) + 0.9 \cdot \max_{a \in \mathcal{A}(C)} Q(C, a) - Q(B, E))$$

$$= 0 + 0.5(0 + 0.9 \cdot 50) = 24.5$$

If the system now attempts to move to the north, as before, the update will result in an actual Q-value change:

$$Q(C, N) \leftarrow Q(C, N) + 0.5(r(C, N, C) + 0.9 \cdot \max_{a \in \mathcal{A}(C)} Q(C, a) - Q(C, N))$$

$$= 0 + 0.5(0 + 0.9 \cdot 50 - 0) = 24.5$$

Finally, let us move the agent one more time into the goal state, to illustrate the effects of a successive update:

$$Q(C, S) \leftarrow Q(C, S) + 0.5(r(C, S, \texttt{exit}) + 0.9 \cdot \max_{a \in \mathcal{A}(\texttt{exit})} Q(\texttt{exit}, a) - Q(C, S))$$

$$= 50 + 0.5(100 + 0.9 \cdot 0 - 50) = 75$$

Table 5.3: Q-value function when simulating Q-learning starting with no knowledge and executing the following environmental interactions: C,N→C,S→exit; B,E→C,N→C,S→exit

Iteration	A,E	A,S	A,W	A,N	B,E	B,S	B,W	B,N	C,E	C,S	C,W	C,N
0	0	0	0	0	0	0	0	0	0	0	0	0
1	0	0	0	0	0	0	0	0	0	0	0	0
2	0	0	0	0	0	0	0	0	0	50	0	0
4	0	0	0	0	24.5	0	0	0	0	50	0	0
5	0	0	0	0	24.5	0	0	0	0	50	0	24.5
6	0	0	0	0	24.5	0	0	0	0	75	0	24.5

Table 5.3 visualizes the update steps, which the system has undergone. Note that the amount of computation from one row to the next is much smaller than when applying dynamic programming, because the agent is actually interacting with the world rather than simulating all possible interactions. We have also assumed that a learning rate of $\alpha = 0.5$ is used. In such deterministic, episodic MDPs, $\alpha = 1$ would have worked and yielded faster convergence. However, when assuming uncertainty about the reliability of the encountered state transitions, a learning rate significantly below 1, such as $\alpha = 0.2$ is commonly used.

Independent of the learning rate, however, it seems somewhat unsatisfactory that learning proceeds so slowly. Given that the state and action spaces are much larger, it would take thousands if not millions of steps to converge to the optimal value function. Therefore, various techniques have been proposed to speed up the temporal difference learning progress.

5.3.3   Speeding up temporal difference learning

To speed up the convergence of the value function estimates, various methods have been proposed and successfully applied. Here, we focus on four fundamental types of approaches. First, *eligibility traces* maintain a memory of the path that was taken before the current action and projects the reward update backwards in time. In this way, the responsibility for the current reward is shared among the recent environmental interactions. Second, the *DYNA-Q* algorithm combines model learning with temporal difference learning, generating a hybrid between temporal difference and dynamic programming-based value function updates. Third, *hierarchical RL* abstracts over individual, single step actions and thus allows RL on multiple levels. Finally, *state factorizations* address the *frame problem* by abstracting the state representations in the attempt to consider only those factors in the environment that are behaviorally relevant.

Eligibility traces

Eligibility traces are particularly useful in RL problems where reward is sparse and the achievement of current rewards not only depends on the current action, but on previous environmental interactions as well. In such problems it is worthwhile to distribute shares of encountered reward not only to the previous state or the previous state-action combination, but also to the whole recent sequence of encountered states and executed actions.

Eligibility traces essentially assume that an update over several interaction steps improves the estimate of the optimal value function faster. In terms of Bellman's equation, this can be formalized as follows:

$$V^{\star}(s) = \max_{a} E\left(r_{t+1} + \gamma \cdot V^{\star}(s_{t+1})|s_t = s, a_t = a\right)$$

$$\approx r_{t+1} + \gamma r_{t+2} + \cdots + \gamma r_{t+x-1} + \gamma \max_{a} E\left(V^{\star}(s_{t+x})|s_t = s, a_{t+x-1} = a\right) \quad (5.17)$$

This approximation not only depends on the state transitions encountered (approximating the expectation operator), but also on the actual behavioral policy that was executed. Nonetheless, in problems in which the approximation can be expected to lie close to the optimum value, the estimation may be useful. In fact, powerful AI programs, referred to as *Monte Carlo tree search* methods, have been generated that implement this principle. With the help of these methods, for example, a powerful AI for the computer version of the board game Go was developed, where the AI executes an informed, but stochastic deep search through future board states, integrating them into the current decision-making process (Gelly & Silver, 2011).

However, when focusing on temporal difference learning, the agent does not have a model about its environment and thus cannot look forward in time. However, the same principle also works backwards in time. In this case we need to maintain a memory of previously encountered states and of the executed actions in these previous states. During each update then, not only the current state-value or state-action value is updated, but also all remembered previous ones. Typically though, not all previous ones should have the same update strength, but more recent states should undergo stronger updates. This is accomplished by determining an *eligibility* of each previous state. The eligibility is easy to determine when defining it using the most recent point in time a particular state had been visited:

$$
e_t(s) \;=\; \begin{cases} (1 - \lambda)(\lambda\gamma)^{t-k} & \text{if } k > 0 \\ (\lambda\gamma)^t & \text{if } k = 0 \\ 0 & \text{otherwise} \end{cases} \;, \tag{5.18}
$$

where $k = -1$ if state s has not been visited at all so far, and $k = \max\{k | s_k = s\}$, otherwise. The factor λ determines the spread of the eligibility, where $\lambda = 1$ corresponds to a normal TD update, while $\lambda \to \infty$ spreads the eligibility uniformly into the past. It is guaranteed that the reward is perfectly spread out into the past, as $\sum_{t'=0}^{t} e_t(s_t) = 1$ when $\gamma = 1$, because the λ factors yield a geometric series.

With the concept of eligibility, the temporal difference update is applied to all states that have been encountered so far until time t, yielding the enhanced temporal difference update equation:

$$
V^\pi(s) \;\leftarrow\; V^\pi(s) + \alpha \cdot e_t(s) \cdot [R_{t+1} + \gamma \cdot V^\pi(s_{t+1}) - V^\pi(s_t)], \tag{5.19}
$$

SARSA learning can be updated accordingly. Q-learning, on the other hand, is not directly applicable because the trace depends on the policy, violating the off-policy principle of Q-learning updates. Nonetheless, initial faster learning can also be achieved in this case, while full convergence to $Q^\star$ relies on proper, off-policy Q-value estimation updates.

Besides speeding up learning in RL and the successes by applying random forward projections using, for example, Monte Carlo tree search, multiple aspects are relevant when viewing eligibility traces in the light of cognitive development, as well as of behavioral learning and behaviorism. First, it has been hypothesized that the dopamine gradient that is generated in the brain upon the encounter of reward declines, while place cells in the hippocampus play out the approximate path that the rat has taken before encountering the reward backwards (Foster & Wilson, 2006). Moreover, the importance of the contiguity of previous stimuli in relation to current reward has shown that the higher the contiguity, the higher the increase in association strength, which is what eligibility traces realize. In general, it thus seems plausible that rewards in the brain are not only associated with the immediately preceding action, but also with those actions that enabled the preceding action in the first place. For example, when consuming food not only the food consumption is rewarding, but possibly the food preparation as well.

Model-based RL

While eligibility is a powerful tool to speed up reward-oriented learning, the updates that take place are still limited to the actually encountered interactions. Model-based RL offers a technique with which reward can also be spread to actions and states that have not been executed in the current interaction episode (Littman, 2015; Sutton & Barto, 1998). To do so, model-based RL updates simulate interactions with the environment by means of an environmental model, which models the state transition function $\mathcal{P}$ and the reward function $\mathcal{R}$ of an RL problem. Due to the application of such *indirect* updates of value estimations, model-based RL is also sometimes referred-to as *indirect RL*.

However, model-based RL relies on the existence of a model, or on additional learning mechanisms, which learns an approximate model. Richard Sutton's *Dyna-Q*-architecture has put forward a general algorithm that combines model-free with model-based RL:

1. Observe the current state s and choose an action a according to the agent's behavioral policy pi.

2. Execute a and observe the resulting state s', as well as the resulting reward R.

3. Apply direct RL, such as Q-learning, given $\langle s, a, s', r \rangle$.

4. Update the internal model given $\langle s, a, s', r \rangle$ – in the simplest form by updating, for example, the action-respective state transition matrix, which estimates the probability of reaching s' when executing a in state s.

5. Also update the internal estimate of the reward when encountering the specific transition in a reward-based state transition matrix.

6. Execute several – say N – model-based RL steps as follows:

   - Choose a known state s and a possible action a at random.
   - Use the internal model to determine the outcome of this state action combination, predicting the resulting s' and r.
   - Apply temporal difference learning, such as Q-learning, with respect to this sample.

7. Repeat steps (1–6) until convergence.

Dyna-Q thus iteratively executes normal temporal difference updates and combines these updates with simulated environmental interactions using the developing environmental model. In the beginning, where the model is basically empty, Dyna-Q will not bring any learning advantages. However, if model learning works well, soon Dyna-Q can speed up the learning process tremendously by essentially spreading encountered rewards into regions that have been explored previously.

Hierarchical RL

When considering value function learning by means of temporal difference learning – even if combined with eligibility traces and Dyna-Q updates – it soon becomes obvious that this learning technique works on small environments only. The more states an environment has and the more actions are possible, the longer it will take to converge toward the optimal Q-value or state-value function. Thus, RL can be successfully applied to small MDP problems only. Larger MDPs still pose a huge challenge to RL. Possibly the most promising approach to tackle such problems is to apply *hierarchical RL* (Littman, 2015; Sutton, Precup, & Singh, 1999).

An example of a very suitable environment for hierarchical RL is the four-rooms problem. Figure 5.2 shows the problem: a maze consists of four rooms, which are connected to each other via doorways in a circular manner. Depending on the number of states in each room, RL will soon reach its limits, requiring thousands if not millions of steps to approximate the corresponding value function. On the other hand, a hierarchical representation of the problem can greatly simplify learning.

Here, we only generally define the hierarchical RL problem to give an idea of its functionality. The main idea lies in extending the actions possible in an MDP to *options*, which

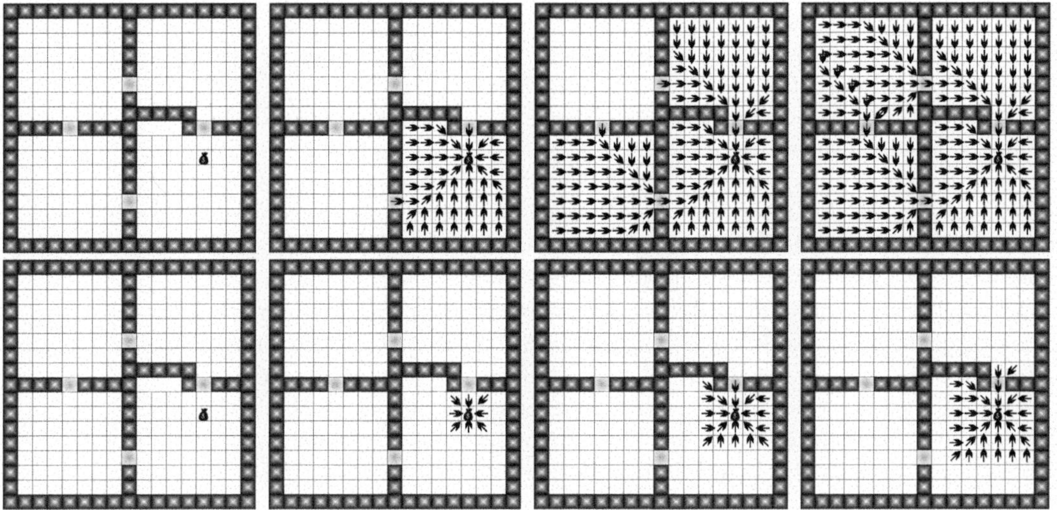

Figure 5.2: Hierarchical RL in a four-rooms problem.

are *behavioral primitives* that can be executed over an extended period of time (Sutton et al., 1999). In the four-rooms problem, for example, an option may be a behavioral primitive that enables an agent to reach one of the two doorways in a specific room. Given that we have eight options – two for each room to reach either doorway – planning can be sped up tremendously. Given, for example, that a particular position in a room triggers positive reward, DYNA-Q-based hierarchical RL updates with a hierarchical model – or dynamic programming techniques given the complete hierarchical model – can generate the $Q^\star$ function with respect to the reward position much faster than a non-hierarchical representation. The two options applicable in the room where the reward was found allow the discounted propagation of the reward from the reward position immediately to the doorways of the room. The four options of the neighboring two rooms then enable the spreading of this reward over the neighboring rooms in the next step. Finally, the opposite room is filled with the corresponding $Q^\star$ values. The options essentially need to specify the discount factor or costs that are encountered when reaching one of the doorways for each possible state in a room. Once accurately specified, discounted reward can be transferred from any state of the room to the doorways, as well as from doorway to doorway across each room.

The four-rooms problem is only a metaphor for many other hierarchically-structured problems. Due the spatial and temporal constraints in our world, it may be hypothesized that most real world problems exhibit particular hierarchical structures. Thus, the concept of hierarchical RL seems to be quite promising. The hard challenge lies in the identification of the hierarchical structure and the development of behavioral primitives that generate behavior and appropriately discount reward estimates on the upper levels of these structures. The challenge is how to develop learning techniques that can robustly detect hierarchical structures in MDP problems as well as in our world. While numerous techniques have been proposed, none of them has established itself as a commonly used technique throughout the RL research community (Barto & Mahadevan, 2003; Botvinick & Weinstein, 2014; Dieterich, 2000; Vigorito & Barto, 2010).

Partial observability and state factorization

So far, we have taken for granted that an MDP can accurately model our world. States have been defined as discrete, symbolic entities, where each state essentially characterizes a complete world state. State transitions fully depended on these symbolic states, assuming that the world obeys the *Markov property*. Two rather straight forward observations stand in sharp contrast to such a formalization:

First, our world is not fully observable. For example, we cannot be completely certain whether a door is locked or not. Rather, we have to rely on assumptions and estimations (such as that we have not locked the door and nobody else could have while we were inside the room). Thus, at best, our world is *partially observable*, where our sensory abilities give us hints about the actual state of the world, but never the true state. In fact, seeing Heisenberg's Uncertainty Principle, our world appears to be intrinsically only partially observable. Thus, RL should also focus on partially observable environments – and in fact, *partially observable Markov decision processes* (POMDPs) have been defined and studied in some detail. A detailed discussion of this matter, however, goes beyond the aims of this book.

Second, and possibly even more important, when encountering reward only a few aspects in our world are typically responsible for the actual reward encounter. For example, when drinking a cup of tea it is not particularly relevant that the room may be painted white, that it may be nine o'clock, or that a particular type of music is playing in the background. Most relevant is the actual state of the body, for example, longing for something warm to drink and the fact that tea is currently available and in reach. The fact that it was prepared by oneself is also of relevance and should be indirectly rewarded – possibly using a suitably adapted eligibility trace or a hierarchical structure. The reward-relevant factors thus play the most crucial role and reward learning should be tuned to focus on these reward-relevant factors.

Factored RL, as well as *anticipatory learning classifier system* approaches, have been developed over the last two decades to address this problem (Sigaud, Butz, Kozlova, & Meyer, 2009). In these cases, temporal difference learning is combined with the learning of a generalized environmental model that focuses on those perceptual factors that are relevant for predicting reward, that is, a particular value function, accurately. In various problems these approaches have shown tremendous success. When the actual, hidden RL problem is reasonably small, such approaches have been shown to solve problems with billions of states in a reasonable amount of time (Butz, 2006). In a later chapter, we will relate other problems and algorithms to such factored RL approaches and anticipatory learning classifier systems (cf. Section 12.3.3).

5.3.4　Behavioral strategies

Regardless of which approach is chosen and how the actual problem is formalized, all the mechanisms we have addressed focus on learning a value function to derive a behavioral policy. If we are approximating the optimal $Q^\star$ value function by means of Q-learning, we have seen that the derivation of an optimal policy is straightforward. When we behave according to the current Q-values, we speak of a greedy behavioral policy, which can be formalized by:

$$\pi(s) = \arg\max_a Q(s, a) \tag{5.20}$$

This behavioral strategy essentially executes in each state the action that is expected to yield the maximum accumulated future rewards.

However, a greedy strategy can have disadvantages. Once a suitable path is found to a rewarding state, this path will always be executed, without considering alternatives and thus without being able to detect potentially shorter routes. Additionally, the Q-learning theory specified earlier stated that in order for Q-learning to converge to $Q^\star$, it is necessary to execute all possible actions in all states infinitely often in the long run. A greedy behavioral policy does not accomplish this.

A simple alternative is the generalization of the greedy policy to an ϵ-greedy policy, where the parameter $\epsilon \in [0, 1]$ specifies the level of exploration, that is, the probability of choosing a random action instead of the currently seemingly optimal action. With $\epsilon = 0$ we end up with the greedy behavioral policy, while with $\epsilon = 1$ we end up with a fully random behavioral policy. Thus, typically a compromise is chosen by setting exploration to $\epsilon = 0.2$, for example. The actual behavioral strategy can be formalized as follows:

$$\pi(s) = \begin{cases} \arg\max_a Q(s, a) & \text{if } \rho > \epsilon \\ \text{rnd } \mathcal{A}(s) & \text{otherwise} \end{cases}, \tag{5.21}$$

where $\rho \in [0, 1)$ stands for a uniformly randomly sampled number and **rnd** denotes a random choice amongst the set of actions. This behavioral policy ensures that all behavioral options will be probed in all states infinitely often in the long run given $\epsilon > 0$. However, it does not consider the other current Q-value estimates. Thus, it may choose an action that is expected to yield negative reward equally likely as an action that is expected to yield only slightly smaller reward than the current best action.

The *soft-max strategy* addresses this problem by determining probabilistic action preferences dependent on the current Q-value estimates:

$$\pi(s) = \pi(a|s) = \frac{e^{Q(s,a)/\tau}}{\sum\limits_{\forall b \in A(s)} e^{Q(s,b)/\tau}}. \tag{5.22}$$

Note how soft-max also accounts for negative reward values: by taking the exponent of the Q estimates, negative values are converted into small positive numbers. Moreover, by taking the proportion of the exponential function with respect to all other exponential action-respective Q-values, a probability of choose a particular action a is determined. The quotient $\tau > 0$ scales the range of Q-values and essentially determines the greediness of the soft-max strategy: When $\tau \to 0$ then the strategy becomes progressively more greedy, increasingly emphasizing the differences between high and low Q-value estimates, converging to a fully greedy strategy in the limit. When $\tau \to \infty$, on the other hand, the behavioral strategy tends toward a random strategy because the Q-value differences are annihilated, yielding exponents that tend toward zero, and thus values that tend toward one for all actions after the application of the exponential function.

Interestingly, the soft-max strategy can be related to a kind of *curious behavior*, where curiosity may be defined as the tendency to choose suboptimal, but promising alternatives, fostering a reward-oriented exploration. Other curiosity definitions typically focus on decreasing uncertainty about reward predictions, or state or perceptual predictions. To minimize uncertainty, however, reward estimates need to have uncertainty estimates. In various problem domains it has been shown that such approaches can increase the learning speed enormously and can prevent the learning system from overlooking certain subareas (Butz, 2002b; Oudeyer, Kaplan, & Hafner, 2007; Schmidhuber, 1991). This is initially because the environment is typically explored in a more distributed fashion, ensuring that important environmental properties are not overlooked. Later on, changes in the environment can be detected much more effectively by tending toward environmental regions that have not been visited for quite some time. Combinations of curiosity-driven behavior, which is also sometimes called *intrinsically motivated behavior*, and reward-driven behavior, which is also called *extrinsically motivated behavior*, are still the subject of research. In Section 6.5 we will explore the challenge of balancing intrinsically and extrinsically motivated behavior further, introducing the principles of motivations and curiosity more generally.

5.3.5 Actor-critic approaches

Behavioral strategies can be directly inferred when approximating the optimal Q-value function $Q^\star$. However, when the action space is large or even continuous, then it is often better to attempt to approximate the state-value function $V^\star$, rather than $Q^\star$. As we have seen though, it is difficult to approximate $V^\star$ directly, especially when knowledge about a model of the MDP in unavailable. In this case, V^π can be approximated by applying temporal difference learning updated given the experiences gained, while interacting with the environments and pursuing policy π.

To optimize both the policy π and the state value function estimates toward the optimal policy $\pi^\star$ and the optimal value function $V^\star$, interactive *actor-critic* updates need to be executed. The "actor" is specified by the behavioral policy. The critic is the value function approximation. While the update of the critic can be computed with temporal difference learning, policy updated need to convert the temporal difference signal δ_t:

$$\delta_t \;=\; R_{t+1} + \gamma \cdot V^\pi(s_{t+1}) - V^\pi(s_t), \tag{5.23}$$

into suitable "actor" updated. Positive values of δ_t indicate that a good action was chosen, so that the likelihood of choosing this action in the respective situation should be increased. In contrast, a negative value indicates that the action was worse than expected, so that the likelihood of the action choice in the respective situation should be decreased.

Q-learning can be viewed as an actor-critic approach, where the actor is directly determined by the critic and the behavioral policy. Taking, for example, the soft-max policy, a positive δ_t (which is defined in this case with respect to Q-value estimates) results in an increase of the respective Q-value. This increase also increases the likelihood of executing the just executed action again, because the soft-max likelihoods directly depend on the Q-value estimates.

In the light of Q-learning, other actor-critic approaches may seem to be superfluous and tedious. However, especially when Q-values or state-value estimates, as well as the behavioral policy, are not represented in symbolic, tabular form, but rather by other generalized, approximate forms – such as by an artificial neural network – actor-critic approaches have often been shown to yield much faster learning progress than standard Q-learning approaches. In such cases, typically a policy is represented by generating a probability density over the action space and choosing an action according to this density. This is similar to the soft-max strategy, which distributes a probability mass of 1 over the discrete, possible actions. Several actor-critic based neurocognitive models suggest that similar learning mechanisms may indeed be at work in our brain when optimizing particular behavioral skills (Herbort, Ognibene, Butz, & Baldassarre, 2007; Lonini et al., 2013; Ognibene, Rega, & Baldassarre, 2006).

5.4   Policy gradients

So far, all RL techniques were based on value function estimates with which a behavioral policy was derived. Thus, policy learning is accomplished indirectly via the value function such that all these approaches rely on good value function approximations. Moreover, we have seen that when the state and action spaces grow, and even more so when they become continuous, learning good value function approximations takes increasingly longer. Remedies to this dilemma, such as value function approximations, factorizations, or hierarchical problem structuring exist, but it is unknown how to apply them in the general case. For example, function approximations by an artificial neural network are certainly possible, but they often suffer from unstable gradients and resulting disruptive fluctuations in the value function approximations. Factorizations can be helpful, but there is no algorithm known that is generally accepted as the one that is most suitable for detecting the features most relevant for estimating state-values or Q-values accurately. Hierarchical structures are definitely helpful, but a widely accepted algorithm for learning such hierarchical structures is also missing.

Policy gradients were developed as an alternative to traditional RL approaches, sidestepping the problem of estimating the value function. As the name suggests, policy gradients attempt to estimate a behavior policy-specific gradient directly. The gradient can be characterized as the direction toward a better behavioral policy given a current policy. By climbing the gradient, then, policy gradients attempt to optimize the policy iteratively to an optimal policy. Policy gradients do not estimate the value function at all. Rather, they optimize the behavioral policy directly by projecting reward-like feedback onto the behavioral policy.

5.4.1 Formalization of policy gradients

Policy gradients mostly focus on episodic RL problems, where the quality of an executed episode can be quantified by a reward value, which reflects the quality of an interaction episode. The example of the simple maze discussed previously can be viewed as such a problem, but typical tasks are found in continuous state and action spaces, such as steering a car, shooting or batting a ball, or attempting to optimally grasp an object.

In order to be able to derive a policy gradient from the quality feedback directly, it is mandatory to have a *parameterized policy*. Given some policy parameters $\boldsymbol{\theta}$ (for example, a vector of real numbers, that is, $\boldsymbol{\theta} \in \mathbb{R}^L$), the behavioral policy in policy gradients may be formalized as follows:

$$\pi_p : \mathcal{S} \times \boldsymbol{\theta} \to \mathcal{A} \tag{5.24}$$

$$\pi_p(\boldsymbol{\theta}) : \mathcal{S} \to \mathcal{A}. \tag{5.25}$$

The policy thus depends on parameters $\boldsymbol{\theta} \in \theta$ and determines an action for each possible input state $s \in \mathcal{S}$. A particular policy given a particular parameter vector $\boldsymbol{\theta}$ is the instantiation of the general policy π_p with that vector, denoted by $\pi_p(\boldsymbol{\theta})$. By thus having parametrized a policy, the gradient on the policy parameters $\boldsymbol{\theta}$ is estimated and used to develop progressively better actual parameterized policies $\pi_p(\boldsymbol{\theta})$.

To formalize this process, it is useful to define continuous MDP problems and to derive the parameter optimization from this definition. A continuous MDP may be defined by a real valued state space $\mathcal{S} = \mathbb{R}^N$ and a real valued action control space $\mathcal{A} = \mathbb{R}^M$. As in the discrete case, we can thus define a state-transition function, which essentially determines continuous probability densities given the previous state s and action a:

$$\mathcal{P} : \mathcal{S} \times \mathcal{A} \times \mathcal{S} \to \mathbb{R} \text{ with } \int_{x \in \mathcal{S}} \mathcal{P}(s, a, x) = 1, \tag{5.26}$$

denoting that the transition from one state $s \in \mathcal{S}$ given action $a \in \mathcal{A}$ to the next state $x \in \mathcal{S}$ is specified by a probability density over the (resulting) state space $\mathcal{S}$.

An interaction episode can then be viewed as a roll out τ of sensorimotor interactions, where each roll out may have a particular length H. The roll out then essentially consists of a sequence of states and actions, that is:

$$\tau = [s_0, a_0, s_1, a_1, \ldots, s_H, a_H]. \tag{5.27}$$

Given interaction experiences in the form of interaction episodes that are specified by such roll outs, the goal is to optimize the expected reward of behavioral interactions with the environment $K(\boldsymbol{\theta})$, that is:

$$K(\boldsymbol{\theta}) = E_{\pi_p(\boldsymbol{\theta})} \left(\sum_{k=0}^{H} a_k R(s_k, a_k) \right) \text{ with } a_k = \gamma^k, \tag{5.28}$$

with respect to the behavioral strategy parameters $\boldsymbol{\theta}$, which determine the behavioral policy π_p. Often in these cases, the discount factor γ is set to one, since usually the whole episode should be equally well optimized.

With these definitions, it is now possible to define the policy gradient **g** mathematically. The gradient is essentially the derivative of the expected reward function with respect to particular policy parameters $\boldsymbol{\theta}$:

$$\mathbf{g}(\boldsymbol{\theta}) = \nabla_{\boldsymbol{\theta}} K(\boldsymbol{\theta}) = \left(\frac{\partial K(\boldsymbol{\theta})}{\partial \theta_1}, \frac{\partial K(\boldsymbol{\theta})}{\partial \theta_2}, \ldots, \frac{\partial K(\boldsymbol{\theta})}{\partial \theta_L} \right)', \tag{5.29}$$

where the Nabla-operator ∇ yields the vector of partial derivatives with respect to the individual dimensions of the parameter vector. Assuming that we can estimate the gradient in some way (see Section 5.4.2), the behavioral policy can be adapted accordingly, by climbing the gradient, that is:

$$\boldsymbol{\theta} \leftarrow \boldsymbol{\theta} + \alpha \cdot \nabla_{\boldsymbol{\theta}} K(\boldsymbol{\theta}), \tag{5.30}$$

where $\alpha \in (0, 1]$ once again denotes the learning rate. Learning then proceeds by iteratively estimating the policy gradient, adapting the policy parameters accordingly, and repeating these two steps until a satisfactory policy is found, or at least until no further improvements are registered. In its general form, the algorithm looks rather simple:

1. Input: initialize policy parameters $\boldsymbol{\theta}$.

2. REPEAT

3. Estimate the gradient $\mathbf{g}(\boldsymbol{\theta})$

4. Change the current policy parameters by means of Eq.(5.30).

5. UNTIL no significant change in the policy parameters occur.

6. RETURN $\boldsymbol{\theta}$

The hardest part of the algorithm is the estimation of the gradient, that is, determining $\nabla_{\boldsymbol{\theta}} K(\boldsymbol{\theta})$.

5.4.2 Gradient estimation techniques

In closed form, the determination of this derivative is generally very difficult and requires a full mathematical description of the system (body, environment, and reward function). Therefore, the policy gradient literature has developed various approaches to estimate the policy gradient. Here, we only introduce the simplest form of estimation in further detail. This is the *finite difference method*.

All gradient estimation techniques typically rely on playing out roll-outs, that is, interacting with the environment in episodes, and observing the resulting reward value gained. When such roll-outs are played out with different policy parameters, the resulting different reward values obtained can be projected on the respective differences in the policy parameters, essentially adapting the parameters in that direction in which higher reward was gained. The algorithm that applies this intuitive method most directly is the *finite difference method* (for further details, cf., for example, Ijspeert, Nakanishi, Hoffmann, Pastor, & Schaal, 2013; Kober & Peters, 2011; Peters & Schaal, 2008).

Starting with some set of parameters $\boldsymbol{\theta}$, small changes in these parameter values may be denoted by $\Delta\boldsymbol{\theta}_i$ for one particular variation i. We can then evaluate the respective performances by simulating or acting out roll outs with these particular strategy parameter values (that is, $\boldsymbol{\theta} + \Delta\boldsymbol{\theta}_i$). With the respective reward values achieved in each roll out, we can then estimate the reward difference caused by the parameter variations:

$$\Delta\hat{K}_i(\boldsymbol{\theta}) \approx K(\boldsymbol{\theta} + \Delta\boldsymbol{\theta}_i) - K(\boldsymbol{\theta}) \tag{5.31}$$

With the help of these difference estimates, it is then possible to approximate the parameter-respective gradient $\mathbf{g}(\boldsymbol{\theta})$:

$$\mathbf{g}(\boldsymbol{\theta}) = \left(\Delta\boldsymbol{\Theta}^T \Delta\boldsymbol{\Theta}\right)^{-1} \Delta\boldsymbol{\Theta}^T \Delta\hat{\mathbf{K}}(\boldsymbol{\theta}), \tag{5.32}$$

where $\Delta\boldsymbol{\Theta} = (\Delta\boldsymbol{\theta}_1, \ldots, \Delta\boldsymbol{\theta}_i)^T$ specifies the applied parameter variations, and the respective differences in reward outcomes are denoted by $\Delta\hat{\mathbf{K}}(\boldsymbol{\theta}) = \left(\Delta\hat{K}_1(\boldsymbol{\theta}), \ldots, \Delta\hat{K}_i(\boldsymbol{\theta})\right)^T$.

This method is easy to apply. However, experiments with it have shown several drawbacks. First, the respective gradients may overshadow each other, thus climbing down one gradient (the steeper one) while almost completely ignoring an equally important gradient, which may be shallower at the current policy parameter subspace. Second, the resulting parameter updates based on equation (5.30) do not consider varying the learning rate, although it is well known that this can yield a much better performance. Finally, the sampling of the parameter subspace around θ is by no means trivial. If parameters are sampled very close to θ the system may easily get stuck in local optima without any gradient information in the local area. On the other hand, if parameters are sampled too far away from θ, then the sampling may jump over optima and possibly even the global optimum, thus overlooking further options for behavioral parameter optimization.

As a result of these observations, during the last decade advanced policy gradient algorithms have been developed. These algorithms approximate the gradient in a more robust fashion and some of which also optimize the sampling around the current best policy parameters. The *likelihood-ratio* method does so by estimating likelihoods of improvements. The *natural policy gradient* method improves the sampling by changing the adaptation of the behavioral strategy parameters dependent on the number of successes that have been encountered in the respective parameter change directions. Interestingly, the latter method has been shown to be very closely related to the *covariance matrix evolution strategy*, which we had shortly touched upon in the evolutionary computations method section in the previous chapter (cf. Section 4.4.1, "Genotype variations"). In fact, two avenues of scientific investigation developed very similar optimization techniques, one focusing on general optimization problems and the other one coming from the RL side, motivated by the challenge to optimize behavior in robotics. Natural policy gradients is in fact the typical choice when a particular behavioral routine or behavioral primitive needs to be optimized on a robot platform.

5.4.3   A racing car example

Let us consider an example where policy gradients are very well suited. A typical task, for example, could be to teach a car to drive around a race track as fast as possible without crashing. The car has no global map information but it drives along equipped only with fast laser sensors, which scan the area ahead. The sensors return the longest distance registered along the track ahead as well as the angular direction of this distance. Figure 5.3 shows an example of such a scenario.

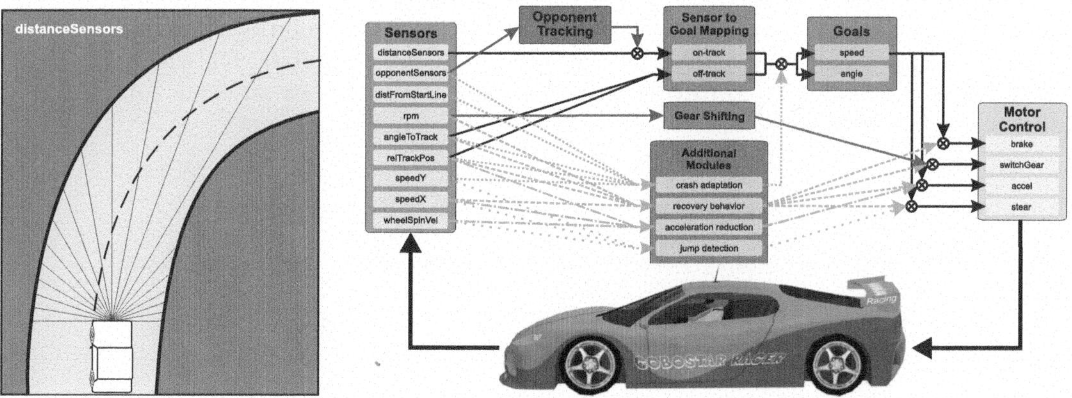

Figure 5.3: Policy gradient example. In a racing car simulation, the racing car is equipped with suitable sensors. The control architecture illustrated on the right was successfully used in several simulated racing car competitions. [© 2011 IEEE. Reprinted, with permission, from Butz, M. V., Linhardt, M. J., & Lönneker, T. D. (2011). Effective racing on partially observable tracks: Indirectly coupling anticipatory egocentric sensors with motor commands. *IEEE Transactions on Computational Intelligence and AI in Games, 3*, 31–42.]

To apply a policy gradient method it is first necessary to define a behavioral policy π_p that maps observations onto actions. For example, one may aim to develop a function that maps the longest free distance and direction information from the laser sensors onto a desired velocity and a steering angle. The desired velocity may then result in a velocity increase or decrease, using the gas-pedal and the brakes accordingly. This latter mapping to actual motor output may even be hard-coded.

Various parameters can now be defined that may influence the behavioral policy. To determine the desired speed, a speed offset value may specify a minimum speed the car always "wants" to drive, a value that linearly maps the registered distance onto the desired speed (the further ahead the next barrier the faster the car should drive), and possibly a value that considers the angle and maps it onto the desired velocity as well (the more to the side the longest distance is, the slower the car should take the apparent curve). Another parameter may influence the steering, steering more to the right/left the more the registered direction lies to the right/left, respectively. Figure 5.3 shows the architecture from (Butz, Linhardt, & Lönneker, 2011), which was developed in 2010 and successfully won several simulated racing car competitions.

Next, we need to specify how the behavioral policy may be evaluated. The policy should be attempted on a range of race tracks, which reflects the types of tracks for which the policy should be optimized. Specific roll-outs should then, for example, allow the policy to race on a track for, say, 5 minutes. The reward function may then reward those strategies that cover the longest distance, but it may also punish the strategy if it caused the car to crash. Clearly, the strength of reward and punishment matter – if the punishment is severe when compared with the reward achieved by the distance, soon very passive strategies are likely to develop – avoiding all crashes, but driving slowly. The development of progressively faster strategies will then likely take a while. On the other hand, if the punishment is mild, a highly aggressive strategy may develop, which may however fail on many other tracks because it crashes too often.

With this example we see that it is far from easy to generate a good policy gradient approach. First, the actual behavioral policy definition is a crucial design choice: which sensory information about the world should be considered? How should the sensory information be abstracted before using it in the behavioral policy? Which mapping should be used to map the processed information onto motor commands? How flexible should the involved mapping functions be (for example, constant, linear, polynomial, exponential, etc.)? Thus, how flexible should the behavioral policy be with respect to its modifying parameters? The choice of the reward function is also critical to guide learning toward an optimal solution, where the reward function can be viewed as a fitness function in relation evolutionary computation (cf. Section 4.4). Finally, the actual roll outs need to be done carefully. On which tracks should the strategy be evaluated? If there is randomness involved, how often should the strategy be run before considering the reward estimate for the policy gradient?

5.4.4   Conclusions and relations to cognition and behavior

Policy gradients have been derived from RL techniques. However, it may be debated how closely related they are to them.

As we have seen, in policy gradients many design choices determine the success of the approach. First, the policy itself needs to be defined, that is, which mappings from state information to action are computable with the chosen policy π_p? In all other RL approaches these mappings are typically much less constrained. Second, the parametrization of the policy, the initial parameter values, and their initial ranges need to be specified with care to prevent overly large, as well as overly small, policy parameter variations. Third, the evaluation, that is, the roll-outs, need to be defined and the quality of a roll out needs to be characterized by a proper reward-like function. Finally, the optimization technique needs to be chosen.

Policy gradients thus rely not only on a good optimization algorithm, but even more on good definitions of a behavioral policy, of behavioral episodes, and of a reward function. In contrast to Q-learning, but similar to evolutionary optimization approaches, policy gradients cannot be guaranteed to converge to a global optimum. With a proper optimization technique, however, at least the convergence to a local optimum can be guaranteed under certain circumstances.

A critical aspect in policy gradients is the choice of the initial behavioral policy, including its mathematic formalization and parametrization. *Dynamic movement primitives* (DMPs) have been proposed and applied in numerous robotics behavioral optimization tasks successfully (Ijspeert, Nakanishi, & Schaal, 2002; Ijspeert et al., 2013). DMPs are essentially artificial *central pattern generators*, which can generate either an attractor dynamic toward a final state or a cyclical dynamic attractor. As a result, DMPs have been successfully employed when optimizing a particular motion sequence - such as hitting a ball – and when optimizing a dynamic motion pattern, such as walking stably (Ijspeert et al., 2013; Kober & Peters, 2011; Schaal, Ijspeert, & Billard, 2003).

Most interestingly, recent theoretical efforts have shown that policy gradients are closely related to CMA-ES evolution strategies (cf. Section 4.4.1, "Genotype variations") in that the gradient estimation and sampling techniques function nearly identically (Stulp & Sigaud, 2013). Moreover, further investigations along these lines have shown that DMPs dramatically simplify the search space and thus the policy optimization problem.

With properly designed, easily optimizable behavioral policies and proper gradient estimation techniques, policy gradient applications have shown much success in robotics applications, including batting a baseball, hitting a table tennis ball effectively, driving a simulated racing car, or grasping objects with a robotic arm. Due to the focus on particular behavioral options defined by the parametrized policy function π_p, learning needs to search through a much smaller problem space, enabling faster learning and convergence. When facing a large parameter space $L >> 10$ and highly flexible policy functions π_p, however, policy gradients will not be successful – or, at least, will take a long time to converge to some local optimum.

When viewing the insights gained by research on policy gradients in the light of cognitive science, cognition, and cognitive development, various associations spring to mind. Infants are equipped with various reflexes, which bootstrap further cognitive development. The grasp reflex especially may not only be seen in the light of evolution as a remainder of the time where the infant had to hold onto the mother tightly while being carried around, but also in the light of ontogenetic development: equipped with the grasp reflex, infants will grasp objects initially nearly by accident, but then soon learn that different objects behave differently when grasped and that grasp variations may be particularly suitable for particular objects. Thus, the motion primitive "grasp reflex" may be optimized by a policy gradient-like algorithm to make manipulation more effective and object-specific. Reward for optimizing different grasps may then come in a positive form from the gathered experiences while manipulating an object, and in a negative form when object manipulation fails and the object is lost. Thus, the grasp reflex may be compared with a DMP that is optimized (and differentiated) during ontogenetic development for accomplishing particular object manipulation tasks.

In fact, neuroscientific research insights suggest that simple *central pattern generators* are pre-wired in spinal cord networks (Hultborn & Nielsen, 2007). These generators are able to generate dynamics, such as a rhythmic activation and deactivation of muscle groups, inherently, that is, without the need for external rhythmic stimulation. As a result, such central pattern generators can be closely related to DMPs and can be assumed to foster the development of crawling and walking, once the body's morphology is sufficiently developed. Similar to the grasping case, also for developing locomotion capabilities it seems rather straight-forward to identify the reward function, which is to successfully bridging space while maintaining stability. Moreover, the task is episodic beginning with the goal and the motion initialization and ending when the targeted location in space is reached or locomotion failed. Thus, policy-gradient-like optimization principles are applicable.

Even speech development is supported by infants starting to babble in the first year of life – apparently experimenting with generating speech sounds and probably optimizing that generation to the sounds produced by their caretakers. Thus, our body and brain seems to be evolutionarily equipped with motor programs that are optimized in a manner closely related to contemporary policy gradient techniques. Higher level planning and reasoning, on the other hand, seems to be more closely related to model-based and hierarchical, factored RL techniques.

5.5 Exercises

1. Relate the Rescorla–Wagner model to temporal difference learning.

2. Construct an episodic simple maze with five states and one reward state. Derive the optimal Q-value and state-value functions and simulate some iterations via Q-learning.

3. Imagine a blocks world with N available stacks and N blocks. Assume furthermore that the world is episodic and the blocks are indistinguishable. Finally, assume that transport actions are available that can transport a block from one stack to any other stack.

    (a) How many possible states does this world have?

    (b) How many actions are possible?

    (c) What could a good problem representation look like?

    (d) Imagine a kind of visual, grid-oriented problem representation with $N \times N$ binary grid positions, which indicate the presence or absence of a block. Why is this problem representation not very suitable?

    (e) Given that the goal is to transport all blocks onto the first stack. How may a system without any state access still solve the problem?

    (f) Specify a maximally compact representation of a value function:
    - when the goal is to transport all blocks onto the first stack;
    - when the goal is to reach a particular block constellation.

    (g) Discuss why "factorized" value functions, that is, value functions that focus on the goal state and its difference to the current state, are particularly suitable in such blocks worlds.

4. Humans are very good at focusing on those aspects of the environment that are currently behaviorally relevant. In which way is this behavior related to state factorizations in RL.

5. Proof that the distribution of rewards by means of eligibility traces equals to 1.

6. In which manner does DYNA-Q combine dynamic programming with temporal difference learning?

7. Contrast the $\epsilon-$greedy behavioral policy from the soft-max strategy.

8. Why does an intrinsically motivated RL agent typically learn faster than a randomly exploring agent?

9. Why is the development of the grasp reflex into pincer, power, and scissor grasps more complex than the optimization of a dynamic motion primitive?

10. Construct a potential mapping function between the distance sensor signals and the target speed in the car racing problem. Choose a maximum of three parameters that may be optimized. Explain the chosen mapping function.

11. In which manner are central pattern generators in our brain related to dynamic motion primitives?

Chapter 6

Behavioral Flexibility and Anticipatory Behavior

6.1 Introduction

In the last two chapters we became acquainted with several optimization and learning processes. When reconsidering these algorithms in the light of cognition, it soon becomes obvious that none of them on their own can lead to the development of higher-level cognitive processes. In essence, none of the mechanisms considered so far can foster a form of "understanding" of body and environment.

Evolution is essentially "blind", meaning that it does not consider explicitly any estimates about how genetic changes may affect the actual phenotypic organism. Evolution optimizes by means of the interplay of the principles of survival of the fittest and genetic variation and recombination. While thus clever behavioral capabilities have evolved, in implicit anticipation of the environmental circumstances that the organism will typically face, no explicit considerations about the future take place. Forms of "understanding", however, typically involve the capacity to simulate a process or reconstruct a process, thus enabling explicit predictions about the process's behavior. Evolution can thus be said to have no "understanding" of what it is actually doing. Note, however, that evolution may and apparently has developed genetically-encoded learning architectures (that is, brains), of which at least some of them enable forms of understanding. The question is, which mechanisms enable such forms of understandings?

Model-free RL and policy gradients do not have any representation about the actual consequences of their actions other than reward predictions. Such systems may very well be able to optimize their behavioral repertoire to the encountered circumstances. The behavioral repertoire itself, however, comes from the designer or, in biological systems, from the evolutionarily shaped body and the ontologically developing reflexes. The systems and organisms solely equipped with such mechanisms may be highly effective in reaping particular energy resources in the world, thus establishing themselves as an effective species. However, as that behavior is optimized solely based on estimations of policy gradients, actor-critic based gradient derivations, or the Q-learning mechanism, once again no real "understanding" about how the world actually works is present.

As we discussed in the last chapter, model-based RL mechanisms use a model about their world to reason and plan goal-directed actions. Model-based hierarchical, factorized RL approaches may be the most powerful ones to optimize behavior reward based. Such mechanisms do have a model and thus the capacity to simulate events and situations, which may be equated with forms of understandings. However, we have not addressed where such models come from except that they may be learned by gathering experiences about sensorimotor interactions. In this chapter we focus on the principles behind learning predictive models of body and environment beyond reward prediction models.

Another consideration motivates this chapter in a related, but different manner. All the mechanisms introduced so far are able to optimize and adapt behavior as a result of reward-based experiences. What happens, though, when the environmental circumstances change? What if my dominant arm is in a cast or I am holding something in my hands and thus cannot open a door in the usual manner? What if the store closes where I usually go shopping? Surely we are still able to interact with the world effectively. We may use the non-dominant arm. We may use the knee to open the door, or temporarily put-down the things we hold in the hand and then hold the door open with the foot. We will usually not walk to the closed store more than once, but look for an alternative.

This knowledge about alternatives, and thus the flexibility to adapt behavior in a one-shot manner to the available alternatives is hardly present in RL. Value functions may signal alternatives to a certain extent, but the adaptation of a value function to a new goal typically takes a significant amount of time. Policy gradients are possibly the least flexible mechanisms, only focusing on the optimization of one particular behavioral interaction. Only some knowledge about alternatives makes it possible to choose between the currently available alternatives and to quickly adapt behavior when the dominant alternative is currently not available.

An important additional aspect when considering the concept of alternatives is the fact that our bodies – and also those of many other animals for that matter – offer partially redundant and partially complementary means to interact with the world. We can, for example, identify objects, such as our keys, by seeing them, naming them, touching them, and often even when hearing them being touched or when hearing them fall onto the floor. We have thus a manifold of sensors that give information about particular things and aspects of the world, and all these redundant sources of information encode aspects of an object. In addition to being redundant in that one alternative often suffices to identify an object (or any other aspect of the world), the redundant alternatives complement each other. Usually, two information sources about an object allow us to increase our certainty about the state of the object.

Interestingly, a similar situation regarding alternatives can be identified with respect to behavior. We can walk, hop, stride, do a silly walk (cf. Monty Python's Ministry of Silly Walks), run, strut, etc., to reach some destination. We can grasp an object with one hand, two hands, or sometimes even with a foot or the mouth. We can even communicate in various ways, for example, by means of pointing, pantomiming, writing, and of course speaking, but also laughing, screaming, and crying, to name only a few possibilities. Again, the alternatives are somewhat redundant and somewhat complementary. Sometimes it seems to hardly matter which redundant alternative is chosen to accomplish a certain task. In other cases, however, it may be that circumstances lead to the choice of a very distinct alternative, but typically this alternative will not always be the same.

Note how in the previous paragraph we have touched upon bodily behavior, as well as communicative behavior in a similar manner. In social interactions, including communication, particularly many behavioral alternatives are available. Assuming that humans evolved at least partially due to the evolutionary advantage of cooperation, the human brain may be particularly well capable of considering the thoughts of others and to "tune into" these thoughts to make cooperation as effective as possible. Alternatives thus exist not only about our own behavior, but also about the behavior of others, potentially opening access to completely different thoughts and ideas. We will discuss these considerations in more detail in the final chapters of the book.

Knowledge about alternatives, however, cannot come from nowhere, but must be learned by experience. An important concept along these lines is the *ideomotor principle* of motor development and, for this matter, also of cognitive development. The principle essentially states that our brains learn from the sensorimotor experiences gathered while interacting with the world, and that the interactions very soon become goal-oriented, choosing behavior by means of the desired and anticipated effects, that is, the current goals. Goals have a dual characteristic: they are desired, that is, they have some reward associated to them, and they are anticipated, that is, they seem to be achievable. Given sensorimotor knowledge, we show how goals can be chosen based on an internal motivational system, which gives potential goals their reward-associations.

To summarize, this chapter emphasizes that knowledge about redundant alternatives for interacting with and manipulating the environment holds the key for developing higher levels of cognition. To be able to choose among alternatives, goal-directed behavior needs to be possible, flexibly choosing among currently achievable and desirable goal states, given the current state of the system. By describing a general cognitive architecture that can yield self-motivated, curious cognitive systems, we show how RL principles can be combined with flexible, goal- and reward-oriented mechanisms, developing encodings that further the understanding of the functionality of the experienced world.

In the reminder of this chapter, we first take a look at how flexible adaptive behavior may have evolved, and which redundant and complementary alternatives we, and also other animals, have to perceive and interact with the world. We will see that only when redundant alternatives are available does it make sense to develop forms of understanding. Next, we will take a closer look on how knowledge about redundancies may develop and how it may be employed to act goal-directedly. Finally, a basic functional motivational system is suggested that can result in goal-directed action choices based on the learned sensorimotor knowledge.

6.2 Flexibility and adaptivity

Behavior and cognition in humans can be considered to be driven by flexible and highly adaptive processes. Humans have managed to survive in remote areas of this planet and have consequently spread out to nearly all inhabitable locations on earth. However, where does this capacity to act flexibly and adaptively come from? Are similar abilities present in other animals? What can be said about redundant versus complementary sensory and motor capabilities?

6.2.1 Niches and natural diversity

When taking a closer look at evolution once again, it soon becomes apparent that evolutionary niches have fostered the evolution of diverse species, each one equipped with particular sensory, motor, and behavioral capabilities to ensure survival and reproduction. As a result of the interaction of individual species, a highly complex ecosystem has evolved, where different species not only compete with each other, but often complement each other. For example, our digestive tract could not work without the help of trillions of bacteria that help our body to digest our food. In biology, there are many more such examples, which are typically referred to as symbiotic interactions between different organisms yielding mutual benefits. However, besides these positive, symbiotic interactions, negative, parasitic interactions are also ubiquitous. Thus, there are cooperative, as well as competitive interactions that drive the evolutionary process.

Besides these interactions, however, evolution has also managed to evolve highly specialized species for very particular environmental niches. It is not necessary to consider animals as bizarre as the aye-aye in Madagascar to see this. A look at the animals found in different regions of this world soon makes it obvious that each animal is equipped with bodily, sensory, and motor capabilities that are particularly well-tuned to the surrounding in which it lives. Evolutionary advantages over other species and individuals within a particular species typically need to be viewed or analyzed in light of particular properties of the ecological niches within which each individual lives. These include:

- The *physical properties* of the environment, including gravity, dynamics (for example, water versus air), temperature, of terrain properties.

- The *ecological, environmental properties*, including the availability of resources, such as food and water, and the presence of other species with their means to consume energy.

- The *bodily properties* of the individuals including their sizes, as well as their behavioral and sensory capabilities.

- The *competitive properties*, such as being predator or prey, competing for scarce resources on either side, the hunting abilities of the present predators, and the hiding, fleeing, and defensive abilities of the prey.

Each individual strives to optimally reap the resources that are available and suitable. Because various resources are typically available in a particular niche, a large diversity of species has evolved that complement each other in reaping the available resources somewhat optimally. This *diversity* in the different species is very important to foster life in its present form. Ecological niches are worked upon by different species and thus by different mechanisms, continuously processing and changing energy resources in various manners. As a result, highly effective biological "super-organisms" are at work in the world, fostering life, including its further evolution.

Besides across species, diversity helps also within a species to improve the chances of survival of that species. Particularly when considering diseases, and especially resistances against diseases, it has been shown that sufficient genetic diversity improves the likelihood that new resistances can evolve and that a species cannot be fully wiped-out by a particular disease.

Let us consider two examples of niche properties and how they are mastered. First, consider locomotion: plants do not posses the ability for locomotion. They do not have brains or even a rudimentary form of nervous system – most likely because without the ability of locomotion and environmental manipulation, brains are of not much use. Animals, on the other hand, all have a kind of nervous system or at least a behavioral coordination system. The challenges for realizing locomotion indeed appears to require some form of coordination. In simpler forms of animals, this coordination is often realized by highly simplistic control mechanisms, very much similar to a Braitenberg vehicle (cf. Section 3.6.2).

Locomotion not only requires bodily motion, but coordinated bodily motion, which also depends on the body's complexity. The body needs to be held somewhat stably, while legs, fins, wings, or other suitable means are moved in way to cause forward propulsion. Thus, not surprisingly, even when only considering legs, many forms of locomotion have evolved, including jumping, running, crawling, climbing, swimming, diving, digging, etc., as a result of the different surroundings within which the respective animals have evolved.

Second, consider speech – social communication by means of a complex auditory communication system can be found in its most complex form only in humans. The challenge is to actually successfully communicate, that is, to transfer information from one individual to the other one by means of *speech acts*. This niche is the social communication and individualized cooperation niche, because we benefit from mutual interactions, information exchange, and effective collaboration for achieving certain tasks (cf. Chapter 13). In order to enable communication, however, the auditory and speech production systems must be sufficiently evolved and must have enough time to develop during ontogenetic development. Thus, a mutual process must have been and still is at work when learning to communicate socially. The result is a lot of flexibility in communicative options, enhanced by grammatic and syntactic principles of human languages, in addition to the environmentally grounded, embodied semantics.

Before concluding this subsection, it is worthwhile to consider the idea of niches and flexibility beyond natural evolution. In our modern world, cultural and economic niches develop, and are continuously in flux. Merchandizing is an obvious example, where economic niches temporarily emerge and then sooner or later disappear, given the rise and fall of a popular movie or computer game. When considering robotics, niches develop dependent on the current and anticipated available robotics technologies, as well as the estimated future demands for such robots. The autonomous car, for example, appears to be very appealing to many of us, so economic evolution invests more money in the development of such technologies, compared with others that seem less appealing. Thus, cultural and economic evolution partially undergo directed evolutionary progressions, which may, however, also lead to local optima.

While the last two paragraphs have focused on communication and intelligent systems, it should not be forgotten that evolution is inevitably embodied in the environment within which it takes place in the first place. In contrast to traditional AI and many current robotics approaches, our brains do not seem to compute exactly what the body is to do next. Approximations and generalizations are at work. The world, as its own best model, is only probed on the fly where and when necessary, focusing on those aspects of the world that seem to be currently behaviorally relevant. Bodily morphologies also support the development of complex control processes, such as walking or grasping.

Thus, behavioral flexibility viewed from an evolutionary perspective can be found in various manners in different species, within a species, and even on cultural, economic, and probably also scientific levels within human cultures. All the niches – and the developed behavioral flexibilities within each niche – exist only due to tight couplings between the niche, the environmental circumstances, the species with their embodied minds and their sensory and motor capabilities, and the interactions between and across the involved species. In the following, we focus on the development of behavioral and cognitive flexibilities within individuals of a species. To do so, we first consider some fundamental insights from cognitive psychology. Then we focus on how redundancies and complements actually can help humans to interact so flexibly and adaptively with the environment.

6.2.2   Beyond behaviorism

In Chapter 5, we mainly focused on behaviorism as the precursor that led to the development of RL and the understanding that animals are capable of adapting their behavior reward orientedly. Behaviorism essentially reduced cognition to a minimum, and focused on direct couplings between stimuli and responses. At the time, this approach was certainly helpful, as there was no room for introspective ideas about how cognition may actually work. In its radical form, behaviorism completely reduced behavior as adapting to reward and punishment alone, denying any other forms of control systems or inner, mental states.

Seeing that this radical form is too narrow-minded to account for our behavioral and mental flexibilities, even during the zenith of behaviorism important insights suggested that there is more to animal behavior than mere reward-based adaptations. Various research results suggested that internal states and forms of knowledge need to be assumed to be able to explain all behavioral observations made with animals including insects. In particular, the results suggested that animals appeared to have expectations about concrete action outcomes, far beyond reinforcement estimates.

We have already heard about Edward C. Tolman and his groundbreaking experiments with rats in T-mazes (cf. Section 2.4). These experiments clearly showed that rats appear to learn a map of their environment even without the provision of an explicit reward. Once a particular location in the maze became relevant, the rats that had explored the maze before without the provision of reward could outperform other rats that had received reward from the beginning. Due to these and similar observations, Tolman fostered the term *expectancies* for explicit forms of expectations about the consequences of actions. The latent learning of cognitive maps in rats is one such example where expectancies are formed and later used to adapt behavior on the fly. However, there are certainly many other situations where forms of expectancies are learned without the provision of reward.

Even Rescorla conducted experiments – after having published the Rescorla–Wagner model – which suggest that rats learn far more than mere reward expectancies (Colwill & Rescorla, 1985, 1990). The results showed that rats do learn more than just reward value estimates with an object manipulation paradigm. By satiating a rat with sugar water, for example, the rat selects the behavior that previously has led to receiving food pellets, and vice versa. In this way, Rescorla demonstrated that the rats learned context-dependent response–outcome expectations and that these expectations co-determined which behavior was chosen. Thus, at least in rats, it has been shown that flexible behavior comes from the ability to decide on a behavior dependent on the current goals and their associations with the currently active context-dependent response–outcome associations.

By now there probably exist thousands of examples of particular animal behavior that exhibit latent learning, knowledge about behavioral outcomes, and explicit goal-oriented behavior, where the behavior is generated due to the desired outcome of the behavior. In the following, we further explore how such goal-directed behavior can come about.

6.2.3   Redundancies and complements

While natural evolution has brought about many species that exhibit a large variety of body morphologies and behavioral capabilities, many of them do not appear to be particularly flexible. However, in rats and many other mammals, as well as specific species of birds, and even in octopi, flexible behavior was observed that can be termed innovative for solving a particular problem. From an evolutionary perspective, one can say that evolution has evolved behavioral capabilities that go beyond reward-driven optimization and inborn, morphological behavioral patterns. These cognitively-driven behavioral flexibilities allow online behavioral adaptations to the current bodily and environmental circumstances.

If we had only one means to manipulate our environment in a certain manner, however, flexibility would be impossible. To enable online adaptations, redundant and partially complementary alternatives must be available. A fundamental distinction in this respect can be drawn between redundancies in perception and action.

Redundant and complementary perceptions

Our sensory systems provide information about our individual bodies and the outside world. Redundant sensory systems thereby provide alternative means to access particular forms of information. For example, when putting down a glass one may visually monitor this action and thus successfully position the glass on the table. Alternatively, one may not look at all and still succeed by focusing on the registered tactile and proprioceptively perceived feedback about the interaction, such as tactilely registering the impact of the glass on the table, as well as the stability of the glass after being put down. Auditory information is also very helpful in this respect, as contact with the table typically produces a distinct event-specific sound. While sensory information is thus partially redundant, offering several alternatives for registering the same event or recognizing the same object, the information sources are also complementary in that several sources of information about the same event typically increase the certainty that the event has actually occurred.

To be able to effectively integrate several sources of information about an object, an event, or about any other property of the surrounding, however, computational mechanisms need to be available that enable the *binding* of these different sources of information. Particularly when considering different sensory systems, the individual bits of information are typically registered in different frames of reference, such as the retinotopic frame of reference of the eyes, the body-surface topology provided by the skin, the auditory topology from the ears, or the muscular topology. Binding is only possible if these different frames of reference can be properly related to each other, expecting that the different information sources are about the same cause, which generates the respective sensations.

On top of this sensory binding problem, the reliability of the different information sources also needs to be taken into account. False information should be identifiable and should thus be ignored or at least devalued. Complementary information, on the other hand, should be fused leading to information gain and thus a more precise estimate about the observed situation. Due to these cross-modal interactions, abstractions about the actual sensory information and sensory-grounded frames of reference into an abstracted and generalized template representation should develop.

For example, when putting down a glass and needing to know exactly when the glass touches the table, the tactile information may be considered to be most reliable. However, when disruptive tactile information occurs or which it cannot be registered at the moment, because of, for example, currently wearing thick gloves, it may be deemed unreliable and other information sources may be considered. Because the impact of the glass on the table can also be registered auditorily as well as by the arm joints due to a significant change in force dynamics, an abstract encoding about establishing contact between two things can develop. This representation will, for example, encode that when an object starts to touch another, typically (i) some auditory signal can be registered; (ii) the two objects appear to be visually very close to each other; (iii) one object can be reached from the other one quickly; and (iv) it can be expected that both objects will be touched when approaching them close to the area of contact. Finally, (v) the result of the touch, that is, the object impact, may be predicted, distinguishing, for example, repulsion, adherence, and moving along. Moreover, when establishing the touch by means of the own body, tactile and proprioceptive touch feedback can be expected.

When the task is to identify an object, such as a glass from which we may want to drink, we can look at it, touch it, knock at it, lift it, taste it (like babies love to do), smell it, and so forth. Visual information may often be sufficient to succeed in such identification processes. However, congenitally blind people, who never had access to visual information, are well able to identify objects by other means. Thus, object identification can be accomplished by various, redundant means. When object identification is accomplished by combining multiple redundant or complementary sensory information sources, as is typically the case, the different information sources need to be temporarily bound together and, to optimize sensor fusion, should be combined taking their estimated information reliabilities into account. If a particular source of information is not available at a certain point in time, it can be easily substituted by other information sources. If the particular source is very noisy at the moment (such as vision when ones glasses are fogged up), it can be complemented by redundant and complementary other available sources.

Thus, perceptual redundancies enhance our knowledge about the world in various respects:

- Features of the environment can be perceived more accurately by fusing multiple sources of redundant, independent sources of information, resulting in information gain.

- Temporarily unavailable sensory information can often be substituted by redundant alternatives.

- Entities in the environment can be perceived in various ways and in various detail – enabling the choice of the currently most-informative alternative where possible.

- The development of abstractions into multimodal, integrative encodings are fostered when learning to bind different information sources temporarily with each other, as they currently provide complementary or redundant information about the same stimulus cause.

As a result, by knowing about redundant and complementary alternatives to perceiving aspects of our world, an organism is able to interact with it in a much more flexible manner because it can perceive the world under various circumstances, can confirm its current perception using alternatives, and can even choose to perceive the world in a certain manner, depending on the circumstances.

Redundant and complementary motor activities

While perception is mostly about gathering information about properties of and entities in the environment, such as our own body, objects, free and blocked space, etc., the motor system is used for manipulating the environment. As different circumstances require different motor actions to maximize the probability of success of achieving a particular goal, redundant and complementary motor activities enable the flexibilization of environmental interactions. In the simplest case redundant alternatives are provided by our different extremities, such as our two hands. In more complex cases, however, different motor programs may be activated to accomplish a certain task – think about the locomotion example mentioned previously. However, object manipulations are not always accomplished in the same manner, suggesting that the brain chooses among available behavioral alternatives.

An interesting observation in this respect is the fact that humans – and also several other animals – exhibit anticipatory, end-state oriented behavior when manipulating objects. For example, we may grasp a glass that currently sits on the table upside down with a $-180°$ rotated hand to be able to rotate it right-side up and, consequently, be able to pour a drink (cf. *end state comfort effect* in Section 12.3.2 and Figure 12.7).   On the other hand, if we want to put the upside down glass into the dishwasher, we may grasp it from the top. Finally, if we want to put it on the top shelf, we may use the standard grasp orientation and transport it upwards accordingly. This example illustrates another form of redundancy that has not been mentioned thus far: the fact that our body is equipped with redundant degrees of freedom. Each one of our two arms, for example, is equipped with seven degrees of freedom – three in the shoulder, one in the elbow, one in the forearm (enabling us to rotate the hand between a prone and supine position, that is, the palm facing downwards or upwards), and two in the wrist. To reach a point in space, however, technically speaking only three degrees of freedom are necessary. To reach a point in space with a particular hand orientation, six degrees of freedom are necessary. Thus, the seven degrees of freedom offer redundant alternatives to manipulate the environment, and give us great behavioral flexibility.

While in a very dull and never significantly changing world, one could always accomplish the same manipulation with the same motor sequence – as, for example, most current industrial robots do. Our real world is much more complex than that, though. Circumstances change all the time and some of these changes require behavioral adaptations. Bodily changes can be relevant in this respect, including muscle fatigue as well as limb or joint unavailabilities, due to, for example, injury. More obvious, though, are other changing environmental circumstances, such as the presence of obstacles, tool properties, or object types. All such circumstances require the versatile adaptation of the behavioral system – and the more elaborate the knowledge about behavioral alternatives, the more versatile and dexterous behavior can become.

However, there are some costs involved. Studies indicate that the brain does not fully consider all currently available behavioral alternatives. Rather, habitual behaviors dominate our interactions with the environment and these behaviors are only modified when necessary. In this way, behavioral interactions often do not seem to be fully behaviorally optimal, but mostly suffice to succeed in the intended environmental interactions. Moreover, the learning of such behavioral alternatives makes life initially harder. It comes as no surprise that humans are the animals where the child typically stays longest with the parents. Our children have to learn to manipulate their environment in a highly dexterous manner, using, for example, the same tool in various ways to accomplish particular tasks – or using the same set of blocks to build completely different objects – or using the same communicative means, for example, speech, to communicate a seemingly infinite variety of ideas and thoughts. Thus, redundancies, including perceptual redundancies, need to be fine-tuned, learned, and differentiated over an extended period of time.

To summarize, robust sensor-based perception as well as motor control rely on knowledge about sensory and motor alternatives, including their partially redundant and partially complementary status. Perception thus becomes more fail-safe and accurate. Behavior becomes more versatile and flexibly goal-oriented. When generalizing behavior to communicative acts and abstract thought, alternatives play a crucial role as well. For example, when empathizing with, or when attempting to understand the view-point, of another person, we are forced to consider alternative interpretations and particularly those that we think the other person is most likely to have about a particular situation. Thus, redundancies and complementaries are highly important when considering social interactions, cooperation, competition, and communication. We will re-address these points in later chapters in further detail.

6.3 Sensorimotor learning and adaptation

To develop behavioral flexibilities and to know about alternatives for perception and action, learning needs to take place. How and when does this learning commence? This is a question that has been addressed by many researchers over the last two centuries and has led to the formulation of the *ideomotor principle* (*ideo* is Greek and means "idea", or mental representation, thus "idea-based motor principle").

We introduced the ideomotor principle in Chapter 2, but here we discuss it in the context of redundancies and behavioral flexibility. As the reader my recall, the ideomotor principle essentially postulates that learning starts with self-generated reflex-like behavior and the registration of the sensory effects caused by the behavior. In consequence, first sensorimotor contingencies are learned. Soon thereafter behavior becomes progressively goal-oriented, where the sensory effects of motor actions are the goals that cause the execution of the associated motor activities. Thus, starting with simple reflexes, the ideomotor principle proposes that predictive, sensorimotor structures are learned and that these structures can be used to choose actions goal-oriented, by choosing to execute those actions that have previously generated the currently desired effects.

In contrast, classical artificial intelligence and classical cognitive science approaches have assumed that sense→think→act cycles unfold in our brains. When "sensing", sensory information is processed and integrated into abstract knowledge. This knowledge is then used to "think", plan, and to make behavioral decisions, which are then acted out in the "act" stage. After that, the cycle repeats, sensing again the next state of affairs in the environment. This cycle is not only very imprecise, it neither solves the symbol grounding problem nor the frame problem and it is too slow to be plausible. Our behavior needs to be much more flexible than iteratively analyzing sensors, processing them to higher-level forms of representation, and then acting according to these forms. Thus, a bidirectional or rather multidirectional cognitive processing cycle appears to be more plausible.

The internal state of a multidirectional information processing and behavioral control system is influenced not only by the sensory information but also by the sensory expectations, which are generated due to the current behavior. Figure 6.1 shows this enhanced point of view in comparison with the classical sense→think→act cycle: Sensory processing is fed not only by sensory information, but also by predictions about the sensory information stemming from higher-level, abstracted forms of representations, as well as from temporal predictions given the current motor activity. Such sensorimotor embodied approaches to cognition thus assume that cognition is accomplished by the controlled unfolding of cascades of sensorimotor coordination on multiple levels. Motor control is, in fact, realized by a control cascade of feedback loops down to the primary control mechanisms, which come in the form of *muscle spindles* within each of our muscles. As a result, a division of labor takes place where various control processes on several time scales influence and control behavior over time. Moreover, disturbances during motor control can be reacted to much faster, as motor activity is directly coupled with sensory feedback.

A very early description of the ideomotor principle was formulated by the psychologist and pedagogue Johann Friedrich Herbart, who wrote in 1825:

> Right after birth of a human or an animal, independent of the soul but just
> due to organic reasons, joint movements take place; and each movement results

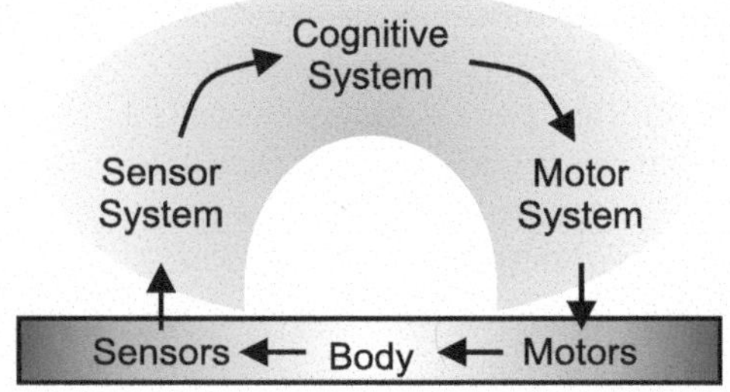

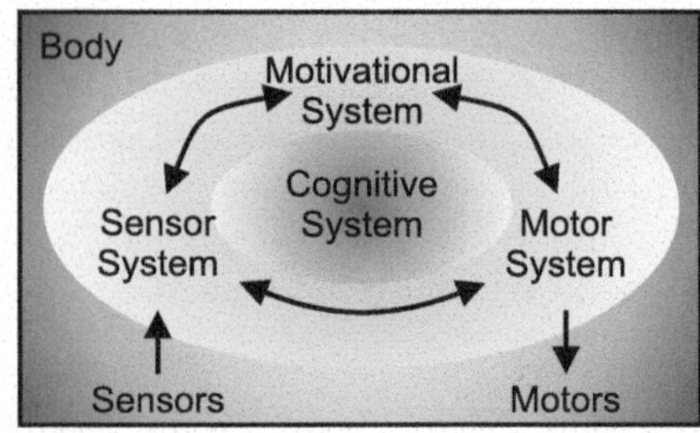

Figure 6.1: In the classical sense-think-act cycle, information was assumed to flow feed-forward only. Modern theories of cognition, on the other hand, assume bidirectional information exchange and control processes at all levels. Moreover, the body somewhat becomes a peripheral part of the cognitive system, by means of which the outside environment is experienced.

in a particular feeling in the soul. Because in the same instant the external senses perceive the change that occurred;

Later, a desire for a particular previously observed change arises. As a result, the associate feeling is reproduced and the nerves and muscles are activated by means of which the desired change can be brought about in the senses. What is desired thus actually takes place; and the success is perceived. Thereby, the previous association is strengthened; a once successful behavior facilitates the next one, and so forth.

[*Gleich nach der Geburt eines Menschen oder eines Thieres entstehn aus bloß organischen Gründen, unabhängig von der Seele, gewisse Bewegungen in den Gelenken; und jede solche Bewegung erregt in der Seele ein bestimmtes Gefühl. Im nämlichen Augenblicke wird durch den äußern Sinn wahrgenommen, was für eine Veränderung sich zugetragen habe;[...]*

In einer späteren Zeit erhebt sich ein Begehren nach der beobachteten Veränderung. Damit reproducirt sich das zuvor mit dieser Beobachtung complicirte Gefühl. [Diesem entsprechen] in den Nerven und Muskeln alle die inneren und äußeren Zustände, vermittels deren die beabsichtigte Veränderung in der Sinnensphäre kann hervorgebracht werden. Das Begehrte erfolgt also wirklich; und der Erfolg wird wahrgenommen. Hierdurch verstärkt sich sogleich die vorige Complexion; die einmal gelungene Handlung erleichtert die nächstfolgende, und so fort.] (Herbart, 1825, p. 464f, own translation.)

In essence, Herbart proposes that initial reflex-like behavior is executed and results in contingent, sensory effects. These sensorimotor contingencies are thus encoded, enabling forward, motor-dependent predictions, as well as inverse, goal-oriented behavior control. The inversion particularly enables flexibility and goal-directedness in behavioral control. However, forward predictions additionally enable the anticipation of the current potentially achievable effects, and thus the bidirectional choice of achievable effects as desirable goals.

Although the ideomotor principle may initially be viewed as a process that works completely subconsciously, it is not restricted to subconscious processes. On higher levels of encodings the principle enables the striving for abstract goal states and to make choices between potential, seemingly achievable goal states. On this conscious level, William James proposed that (cf. Section 2.4.2):

An anticipatory image, then, of the sensorial consequences of a movement, plus (on certain occasions) the fiat that these consequences shall become actual, is the only psychic state which introspection lets us discern as the forerunner of our voluntary acts. (James, 1981, p. 501.)

where the *fiat* in this respect refers to an act of will, which desires that the specific anticipated consequences become actual, typically because they are motivationally desirable. This anticipatory behavior makes explicit not only that we act in anticipation and by anticipating the behavioral consequences, but also that we actually desire those consequences. Learning then also focuses on developing sensorimotor codes – regardless of which level of abstraction – that encode expectations about the effects of particular motor behavior. Once again, when abstracting motor behavior to attention control, anticipatory behavior becomes anticipatory mental processing, leading to anticipatory cognition. The ideomotor principle and anticipatory behavior thus open up new horizons with respect to cognition: on the one hand, flexible interactions with body and world become possible because goal-oriented, situated behavioral decision making and control become possible; on the other hand, invertible sensorimotor structures enable the development of an understanding of the world's functionality and causality – enabling the anticipation of interaction consequences and the consideration of alternative environmental interactions. In the following section, we look in further detail into the cognitive processing capabilities that open up when implementing the ideomotor principle and anticipatory behavior.

6.4 Anticipatory behavior

Given a system that develops sensorimotor, predictive structures, which are sometimes also referred to as *temporal forward models*, it becomes able to generate expectations about future environmental situations. These expectations can be used in various ways to improve behavior and environmental interactions, which is then called *explicit anticipatory behavior* (Butz, Sigaud, & Gérard, 2003), that is, behavior that takes explicit forms of potential future state representations into account. Two types of explicit anticipatory behavior can be distinguished: in *forward anticipatory behavior*, also called state anticipation, current behavior and information processing is influenced not only by the current sensory information and the internal state of the system, but also by the forward predictions about the current state of the environment; in *inverse anticipatory behavior*, also called active inference, the inversion of potential future states leads to modifications of the current system behavior, enabling the execution of explicit goal-directed behavior. The goal-oriented anticipatory direction of attention is also part of this category. Figure 6.2 contrasts these two forms of explicit anticipatory behavior.

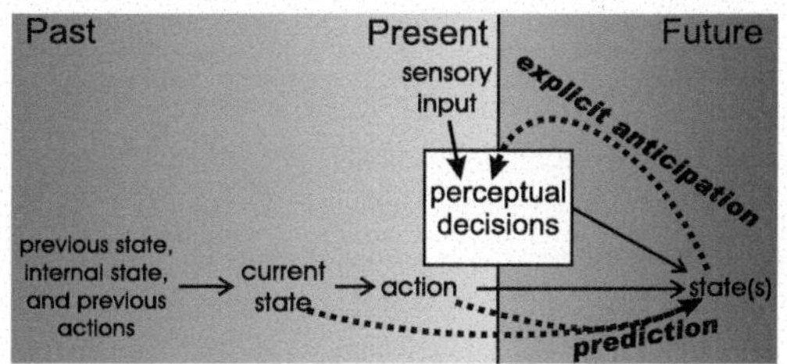

(a) Forward anticipatory behavior

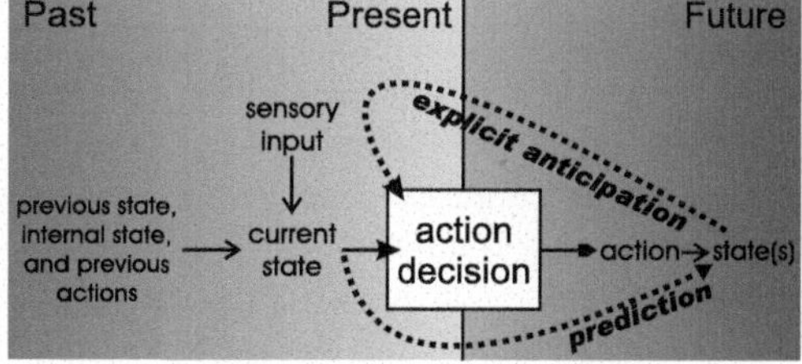

(b) Inverse anticipatory behavior

Figure 6.2: Explicit anticipatory behavior can be separated further into forward anticipatory behavior, where expectations about changes in the world influence sensory processing and actual state inference, and inverse anticipatory behavior, where desired future states co-determine current behavior. [Reproduced with permission from M. V. Butz, O. Sigaud, & P. Gérard (Eds.), *Anticipatory behavior in adaptive learning systems: Foundations, theories, and systems*, Volume 2684 of the series Lecture Notes in Computer Science. Internal models and anticipations in adaptive learning systems, 2003, pp. 86–109, Authors: Martin V. Butz, Olivier Sigaud, Pierre Gérard, © Springer-Verlag Berlin Heidelberg 2003, with permission of Springer.]

6.4.1   Forward anticipatory behavior

Forward anticipatory behavior characterizes behavioral or cognitive processing mechanisms that are influenced by temporal forward-directed expectations about the current state of the world, possibly including the sensory information about that state. While the principle is kept rather general, here we will focus on examples that consider forward predictions of the sensory consequences of actual motor behavior.

Reafference principle

One of the most important and also most well-known forward anticipatory behavior was formulated as the reafference principle by von Holst and Mittelstaedt (von Holst & Mittelstaedt, 1950). It postulates that our nervous system not only sends motor control commands, that is, *efferences*, to our muscles, but also copies of those efferences to our sensorimotor forward models. Via these predictive forward models, expectations of *reafferences* are formed, that is, expectations about the sensory consequences of the motor commands that are currently being executed. The actual reafferences, which are then sensed by the sensory organs, are compared with the expected reafferences. When subtracting the two from each other, the remaining *residual* contains two components: first, error information, which can be used to further adapt the sensorimotor forward models; second, information about other things, which may have moved in the world while the motor command was executed.

The latter component is also referred to as the *exafference*, that is, the external causes that may have changed and thus cause sensory impressions different from those that were expected. Figure 6.3 schematically illustrates this principle.

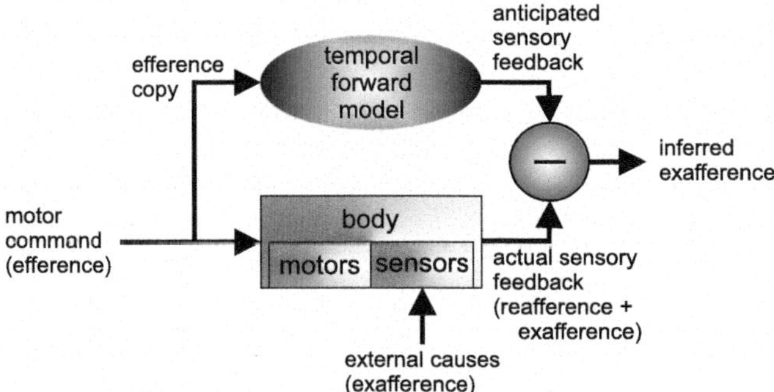

Figure 6.3: The Reafference principle

A very obvious example along these lines are eye saccades. When we want to visually focus on another aspects in the environment, our eyes typically execute a saccade toward that location, thus approximately fovealizing it. Typically, we do not become aware of our eye saccades. Without the reafference principle, this unawareness should actually surprise us – seeing that each saccade causes an immense shift in the image that is projected onto our retina and thus into different visual, sensory stimuli. However, due to the predictive model, the consequences of an eye saccade are anticipated and the resulting sensory information is compared with the expected reafference. If the anticipation was sufficiently correct, no surprise is triggered and the world seems to remain stable. To illustrate this further, attempt to close or cover one eye and lightly touch the eye ball of the other eye from the side. The visual perception is instable, reflecting the unusual manipulation of the eye, for which no sufficiently accurate forward model (about the visual consequences of touching the eye ball) is available.

Many other examples of this kind can be given. Temporal forward models seem to be at work at all times during an awake state – stabilizing the world during locomotion and other bodily motions. Forward models are also available for manual environmental interactions, seeing that we are typically not surprised when we move our hands across our field of vision or in any other manner that is visually perceivable. Even without considering vision, we are not surprised when our body moves and we feel the movement, because it was our own will that produced the movement in the first place.

Adaptive filtering

Adaptive filtering addresses the other residual component of the reafference process, that is, the error component that was not due to unforeseeable changes in the outside environment. The error component, however, can again be considered to consists of two error sources. The first source is due to noise, that is, errors in the sensory readings due to neural fluctuations. The other source stems from inaccuracies in the predictive forward model, which predicted the reafferences.

While it is important to distinguish the two potential error sources, we will not go into details here how this may be accomplished. Rather, we acknowledge that sensory, reafferent signals and predicted, reafferent signals come from two independent sources of information. The one stems from the outside environment, registered via sensors. The other one comes from the internal, sensorimotor forward model, given the previous internal state of the system. Because these two sources of information are independent of each other,[1] they can be effectively fused producing information gain and thus higher certainty in the current perceptions.

The fusion process needs to take into account the reliability of the two independent sources of information, and this reliability needs to be continuously estimated. Sensors typically are noisy. Biological sensors suffer from fatigue and rely on a sufficient supply of nutrients, such as oxygen, water, fat, and proteins. Thus, their reliability is not constant. Similarly, forward model-based reafference predictions will be more or less certain about the current behavioral circumstances, depending on behavioral expertise, fatigue and nutrients, and on environmental circumstances.

Thus, filtering incoming sensory information by means of the sensorimotor, forward predictions needs to be adaptive, continuously taking into account certainty estimates about the sensory content as well as about the forward model predictions. Approximations of such mechanisms are available and are widely in use in engineering applications and robotics. However, without any assumptions about the system at hand, engineering and robotics still struggles to identify a learning mechanism that can robustly produce reliable forward models and, at the same time, filter incoming sensory information with these models (Kneissler, Drugowitsch, Friston, & Butz, 2015).

Anticipatory behavioral adaptation

In addition to being able to improve the perception of the outside world by means of adaptive filtering and to notice other things in the world by means of the reafference principle, ones own behavior can be directly adapted by means of forward anticipatory mechanisms.

Beginning to execute a movement, the forward model predictions essentially lead to an

> [...] anticipatory arousal of the [perceptual] trace, and the feedback from the ongoing movement is compared with it (Adams, 1971, p. 123.)

In this case, comparisons between the sensorimotor forward model-based predictions and the actual sensory feedback can lead to anticipatory behavioral adaptations as a result of the differences between the anticipated trace and the bodily behavior that is actually unfolding. In this manner, changes in the trace can be compensated for much faster, executing online *closed-loop control.*

However, even without sensory feedback, the unfolding behavior can be adapted by continuously comparing the anticipatory trace with the desired goal state. That is, if feedback

---

[1] Independence is not completely true in this case and a full formalization would need to take into account a Markov state assumption, but this is beyond the point here.

is delayed or even completely unavailable, behavior may be controlled by *open-loop motor control* programs, which are tuned to achieve a certain goal state. While the execution unfolds, the sensorimotor forward model alone can be used to produce the anticipated sensory feedback, executing the motor control until the goal state is believed to be reached. Clearly though, only actual sensory feedback can be used to confirm that the behavior was actually successful. Desmurget and Crafton described this type of anticipatory behavioral processing in the following way:

> During the realization of the movement, a forward model of the dynamics of the arm is generated. In its simplest version, this model receives as input a copy of the motor outflow. Based on this information, the end-point of the movement can be predicted and continuously compared to the target location. (Desmurget & Grafton, 2000, p. 426.)

By anticipating the actual sensory feedback without waiting for the actual, typically delayed feedback, goal-directed behavior can still be executed. The system does not wait for the actual sensory feedback, but rather executes the current behavior only taking into consideration the anticipated bodily changes.

Finally, the forward projection of actual behavioral consequences into the future allows us to anticipate undesired events before they actually occur. For example, while jogging we may avoid collision with another jogger much before collision is imminent. Similar behavior can be observed not only when interacting with others, but also when interacting with objects and when avoiding obstacles of any kind. While executing the current behavior, behavioral adaptation occurs in anticipation of a negative event, leading to appropriate modifications.

6.4.2   Inverse anticipatory behavior

Inverse anticipatory behavior is behavior that is controlled by actually desired, anticipated motor consequences. In contrast to forward anticipatory behavior, in inverse anticipatory behavior it is the anticipated future before it actually takes place that influences current behavior. This form of anticipatory behavior is also the one that was formulated in the ideomotor principle. On the sensorimotor level closest to actual sensory readings and motor activations, Greenwald stated that

> For the ideo-motor mechanism, a fundamentally different state of affairs is proposed in which a current response is selected on the basis of its own anticipated sensory feedback. (Greenwald, 1970, p. 93.)

Thus, the anticipated consequences guide behavior, rather than the motor activities themselves.

From a control-theoretic perspective, this may come as a surprise, since the motors are actually the ones that are controlled by motor activities. From a biological perspective, however, this insight is not that surprising: how could a system control its motors without monitoring what they are actually doing? Only precisely engineered motors can function in a fully (non anticipatory) open-loop manner given a precisely regulated energy supply, which we have available by means of electric power. Biological systems grow and change in strength and energy supply all the time. Thus, the focus typically should lie on the control of the monitored sensory effects rather than on the motor control itself.

While Greenwald and others emphasized that the representations of the anticipated sensory feedback themselves are the ones that actually activate motor behavior, it is still under debate to which extent this is the case. *Desired effects* can come in various forms of encodings and thus may not need to be restricted to the actual, direct sensory effects. Additionally, the wiring in the brain suggests that motor control can be activated in various ways, including, but not being limited to, the sensory effects. Moreover, it should be acknowledged that behavioral effects on sensor readings come in various forms and are not necessarily straight-forward. For example, when we flex our elbow, our hand may move toward our body mid-axis, but it may also move away from our body mid-axis when readjusting the shoulder joint appropriately. Thus, some sensory effects are posture-dependent and they are thus slightly difficult to predict. Moreover, sensory effects can be registered in various modalities, where each one is grounded in a specific frame of reference dependent on the type and morphology of the sensor. While some sensory effects can be very directly related to motor activities, others require additional computational capabilities.

Inverse anticipatory behavior, however, is not restricted to inverse, sensorimotor control of actual current bodily motion. Rather, it can be extended to any form of "sensory code" and any form of "motor behavior". As illustrated, sensory information comes in various forms and is grounded in various frames of reference. The brain appears to abstract over different sources of sensory information, integrating them on abstract levels, and projecting the respective information onto each other. As a result, abstract and generalized forms of multisensory codes develop, such as the encoding of a direction, a body-relative location, or a type of object. Directional movements are possible as are particular object manipulations that are suitable for a particular type of object. Thus, sensory effects can be encoded in various frames of reference and on various levels of abstraction.

Inverse anticipatory behavior on multiple levels of abstraction enables the pursuance of effects on various levels. For example, when intending to go to a lecture (and actually going) typically several means of locomotion are pursued to reach the lecture hall – for example, by bike or car – to enter it, to reach the correct floor, to reach the lecture hall entrance, to find a suitable seat, to move to that seat, and finally to sit down. The final goal of attending the lecture is pursued by activating an overall goal, which activates multiple subgoals in an appropriate sequence, which ultimately activates the currently appropriate behavior in the light of the current active goals, subgoals, and the environmental situation. In this manner, we have described a hierarchical, model-based RL mechanism, where the anticipation of rewarding goals leads to the generation of a behavioral plan that is then pursued over an extended period of time. Actual sensorimotor control is guided by all these goals and it is continuously adapted in light of the current circumstances, for example, stopping the car at a red traffic light or making way for other people while entering the building.

When we move even further away from manipulating the environment or reaching a certain destination with the own body, we enter the cognitive, mental world. Imagine the simple task of adding a few numbers – say adding up all numbers from 1 to 10. With this goal in mind, we may go about it step by step, going through the numbers, mentally adding them, and storing the sub-sums, $1, 3, 6, 10, 15, 21, 28, 36, 45, 55$. Or, we may emulate young Gauss, thinking we are clever, and remember that $10 + 1 = 11$ as is $9 + 2$ and so forth, and thus more quickly calculate $5 \cdot 11 = 55$. Regardless, the point is that we can mentally represent the goal of summing up a few numbers, make a plan to get that sum, and then do the calculation (possibly with the help of paper and pencil to avoid making working memory errors).

Note how both inverse, goal-oriented anticipatory behavioral examples are very similar: In both cases, alternative plans reaching the goal offer themselves: "Should I take the car or the bike to reach the lecture hall", and "Should I do a straight-forward summation or a slightly more intricate but faster computation?" Moreover, an overall goal leads to the activation of successive sub-goals: the sequential means to reach the seat in the lecture hall, and the sequential means to compute the overall sum. Finally, the inverse anticipatory episode is concluded when the formulated final goal is reached — the lecture hall or the result of the summation.

While these similarities are somewhat striking, as in the case of model-based RL, the crux lies in learning the necessary hierarchical encodings. Behavioral or mental goals need to activate sub-goals, which ultimately activate the unfolding goal-directed behavior.

Anticipatory learning

The ideomotor principle contains considerations about actual learning mechanisms, which may develop sensorimotor encodings. As Herbart had put it (Herbart, 1825), the experiences of sensorimotor consequences are registered and may lead to the further differentiation of the hitherto available sensorimotor encodings.

On an abstract level, learning needs to consider the sensorimotor contingencies experienced while interacting with the environment. Hoffmann (1993) has proposed an *anticipatory behavior control* principle, which suggests how learning progresses. He postulated that first action-effect relations are formed and differentiated based on the encountered sensorimotor experiences. Later, when the learning system experiences situations in which the effects are not achieved by the correlated action, these action-effect relations may be further differentiated, taking into account the conditions under which the effects may become actual.

For example, we may learn how to open a door by pushing down the handle and pulling or pushing the door open. When for the first time we are confused by a door that does not open in this manner, however, we may learn to distinguish situations in which a door is locked and in which it is not locked. Locked doors require unlocking before opening can be successfully executed. Thus, an "opening a door" behavior may be conditioned on the unlocked state of the door – requiring "unlocking" behavior before the actual "opening" behavior when the door is locked.

In sum, anticipatory learning, that is, the differentiated learning of behavioral consequences may yet be the hardest task in the development of higher levels of anticipatory, cognitive processing. It is apparent that this learning starts at a very young age and most likely before birth. In the womb, the infant faces a rather safe environment within which it can explore its own body and the environment surrounding its body. How learning actually works, however, is still debated. Recent considerations of predictive encoding and free energy-based minimizations of these encodings, which are closely related also the principles of a Bayesian brain, are closely related to the learning mechanisms formulated in a highly abstract manner in the ideomotor principle (Doya, Ishii, Pouget, & Rao, 2007; Friston, 2010; Rao & Ballard, 1998). We will re-consider these relations in several later chapters.

6.5 Motivations and curiosity

While we have characterized and differentiated various forms of anticipatory behavior and anticipatory learning, what has not been addressed is how goals are chosen in the first place to initiate inverse, goal-directed anticipatory behavior. Behaviorism and reinforcement learning, which were addressed in the previous chapter, have assumed that organisms can register reward and punishment values, and directly adapt their behavior and reward estimates based on these values. In anticipatory behavior, goals are chosen on various levels of abstraction. Closely related to hierarchical, model-based RL, the goals in anticipatory behavior may indeed be chosen based on the estimated reward, which the goals trigger once achieved. Thus, decision making in anticipatory behavior is closely related to decision making in RL – and most particularly in model-based RL – because in both cases behavior is strongly determined by the expected future reward.

However, in addition to actual *extrinsically* gained reward and punishment, another form of rewarding interactions can be identified when considering the sensorimotor learning task. *Intrinsic reward* characterizes reward that is gained due to information gain, that is, as a result of an improvement in the internal knowledge structure. Intrinsic reward is thus closely related to curiosity, that is, inquisitive or *epistemic behavior*. Both forms of reward seem to be at work when making behavioral decisions. For example, when something unexpected happens, such as hearing a noisy thump close by, we direct our attention toward that thump because we want to know what has happened. Children more than adults show curious, information seeking behavior all the time, such as when they explore a new object. However, sometimes extrinsic rewards dictate our goal-oriented behavior, as for example, when we indulge ourselves in a piece of cake.

Both forms of reward appear to influence our behavior – depending on the current circumstances, more the one or the other. In the following, we detail how these two forms of reward actually work from a functional perspective.

6.5.1 Intrinsic reward

Sometimes intrinsic reward is equated with novelty, where everything that is novel triggers intrinsic reward. When an organism searches for novelty, it can be shown that learning proceeds more robustly and is less prone to converge to local optima. Remember the latent learning experiments of Tolman: rats that had not received extrinsic reward in the first maze trials showed that they had learned a more suitable cognitive map, which they later were able to exploit in order to reach a rewarding position faster. The rats had explored the maze more thoroughly because they were not distracted by extrinsic rewards during the initial trials. Nonetheless, *latent learning* took place and, driven by *intrinsic reward*, the rats explored the maze in more detail during these trials.

In evolutionary computation and RL algorithms, mechanisms to foster exploration have received much attention. In evolutionary computation, for example, often an adaptive fitness function is applied that provides reward for novel phenotypes, that is, when genotypes evolve that exhibit a novel type of behavior. Phenotypes that generate exploratory behavior also sometimes receive higher fitness. In this way it is assured that the evolutionary process does not converge to a local optimum overly quickly, potentially overlooking promising optimization alternatives. Additionally, in RL the reward function is sometimes initialized in such a way that the values in the Q-table, for example, are initiated to rather high values, thus fostering the exploration of those state-action combinations that have not yet been explored.

When equating intrinsic reward with novelty, however, a big problem arises when considering situations in which novelty leads to highly undesired states. Imagine a child that behaves overly exploratory. The child will typically fall and get hurt more often, because he or she does not sufficiently consider the danger in the behavior. Additionally, while novelty may be interesting, it seems useless to explore novel situations in which it appears impossible to learn anything. Thus, while the novelty concept goes in the right direction, it needs to be further differentiated.

A more advanced intrinsic reward concept is that of uncertainty. Given a developing sensorimotor model, uncertainties may be stored along with each model prediction. Intrinsic reward may then be associated with those actions in particular situations that predict the highest uncertainty in the action outcome. This concept essentially enhances the novelty concept when assuming that actions with totally unknown sensory effects are the ones with highest uncertainty. However, uncertainty-based intrinsic reward suffers from particular problems. As in the novelty-based reward, uncertainty-based intrinsic reward does not consider extrinsic reward. Thus, it does not avoid dangerous situations. We will address this problem once we have introduced extrinsic reward.

More importantly though, uncertainty-oriented behavior may fall into another trap: the system would love those situations most where the uncertainty continuously stays high. For example, the system may end up running around on clear ice, because it simply cannot predict if it will manage to stand or fall in the next second. While walking on clear ice may certainly be fun for a short while, it sooner or later becomes dull and boring. One reason this is the case is probably because our learning progress is saturated. We realize that there is nothing more to learn, thus we stop finding the situation interesting. This applies not only to clear ice, but to many situations in which the system – or us for that matter – no longer encounter much learning progress. Once uncertainty does not decrease, things become boring.

Computationally, this concept can be characterized as uncertainty decrease, or *information gain*. As a system that strives to maximize information gain will act curiously when detecting novel things, but will stop being curious once the behavior of this novel thing has been sufficiently explored, that is, once the thing-specific sensorimotor forward models no longer improve significantly. Intrinsic reward based on information gain can thus mimic novelty-oriented and uncertainty-oriented reward, but it makes these concepts more generally applicable. Novel things are intrinsically associated with high uncertainty, thus expecting large information gain. However, once starting to explore the novel thing, the information gain can be monitored, such that things that just behave strangely are soon left alone while other things that behave in a complex but predictable manner are most interesting. It is now believed that curiosity is realized by mechanisms that strive to maximize information gain. Information gain expectations in novel situations are typically initialized to high values, thus fostering curiosity about novel things.

However, all concepts so far have only considered intrinsic reward, completely ignoring extrinsic reward. Thus, all mechanisms will find dangerous novel situations as interesting as non-dangerous novel situations, which is, of course, undesirable. We thus introduce extrinsic reward in the following and relate the two concepts to each other.

6.5.2   Extrinsic reward and motivations

Extrinsic reward addresses forms of reward that are triggered due to bodily interactions with the outside environment. The simplest forms of extrinsic reward are probably triggered while executing consummatory and related behaviors. However, social interactions can also be highly rewarding, including, for example, cooperative actions or the feeling of being protected. Extrinsic rewards are embodied and are typically triggered by an internal, *motivational system*, which is based on the principle of *homeostasis*. Motivations, once satisfied, are thought to trigger extrinsic reward signals akin to the reward signals assumed to exist in behaviorism, and closely related to the many forms of reward signals in RL. In anticipatory behavior, extrinsic reward may be associated with goals, such as "consuming food". Negative extrinsic reward may be associated with the strenuousness of particular behavior, the danger involved, as well as with injury. Thus, positive and negative forms of extrinsic reward can be generated and need to be taken into account to optimize behavioral interactions.

How is extrinsic reward generated, though? From a computational perspective, a mechanisms based on homeostasis seem to be most plausible. This mechanism is closely related to the property of *autopoiesis* in biology, which was put forward by the Chilean philosophers and biologists, Francisco Varela (1946–2001) and Humberto Maturana (*1928). Biological systems are evolutionarily designed to maintain internal homeostasis, thus striving for survival and reproduction. Once the internal homeostasis falls too much out of balance, the system dies and disintegrates. The Austrian physicist and theoretician Erwin Schrödinger (1887–1961) has related this concept to the information-theoretic and physical concept of *entropy*, which quantifies the in-orderliness of a system. He essentially postulated that organisms generally must feed on negative entropy: they consume lower entropy structures (such as food, oxygen, or water) the environment that can be processed by their metabolisms and extract parts of that structure, getting rid of higher entropy structures. The extracted structure, that is, the negative entropy is used to maintain internal homeostasis, that is, internal structure (Schrödinger, 1944). Striving for the maintenance of bodily homeostasis and thus bodily structure may be considered as one of the most fundamental mechanisms that establishes life.

Moving back to a functional level, however, a homeostatic system can be thought of as containing reservoir-like states, which trigger reward when behavior changes the states to one of near saturation. Additionally, over-saturation is again associated with negative reward, thus avoiding overly low reservoir states, as well as overly high reservoir states. Imagine a food reservoir, which may indicate the current state of the stomach. If the stomach is well-supplied with food, it signals a saturated reservoir and further eating behavior will not yield additional extrinsic reward and may even yield negative reward when full saturation has been reached. Thus, a system equipped with such a mechanism will want food and consume food while hungry, that is, while the body signals a low reservoir state. Once food is consumed, however, the hunger goes away because the reservoir fills up – decreasing the motivation for food consumption. As a result, such a self-motivated system will strive for food when hungry, but may do other things when saturated.

Multiple such motivational reservoirs may be maintained, yielding distinct reward values, which may be associated with distinct environmental interactions. Maintaining a balance between these distinct motivations is tricky and was most likely developed by evolution in all species.

Computationally, a homeostatic system that is equipped with a number of N motivational reservoirs $\mathcal{R}$ may be formalized as follows. Each reservoir $i \in \mathcal{R}$ may signal its current reservoir level by the value $r_i \in [0, 1]$. Moreover, each reservoir can be expected to be equipped with a reward mapping function, $f_i : [0, 1] \to \mathbb{R}$, which may be thought to compute an *urgency level* dependent on the current reservoir state. A simple function, for example, may be a linear function that yields its maximum value when reaching a particular reservoir state:

$$f_i(x) = \theta_i - x, \tag{6.1}$$

which essentially yields progressively more positive urgency when the current reservoir state x is smaller then the saturation level θ_i and begins to yield negative values when the reservoir is overly saturated. Clearly, there is much room for optimizing this function further by, for example, modifying the maximum and minimum values reached when the reservoir is empty or overly saturated and by modifying how quickly these values are reached.

Given several such reservoirs with their respective urgency functions, the respective urgencies still need to be put in relation to each other. For example, particular urgencies may reach such high values compared with others that they fully dominate the others. This can, on the one hand, be realized by the maximum values reachable in the respective urgency functions. On the other hand, the urgencies may be further modified by multiplicative values m_i, which enable the further adaptive balance of the different urgencies. As a result, the different urgencies may be computed, given the current reservoir states $r_i(t)$, by $m_i \cdot f_i(r_i(t))$.

These values can now be viewed as the expected reward value when the respective motivation is being satisfied, for example, when food is consumed. Thus, the reward values can be associated with those environmental interactions that have previously led to the encounter of similar reward. Learning thus not only needs to form sensorimotor structures, but it also needs to associate motivation-based, extrinsic reward with those structures.

The resulting system can generate anticipatory behavior by employing principles of model-based RL where reward comes from the internal motivational system. Given the current motivational system state, the resulting expected reward (or "urgency") values are associated with those sensorimotor codes that may satisfy the respective motivational states. Larger urgencies thus project larger expected extrinsic reward values into the sensorimotor model. Planning mechanisms then propagate the current reward values inversely through the system, yielding appropriate reinforcement-learning-based gradients. Behavior then proceeds to pursue the activated goals, and subgoals, which promise to lead to the currently most desired motivational satisfaction.

Clearly, the involved computations are not trivial and the maintenance of a balanced system that is able to consider many motivational states is a difficult challenge. In fact, research suggests that different motivations may need to be distinguished, such as consummatory motivations from property-based motivations. The former addresses reward situations where one particular interaction triggers reward. The latter addresses motivations that are continuously affected by the environmental situation and executed behavior, such as, for example, behavior-dependent energy consumption, as well as safety considerations. While the former may be associated with particular states and propagate this reward inversely through the sensorimotor model, the latter influences the reward propagation because it is relevant in all states and actions. Moreover, considerations of how quickly a reservoir is actually satisfied by, for example, a consummatory action, needs to be addressed in further detail.

Another consideration is the balance between different motivations, as, for example, manipulatable by the multiplicative biases m_i. These biases may adapt depend on the current mood or emotional state of the system. When in a very "optimistic mood", negative consequences may be disregarded to a certain extent, leading to high confidence and focused, goal-oriented behavior. However, it may also cause unfortunate side-effects, such as not being sufficiently cautious, possibly leading to an accident, or being over-confident, thus, for example, spending time on unsolvable problems. On the other hand, when in a "depressive mood", nothing may seem to be promising because nothing is expected to actually generate sufficiently high reward, thus leading to lethargic behavior, essentially exhibiting behavioral symptoms of depression.

Finally, even if a good balance can be maintained between the urgencies generated by different motivations, intrinsic reward still needs to be balanced with the motivationally-determined, extrinsic drives. When should we act curiously, when motivationally, goal-directed? Interestingly, to a certain extent the two mechanisms may be merged when taking into account that the learned sensorimotor model is endowed with certainty estimates.

An interesting study, which was partially inspired by the Wilhelm Tell story, shows that our brains do indeed consider uncertainty when making action decisions. Wilhelm Tell, so it is told, had to shoot an apple from his son's head in order to save his son. In this case, the apple is the target, which will yield a positive reward, and the head is the close-by region, which will generate a very negative reward. Thus, it can be expected that Wilhelm Tell probably aimed for the top part of the apple to avoid the negative region as much as possible. In fact, an experiment by Trommershäuser and colleagues (Trommershäuser, Maloney, & Landy, 2003a, 2003b) showed that the brain acts according to statistical decision theory principles, aiming approximately optimally away from the negative reward region toward the positive reward region taking into account aiming precision estimates. Thus, uncertainties are taken into account when choosing exact goal locations.

Intrinsic reward, however, additionally should co-determine if a current goal is striven for at all. If the uncertainty about reaching the aimed-at goal state is very high and epistemic behavior is available, which may decrease this uncertainty, the system may choose to first decrease its uncertainty before attempting to reach the goal. When the uncertainty in the sensorimotor model cannot be sufficiently decreased, however, the pursuance of the goal may be dismissed or completely other means may be considered.

A final consideration is that intrinsic reward may be further fine-tuned by a *curiosity motivation*, which may modify the estimated information gain-based reward values. For example, curiosity may interact with the other motivational drives in a way that while the other motivational drives of the system are rather low, curiosity may increase in strength. In this way, a system can develop that is curious when there is time for it and that focuses on its bodily needs when necessary. It still remains unclear with respect to both biological and AI systems how exactly different motivations interact with each other and thus influence behavior interactively in the most effective manner. The later chapters on attention (Chapter 11), as well as on decision making and behavioral control (Chapter 12), will address these aspects in further detail.

6.6   Summary and outlook

This chapter introduced how behavioral and cognitive flexibility develops ontogenetically. At the beginning of this chapter, we saw that diversity plays a crucial role when a system needs to work robustly in different environmental niches. Evolution has therefore developed a large variety of species, each one specialized to reap particular resources in a particular niche. Moreover, evolution has developed redundancies in the populations of the respective niches to harvest the available resources in various manners, thus avoiding missing available resources.

However, evolution has not done so only with regard to the variety of species, which populate environmental niches, but also with regard to the sensory and motor systems available in one species – particularly in cognitively more advanced species. Redundancy and complementarity in the sensory system of a species enable it to sense the environment by different means, thus having a more fail-safe system, as well as having a system that benefits from information gained due to the availability of multiple, independent sensory sources about the same environmental causes. Redundancy and complementarity in the motor system of a species enable it to learn to use and consider alternative behavioral interactions with the environment in order to be able to reach particular goals. These two capabilities are especially useful in dynamically changing environments in which different sensory and motor capabilities are particularly useful for executing particular interactions.

To an even greater extent evolution has evolved more complex brains, which allow behavioral adaptations to environmental circumstances by means of learning and goal-directed, anticipatory behavioral control. These capabilities partially go hand-in-hand with the alternatives in the sensory and motor systems, because only a system that is sensorially and motorically able to consider and to accomplish particular environmental manipulations by different means needs to be endowed with cognitive flexibility. Human life developed in the ecological niche of strong social interactions, including intricate forms of cooperation and communication. In all the particular forms of interaction, various alternatives are usually possible and need to be considered to make effective choices, and to interact with the environment and other individuals effectively. This may be part of the reason why humans have managed to start a cultural evolutionary process, which has led us to develop our current, highly sophisticated environments, such as tool, machines, the computer, or the Internet.

To achieve behavioral and cognitive flexibility, however, anticipatory behavioral mechanisms that choose goals and strive for their pursuance seem to be necessary. We have shown that these mechanisms have various advantages when considering forward-directed anticipatory processing as well as inverse-directed anticipatory behavioral control. Forward-directed anticipatory mechanisms are particularly well-suited to filter sensory information leading to information gain as well as to identify exafferences, that is, other causes of sensory changes. Moreover, behavior can be adapted faster and can be controlled faster, when considering the anticipated sensory effects during the unfolding control process. Inverse anticipatory behavior result in goal-directed behavior and the flexible realization of these behaviors under varying circumstances, essentially enabling the consideration of behavioral alternatives for achieving a particular goal. Furthermore, information-driven, epistemic behavior can be viewed as inverse anticipatory behavior, which strives for information gain.

The ideomotor principle, which dates back to the early 19th century, proposed that sensorimotor structures are the ones first learned by an organism in order to enable goal-directed, ideomotor-based control. Inborn reflex-like behavior probably bootstraps the initial ideomotor learning progress, which monitors sensorimotor contingencies, and thus learns sensorimotor models. The simplest models to learn in this manner are those about the functionality of ones own body, including the available different sensory and motor systems. Further learning progress can then focus on external sensory causes and their behavior.

The developing sensorimotor models essentially allow the effect-oriented, anticipatory execution of behavior. When abstracting sensorimotor models to abstract sensory and perceptual encodings as well as to complex, motor-primitive-like encodings, goal-directed behavior can lead to intricate, hierarchical goal pursuance behavior. Such behavior seems to be closely related to abstract thought processes, such as when solving a mathematical problem by executing a sequence of mathematical calculations.

Finally, to coordinate the goal-selection process, a motivational system has been described which may indicate expected extrinsic rewards when achieving particular states in the environment. This system needs to be able to balance the importance of different extrinsic and intrinsic forms of reward. Interestingly, intrinsically-motivated, epistemic behavior may be triggered when the uncertainty about achieving particular extrinsically motivated goals is high and when epistemic actions are anticipated to decrease this uncertainty. Intrinsically motivated behavior may also be pursued when no extrinsic behavioral motivations are particularly urgent.

In Chapter 7, we look at the human brain from the perspective of a behavior-oriented, functional, cognitive architecture. Subsequent chapters will re-consider the mechanisms explained in the chapters covered so far from a cognitive brain perspective. In doing so, we will differentiate the mechanisms further and provide details about further computational principles, mechanisms, and the developing encoding structures that bring the mind about. We will thus shed further light on the question how predictive models about body and environment may develop and how they may be structured to enable the development of the abstract thought, reasoning, and language abilities in humans.

6.7   Exercises

1. In your own words, describe what "to understand something" may mean from a computational perspective.

2. Why can model-free RL techniques not develop any form of understanding beyond an understanding of reward?

3. To which extent does our ability to adapt our behavior on the fly to novel circumstances indicate that we possess an understanding of our environment?

4. Why is redundancy in sensory and motor behavior essential to enable the development of flexible, adaptive behavior and cognition?

5. Relate the classical sense-think-act cycle to the homunculus problem (cf. Section 2.2.2). Why does closing the loop between the motor and the sensory system offer a solution to the homunculus problem?

6. Predictive models of our world enable us to plan and make decisions in anticipation of the behavioral consequences. Give examples of cooperative, social scenarios in which predictive models are particularly useful or even absolutely mandatory to successfully cooperate.

7. The DYNA-Q algorithm, which was introduced in the previous chapter, can accomplish latent learning. Sketch-out how DYNA-Q may generate behavior that is similar to the one observed in rats by Eward Tolman. Which types of "expectancies" would the algorithm form?

8. Relate the observations of Tolman in rats to the situation when we are confronted with a new environment, such as a different city or a new building. Initially, we typically feel easily disoriented, but sooner or later we feel comfortable and find our way around. What has happened?

9. It is know to happen that we sometimes find an actual shorter route to a certain location, such as a shopping center, after several months of going to that location. Explain how this may happen and relate it to insufficient exploration and curiosity.

10. Gently push one of your eyeballs with your index finger from the side, while keeping your other eye closed. The visual image of the environment shakes. Relate this observation to the reafference principle and missing forward models.

11. Identify the commonalities and differences in the three types of forward anticipatory behavior introduced.

12. Contrast anticipatory behavioral adaptation in a forward anticipatory manner from inverse anticipatory behavior.

13. Formalize a simple cooking recipe and thus show how the very embodied behavior of actually cooking the specified dish is formalized into discrete states, involved entities, and behavioral primitives.

14. Fundamental concepts of living systems can be characterized as striving to feed on negative entropy – thus consuming "structure". How can the feeling of being hungry and thus consuming food be related to this concept via the principle of homeostasis?

15. Intrinsic motivations are concerned with information and knowledge, while extrinsic motivations are concerned with homeostatic states. Give examples of intrinsically and extrinsically motivated human behavior.

16. To which extent may social behavior be intrinsically and extrinsically motivated.

17. In social situations, it is typically useful to develop theories of mind of others – that is, to estimate the current knowledge and abilities of others. Give examples how such theories of mind can be useful to realize effective social cooperations.

Chapter 7

Brain Basics from a Computational Perspective

7.1 Introduction and overview

In the previous chapters we approached the concept that cognition in humans needed to become flexible and adaptive. To enable such adaptations to the environmental circumstances, to enable planning ahead, and even to enable cooperation, we deduced fundamental computational principles and we considered fundamental algorithmic principles. Thus, we have looked at David Marr's (cf. Section 2.5) first and second levels of understanding a system.

What about the third level? What about the human brain and the central nervous system? How does the brain learn and develop the cognitive capabilities we have? As we have suggested, cognition is the result of mental processes, which are generated by the brain-body device. In order to fully understand the computational and algorithmic principles, it is necessary for them to be embedded in the brain. Although research does not yet know exactly how this embedding is accomplished, many novel insights have been gained for the last two decades.

In this chapter, we first give a basic introduction to the brain and then look at its fundamental components, modules, mechanisms, and their development from a computational, algorithmic perspective. We focus on the question of how the brain implements the algorithms and computations necessary to allow our minds to come into being. From a reverse perspective, we also ask the question what can we learn from the brain and current knowledge in the neurosciences about possible algorithmic and computational principles that are implemented by the brain. To do so, we look not only at the brain's modules and neurons, but also at fundamental neural processing pathways and the functionalities of information exchange, temporary processing focus, decision making, and action control mechanisms.

7.2 The nervous system

The nervous system is usually partitioned into two main parts. The central nervous system (CNS) consists of the brain and the spinal cord. The peripheral nervous system consists of all neurons that are not part of the CNS. Despite this partition, it is apparent that the central and peripheral nervous systems are strongly interactive and their interactions are not only mediated by the neuronal information exchange, but also by the body's morphology, sensor and motor properties, as well as by various chemically-mediated interactions.

Clearly, the main purpose of the nervous system is to ensure the survival and the possible reproduction of the organism. To ensure this, the nervous system receives sensory information about the body and the outside environment, and, to some extent, it can manipulate the body and the environment using its bodily and motor capabilities.

For example, Figure 7.1 sketches out important pathways by which visual information can influence manual action decision making and action control. Visual information registered by cells on the retina are transferred via the *lateral geniculate nucleus* (LGN) to V1, where the primary visual information is analyzed and transferred via a ventral pathway to *inferior temporal* (IT) areas. In these areas, neurons selectively respond to particular entities and thus support entity recognition. Moreover, this recognition also supports decision making in the *prefrontal cortex*, such as if one focus further or interact with a particular entity. The decision is then transferred to the *premotor* and *motor cortex* to initiate and control the action, sending the motor commands to the relevant muscle groups via the *spinal cord*.

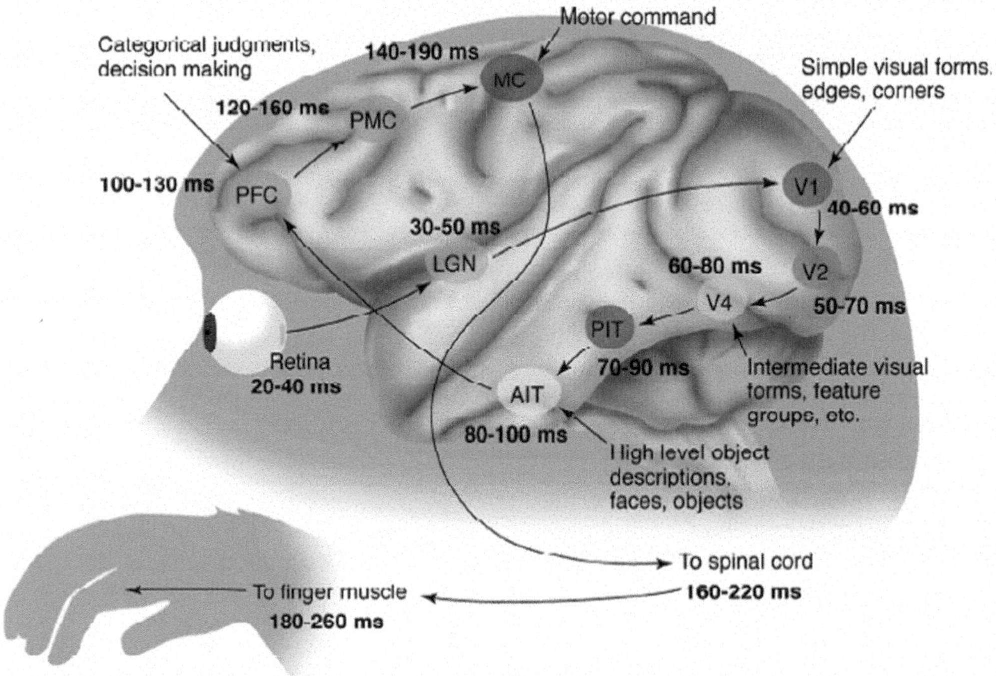

Figure 7.1: A simple sketch of how visual information may influence action decision making and control . [From Thorpe, S. J. & Fabre-Thorpe, M. (2001). Seeking Categories in the Brain. *Science, American Association for the Advancement of Science, 291*, 260–263. Reprinted with permission from AAAS.]

Note that this image is extremely simplified. The most fundamental simplification is the fact that premotor and motor areas are also strongly supported by *parietal areas*, which analyze the visual information via the *dorsal information processing stream*, and which are believed to provide information about entity locations and orientations relative to the own body and relative to each other. Another fundamental simplification is the fact that cortical areas typically communicate bidirectionally, such that the information from the visual area does not really flow in a feed-forward manner toward the decision making and motor control areas. Rather, on its way it is strongly modulated by feedback connections. Moreover, all subcortical interactions are left out and many further cortical-cortical and cortical-subcortical connections are ignored.

While the simplified processing pathway does give an idea about how visual information can influence motor behavior, the brain is not really a "feedforward", reactive information processing system that transfers sensory input via some decision making center to motor output. Rather, the brain should be thought of as striving for maintaining a consistent internal model of the environment and the currently unfolding interactions (Butz, 2016;

Friston, 2009). Later in this chapter and in the remainder of this book, we will provide many more details about this view of the brain and its implications.

Besides interactions of the central with the peripheral nervous system, functional modularizations should be mentioned. These particularly highlight the fact that our body with its manifold control systems can be compared to a *subsumption architecture* (cf. Section 3.6.2). The body contains a *society of mind* – as Minsky put it (Minsky, 1988) – where each organ or body part may be considered an individual, but often highly interactive entity. In other words, the body may be viewed as a society of interactive, but partially self-sustaining systems. The *somatic* or *voluntary nervous system* (from Latin soma = body) enables the recognition of sensory stimulations and the goal-directed control of bodily motions. The *vegetative, visceral*, or *autonomous nervous system* (from Latin visus = intestines) continuously and autonomously works on maintaining bodily homeostasis, such as body temperature, blood pressure, sugar level, oxygen concentration, and energy processing, by controlling the inner organs. This part of the nervous system is often partitioned further into two subsystems. The *sympathetic nervous system* is mostly responsible for maintaining homeostasis, but it also plays a role in fast, sub-conscious fight-or-flight decisions. The *parasympathetic nervous system* is complementary and regulates energy storage, food digestion, and other bodily needs. It is known for stimulating rest and digest activities as well as feed and breed activities among other basic body-oriented activities.

This short overview indicates that the visceral nervous system is a highly complex and modularized system on its own. The stomach alone actually contains more neurons than the spinal cord – indicating that digestion alone is a neurally controlled process, which, however, typically functions fully autonomously without our awareness. The central nervous system thus does not need to take full care of the body with all its functionalities. Many such functionalities are taken care of by the peripheral nervous system, often without any brain interactions. This eases the tasks for which the brain is responsible. It allows the brain to focus on controlling voluntary environmental interactions, communicating with the voluntary part of the peripheral nervous system. However, although the autonomous nervous system functions largely autonomously, it certainly influences the brain by indicating current needs (such as hunger or thirst) and preferences (such as fight-or-flight tendencies). In the remainder of the chapter, we focus on the brain while keeping in mind that the brain is not only supported by, but also influenced by both the vegetative and the somatic parts of the peripheral nervous system.

7.3 Brain anatomy

To get an idea of the complexity of the brain it is first necessary to examine the brain's anatomy. To be able to build a crude brain map, we follow a bottom-up approach, focusing first on neurons and the principles that underlie neural information processing and transition. We then look into particular areas, and finally consider brain modules and the overall brain architecture.

7.3.1 Neurons and neural information processing

The central nervous system consists mainly of two types of cells: *neurons* and *glial cells*. Neurons are the main functional units that process neural information. They are responsible for encoding afferent sensory signals and for producing efferent motor signals. Moreover, they are responsible for information processing and encoding, and ultimately cause cognition. Glial cells, on the other hand, are mainly responsible for maintaining structural integration, for insulating axonal nerve fibers, and for energy supply. Depending on the brain module, there are either very few glial cells when compared to neural cells or up-to about 17 times more glial cells (Herculano-Houzel, 2009). The insulation of neural axons by means of myelin sheaths is especially important to ensure effective *neural action potential* transmissions to other cells. The mutual insulation of neurons against other neurons is also an important neural information processing contribution. Nonetheless, the exact role of glial cells in information processing still remains unclear. We thus focus on neural morphology.

Individual neurons consist of four basic components:

- *The soma* or neural cell body is surrounded by the cell membrane and contains the cell nucleus and cell organelles. The energy supply of the cell is regulated by means of mitochondria. In the nucleus, gene expressions unfold, generating protein syntheses, which generally control the cell's activity (cf. Section 4.3.2).

- *Dendrites* are information fibers, which typically receive information from other cells. They usually come in the form of a complex *dendritic tree*. The received information is integrated in the dendritic tree and transported to the soma. While simple models of dendrites originally assumed that presynaptic activities are simply integrated additively, over the last few decades it has become clear that much more intricate neural integration processes can occur in a dendrite, including neural activity inhibitions and even multiplicative activity interactions between presynaptic activities.

- *The axon hillock* is the part of the soma, which extends toward the axon. Neural activity is transferred over the axon hillock in a threshold-like fashion, leading to the generation of an *action potential*.

- *The axon* of a cell is the cell's projection fiber, which sends neural activity from this cell to other cells. It begins after the axon hillock. An axon has a diameter of 0.5 to $10 \mu m$ and can be up to one meter in length. It is responsible for transporting mainly electrical activities. Axons typically end in presynaptic connections, which connect to dendrites of other cells via synapses. Myelination of the axon enables a much faster information transfer.

Figure 7.2 shows a cut through of a neuron highlighting several other neural components. For our discussion it suffices to acknowledge that neural processing is much more complex than described previously, but it can nonetheless be crudely characterized by these four components.

Nonetheless, it is worthwhile to acknowledge that neural cells come in various forms and shapes. There are many non-exclusive classifications that distinguish particular neural cell types, where each one is taking particular cell characteristics into account. The most important characteristic is the shape of the dendritic tree. *Pyramidal cells* are the primary excitatory cell type, the soma of which has a pyramidal shape. They typically have extensive projection fibers in the form of a complex axonal structure. The dendritic tree can be highly complex. *Stellate cells*, on the other hand, have a star-like dendritic tree that radiates from the soma. Many other cell types have been identified, including *Purkinje cells, granule cells*, and *tripolar cells*.

The main direction of information processing has led to further distinctions. *Projection neurons* have long axons that transmit signals from tissues and organs into the CNS (afferent sensory neurons) or convey information from the CNS to effector cells (efferent motor neurons). *Interneurons* (of various types and forms) convey signals between projection neurons thereby enabling the communication between sensory and motor neurons. Electrophysiological characteristics can also be distinguished. These characterize typical neural discharge patterns, such as *tonic or regular* activity, that is, a persistent, approximately constant discharge, versus phasic cell activity, which is characterized by bursts of discharges. Another distinction characterizes the effect of a cell's activity on the postsynaptially connected neurons. *Excitatory* connections increase the likelihood that the postsynaptic neuron will fire, while *inhibitory* connections decrease the likelihood. Finally, the neurotransmitters that are primarily produced by a cell lead to further distinctions, the most well-known being *cholinergic, GABAergic, glutamatergic, dopaminergic,* and *serotonergic* types.

These various distinctions point out that each neuron has particular characteristics, which most likely also result in distinct functional, computational characteristics. As a

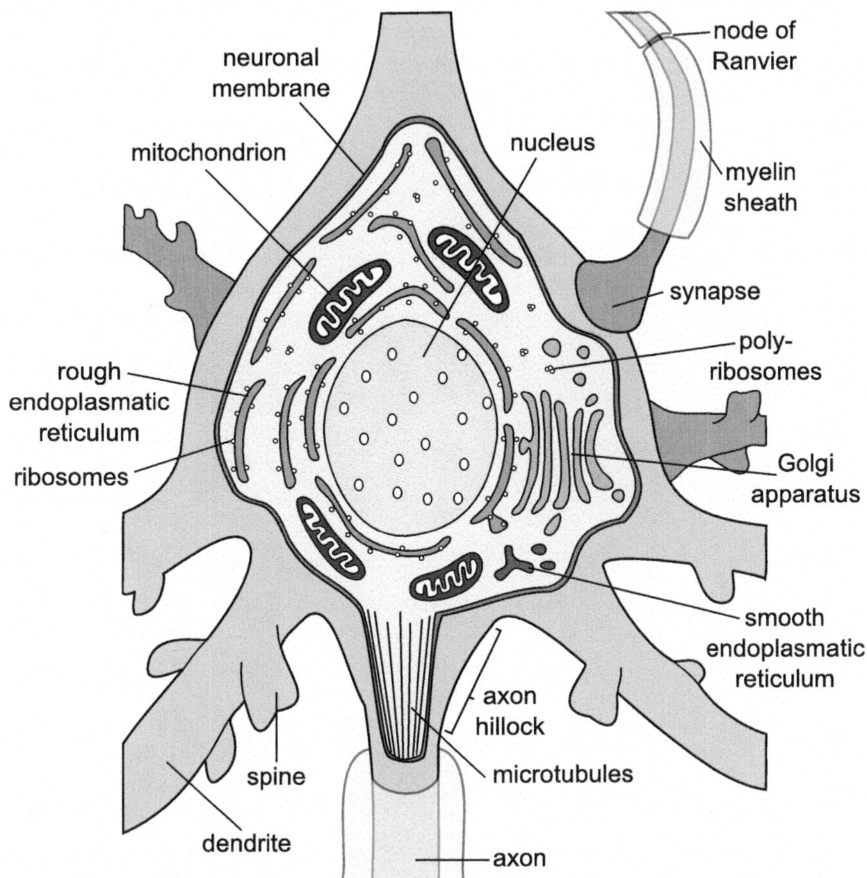

Figure 7.2: A neural cell has a complicated, self-sustaining structure, a dendritic tree, which receives information from other neurons, the axon hillock, which thresholds the integrated received information, and the axon, which transfers the own neural activity to other neurons and other cells. [Adapted with permission from Mark F. Bear, Barry W. Connors, Michael A. Paradiso, *Exploring the Brain*, 3rd Edition, (c) Lippincott Williams and Wilkins, 2007.]

result, neural computation can be assumed to be much more complex than typically assumed by an *integrate and fire mechanism*, which simply adds up incoming neural activity and generates outgoing neural activity in the form of *action potentials* once a threshold is reached. Rather, complex activity integration mechanisms appear to unfold, which are also dependent on current chemical and energy distributions, and their availabilities. Once again, for the purpose of this book, it suffices to acknowledge this complexity and to realize that rather intricate computational processes can unfold even within single neurons as well as between intricately connected groups of neurons.

So far we have considered neurons as the units that process information, but we have not detailed how this processing is actually implemented. The key component in information processing can be considered the *action potential*, which characterizes the manner in which a neuron transmits information to other cells (mainly other neurons, but also, for example, muscle cells).

Action potentials are generated at the neural cell membrane. The cell membrane is made of an impermeable phospholipid layer that separates intra- and extra-cellular fluids, in which different molecules and charged ions are dissolved. The cell membrane controls the flow of charged ions, including Na^+, K^+, Ca^{2+}, and Cl^-, by means of ion channels, which enable the inflow or outflow of particular ions. Electrochemical forces, most importantly concentration gradients and voltage gradients, cause a flux of ions across the membrane thereby establishing a stable equilibrium of unequal ionic concentrations on either side.

During this so-called *resting potential* of a cell, the difference between the inside and outside charge lies at about −65mV. Figure 7.3 shows a cell membrane with distinct ion channels, illustrating the basic principle underlying action-potential generation.

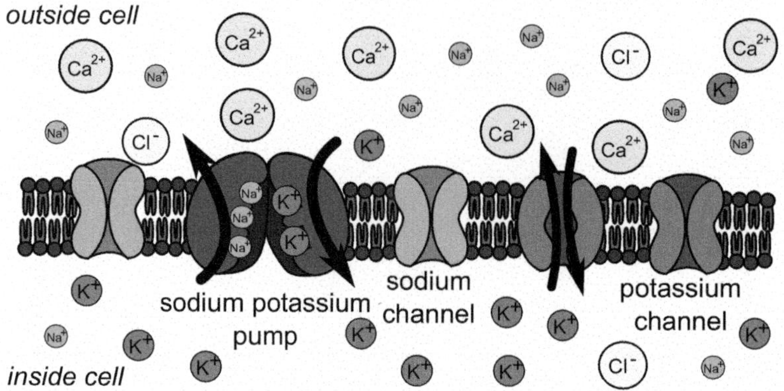

Figure 7.3: Concentrations of extra- and intracellular molecules, which are exchanged via the cell membrane, result in cell polarization and depolarizations, and eventually decide if the cell fires an action potential.

An action potential can be characterized as a sudden, short-lasting reversal of the membrane potential, which is generated by opening voltage-dependent ion channels in the cell membrane of a neuron.   Incoming neural signals invoke the opening of Na-channels, such that NA^+ ions flow into the cell, thereby depolarizing it. Once a particular threshold is reached, the sudden opening of many more Na-channels results in a fast *depolarization,* which typically peaks at about 40mV. This depolarization results in the closure of the Na-channels and the opening of K-channels, which enable the outflow of K^+ ions, resulting in the *repolarization* of the cell. As a result, the voltage temporarily drops below −75mV (which is called *hyperpolarization*) and then proceeds toward the typical resting potential of a neuron (about −65mV), by a slower inflow of K^+ ions. The period before the resting potential is reached again is called the *refractory period*, during which the cell is not able to fire again. The whole process takes about two milliseconds. As a result, action potentials in a cell cannot be generated faster than 500Hz. Figure 7.4 depicts this depolarization and repolarization process.

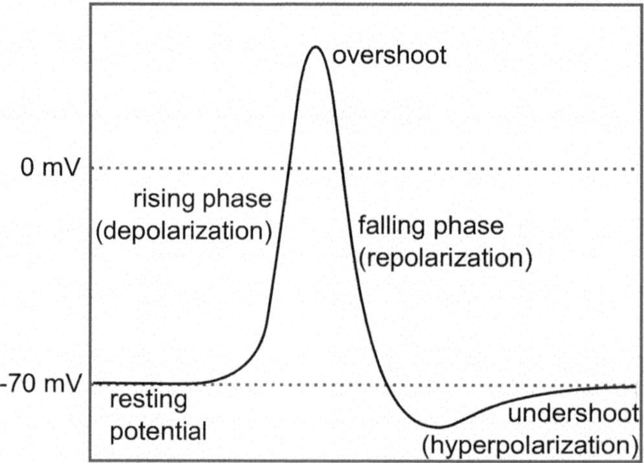

Figure 7.4: An action potential has a typical overshoot and consequent undershoot characteristic.

An important principle of an action potential is the all-or-nothing characteristic, which was originally proposed by the American physiologist Henry Pickering Bowditch (1840–1911) in 1871, who concentrated on the contraction of the heart muscles. The principle says essentially that the magnitude of the action potential is assumed to be constant. That is, given enough exciting stimulation such that an action potential is triggered, it is irrelevant how much more excitation currently is incoming. The action potential will be of equal strength. On the other hand, if the excitation does not reach the threshold, no action potential is triggered. Nonetheless, the frequency of successive action potentials is modulated by the incoming excitatory strength.

Once an action potential is generated, the cell transmits the potential along the cell membrane and ultimately along the axon toward synaptically connected cells. This transmission speeds up when the axon is myelinated, because the myelin sheath insulates the axon transmitting the depolarization until the next cleft in the myelin sheath. At the axon terminals, the potential stimulates the connected presynaptic terminals. Once again, chemical processes unfold, called *exocytosis*. During this process, neurotransmitters, which are stored in synaptic vesicles of the axon terminals, are released into the *synaptic cleft*, which links the presynaptic axon terminal with the postsynaptic dendrite. As a consequence, the postsynaptic part of the information receiving cell is stimulated so that the neurotransmitters dock onto transmitter-specific receptors, influencing the previously described polarization and depolarization processes. Later, the axon terminals reabsorb (that is, reuptake) the released neurotransmitters, which is called *endocytosis*. Figure 7.5 shows the basic process, which takes place in the synaptic cleft.

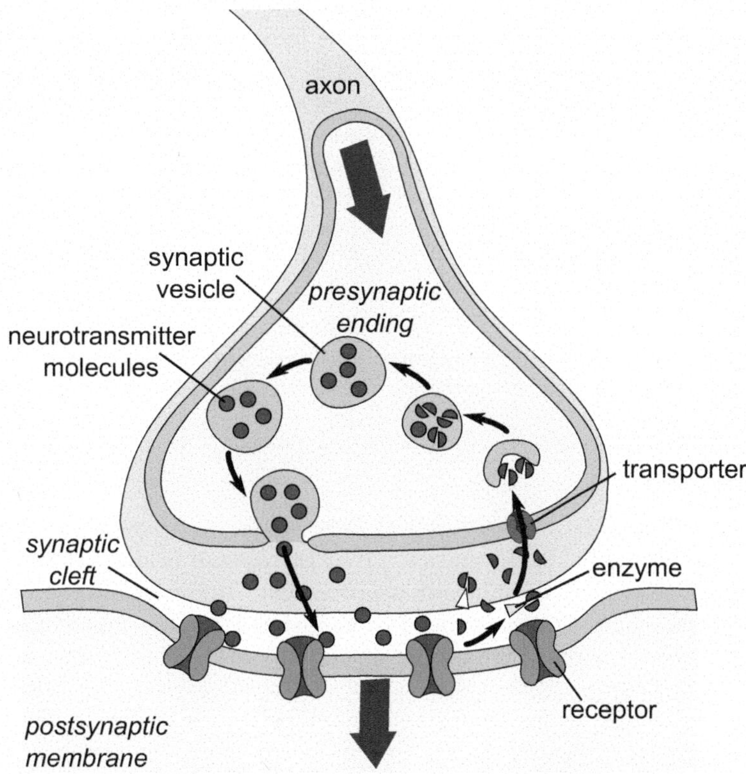

Figure 7.5: Presynaptic activities stimulate via the synaptic cleft the postsynaptic membrane and its receptors, eventually exciting or inhibiting the postsynaptic cell.

The communication between the axon terminals and the stimulated postsynaptic dendrites is controlled by various neurotransmitters, of which we have mentioned only a few. These neurotransmitters selectively activate particular types of receptors, of which two fundamental types can distinguished from a functional perspective. Particular transmitters result in *excitatory postsynaptic potential*, pushing the receiving neuron toward an action potential. In contrast, other transmitters may generate *inhibitory postsynaptic potential*. Moreover, *metabotropic* and *ionotropic* types of receptor determine if the information transmission is indirect, modulatory, slower, and longer-lasting or more direct, faster, and short lasting, respectively. Thus, activities from axon terminals can result in postsynaptic excitation and inhibition, depending on the released neurotransmitters. Moreover, these postsynaptic effects can also differ in their immediacy and duration.

Disruptions in these signal transmission processes are known to potentially generate many influences on our cognitive abilities. Diseases such as Parkinson disease, Attention-deficit/hyperactivity disorder (ADHD), or depression, for example, are associated with disruptions in dopamine concentrations. Also many drugs are well-known to influence the transmission process in various ways. For example, the neurotransmitters may be substituted or imitated by a drug, or the drug may result in the erroneous activation of neurotransmitter receptors, or they may inhibit the reuptake of neurotransmitters.

To summarize, neurons transfer information to other neurons mainly by firing action potentials. The action potentials are transmitted via the axon of a neuron toward synapses. At synapses, intricate electrical and chemical processes unfold, which control the extent of stimulation or inhibition of the membrane of postsynaptic dendrites. The integration of these stimulations again leads to the potential generation of an action potential in the connected neurons, and so forth.

Looking back at the simplified example of the information flow from the eyes to the hands, light-sensitive neurons in the retina fire action potentials when stimulated by photons. The integration of many of these action potentials essentially analyzes the sensory activities, deducing information about the outside world. Based on this information, further neurons integrate this information as well as other information sources, such as the current state of satiation, to come to an object interaction decision. This decision is finally transferred to motor neurons, which cause muscle activations when being neurally stimulated.

Neurons can thus be viewed as units of information processing which, however, also strongly interact with the current energy and chemical distributions. Especially the distributions of neurotransmitters, ion channels, and receptor types strongly influence how neural activities unfold, and thus how neural information processing proceeds. Of course, the general energy supply of a cell by means of, for example, oxygen, also needs to be continuously ensured to enable proper information processing. In later chapters, we will relate fundamental computational mechanisms with these general neural information processing principles.

7.3.2   Modules and areas

While we have seen that individual neurons and neural networks can apparently perform intricate information processes, and thus computations, it is the overall brain architecture that actually enables human-like cognitive information processing. Information about the environment and our thoughts are not so much encoded by single neurons, but much more so by billions of highly interconnected neurons. In fact, the brain appears to contain about 120 billion neurons, with particular neural distributions, which are detailed later. Networks of neurons are likely the units of thought, rather than single neural activities, and redundancies in the neural encodings ensure that cognition does not break down when some neurons die. Let us now take a closer look at the brain as a whole.

To be able to orient ourselves in the three-dimensional brain it is useful to be able to name different sections and directions within it. Figure 7.6 provides an illustrative overview of the most important terms. Similar to an area map, the map of a brain can be partitioned into particular sections. Due to its three dimensionality these sections need to be specified with respect to particular planes and plane-respective brain regions. The planes perpendicular to the x-, y-, and z-axes, are referred to as *sagittal*, *coronal*, and *axial* (or horizontal) planes, respectively. Section-wise then, *anterior* ("toward the front") sections can be contrasted with *posterior* ("toward the back") sections, where anterior and posterior can be used in various frames of reference. Generally, however, anterior areas refer to the brain areas that lie toward the forehead, while posterior areas lie toward the back of the head. In a related, but not identical manner, *rostral* areas are referred to as lying toward the oral or nasal region, whereas *caudal* areas lie toward the tail or the tailbone. Within the axial plane, brain regions can lie *medial*, that is, toward the middle or center, or *lateral*, that is, toward the left or right side of the brain. Within the sagittal plane, one refers to *superior* and *inferior* regions, which are sometimes used synonymously with *dorsal* and *ventral* regions, and *anterior* from *posterior* regions.

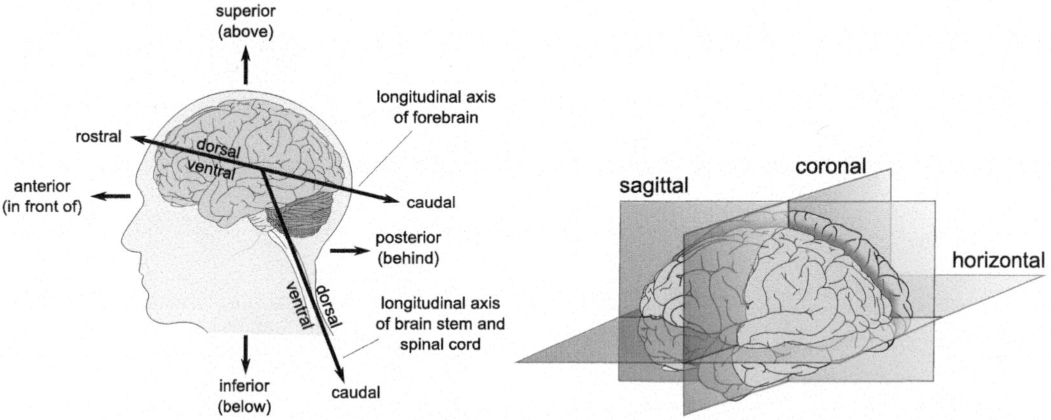

Figure 7.6: To be able to address certain brain regions, the visualized anatomical terminology is typically used. [Reprinted with permission from Purves, D., Augustine, G. J., Fitzpatrick, D., Hall, W. C., LaMantia, A. S., McNamara, J. O. & Williams, S. M. (Eds.) (2004). *Neuroscience*. Sunderland, MA: Sinauer Associates, Inc..]

From a surface and cross-sectional anatomical perspective, the brain can generally be partitioned into three main sections. The *cerebrum* is the largest brain structure, which includes the *cerebral cortex* as well as several subcortical structures, of which the best known are the hippocampus, the basal ganglia, and the olfactory bulb. The largest part of the cerebral cortex is the large, folded brain structure, which covers the rest of the brain. In humans, it is dominated by the *neocortex* or *isocortex*, which has a six-layered neural structure. The brain is furthermore partitioned into two *hemispheres*, which communicate with each other via the *corpus callosum*. The cerebral cortex in humans is folded such that the surface of it can be further characterized by gyri (ridges) and sulci (furrows), some of which are sometimes referred to as *fissures*. Most pronounced and well-known are the *central sulcus* and the *lateral sulcus*. Moreover, the main areas of the cortex are also partitioned into lobes: the *frontal lobe* is the part anterior of the central sulcus, while the *parietal lobe* refers to the posterior part, which extends to the *occipital lobe*, which refers to the most posterior parts of the cortex. Finally, the *temporal lobe* refers to the brain areas that lie ventral of the lateral sulcus and the parietal lobe, and anterior of the occipital lobe. Figure 7.7 provides an overview of these areas and their anatomical position in the brain, from various brain surface-specific orientations.

Over the last century, efforts have been made to address individual cortical areas more explicitly and precisely. The German neurologist Korbinian Brodmann (1868-1918), mentioned in Section 2.3.1, generated a map of the isocortex, based on variations in cytoarchitectonic and histological features. He distinguished, for example, layer thickness and cell distributions. In his book from 1909 (Brodmann, 1909), Brodmann identified 52 distinct cortical regions, which are now called Brodmann areas (BA). Although his numerical nomenclature is only one among several, and his areas have been debated, revised, and refined many times, especially as afferent, efferent, and internal connectivity patterns were not considered by Brodmann, many Brodmann areas have been closely correlated to diverse cortical and

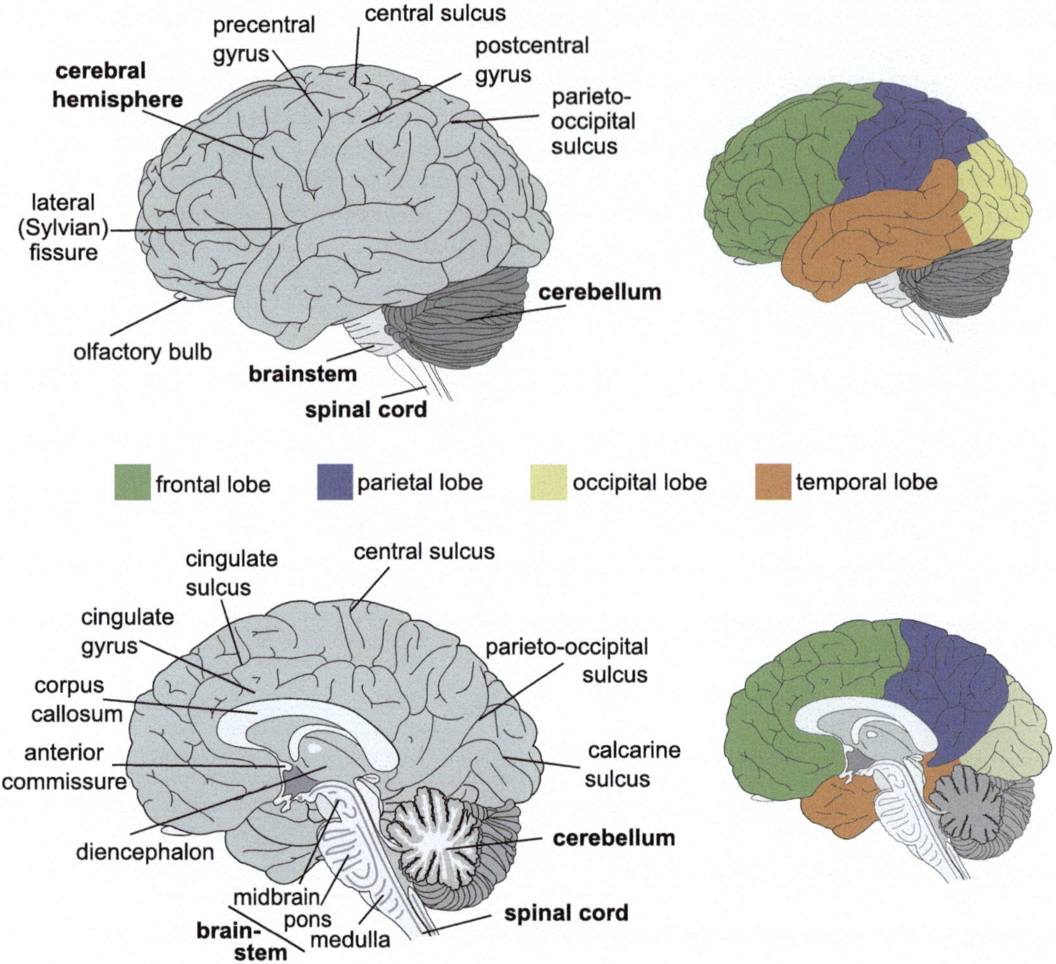

Figure 7.7: Most important brain areas and their names. [Adapted with permission from Mark F. Bear, Barry W. Connors, Michael A. Paradiso, *Exploring the Brain*, 3rd Edition, (c) Lippincott Williams and Wilkins, 2007.]

cognitive functions. For example, the Canadian neurologists Wilder G. Penfield (1891–1976) and Theodore B. Rasmussen (1910–2002) showed that stimulations of BA 1–3 (located in the somatosensory cortex) can lead to the invocation of particular somatosensory impressions. Similarly, particular stimulations of sites in the primary motor cortex (BA 4) result in muscular activity. Even more importantly, the discovery led to the cartography of the *somatosensory homunculus* (in BA 1–3) and a motor homunculus (in BA 4), implying a somatotopic modularization of bodily perceptions – that is, neighboring regions in BA 1–3 typically perceive sensory information of the skin, muscles, and joints in neighboring body parts. As a result, Brodmann's nomenclature is still used as the basis for describing the cortical locations of functional and anatomical findings in humans. Figure 7.8 shows a chart of the Brodmann areas in the original and in a more readable, currently accepted form.

When considering the brain's anatomy from a developmental perspective, not only the neocortex, but the whole brain's anatomy falls into distinct modules, exhibiting modular and hierarchical structures. As we discussed in Section 4.2.1 on prenatal development, the brain develops very early during the embryonic stage and further during the fetal stage. After about four weeks, the developing brain structure is already partitioned into three basic components: the *mesencephalon*; the *rhombencephalon* or hindbrain, which then is further differentiated into *metencephalon* and *myelencephalon*; and the *prosencephalon* or forebrain, which is further differentiated into *telencephalon* and *diencephalon*.   Over

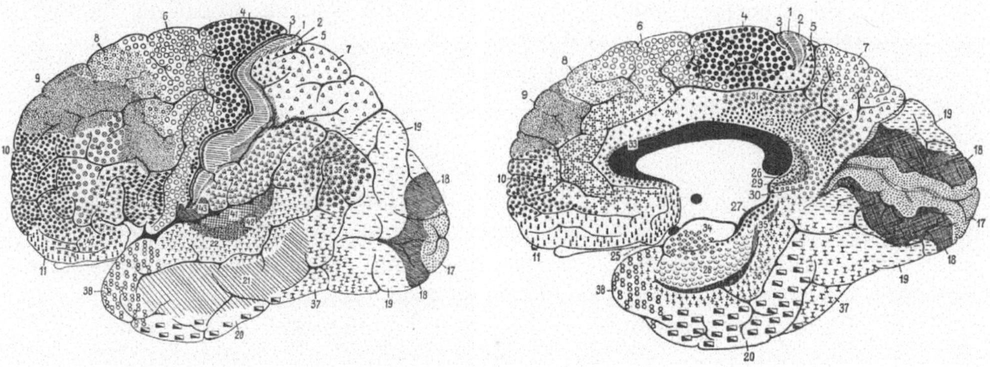

Figure 7.8: Brodmann areas from the original paper. [From Brodmann, K. (1909). Vergleichende Lokalisationslehre der Grosshirnrinde: in ihren Prinzipien dargestellt auf Grund des Zellenbaues. Leipzig: Barth. Republished with permission of Springer-Verlag US, from Brodmann's *Localization in the Cerebral Cortex*, translated by Laurence Garey, 3e, Copyright (c) 2006.]

the next 8 months the brain develops into its well-known, adult-like shape with the folded cortical structures and all relevant, functional components. Table 7.1 gives an overview of the major subdivisions and the most important brain components that are contained in these subdivisions.

7.3.3 Basic brain and body maintenance

The brain is supplied with energy and protected against disruptions, such as external forces by multiple means. Twelve pairs of *cranial nerves*, which originate (mostly) in the brain stem, provide sensory signals (including taste, vision, smell, balance, hearing, and somatosensory information) and enable motor efferences mainly to muscles in the head (for example, face, mouth, oculomotor) and neck. *Meninges* (singular meninx, Greek for "covering") are three membranes that envelop the brain and the spinal cord. The outermost and toughest one is the *dura mater*. The middle meninx is the *arachnoid mater*, so named because of its spider-web appearance. Finally, the innermost, very thin and soft membrane is the *pia mater* (cf. also Figure 7.11 and Figure 12.3). It follows all of the brain's contours (gyri and sulci), while the other two membranes form rather "loosely fitting sacs". The meninges are very useful in protecting the central nervous system. The *ventricular system* provides further protection of the brain. It is a system of hollow spaces inside the brain that are filled with cerebrospinal fluid (CSF), which flows through the ventricles and into the subarachnoid space between the pia and the arachnoid mater. Due to the ventricular system, the actual neural and glial brain tissue actually floats inside the head and is thus further protected against external forces, such as a hit on the head.

The blood supply is provided via paired *vertebral arteries*, which first converge on an unpaired basilar artery. The basilar artery then splits into three large cerebral arteries, which later further divide into smaller internal carotid arteries. These arteries provide oxygenated blood, glucose, and other nutrients to the brain and carry back deoxygenated blood, carbon dioxide, and other metabolic byproducts. The *blood–brain–barrier* prevents viruses and toxins from infecting the brain.

When making cross-sections of the brain, different regions in the section appear to be whitish and others look grayish. Correspondingly, one speaks of *white matter* and *gray matter*. Gray matter regions contain cell bodies and *neuropil*, which is an accumulated mesh of unmyelinated dendrites, axons, and glial cells. These structures primarily form nuclei, that is, clearly distinguishable neuron assemblies deep in the brain, and cortical components, which are collections of neurons that form the thin, layered neural structure at the brain's surface. White matter regions are typically myelinated axon tracts and *commissures*, which

Table 7.1: Taxonomy of brain structures based on anatomy and development, including the main components of the respective sub-structures and some of the main functionalities.

Main structure	Included sub-structures	Main components	Main functionalities
Prosence-phalon (forebrain)	Telence-phalon	Cerebral cortex	Associative learning; main center of the central nervous system; sensorimotor and cognitive control; consciousness
		Limbic system including hippocampus, amygdala, nuclei, olfactory bulb and others	Various functional centers for establishing episodic memory (hippocampus), regulating emotions (amygdala), sensory stimulations (olfactory bulb), neural information exchange (nuclei), and interaction with the vegetative nervous system
		Basal ganglia	Support of motor control, procedural learning, and habits
	Dience-phalon	Thalamus	Relay and control of incoming sensory and outgoing motor information, as well as cortical information exchange
		Hypothalamus	Control of metabolisms and other functions of the autonomous nervous system
Mesence-phalon	Mesence-phalon (midbrain)	Tectum with inferior and superior colliculus, cerebral peduncle	Reflex systems, information exchange between spinal cord and brain
Rhomb-ence-phalon	Metence-phalon (hindbrain)	Cerebellum	Automatization, stabilization, and smoothing of behavior
		Pons	Relay station between brain and cerebellum as well as between sensory afferents and thalamus; control of breathing and circulation
	Myelence-phalon	Medulla oblongata	Regulation of autonomous bodily functions including heartbeat, respiration, and circulation

connect the various gray matter areas. Figure 7.9 shows a particular cross-section where gray and white matter regions are easily distinguishable.

7.4   General organizational principles

Starting in the 19th century (and partially even earlier), the first insights about the functionality of particular brain modules were made by observing the results of brain lesions. Three famous studies from the 19th century are of particular interest to cognitive science.

The first study came about as a result of the peculiar accident the metal worker Phineas Gage had in 1848. A piece of metal flew through a large part of his left frontal brain region, so that he suffered from a severe injury in the area of the orbitofrontal cortex (OFC) and prefrontal cortex (PFC). Gage survived this accident and appeared normal initially with respect to language, intelligence, memory, and reasoning, but not with respect to his social behavior, which was much more impulsive and vulgar after the accident. Gage lived for another twelve years and to a certain extent recovered his social abilities. The question of how exactly OFC and PFC influence our social and personal traits, however, is still an important matter of research.

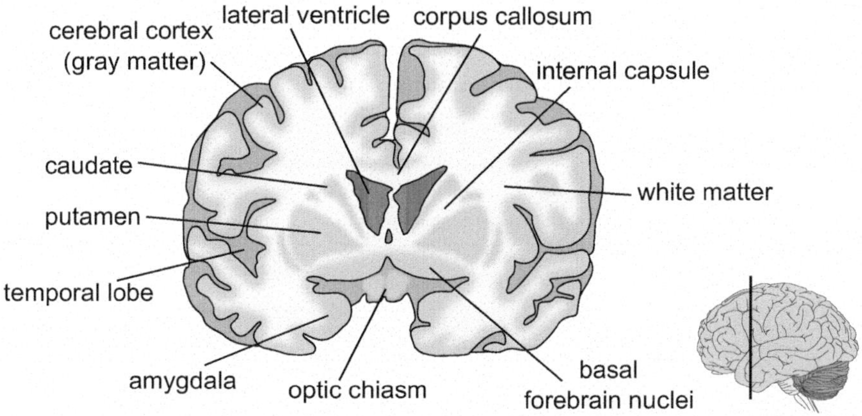

Figure 7.9: Cross-section with most important structures indicated. [Reprinted with permission from Purves, D., Augustine, G. J., Fitzpatrick, D., Hall, W. C., LaMantia, A. S., McNamara, J. O. & Williams, S. M. (Eds.) (2004). *Neuroscience.* Sunderland, MA: Sinauer Associates, Inc..]

The second and third famous lesion studies concern language and the two major language areas in the brain: named after the French medial doctor P. Paul Broca (1824–1880), the *Broca area* in the left inferior frontal gyrus is necessary for speech production and lesions lead to selective impairments in the ability to speak, but to hardly any impairments in the ability to comprehend speech. Named after the German neurologist Carl Wernicke (1848–1905), the *Wernicke area* in the posterior section of the superior temporal gyrus is predominantly responsible for speech comprehension, such that lesions in this area lead to the failure of comprehending speech, but typically leave speech production capabilities intact. These insights suggest that particular areas can be important for realizing specific cognitive functions. However, today neuroscientists often hesitate to assign clear functions to particular brain regions, because observed functionalities or impairments often depend on the actual experiments conducted, on which paradigms are employed, on which stimuli are shown, and even on general health conditions of the patients or study participants. Indeed, over the last decades, Broca's area, for example, has been shown to be also involved in manual communication by means of sign language and homologous areas have been identified in monkeys, which are involved in the production of alarm calls (Corina, McBurney, Dodrill, Hinshaw, Brinkley, & Ojemann, 1999; Gil-da Costa, Martin, Lopes, Munoz, Fritz, & Braun, 2006; Petrides, Cadoret, & Mackey, 2005) (cf. also Chapter 13.3).

7.4.1 Function-oriented mappings

Over the last several decades numerous researchers have assigned particular functions, encodings, and representations to particular brain areas. Beginning in the 19th century, phrenologists tended to assign brain areas particular cognitive functions. A somewhat revealing picture from the end of the millennium provides an even more detailed perspective (cf. Figure 7.10), which certainly has some truth to it. Despite the insights that might be gained from such a map, it is currently generally accepted that individual brain areas are typically not involved in only one cognitive function. Moreover, depending on the paradigm and stimuli used, the functional assignments may be overgeneralized or overly restricted. Thus, assigning functions to individual brain areas, if one considers them at all, should be interpreted with a grain of salt, if not with extreme caution.

Despite this danger, it does come in handy to develop a crude brain atlas in ones mind to be able to have a working hypothesis of most likely critically involved brain areas when particular mental processes unfold. Clearly, primary sensory and motor areas can be linked to particular functions, which is the primary analysis and generation of suitable sensor and motor codes. Next, the temporal lobe can roughly be separated into posterior and

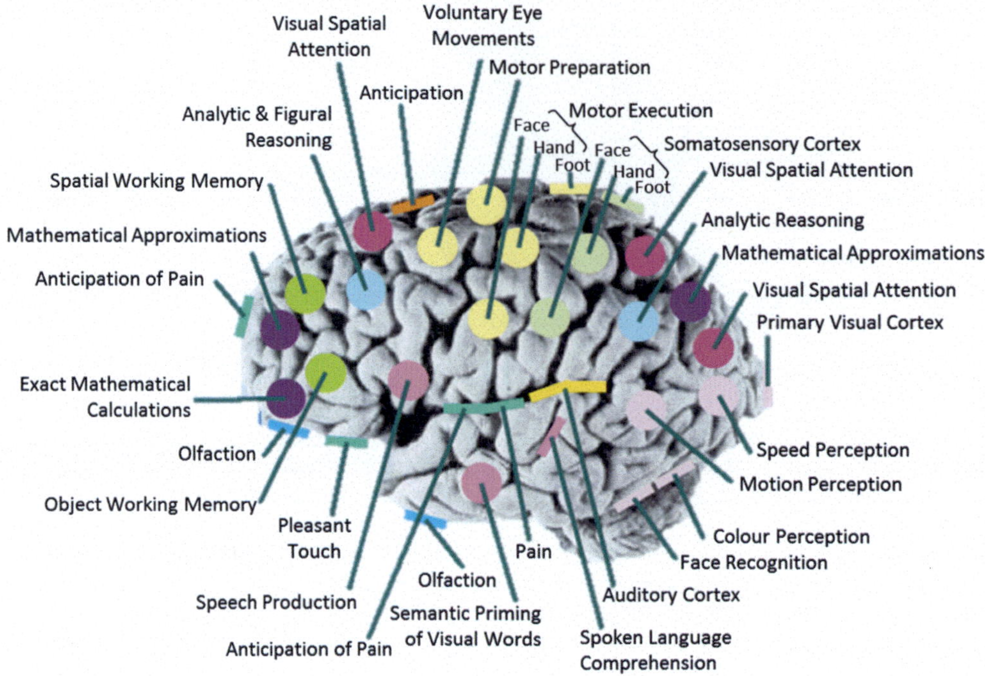

Figure 7.10: An area-function map as illustrated in Nichols and Newsome (1999, p. C36.). [Reprinted by permission from Macmillan Publishers Ltd: *Nature, 402*, C35–C38, The neurobiology of cognition, Nichols, M. J. & Newsome, W. T. Copyright (c) 1999.]

anterior regions. The posterior temporal regions are typically strongly linked to object recognition, including humans, animals, tools, and simple objects. The anterior temporal regions are involved in linking words with meaning and also in supporting meaning. The superior temporal regions focus on dynamics while the inferior regions focus on snapshot-like encodings. Note how the temporal lobe largely abstracts over spatial aspects, focusing on feature and identity encodings independent of space.

In contrast, the parietal areas primarily focus on spatial aspects of the environment. Coming from the visual side, the posterior parietal cortex has often been characterized as the *where-* or *how-stream* of visual processing, while the temporal lobe has been characterized as the *what-stream* (Mishkin, Ungerleider, & Macko, 1983; Milner & Goodale, 1995, 2008). It is generally accepted that posterior parietal areas are largely involved in encoding and processing spatial interactions – regardless of whether these interactions are physical, attentional, or mental (for example, numbers). As physical interactions are experienced by one's own body, it may not come as a surprise that anterior regions contain the somatosensory cortex, and thus generally speaking a map of ones body. In recent years, the superior parts of the posterior parietal cortex have additionally been distinguished from the inferior parts, where the former have been related with motor control and thus continuous changes in space, while the latter have been related to more abstract forms of planning, tool usage, and action observation (Glover, Rosenbaum, Graham, & Dixon, 2004; Turella, Wurm, Tucciarelli, & Lingnau, 2013).

Between these two regions the *intraparietal sulcus* has been closely related to controlling environmental interactions of distinct motor actions, such as eye saccades, manual manipulations, interactions with the mouth, as well as protective actions (Graziano, 2006; Graziano & Cooke, 2006). Intraparietal areas have also been shown to be closely interactive with premotor areas, suggesting the establishment of a recurrent network between the two areas. These interactions have even been related to Wernickes's speech comprehension area and Broca's area, which lie further inferior to inferior parietal and inferior premotor areas

(Graziano & Cooke, 2006). Premotor cortical areas have also undergone further distinctions over the last decade or so, separating inferior from superior and medial areas, where different motor actions seem to be dominantly controlled, such as hand-to-mouth, defensive, reach-to-grasp, and climbing movements. Lower level actions, such as chewing or manipulating the space in front or below the body have been localized more posterior in M1 of monkeys (Graziano & Aflalo, 2007). Thus, it seems that interactions are encoded conceptually distinctly, separating types of potential interactions with the environment.

Decision making and abstract forms of planning have been localized in the frontal lobe. Moreover, the medial prefrontal cortex (MPFC) was shown to be involved in social cognition.

In particular, MPFC was shown to be involved in action monitoring, perception of the intention, self-knowledge, and mentalizing the current knowledge of others (Frith & Frith, 2003). Thus, it seems that here the brain focuses on distinguishing the self from others in social spaces, including the current knowledge of oneself and of others. For example, knowledge and inference processes about, for instance, whether a child thinks that Smarties are in a Smarties box or a pen, which was secretly put inside, is actually in the box, seem to be supported by these areas. Finally, the hidden *cingulate cortex*, which can be found medially centrally covered by the frontal and parietal lobes, is part of the limbic system. It is believed to be strongly involved in assessing values, that is, reward, and co-controlling motivations and emotions.

Finally, the *lateralization* of functionalities in the two brain hemispheres should be mentioned. Although still hotly debated, it appears that the right hemisphere is more strongly involved in spatial processing than the left hemisphere (Suchan & Karnath, 2011). For example, *neglect patients* typically show a neglect of the left part of their body, as well as the left part of the environment (where left can be interpreted in various frames of reference), after a lesion in the right parietal area. A lesion in the left parietal area, on the other hand, typically does not result in an equally strong neglect to the right side, but typically rather impairs linguistic abilities to a certain extent. Broca and Wernicke's language areas are located in the left side, so that grammar and word production are dominantly controlled by the left hemisphere. However, for the realization of most cognitive functions, both hemispheres are typically involved. Thus, while some researchers believe that the brain lateralization is an important aspect of human phylogenetic development, the reason for or purpose of this lateralization remains obscure (Gazzaniga, Ivry, & Mangun, 2002; Suchan & Karnath, 2011).

7.4.2 Cortical columns and topographies

Another interesting general organizational principle of the brain is that the neocortex is structured into six well-separable neural layers (I–VI). Each layer contains a characteristic distribution of particular neuronal cell types. Moreover, the connectivity within and between layers exhibits a systematic structure; only particular layers are connected with other cortical and subcortical areas:

- *Layer I (stratum moleculare)* has the least density of cells. During development, neurons grow here first and tend to die out when the other five layers have established themselves.

- *Layer II (stratum granulosum externum)* is mainly populated by various stellate and small pyramidal neurons.

- *Layer III (stratum pyramidale externum)* contains mainly smaller pyramidal cells and intracortically connected cells. It is the main target of neural projections from cortical areas of the other hemisphere.

- *Layer IV (stratum granulosum internum)* contains stellate and pyramidal neurons. It receives signals via the thalamic nuclei and from other intra-hemispheric areas.

Accordingly, this layer is particularly pronounced in primary sensory areas while it is almost completely missing in primary motor areas.

- *Layer V (stratum pyramidale internum)* contains much larger pyramidal cells, whose axons typically project their neural activity to subcortical structures. In the motor cortex, this layer is particularly pronounced and contains cells that form the corticospinal tracts to generate motor efferences.

- *Layer VI (stratum multiforme)* is populated by few large pyramidal neurons and many much smaller spindle-like pyramidal and other neurons. It projects activities to the thalamic nuclei, establishing very precise interconnections between thalamus and cortex.

Figure 7.11 shows the six-layered structure, which was first described by Santiago Felipe Ramón y Cajal (1852–1934), whom we introduced in relation to the discovery of synapses (cf. Section 2.3.1).

Although knowledge about how and why the neocortex exhibits this structure is still thin, the systematics in the six layers suggest that general, evolutionary principles are at work, which probably facilitate learning, particular types of neural information processing, and thus particular cognitive development. Interestingly, the cerebellum also exhibits a typical, layered structure which, however, contains only three layers and is populated by different types of cells. The bottom layer is mainly filled with small, granule cells. The middle, thin layer is populated mainly by *Purkinje cells.* Finally, the top layer contains the dendritic trees of the Purkinje cells and huge arrays of parallel fibers, which penetrate the dendritic trees of the Purkinje cells at right angles. Without going into further detail, it should not come as a surprise that the main function attributed to the cerebellum is quite different from that of the isocortex. In particular, while the isocortex is typically referred to as the association cortex and is known to be mandatory for cognition and human consciousness, the cerebellum is most relevant for smoothing and dynamically controlling motor behavior (Barlow, 2002; Fleischer, 2007; Shadmehr & Krakauer, 2008; Wolpert, Miall, & Kawato, 1998).

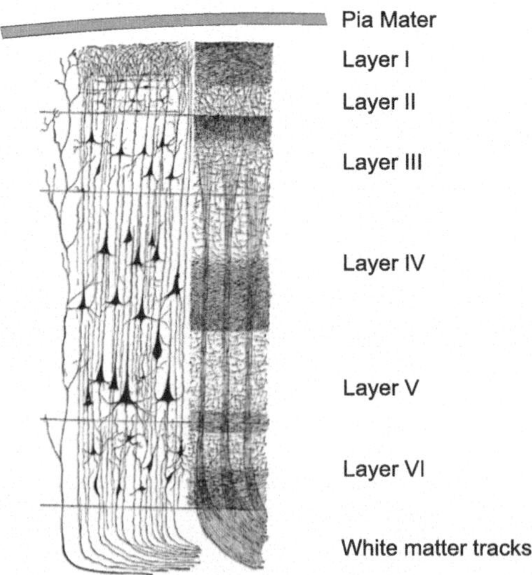

Figure 7.11: Illustration of the typical six-layered structure found in the neocortex [Adapted and annotated from Gray, H. (1918). *Anatomy of the Human Body.* Philadelphia: Lea & Febiger.]

While the cortical columnar structures suggest that cortical information processing obeys particular systematics, it is still unclear what these are exactly. Nonetheless, certain insights have been made and theories proposed.

With the advent of the single-cell, electrophysiological recording technique, the neurophysiologists David H. Hubel (1926–2013) and Torsten N. Wiesel (*1924) worked on neural recordings in the occipital area of cats – known also as the visual cortex (Brodmann area 17 in humans). In 1959 they discovered that individual neurons selectively responded to particular edge orientations when presented visually either statically or dynamically. In 1981 Hubel and Wiesel received the Nobel Prize in medicine and physiology for the discovery of the information processing principles underlying the visual cortical system in the brain. In addition to the particular tuning of individual cells, Hubel and Wiesel uncovered the systematic, columnar arrangement of cells in the visual cortex, which was originally discovered in the 1950s by the neuroscientist Vernon Benjamin Mountcastle (1918–2015) in the somatosensory cortex of cats.

It has been theorized that columnar structures can be found ubiquitously throughout the neocortex, tending to form systematic, somewhat topographically organized encodings. A topographic organization essentially refers to a neural organization where spatially adjacent stimuli on some sensory receptor surface, that is, with respect to a particular frame of reference, are encoded in adjacent neural positions within the cortex. The hypothesis is that such *topographic maps* continuously and completely represent their relevant sensory or motor dimensions. Additionally, topographic and anatomical boundaries align with each other (Patel, Kaplan, & Snyder, 2014). The most clearly established examples are the encodings in the primary sensory, motor, and somatosensory areas. A retinotopic map can be found in the primary visual cortex, the properties of which we will detail further in Section 8.4. In the motor cortex (M1), a motor- or muscle-topographic organization can be found, which maps the muscle-topography of the body. Additionally, in the neighboring somatosensory cortex (S1), a body-surface grounded topography can be identified, which essentially reflects the sensitivity of the skin and bodily joints in a body topography. Figure 7.12 shows the homunculi of M1 and S1.

In all these three areas, the topographies are sensor-grounded and reflect the sensitivity of the respective spatial areas. While in V1 the fovea is encoded with more neurons, in S1 more neurons process sensory signals from tongue and fingers than from a leg or the belly. Additionally, in the auditory system, a tonotopy has been identified, which encodes similar tones (in terms of frequency) in adjacent regions. Note how also in this case, embodiment supports the development of the tonotopy, because the tonotopy begins already in the cochlea, where the basilar membrane vibrates at different sinusoidal frequencies depending on the incoming tones.

Besides these strongly sensory- and motor-grounded topographies, however, it appears that deeper cortical areas also exhibit topographies which, however, are typically neither fully sensory- nor motor-grounded. For example, Patel et al. (2014) investigated the topography in the lateral inferior parietal area (LIP), which responds both to visual stimuli as well as to eye saccades. Functionally, theories suggest that LIP is involved in integrating a saliency map to plan eye saccades. LIP has also been shown to be involved in object categorization, reward estimation of eye saccades, and deeper oculomotor planning. Deeper investigations suggest that the topography in LIP can be divided into a ventral area (LIPv) and an anterior LIP area. LIPv is hypothesized to encode a polar angle map, which has been shown to be involved in both, oculomotor planning as well as orienting spatial attention. The anterior part of LIP, on the other hand, seems to be mostly involved in inspecting the currently fovealized stimulus. Thus, LIP violates the principles of a sensory- or motor-grounded topographic map, because its anatomical structure contains several topographic maps, which additionally appear to support several functions. Nonetheless, the organization of LIP still seems to be still generally topographically, but – probably because LIP processes multiple sources of information (visual and oculomotor) – it appears to integrate these sources in maximally suitable topographies – focusing on the fovealized stimulus in the anterior part while planning the next focus in the ventral part. Somewhat similar insights exist for the frontal eye field (Patel et al., 2014, cf.) and even a numerosity-differentiating topography

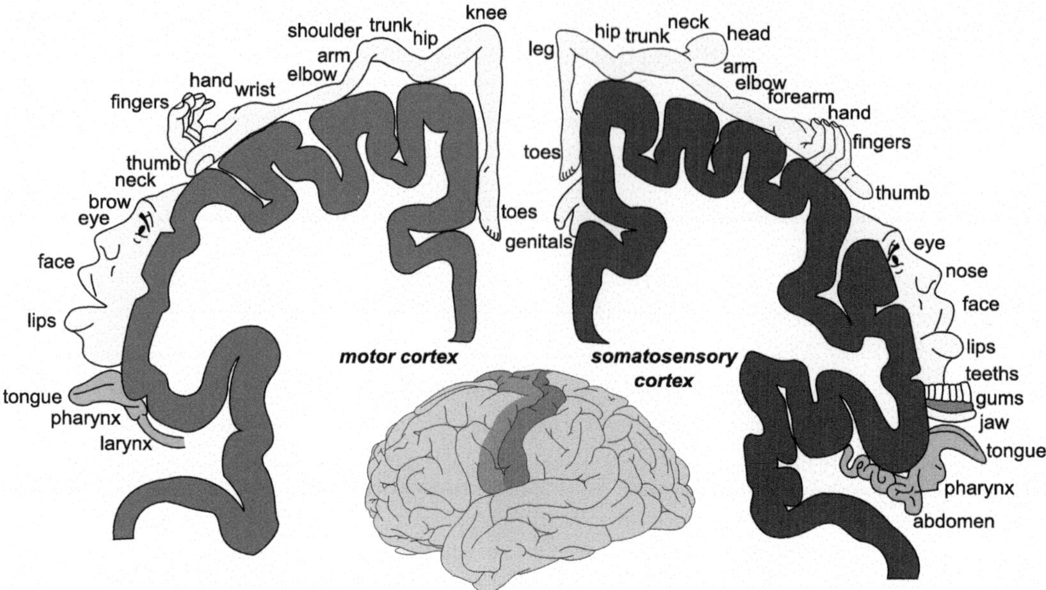

Figure 7.12: The somatosensory- and motor-homunculi beautifully illustrate sensory-grounded topographic encodings.

was identified in a distinct posterior superior parietal area (Harvey, Klein, Petridou, & Dumoulin, 2013).

While these insights are still sketchy, they suggest that the principle of a topography may be implemented in many if not all cortical areas. However, still it is not well-understood what these topographies are when considering deeper neural areas. As a general principle, it might be the case that the brain strives to minimize wiring lengths, encoding neighboring stimuli or neighboring abstract concepts neurally close to each other. Because neighboring information in a topography may complement each other or may contrast with each other (think of an edge or a surface), neighboring very short-range neural interactions may support such computations. The six-layered structure of the cortical surface additionally suggests that this principle may apply generally. In deeper layers, however, it still needs to be shown, which topographies – which may even not be spatial at all, but may encode particular feature or conceptual spaces – are actually being encoded. Clearly, further brain imaging studies are necessary to verify or falsify these claims.

7.4.3   Neural tuning and coordinated communication

In addition to the spatial aspect of a topographic encoding, the term also implies that particular properties are encoded in a spatially distributed manner. That is, single cells are *neurally tuned* to respond to one particular type of sensory, motor, or associative information. In the simple case of V1, neurons are known to respond to particular visual stimulus orientations. Thus, neurons in V1 are tuned both, retinotopically and feature-specifically such that a neuron that is maximally responsive to a vertical edge will respond progressively less strongly to progressively less vertically-oriented edges, as well as to vertical edges that are presented progressively further from the center of the neuron's retinotopic receptive field. Figure 7.13 illustrates this tuning property.

Besides such tuning to rather simple sensory or motor features, however, highly complex features have also been shown to be encoded by single neurons, which are then referred to as *grandmother neurons*. Such a neuron essentially is maximally responsive to a particular entity, such as your grandmother. In 2005, a widely cited study was published that reported about

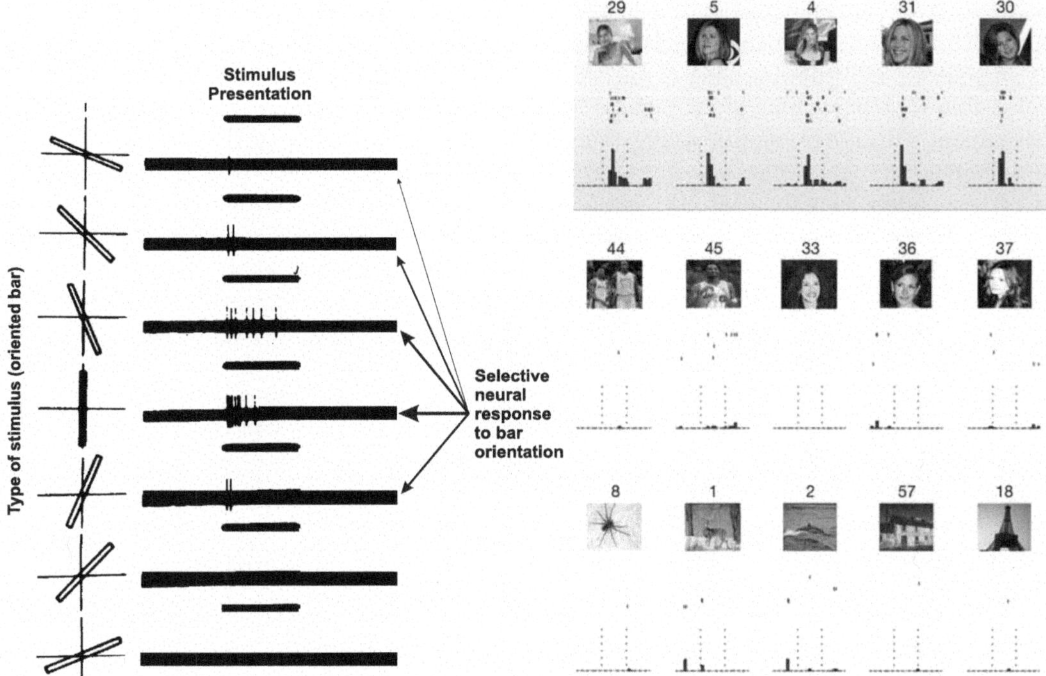

Figure 7.13: Neuronal tuning cells in the visual cortex according to Hubel and Wiesel (left) as well as more recent findings of person-selective cells in the hippocampus within the medial temporal lobe (right) (left: adapted from Hubel, 1993, p. 32; right: Quiroga et al., 2005, p. 1103). [left: Reprinted from *Nobel Lectures, Physiology or Medicine 1981–1990*, Editor-in-Charge Tore Frängsmyr, Editor Jan Lindsten, World Scientific Publishing Co., Singapore, 1993. Copyright © The Nobel Foundation 1981. right: Reprinted by permission from Macmillan Publishers Ltd: *Nature, 435*, 1102–1107, Invariant visual representation by single neurons in the human brain, Quiroga, R. Q., Reddy, L., Kreiman, G., Koch, C. & Fried, I. Copyright (c) (2005).]

> [...] a remarkable subset of MTL [medial temporal lobe] neurons that are selectively activated by strikingly different pictures of given individuals, landmarks or objects and in some cases even by letter strings with their names. (Quiroga et al., 2005, p. 1102.)

For example, a neuron was shown to be highly responsive to the actress Jennifer Aniston (famous for her role in the sitcom "Friends"), such that the neuron responded to various different pictures of Ms Aniston, but not to pictures of other persons or even to completely different pictures, such as landscapes or animals (Figure 7.13). Seeing that the selective response of some of these neurons could partially even be extended to the encoded person's name in the form of letter strings, these encodings link auditory and visual information sources (and probably others), merging them into one distinctive code – offering a partial solution to the *symbol grounding problem* of naming entities (cf. Section 3.4.1). While there appear to be grandmother neurons, it should not, however, be concluded that when a grandmother neuron dies then the grandmother cannot be recognized any longer. Most likely redundancy in the encoding, supported by a neural network of corresponding encodings, prevents dependencies on individual neural activities.

Neural tuning appears to be additionally supported in the brain by the principle of *sparse coding*. In fact, there is accumulating and striking evidence that the brain encodes particular stimuli by a rather small subset of maximally active neurons. Most of the other neurons remain silent. With respect to the insights noted previously, for example, grandmother cells appear to encode particular persons, and most likely similar neurons exist for particular objects and other particular entities. Similarly, when focusing on a particular spatial location, neurons selectively encode this spatial location, typically in sensory, sensorimotor, or motor-grounded frames of reference. Given the large number of possible inputs and the huge amounts of sensory information that is gathered from the environment, it seems to be generally plausible that particular brain regions focus on encoding particular aspects of body and environment. Moreover, modularizing the neural encodings enables to focus on a particular spatial subspace, feature subspace, entity subspace, or even a motivational sub-space. In fact, sparse coding may facilitate the *binding problem* (cf. Section 3.4.3), enabling the binding of several stimulus sources temporarily to each other, essentially focusing on those features, entities, or other aspects that currently belong together and are currently relevant.

Together with the topographic encodings in the form of population codes, which encode a particular (possibly loose) topography as described previously by means of neural cortical columns, sparse coding may selectively activate those aspects in a neural topography that belong together, thus binding them together.   Meanwhile, other possibly contradictory facets may be inhibited, enabling the inhibition of disruptive influences. If the brain knows which topographically encoded information in different brain modules typically co-occur with which other topographically encoded information, it may establish a temporarily active network of relevant encodings, which essentially constitutes the current focus of attention. In fact, this may be the brain's approach to solve the *frame problem* (cf. Section 3.4.2), that is, the problem of being able to focus on those aspects of the world that are relevant for an organism or a system, essentially making planning and even abstract reasoning possible.

How does the brain accomplish the sparse encodings in topographically organized pop-ulation encodings and probably also other neural encodings? In other words, how does our brain enable us to focus on particular aspects of the world and to largely ignore other as-pects? Where this capability comes from is still under debate. However, it seems clear that there is no distinction between hard-disc memory and RAM, that is, random-access memory, as is the case in most computers. That is, there is no central working memory unit. Rather, working memory, that is, what we currently focus on and process actively in our minds, is solely determined by the neural activities themselves.

In the past decade or so, it was proposed that neural activity is coordinated and brought into coherence by particular rhythms or neural activity (Fries, 2005; Fries, Nikolic, & Singer, 2007). In particular, it was suggested that a fast *gamma-band rhythm*, which lies between 30 and 90Hz, may coordinate current neural binding, where the currently bound activities fire selectively at the peak of this rhythm. Meanwhile, a much slower *theta-band rhythm*, which typically has a speed of about 7–8Hz, was shown to be able to reset the gamma-band rhythm. It was thus proposed that this rhythm enables the selection of the next focus of attention. Combined with inhibitions of the most recent focus of attention then, these two rhythms may enable progressions in the focus of attention, preventing to get stuck on one particular point of focus. Finally, an intermediate *alpha-/beta-band rhythm*, which lies between 8 and 20Hz, was proposed to coordinate top-down influences of focus, possibly enabling to maintain the focus on particular entities, items, or even thoughts over an extended period of time (Fries, 2015). While it has thus been proposed that the rhythms establish neural communication through coherent neural firings with respect to particular neural rhythms, the details of the involved mechanisms as well as their functional implications are still being debated.

7.5   Brain mechanisms and brain imaging

The surveyed insights previously, as well as current research in cognitive neuroscience, are based upon, and dominantly rely on, a diverse toolbox of brain imaging techniques. Indi-vidual techniques, however, only give a glimpse at what is actually happening in the brain.

Think for a moment of a computer. A computer also has various modules and devices, which are interconnected in rather intricate ways. There is a hard disc and RAM, a mother board, and the central processing unit (CPU). Depending on at which level we investigate these modules, assuming no further knowledge, we will detect highly intricate wirings. On a molecular and atomic level, we may be able to identify semiconductor material, which is arranged in highly complex, but systematic structures. On the cellular level, we may be able to identify transistors and other basic electronic processing units. At the next level, electrical circuits may be identifiable and one level more coarsely grained, the arithmetic logic unit (ALU) of the CPU may be detected. When we monitor the CPU level, streams of bits may be registered passing through it, seemingly without any actual higher-level meaning or reason.

As in the computer, it is the actual encoding that is critical: neural encodings need to be deduced and understood, including the involved activity inducing and maintaining mechanisms, learning and memorization mechanisms, information exchange mechanisms, and so on. Depending on which level of granularity is investigated, the brain will reveal different aspects and components of its mechanisms. However, the analogy to the computer should not be taken overly literally. Although it is probably the case that the brain has the same computational capacity that the Turing machine has – and thus any computational device that is equivalent to a Turing machine – *how* these computations unfold seems to be radically different from any currently available computer. Essentially, it seems that the brain activates its working memory directly within its long-term memory structures, thus integrating its CPU into RAM and hard disc, where the latter two are not directly separable in the brain.

The comparison with a computer points out that the right level of granularity needs to be identified for a particular purpose, that is, for understanding a particular functionality of the brain with its neurons and other cells. Table 7.2 shows the different levels and the approximate explanatory power that can be gained at each level. In light of these considerations, we will provide a short overview of the current most prominent neuroscientific brain imaging techniques.

7.5.1   Brain lesion studies

Although not a brain imagining technique, the study of brain lesions, which reaches far back in history, may be considered as the first means of studying brain mechanisms. From the clinical neuroscience side, brain lesion studies have revealed important functions of particular brain areas, although the attribution of particular functions should be considered with care. We already mentioned Wernicke's and Broca's language areas, as well as the case of Phineas Gage (1823–1860) and his rather dramatic changes in personality after a large, prefrontal brain penetration (cf. Section 7.4).

In contemporary research, many advances have been made. The focus typically is on stroke patients, patients that suffer from brain degeneration (for example, Alzheimer patients), although tumor patients and patients with traumatic brain injuries are also studied. In the latter cases, though, regional assignments of functionalities are even more difficult, because damages may be more widely distributed. In all cases the hardest problem may be the correct localization of the brain tissue that is disabled. MRI data (discussed later) typically helps in this localization. The comparison of behavioral data from patients with control groups is also important. Finally, it is typically rather difficult to identify several patients with nearly identical lesions, such that observations in individual patients may be caused by other mental predispositions and not by the actual brain lesion.

Given a lesion occurred, patient studies can yield different results dependent on the length of the recovery period. For example, in acute cases sudden loss of cognitive functions have been reported in a region that is only connected to the actual damaged area, but that was not damaged itself. In this case, the loss just indicates that the cognitive function somewhat relies on neural information from the damaged area. An assignment of the lost cognitive function to the damaged area would be false. In chronic patients, on the other hand, functions that are typically assigned to a damaged brain area may be compensated for by other areas, thus obscuring the typical functional properties of the damaged area.

Table 7.2: The brain's functionalities and mechanisms can be investigated at several different levels, starting with the very fine-grained, atomic and molecular levels up to the organismal, human level. For a computer, similar levels of granularity can be contrasted.

Level of organization	Explanation	Example in the brain	Computer analogue
Organismal level	Several organ systems that function together	Human	Personal computer
Organ system level	Group of organs that carries out a more generalized set of functions	Neuronal system, digestive system	Central processing unit (CPU), main memory
Organ level	Two or more types of tissues that work together to complete a specific task	Heart, stomach, brain	Control units, processor registers, arithmetic logic unit (ALU)
Tissue level	Groups of cells with similar functions	Muscle, epithelial, neuronal tissue	Electrical circuits (logic gates) constructed with transistors and passive components
Cellular level	Smallest unit of life; membrane bound structure of biomolecules	Muscle cell, skin cell, neuron	Transistors, resistors, capacitors
Molecular level	Combination of atoms that can have entirely different properties than the atoms it contains	Water, DNA, carbohydrates	Semiconductor material, for example, gallium arsenide
Atomic level	Smallest unit of an element that still maintains the property of that element	Carbon, hydrogen, oxygen	Germanium, silicon, gallium

Nonetheless, careful studies, which keep these difficulties in mind, have shown that valuable insights can be gained.

Interestingly, bodily lesions also allow for rather intricate deductions about brain functionalities. The most prominent example comes from patients who had an arm or leg amputated. Some of these patients report the existence of a *phantom limb* (Ramachandran & Blakeslee, 1998). Although the limb is gone, their brains seem to indicate its presence, which, understandably, may lead to very uncomfortable feelings at best, but often even to excruciating pain. Studies with such patients have shown that the presentation of a fake arm, for example, by mirroring the other arm, can temporarily ease this pain. It appears that somatosensory brain areas are partially responsible for these symptoms, in that neighboring areas expanded into the lost arm area – signaling false information about its presence. Thus, neural plasticity in this case can lead to the effect of feeling the presence of phantom limbs.

7.5.2   Active methods

In its underlying principle somewhat related to lesion studies, *transcranial magnetic stimulation* (TMS) uses a magnetic field generator (called a *coil*) to temporarily inhibit or enhance neural activities in particular brain regions. Placed near the head of the subject, small electric currents are produced in the brain region directly under the coil by electromagnetic induction. In effect, a kind of temporary, reversible lesion is induced and the behavioral effects of the lesion can be investigated.

A clear advantage of TMS is that it provides insight into the causal relationships between brain areas. This stands in contrast to all "passive" neuroimaging techniques, which are surveyed in more detail later. Disadvantages lie in the limited stimulation depth and the difficulty of adjusting the stimulation strength accurately. It needs to be high enough to evoke an effect, but not so high that multiple effects (that probably also involve other areas) may occur. Despite these difficulties, over the last decade TMS has established itself as a valuable paradigm. Related techniques have also been used to selectively excite particular brain areas or to enhance communication between selective areas.

A related perturbation technique is called *microstimulation*, which is used to stimulate small cell clusters and nuclei. Luigi Galvani (cf. Section 2.3.1) in 1780 was one of the first to use electrical stimulation to produce movements in frog legs. Later, neurons were stimulated by means of cortical microstimulation. In this case, small populations of neurons are stimulated by passing a small electrical current through a nearby microelectrode. For example, perceptual judgments of motion direction were manipulated (Salzman, Britten, & Newsome, 1990) and complex movements, such as hand-to-mouth movements, were invoked by stimulating the premotor and motor cortex over an extended time of about 500ms (Graziano, Taylor, & Moore, 2002). As is the case with TMS, microstimulation yields causal relationships. However, it is clearly more invasive than TMS and thus mostly used in animals. More recently, deep brain stimulations have been applied to human patients, though, by implanting electrodes to specific brain nuclei, eliciting electric impulses for treatment of movement or affective disorders, such as Parkinson's disease. Although several of these treatments have been very effective, the underlying reasons for this effect are still unclear. Furthermore, also in patients that are suffering from severe epilepsy as well as in brain tumor patients microstimulation and single cell recording techniques are applied partially.

Most recently, *optogenetics* has been shown to be successfully applicable to modify the activity of neural cells. In this case, neurons are genetically modified so that they develop light-sensitive ion channels. After the modification, light stimulation can activate these channels in real-time. Thus, the development of complex brain interfaces may be possible, by controlling biochemical events at the milliseconds scale in temporal precision within normally behaving subjects. Chosen as the "method of the year 2010" by the journal *Nature Methods*, combining insights from optics and genetics, this method seems to have quite some potential for gaining new insights and even to develop functional brain interfaces.

Somewhat more relevant in the medical domain, we should lastly mention one more active method. Various kinds of drugs are currently available to treat neuropsychological disorders, ranging from standard pain killers to potent sedatives, drugs to increase concentration capabilities, as well as drugs to fight depression – to mention only a few. We already touched upon some of their functionalities in Section 7.3. Everyday drugs, such as coffee, alcohol, tobacco, or even chocolate also have obvious effects on our mood, and, alcohol most obviously, on our cognitive abilities. While we do not address these substances and the current knowledge about how they affect our cognition and mood in further detail, the following insight derived from their effects on the mind should not be forgotten. Our brain is not an electrical computer. It is a biological system where the chemistry is at least as important as the neural connectivity and the firing of action potentials. Changes in the chemical balances by means of, for example, everyday drugs, can influence cognition in systematic ways, indicating that evolution has developed means to maintain a particular balance, but also the flexibility to adjust this balance based on external circumstances.

7.5.3   Passive methods

In addition to active methods to change neural activities in the brain, which essentially stimulate neurons or clusters of neurons actively, "passive" brain imaging techniques record neural activities. They are passive in the sense that they only measure ongoing activity, but do not manipulate this activity actively. Thus, they can measure correlations, but not causal relations. Each technique has particular advantages and disadvantages, since each technique records at different resolution levels in space and time, monitoring different aspects of the neural activity.

Electroencephalography (EEG) records electrical signals of neural activities at a rather coarse-grained spatial resolution, but rather fine-grained in time. EEG records the voltage fluctuations resulting from ionic currents within thousands of synchronized neurons in the brain. This may imply that EEG sums over action potentials, but this does not seem to be the case. Rather, EEG measures postsynaptic potentials in dendritic trees, which can be best related to input to, rather than output of, groups of neurons. Different frequency spectra or wave patterns can be differentiated in such recordings. For example, from the signal it is easy to detect if the person was awake or asleep. Accordingly, medically EEG is being used to monitor the depth of anaesthesia or to detect and characterize epileptic seizures. There have also been published attempts to use EEG as a brain computer interface in the hope of establishing communication channels with *locked-in patients*, who progressively lose the ability to control their body and thus to communicate with the outside environment.

The psychophysiologically most relevant variant is the recoding of *event-related potentials (ERPs)*, which average EEG signals time-locked with respect to a particular stimulus presentation in a particular trial or task.   By means of ERPs, it is possible to interpret the progression of the EEG signal in an experiment, contrasting, for example, expected from unexpected stimulus presentations. These signals are characterized by names, such as P300, indicating that a positivity is expected at about 300ms after stimulus onset, which has been correlated with the recognition of an unexpected or improbable, but relevant stimulus. Higher positivity in the P300 has thus been interpreted as signals of increased surprise or awareness.

Big advantages of EEG are that it is relatively cheap and it is relatively easy to conduct. EEG electrodes are simply placed across the scalp, thus being absolutely non-invasive. Another advantage is the high temporal resolution, enabling the interpretation of the EEG signals directly with respect to a particular stimulus presentation. Moreover, EEG is rather robust against disruptions, which may be caused by movements of the person from whom EEG signals are recorded. Most disadvantageous is the low spatial resolution and the fact that EEG records only from the head's surface. As a consequence, very similar EEG signals can be recorded even if the neural activity sources, which caused the recording, may vary significantly. Although signal localization methods have been developed, their accuracy is limited.

Magnetoencephalography (MEG) measures the magnetic fields induced by neuronal currents, such that the signals are generally believed to originate from the same neurophysiological processes as EEG signals. Similar to EEG, MEG integrates neural activities and is non-invasive, but much more expensive. Its main advantage is that it has better spatial resolution compared to EEG, enabling localization of particular signal sources within millimeter precision. As EEG, MEG has a very high temporal resolution. Apart from the surface recording disadvantage, MEG is quite sensitive to magnetic signals – a car driving by at a distance of 100 meters can be detected! As a consequence, expensive equipment and shielded rooms are mandatory, making MEG an expensive technique both for its initial purchase as well as for its maintenance.

In contrast to MEG and EEG, *positron emission tomography (PET)* is invasive. PET detects (pairs of) gamma rays, which are emitted in opposite directions by a positron-emitting radionuclide, called a "tracer". This tracer is introduced into the body and then accumulates in specific brain regions by binding to specific receptors. The reconstruction of the resulting brain image (using, for example, expectation-maximization algorithms or more recently Bayesian methods) depending on the tracer signals allows the inference of selectively enhanced activities in particular brain regions. A few decades ago, PET was used in various neuro- and cognitive science laboratories because it was believed to be harmless. Currently, however, the harmlessness has been questioned seeing that the brain is exposed to ionized radiation. As a consequence, recent PET studies in cognitive science are rare. Medically though, the technique is still being successfully applied to identify diseases.

Much more common and currently used is the *[functional] magnetic resonance imaging* ([f]MRI). MRI produces a strong magnetic field (for example, 3T = 60.000 times the Earth's magnetic field), within which orthogonally applied radio frequencies interact with cell tissue and allow the detection of neural activities. By monitoring temporal de- and re-alignments in response to the radio frequencies, body tissue and brain tissue can be probed. More importantly, selectively enhanced brain activities can be detected. In particular, fMRI measures the neural activity indirectly by focusing on its energy consumption. Most of the cell's energy is provided in the form of oxygen, which is released from the blood (a hemodynamic response) to neurons. This oxygen supply results in changes of the relative levels of oxyhemoglobin and deoxyhemoglobin as a result of increased blood flow (neurovascular coupling) to more strongly activated areas. The differential magnetic properties of oxyhemoglobin and deoxyhemoglobin interfere with the MR signal so that the recorded signal reflects a *blood-oxygen-level dependent* signal, which is often referred to as *BOLD* signal. As a result, areas of higher and lower activity become visible. Apart from the magnetic field, which is believed to not affect brain or body, fMRI is completely non-invasive and has a relatively high spatial resolution of about $1mm^3$ (*voxels*). Due to the three dimensional localization capabilities, it outperforms both EEG and MEG by far. Unfortunately, one full fMRI scan, which is generated by means of radio frequency sheets, takes a little while. As a result, there is a relatively low temporal resolution, which is in the order of seconds. Additionally, fMRI does not measure cell activity. It reflects brain metabolism, that is, it highlights those areas most in which blood flow increases. Thus, activated neurons are not measured directly, but only indirectly, which can result in false area localizations. Finally, the statistical analyses, which are carried out with the data, are still being debated – such as when an area can be said to be significantly more activated under certain experimental conditions. Nonetheless, fMRI is widely used and has produced various very revealing insights.

Finally, *single cell recordings* have been applied – mainly in animals, but also in humans, as, for example, before a brain tumor removal operation. In this case, very small electrodes are placed close to the neurons, measuring the action potentials elicited by the neurons in subjects that are being studied. Simultaneous recordings at several sites are possible and are now accomplished by implanting multiple 3D electrode arrays by means of an operation. Such recordings allow the study of neuron populations and even interactions between multiple neural populations.

Important insights have been gained by means of single cell recordings. Hubel and Wiesel (1959) have characterized the columnar, mostly edge-encoding receptive field maps in the visual cortex (cf. Section 7.4.2). More recently, the technique revealed *mirror neurons* in premotor and parietal areas in monkeys, where some neurons fire not only when the monkey executes a particular action, but also when the monkey watches a human executing a similar action (cf. Section 10.3.2). Equally interesting is the fact that different neural groups communicate with each other in a rhythmic manner (cf. Section 7.4.3).

Single cell recordings have high spatial and temporal resolution and thus allow much more detailed insights into the functional and topographical mapping of the cortex, as well as into the way neurons principally communicate with each other. The most obvious disadvantage is that the necessary electrodes are physical devices that need to be implanted in the brain, requiring an operation. Thus the approach is very invasive. Additionally, the placement of the electrodes is important and mostly the neural activities of larger cells are recorded. Finally, only the spark of a cell is recorded, but no information about postsynaptic potentials, resting membrane potentials, or chemical gradients.

Besides yielding neuro-processing insights, microelectrodes have recently also offered potential *brain-computer interfaces* (BCIs), opening a way for developing neuro-prosthetics. In this case, the microelectrodes are implanted in the skulls of patients with motor disabilities (for example, suffering from tetraplegia or amyotrophic lateral sclerosis), where they capture neuro-electrical signals from motor areas to decode movement intentions to enable the control of prosthetic limbs. This technology, however, is in its infancy.

7.5.4   Summary

Brain mechanisms and processes take place on multiple levels of granularity in parallel and interactively, as for that matter does cognition. When investigating these mechanisms and processes, it is impossible to analyze all of these levels concurrently and to record both in high spatial and temporal resolution. As a result there is no optimal method for analyzing or recording the brain, but individual methods analyze particular aspects of brain and cognition. A summary chart of the temporal and spatial resolution levels covered by these methods can be found in Figure 7.14. Additionally, we have emphasized that active methods can typically reveal causal relationships between artificially induced manipulations and neural and cognitive effects, while passive methods reveal correlations, but not necessarily causal relationships.

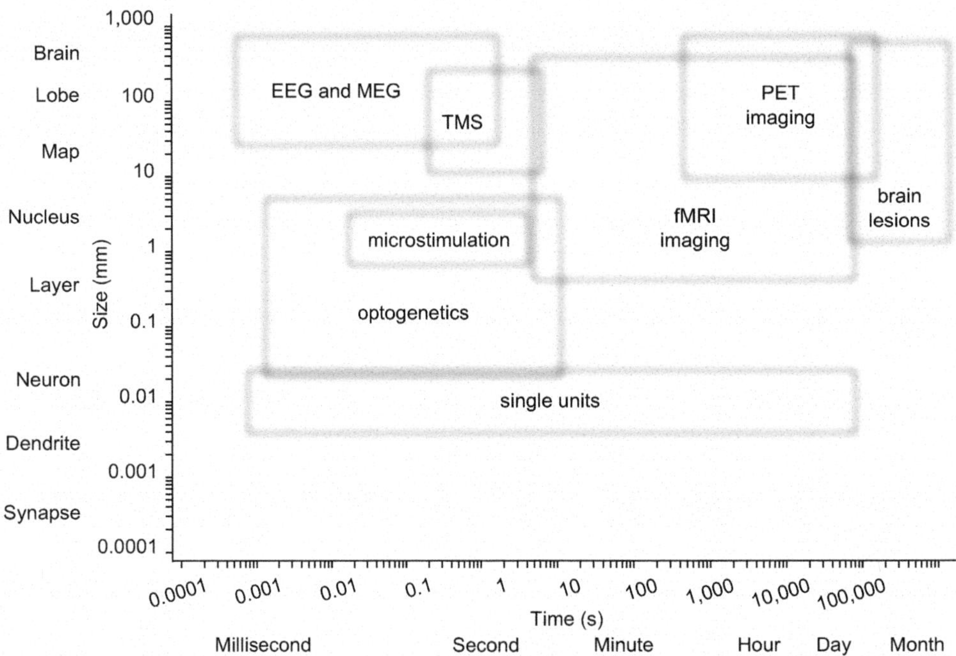

Figure 7.14: Brain imaging techniques vary significantly in their temporal and spatial resolutions, thus revealing different aspects about the brain's functionality [Reprinted by permission from Macmillan Publishers Ltd: *Nature Neuroscience, 17*, 1440–1441. Putting big data to good use in neuroscience. Sejnowski, T. J., Churchland, P. S. & Movshon, J. A. Copyright (c) 2014.]

Due to the different advantages and disadvantages of each neuroimaging technique and of brain lesion studies, it is important to put together the insights gained from several techniques and studies like a jigsaw puzzle. The result offers a progressively improving picture of the brain's functionality – although sometimes puzzle pieces are temporarily placed at the wrong location. Clearly, though, the images of the individual jigsaw puzzle pieces are determined by the particular technique used, the particular experimental paradigm pursued while applying the method, as well as background assumptions and other prior conceptualizations. To develop an overall image of brain functionalities, standardizations, and normalizations are necessary to enable the derivation of meaningful relationships and brain functionalities across the different techniques. Machine learning techniques are immensely helpful when analyzing the data from the individual techniques as well as when combining data from several techniques. Moreover, simulations and computer models help to further corroborate evidence for the validity of particular interpretations. Much additional work seems necessary to really gain a complete, functional understanding of the brain at all levels of granularity.

7.6 Summary and conclusions

To be able to proceed with the book's overall goal of providing an introduction to how the mind comes into being, this chapter provided a very broad overview of the crucial components of body and brain. As with cognition, the brain cannot be viewed alone or fully detached from the body, since it co-develops ontogenetically with the body and since it has co-evolved with the body over millennia. As a result, the nervous system consists of highly interactive peripheral and central nervous systems, both of which can be further modularized into various subcomponents.

The brain in particular exhibits a hierarchical, tree-like structure from a phylogenetic, ontogenetic, and functional perspective. While our focus was on the cerebral cortex, it is clear that all other brain components – including subcortical structures, the peripheral nervous system, and the cerebellum – play important roles in keeping the organism alive, coordinating behavior, information exchange, and learning. To a certain extent, a strongly interactive *society of mind* (Minsky, 1988) may be a good analogy (cf. also Section 3.6.2). However, due to the strong interactivity, the brain appears to control the body not by means of a subsumption-like architecture, but rather by a highly versatile, modularized network of control components. The control components – such as different sensory analysis, integrative spatial, motivation-based, or motor-oriented components – appear to continuously exchange information.

Currently, a widely debated idea is whether the brain works by minimizing *free energy* as its core control, adaptation, and learning principle (Friston, 2009, 2010). Slightly simplifying, free energy is defined as the difference between neural activities and predictions of these activities. At the sensory level, for example, deeper neural levels attempt to predict incoming sensory information, minimizing surprise about sensations. At the motivational level, deeper neural layers attempt to keep the bodily demands within their predicted parameters, generating actions that are believed to address bodily needs (such as eating when hungry) by means of active inference. At the motor control level, forward models minimize surprise during action executions, enabling their smooth control. In later chapters, this perspective on the brain's functionality will be elaborated in further detail.

With the knowledge of the brain and the nervous system's fundamental components in mind, the remainder of the book proceeds with analyzing and detailing the functionality of sensory processing, multisensory integration processes, attention, and motor control in the brain from a computational perspective. Moreover, we detail how abstractions can build upon the involved information processing and control mechanisms, yielding reasoning and conceptualization abilities, and ultimately human language.

Chapter 8

Primary Visual Perception from the Bottom Up

8.1   Introduction

With a sketch of our modular brain in hand, we now look into one particular sensory processing pathway in detail. We will focus on the visual pathway because this is the one that arguably may be considered to be best understood today. However, the principles that we uncover for the visual pathway generally appear to be applicable for other primary sensory information processing areas as well. Seeing the general systematicity of the six-layered structure all the neocortex, one can deduce that some of the principles uncovered for the visual system may also hold not only for other primary sensory systems, but possibly even also for deeper neural processing modules.

To approach the visual system we first look at the information that is actually registered by our eyes, or rather by the neurons distributed on the retina of our eyes. We also consider a couple of general properties of light to better understand what information about the world can be assumed to be contained in light. Next, we detail several visual pathways and their suspected functionality. Then we focus on cortical processing and highlight different, redundant, and complementary sources of information that are contained in light and that are also explored in the primary visual cortical areas. We emphasize that these different sources provide information about particular, distinct aspects of objects, other entities, and states in the world. Finally, the redundant information sources are put together again – at least to a certain degree – to recognize particular causes for particular sensory perceptions, where such causes may be objects or other entities that have particular visual signatures. Additionally, spatial relationships between the causes – such as between ones own body and a particular object – appear to be extracted.

In subsequent chapters, we show how the sketched-out bottom up information can be combined with top-down, predictive processing and with other sensory and motor sources of information. We will particularly show that interactions between bottom-up and top-down information sources can yield suitable stimulus abstractions, which conceptualize space and time, objects and other entities, and goal-directed behavior.

8.2   Light and reflections

Atomic fusion, which progressively unfolds in the stars of our universe due to quantum-determined particle interactions, releases energy in the form of electromagnetic waves of a large spectrum of wavelengths. After traveling through the vacuum of outer space with a speed of 3×10^8m/s and being partly absorbed by the earth's atmosphere, a small part of radiation reaches the surface of our planet.

As life evolved from single-cell to multicellular organisms, evolution managed to develop specific photo receptors that are sensitive to a particular spectrum of this electromagnetic radiation, which generate different color impressions in our brains. In primates, this so-called *visual spectrum* covers wavelengths of about 400–700nm that are perceived as the rainbow colors being distributed in a continuous color-changing manner starting with blueish, short wavelengths and ending up with reddish, longer wavelengths. Figure 8.1 shows this color spectrum and its embedding into the overall electromagnetic wave spectrum. Note that other animals, such as some beetles or snakes, have evolved different photo receptors, which can also be sensitive to infrared light.

Direct light sources, as the sun or a light bulb, contain many diverse wave lengths covering more or less uniformly the whole visual spectrum, which generates the visual impression of a bright white light. A prism can be used to make the actual color spectrum visible, because the prism re-directs the light wave-length dependently, thus producing the rainbow-spectrum, which is also shown in Figure 8.1.

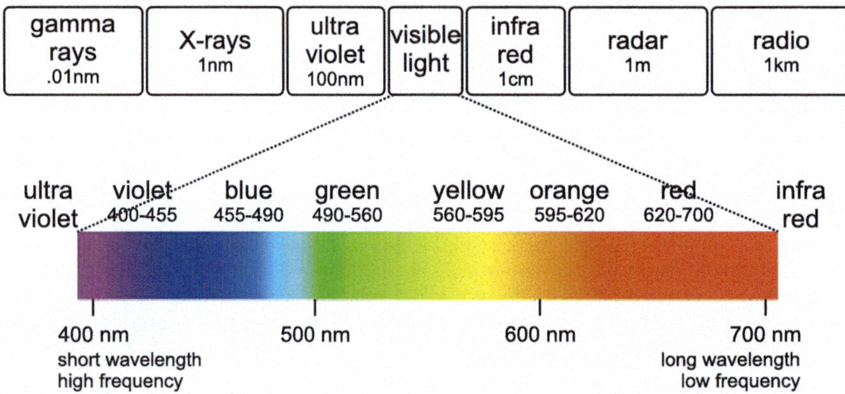

Figure 8.1: Only a small part of the spectrum of electromagnetic waves is visible to humans. Within the visible light spectrum, we perceive a continuous color spectrum starting with violet light from about 400nm wavelength and changing, like the rainbow colors, to red light at about 700nm wavelength.

What we most often perceive with our eyes, however, are not direct light sources (do not look directly into the sun or a glowing light bulb!), but *indirect light sources*, which are *reflectances* of light from surfaces in the environment, such as the ground, objects, and other entities, the atmosphere, or the moon. Such surfaces typically absorb some fraction of the light spectrum and reflect others or transmit others through a transparent surface, such as glass. The light that is reflected is the one we perceive, such that distinct surfaces give distinct light and color impressions.

Two kinds of reflections should be distinguished further: first, *specular reflection* is most obvious in the mirror, but it is present – at least to a small extent – in all surfaces. Specular reflection essentially refers to light that is reflected on a surface by maintaining the same angle, that is, the incoming angle is equal to the outgoing angle. Second, *diffuse reflection* refers to the parts of the wavelengths that are reflected diffusely when hitting a surface, that is, the light is reflected in all possible directions approximately uniformly. The swiss polymath Johann Heinrich Lambert (1728–1777) modeled this diffuse reflection mathematically, proposing that the apparent brightness of a surface for an observer is the same regardless of the angle from which the surface is being watched. The moon is a good example of diffuse, Lambertian reflection because we see, for example, the full moon as a nearly equally bright disc, even though the surface on the sides of what we see faces earth progressively less directly. This Lambertian reflection property of surfaces is essential to enable the perception of uniformly colored object surfaces in an approximately uniform color. Imagine all surfaces in our world would only produce specular reflections – we would face a house of mirrors and probably could not use light as a useful source of information about things in the world at all.

Because different objects typically have different absorption properties, it becomes rather easy to distinguish different surfaces. Figure 8.2 shows a sketch of some common food objects with their characteristic selective color spectrum reflection properties. Accordingly, a lemon is typically perceived rather yellowish, while a tomato is more reddish, while cabbage comes across light greenish. For extreme cases, white surfaces reflect most of the visible wavelengths, while black surfaces absorb most of them, which is easily perceived by the fact that black surfaces heat up much faster under direct sunlight than white surfaces.

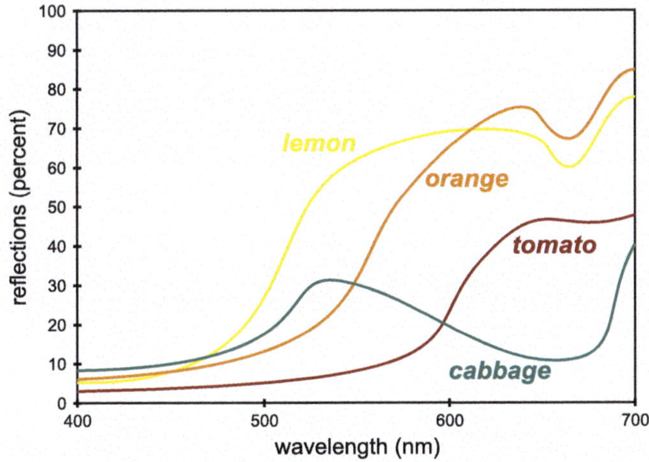

Figure 8.2: The perceived color of an object's surface is determined by its light reflection and absorption properties. The figure shows some exemplar, approximate reflection distributions over the color spectrum..

8.3 The eye

The light that is sent out by a light source – regardless if it is a direct light source or an indirect light source – is the one that we sense with our eyes. Essentially, the incoming wave lengths pass through the eye's pupil, which is controlled by the iris, then further through the lens, and finally onto the retina. The *ciliary muscle* of the eye controls the form of the lens to focus the incoming visual information on the retina. In the focus point, which typically lies in the fovea centralis and which has a diameter of about $0.5mm$, the eyes sense visual information most accurately. Toward the periphery, this acuteness degrades so that the visual information becomes progressively fuzzier. Figure 8.3 sketches the eye, the path of the light onto the retina, and the main nerve tracks.

Viewed from its basic physical functionality, the principle behind the human eye can be compared with a classical camera. In a pinhole camera without a lens, light information coming from the three dimensional coordinates (X_o, Y_o, Z_o) centered on the eye, will be projected onto two-dimensional photographic paper, which is the retina in our eyes, to the corresponding point (X'_i, Y'_i). The distance between the pinhole and the retina can be assumed to be constant d. As a result, geometry tells us that

$$\frac{X_o}{Z_o} = \frac{-X'_i}{d} \quad \text{and} \quad \frac{Y_o}{Z_o} = \frac{-Y'_i}{d}, \tag{8.1}$$

given that the respective two X and two Y axes are parallel to each other. Given a known object position, it is thus possible to determine the corresponding position on the photographic

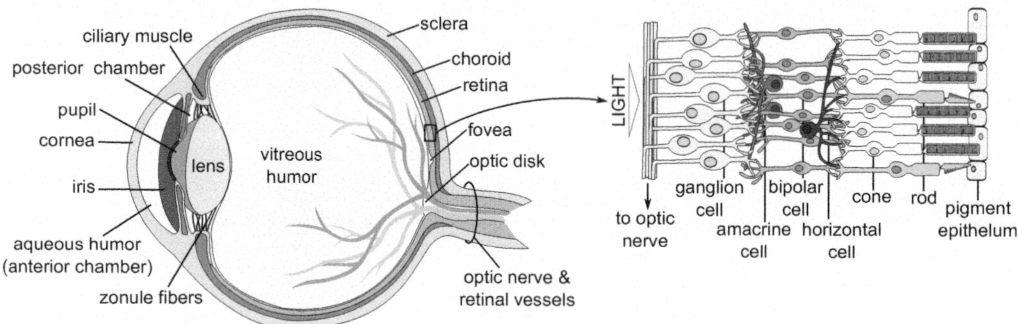

Figure 8.3: General anatomy of eye and resulting projection onto the retina. The lens of the eye continuously adapts to the incoming light sources such that a focused image is projected onto the retina. Interestingly, the light-sensitive rods and cones in the retina are reached by the incoming light only indirectly passing through the optical nerve fibers and a layer of ganglion cells. [Adapted with permission from Mark F. Bear, Barry W. Connors, Michael A. Paradiso, *Exploring the Brain*, 3rd Edition, (c) Lippincott Williams and Wilkins, 2007.]

paper in image coordinates where the object will be perceived:

$$X_i' = \frac{-d \cdot X_o}{Z_o} \ \text{ and } \ Y_i' = \frac{-d \cdot Y_o}{Z_o} \tag{8.2}$$

This principle is also shown in Figure 8.4 for illustration purposes. While the lens of the eye – and also the lens in a camera for that matter – bundles more light in the center and influences acuity, the geometric relationships still hold approximately. Thus, we have seen how three dimensional light sources determine a two-dimensional retinotopic image, which is then further processed by the photo-receptors in the eye. Note how the image on the retina is upside down and the brain needs to learn to interpret the information accordingly.

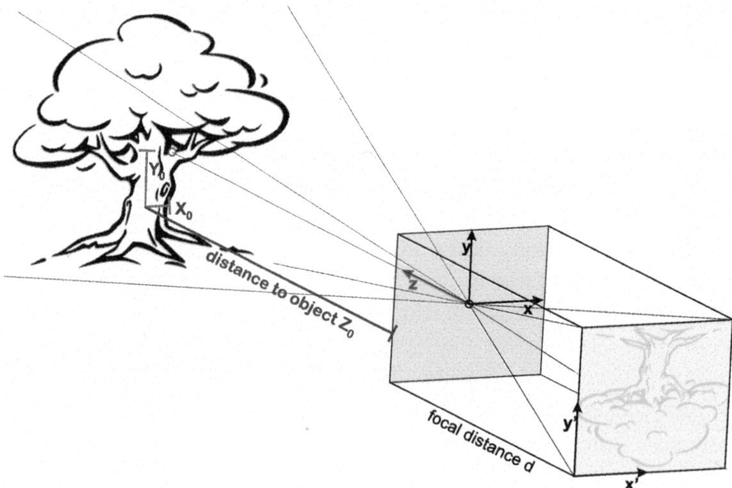

Figure 8.4: Basic geometry allows to determine where on the retina – or on the photographic paper – which light source or surface will be perceived.

The information actually perceived by our eyes then depends on the distribution and sensitivity properties of the photoreceptors, which are found in the retina. Two major types of photoreceptors can be found in the human retina, both of which react to the incoming light by means of light-sensitive photopigments, which are chemically changed when absorbing light. Monochromatic *rod cells* are color-insensitive and mainly react to ≈500nm light wavelengths. Color-sensitive *cone cells* come in three types, being sensitive to longer-range wave lengths (L cones have their maximum sensitivity at $\approx$570nm, which is yellow-greenish light, and have a sensitivity range of $\approx$500–700nm), middle-range wave lengths (M cones, maximum: $\approx$530nm, greenish, range: $\approx$450–630), and short-range wave lengths (S cones, maximum: $\approx$430nm, blueish, range: $\approx$400–500nm). In accordance with evolution, which has evolved three color-sensitive receptors, the German physiologist Hermann L.F. Helmholtz (1821–1894) has shown that it is possible to produce any color visible with the human eye by mixing the three primary colors. Figure 8.5 shows the two types of photoreceptors in our eyes.

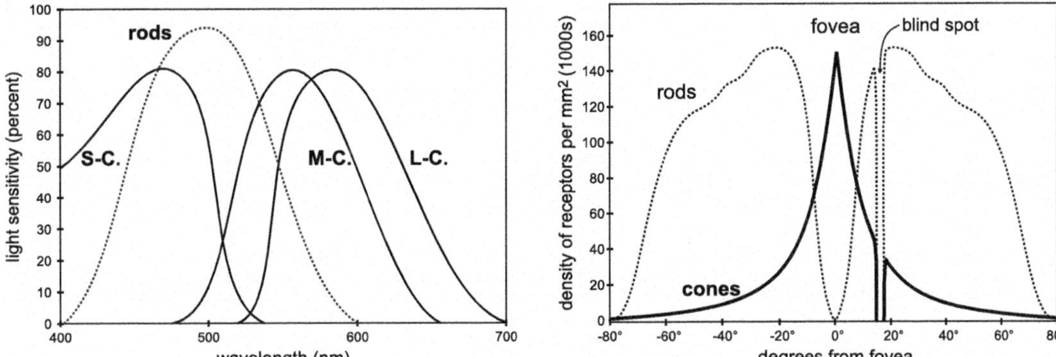

Figure 8.5: The sensitivity of a particular photoreceptor (left) depends on the wave length of light. Cones come in three types, being dominantly short-wave S, middle-wave M, and long-wave L sensitive. Rods are much more light sensitive, but are much less spectrum selective. While cones are mainly found in the fovea, the periphery of the retina is mainly populated by rods. Interestingly and counterintuitive to the fact that we seem to perceive a full image of the outside world on each retina, in the blind spot, where the nerve bundle from the rods and cones leaves the eye, no visual perception is possible.

While the color-sensitive cone cells are densely found in the fovea centralis and their density quickly decreases toward the periphery, the rod cells populate the periphery much more densely and are less densely found in the fovea centralis. Figure 8.5 shows the distribution of rods and cones along an angular axis centered on the fovea centralis. Rod cells are more light sensitive than cone cells and are thus the ones mainly responsible for night vision, which is the reason why colors are much less accessible at night. Somewhat surprisingly, the light-sensitive parts of rod- and cone-cells do not face the lens, but face away from the lens, such that other cells, and nerve fibers from the cells, are in-between (cf. Figure 8.4). Thus, the visual information perceived by the rods and cones could have been arranged even better – albeit unknown physiological reasons may have caused evolution to favor the actual orientation.

Because the nerve fibers have to leave the eye somewhere, there is also a hole in the visual image of the retina, which is termed the blind spot. Because the nerve fibers leave the retina at this location, no photoreceptors are present and thus no visual perception is possible here. Figure 8.6 illustrates this "blind spot" phenomenon. Considering the blind spot as well as the fact that the visual information that hits the retina is not perceived by a uniform distribution of light-sensitive cells, it comes as a surprise that we notice neither a hole in our visual field (even with one eye closed we usually hardly notice the blind spot) not that the visual information in the periphery becomes grayish. Both of these observations suggest that our brain is not a passive perceiver of visual information, but actively processes the information, filling in gaps and augmenting the incoming visual information with color estimates.

Figure 8.6: The blind spot can be easily noticed when keeping the image horizontal, closing the right eye, and fixating the star with the left eye. At about the 50cm and in a range of about 10cm distance, the left cone is not visible.

8.4   Visual processing pathways

The visual information perceived by the retinotopic distribution of rods and cones on the retina is then transferred via the blind spot to the *optic chiasm* where the visual information from the two eyes is divided in such a way that each visual *hemifield* is further projected toward the contralateral side in the brain (cf. Figure 8.7).

The next processing stage is accomplished by the *lateral geniculate nucleus* (LGN), which is a thalamic nucleus. It normalizes the visual information and even first considers top-down visual expectations. An evolutionarily older visual processing path enters the *superior colliculus*. The superior colliculus can induce, amongst other things, fast orienting eye saccades and thus complements the slower, visual pathway via the cortex. Seeing that the superior colliculus is much more in control of the eyes in other animals, such as barn owls, than it is in humans, we do not consider it further in this book.

Proceeding from the LGN, the processed visual information enters the primary visual cortex (V1), where the incoming sensory information is still organized retinotopically. Figure 8.7 shows these processing stages and some aspects of the structuring, which the visual information undergoes along the main visual pathway.

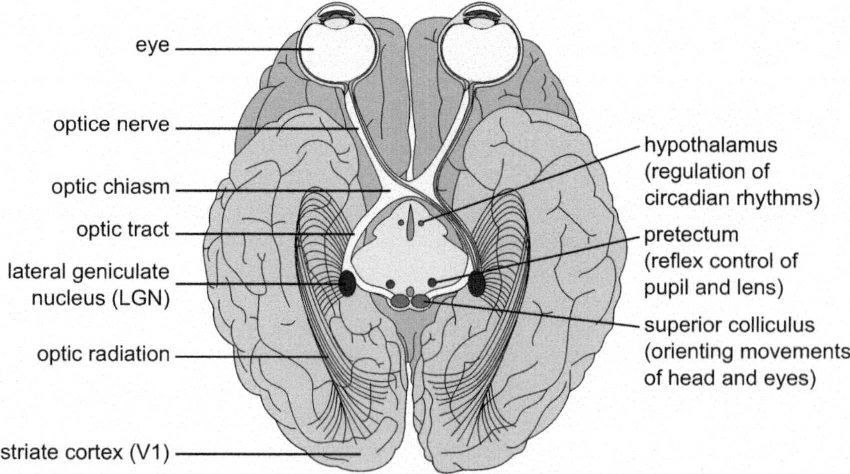

Figure 8.7: Visual information from the eyes is projected via the optic chiasm into LGN and then further into primary visual cortex. An evolutionary older path projects in parallel some of the visual information into the superior colliculus. [Reprinted with permission from Purves, D., Augustine, G. J., Fitzpatrick, D., Hall, W. C., LaMantia, A. S., McNamara, J. O. & Williams, S. M. (Eds.) (2004). *Neuroscience.* Sunderland, MA: Sinauer Associates, Inc.]

In V1 the visual information is then further analyzed and structured. Essentially, V1 and neighboring areas appear to extract particular aspects of the visual information, including edge, motion, color, and depth information. Viewed from a pure bottom-up perspective, the information extraction neurons essentially act as filters in local space and time, being maximally sensitive to particular local visual distributions. As we saw in the last chapter, the visual information appears to be processed in a topographic and thus retinotopic manner at this very early visual processing stage. Cells in V1 and neighboring regions typically exhibit local receptive field properties, that is, the cells are sensitive only to a relatively small, local region of the retinotopic space. Moreover, the cells within this region are sensitive to particular spatial and temporal stimulus properties, such as to a particular color, a particular directional motion pattern, or a particular edge orientation.

Deeper visual areas, including V2–V4, IT, V5/MT, receive as bottom-up input the processed information from V1 and further analyze this information focusing on particular feature aspects and feature combinations, such as corner detections, extended line detections, curve detections, etc. Figure 8.8 shows the distribution of these areas. V2 is mainly sensitive to more complex edge, corner, and basic shape properties, where the neurons also exhibit larger receptive fields. V3 is mostly sensitive to local motion signals. V4 yields activities that are sensitive to even more complex form and shape properties. The inferior temporal cortex (IT) is known to encode objects, faces, and, generally speaking, complete shapes and forms. The middle temporal (MT) area and particularly V5, on the other hand, are most sensitive to complex motion signals.

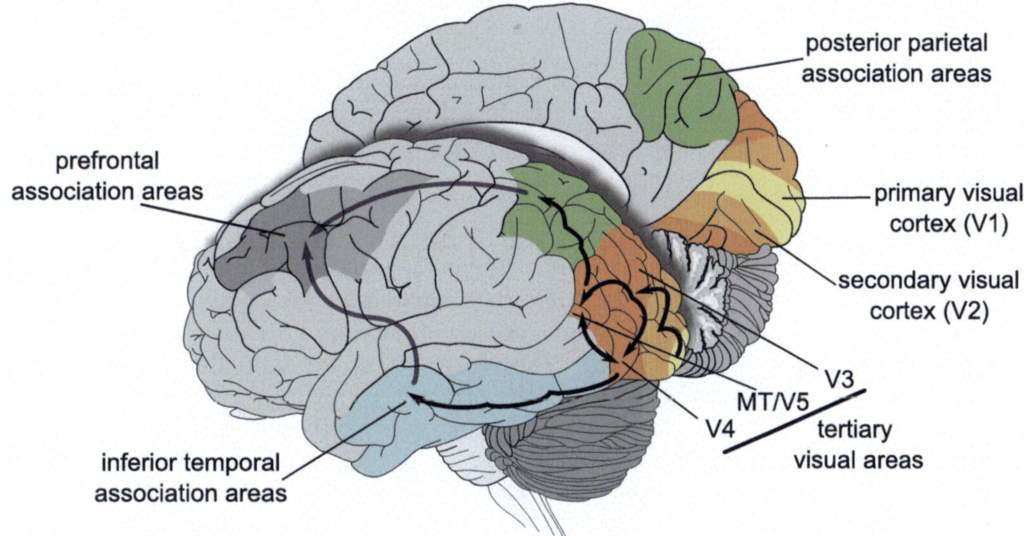

Figure 8.8: Beginning with the primary sensory processing area V1, visual retinotopic information from LGN is analyzed in further detail, extracting basic edge and motion signals. Deeper visual areas (V2–V4, IT, V5/MT) extract even more complex visual information. While the visual analysis in primary visual areas considers all incoming visual information, later modules split the information toward a dorsal pathway ("where" or "how" pathway), which is believed to focus on spatial and motion information, and a ventral pathway ("what" pathway), which is believed to focus on object identification.

Further information processing splits into two pathways, where the interaction between the pathways is much lower than the interaction within each pathway (Goodale & Milner, 1992; Milner & Goodale, 2008; Mishkin et al., 1983; Ungerleider & Haxby, 1994). The *ventral pathway*, which is also sometimes referred to as the "what" pathway, leads from V1 into IT and further into the temporal lobe. It is believed to be dominantly responsible for object recognition. Seeing that object recognition depends on rather acute visual information, it is particularly sensitive to higher spatial visual frequencies. The *dorsal pathway*, which is also sometimes referred to as the "where" pathway or the "how" pathway, leads from V1 over the occipital lobe into the posterior parietal cortex. This pathway appears to focus on motion, and body- and object-relative spatial positions, and orientations. It is much more time-sensitive, exhibiting much faster activity fluctuations, thus enabling a faster perception of, for example, sudden movement onset stimuli, even without recognizing their identity.

Table 8.1 gives an overview of important regions that are typically assigned to one of the two deep visual processing streams. The reader should be aware, however, that the assigned functionalities described here and throughout this book, are still considered hypothetical. In particular, the functionalities probably only characterize some particularly strongly encoded aspects of the actual information that is neurally encoded in the particular brain region. Further differentiations and more exact characterizations based on future research insights will be inevitable.

Table 8.1: Important regions of the two visual processing pathways

Ventral stream		
LO	Lateral occipital	Object analysis
FFA	Fusiform face area	Face analysis
EBA	Extrastriate body area	Body analysis
FBA	Fusiform body area	Body analysis
STS	Superior temporal sulcus	Analysis of biological motion
STSp	Superior temporal sulcus (posterior)	Moving-body analysis
PPA	Parahippocampal place area	Analysis of landmarks
Dorsal stream		
LIP	Lateral intraparietal sulcus	Voluntary eye movement
AIP	Anterior intraparietal sulcus	Object-directed grasping

8.5   Redundant and complementary visual processing

When we consider visual information from the bottom-up, starting from the sensory information without any prior expectations, one has to acknowledge that it seems rather messy. The distribution of rods and cones is not uniform. The sensitivity of the individual cone types should be integrated to enable color vision, but the brightness information from the rods should also be considered. The visual field is covered progressively coarsely when moving away from the fovea. And to make things even worse, usually our head and eyes do not stare at something without motion, but microsaccades keep our eyes nearly always in motion. Actual full saccades yield huge shifts in the retinotopic image, which – also due to the uneven distribution of rods and cones – results in non-uniform activity transformations.

A constant in these dynamics, however, is the retinotopy, that is, the topological distribution of rods and cones. Although their distribution and their activities are nonlinear, there is still a certain neighborhood relationship present between the cells. Thus, localities can be identified and a cell's activity allows the refinement of a prediction about the activities in neighboring cells.

Starting from this retinotopically distribution of particular light-sensitive cells, evolutionary and ontogenetic predispositions guide the neural wirings to the LGN and further into V1, preserving the retinotopic topology. Natural images, which are neurologically processed in a topology-preserving manner, may generally be characterized by stating that the light or color intensity $I(x,y)$ at a particular retinotopic location (x,y) is most likely similar to the light or color intensity in neighboring locations, that is, $I(x,y) \approx I(x \pm 1, y \pm 1)$. Large changes within a natural image can be characterized by edges, which produce discontinuities in the image, that is, exceptions from the similarity rule.

Moreover, natural images have the tendency to stay the same when comparing successive images over short temporal intervals, that is, the light or color intensity at a certain location

(x, y) at a certain point in time t can be expected to stay almost the same over a short time period δt: $I(x, y, t) \approx I(x, y, t + \delta t)$. Motion in the image, especially when maintaining a rather stable stare into the world, is the exception to this rule, essentially predicting discontinuities in the form of light or color intensity changes. Motion will typically lead to a transfer or continuous shift of particular light or color intensities. New colors or intensities may also become visible when, for example, an obstacle is moved aside to show a clear view of something.

It turns out that these principles largely characterize early, bottom-up visual processing. The visual cortex appears to essentially analyze the visual information seeking regularities and exceptions. The types of regularities and exceptions are processed in various modules The different analyzed aspects may sometimes be viewed as redundant information processing – to ensure that fail-save mechanisms are available. These fail-save mechanisms also enable much flexibility in the involved information extraction processes. An object, for example, may be identified from its contour alone, focusing on edges only, or it may be identified mainly by color properties, or even solely by motion cues. Thus, in vision information is processed redundantly, but also complementarily, in that color plus contour typically increases certainty and allows for flexible information source substitutions and information-content-dependent fusion.

Early bottom-up processing typically starts with an analysis of the basic retinotopic image properties. To do so, the brain first appears to apply some normalization and smoothing operators, which help to ignore temporary failures of particular neurons as well as neural noise. Next, edges, motion, and other information aspects are extracted. Later on, the redundant information is partially reintegrated, to enable, for example, object recognition. Before we go on to the functional details of these mechanisms, however, we give some background about the initial discoveries of the basic structures in LGN and the cortical columns, which can be found in V1.

8.5.1 Receptive fields, columns, and hypercolumns

In 1957, the neuroscientist Vernon Benjamin Mountcastle (1918-2015) published in an edited volume the hypothesis that columnar structures are found throughout the primary somatosensory area. Inspired by Mountcastle's work, David H. Hubel (1926-2013) and Torsten Wiesel (*1924) then showed that a similar structure could be found in V1, where the columns are retinotopically organized. That is, each column reacts to visual stimulation at a certain small local area of the retina, which is the common *receptive field* of the neurons in the column. Each neuron, however, reacts selectively to different properties of the visual stimulus in the particular receptive field. For example, the neurons may be selective for particular edge orientations (cf. also Section 7.4.2).

In further studies, Hubel and Wiesel showed that the systematicity in the encodings of the visual pathway can be characterized in even further detail. In the LGN, cells could be identified that responded most to light blobs, that is, bright spots with dark surroundings or vice versa. In V1, this information appears to be combined yielding responses to dark or light lines of particular orientation and thickness, or to edges with a bright area on one side and a darker area on the other side. Thus cortical columns were identified in V1, where the cells in one column respond exclusively to one particular retinotopic area, but selectively for particular, edge-like visual distributions in that area. Hubel and Wiesel called these cells *simple cells* and showed in early neural wiring diagrams, how these cells may generate their activity by integrating cells from V1. Figure 8.9 shows the key results from the authors, published in the Nobel Laureate collection (Hubel, 1993).

They furthermore discovered *complex cells* with larger receptive fields, which exhibited a sensitivity to a particular edge anywhere in the larger receptive field. Thus, it appeared that a maximum operation is at play, which takes as input simple cell activities from several columns, but with the same orientation. Many of these complex cells showed additional movement selective properties, being particularly sensitive to an edge that moved in the

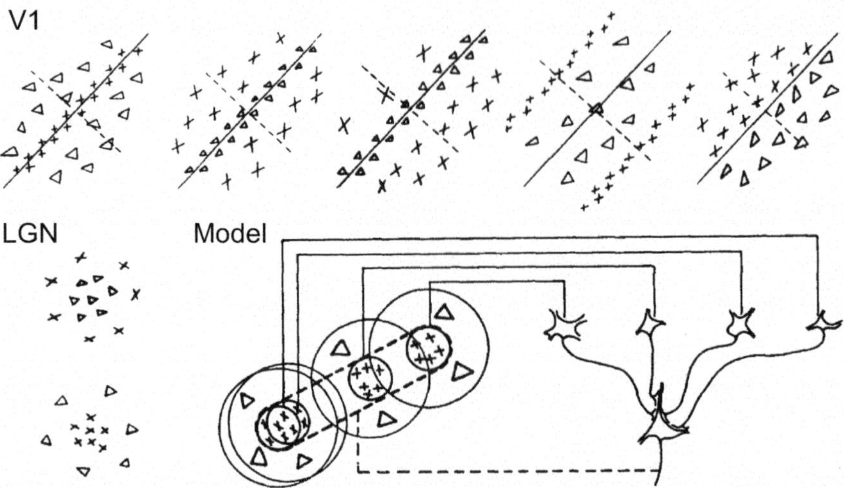

Figure 8.9: In the lateral geniculate nucleus, cells typically exhibit spot-light sensitivities. On the other hand, simple cells in V1 exhibit small, local receptive fields within which they are selectively active for particular visual edges with particular orientations. Complex and hypercomplex cells exhibit larger receptive fields, within which they additionally detect edges, motion, and/or end-stopping properties. [Adapted from *Nobel Lectures, Physiology or Medicine 1981–1990*, Editor-in-Charge Tore Frängsmyr, Editor Jan Lindsten, World Scientific Publishing Co., Singapore, 1993. Copyright © The Nobel Foundation 1981.]

one or the other perpendicular direction with respect to the edge. Moreover, *end stopping* behavior was found in *hypercomplex cells*, which were selective to lines of limited lengths in addition to the line's orientation. Besides these property-selective characterizations, Hubel and Wiesel also looked in detail at how the columns cover the whole retinotopic space, exhibiting general systematics in the columnar distribution and their right and left-retinal dominances within these distributions.

These differentiations have now been much further explored, confirmed, and differentiated. It is now believed that later visual areas focus on particular aspects of the information that is encoded in V1, combining these aspects in particular manners. For example, several edges may be combined into corner or roundness detectors; and the motion selective information may be combined into more complex combinations of dynamic motion detectors. Nonetheless, the basic principles of columnar and hypercolumnar structures as well as of topological neighborhood preservation seem to be maintained in deeper visual areas. In the following, we explore from a functional-computational perspective, how these selective sensitivities may come about and for what they may be useful.

8.5.2   Smoothing

Biological vision is driven by neurons, which depend on the current oxygen level, the availability of other important chemical substances, a proper blood flow, etc. This very crude characterization shows that neurons are noisy and sometimes will fire spontaneously without any actual stimulation, or they may fire delayed or in an otherwise noisy fashion. To get rid of this noise before proceeding with the actual information extraction process, it is useful to first smooth the neural information. Due to the retinotopic distribution, smoothing works by considering local neural neighborhoods.

Figure 8.10 illustrates the "smoothing" principle: Given a neural activity map I and particular neurons within that map (x, y), smoothing is achieved by propagating the average activity value of the local neighborhood, rather than of the neuron only. To maintain locality, of course, this neighborhood should typically not be overly large and should integrate more distant neural activities much less strongly than the activity of immediate neighbors. Such a smoothing image processing is in fact apparently realized by the LGN.

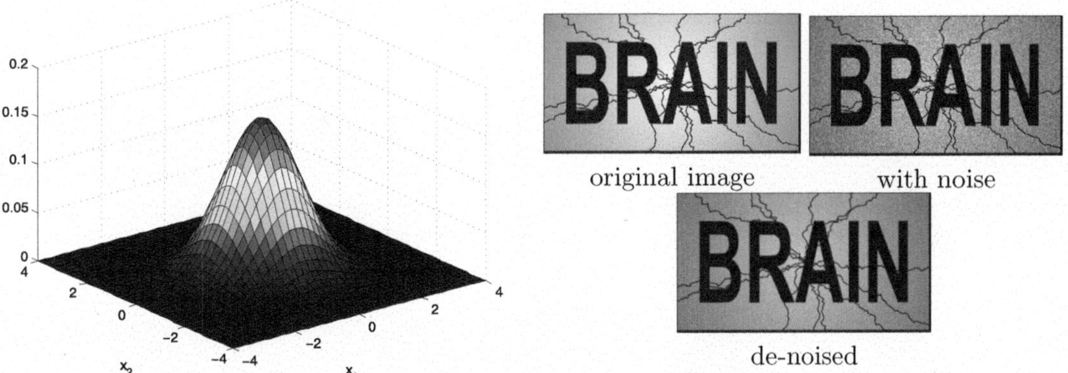

de-noised

Figure 8.10: Noise in an input image can be reduced by applying a Gaussian filter using convolution. In consequence, the image with unsuitably distributed dots is converted into a smoother image with homogeneous light-gray background.

Computationally, an output neuron that is responsible for position (x, y) may signal the average activity $h(x, y)$ of a local neighborhood of input neurons $I(x + u, y + v)$. The neighborhood may, for example, be computed by means of a two-dimensional Gaussian filter $G_\sigma(u, v)$, which yields the maximum value for $u = v = 0$. To determine the activity in the local neighborhood then, the average weighted activity can be calculated by summing up the weighted input activities:

$$h(x, y) = I(x, y) * G_\sigma = \sum_{u=-\infty}^{+\infty} \sum_{v=-\infty}^{+\infty} I(x + u, y + v) \cdot G_\sigma(u, v), \qquad (8.3)$$

where

$$G_\sigma(u, v) = \frac{1}{\sigma \cdot \sqrt{2\pi}} \cdot \exp \frac{-(u^2 + v^2)}{2\sigma^2}, \qquad (8.4)$$

determines the Gaussian receptive field of the neuron with activity $h(x, y)$. The important parameter σ, which specifies the standard deviation of the Gaussian distribution, determines the breadth of the filter. Larger values yield larger receptive fields, an integration of a larger input area, and thus a stronger smoothing of the image. With a proper σ, the result is a smoothed image. The operator $*$ denotes the convolution operator, where in this case the Gaussian kernel is convolved over the whole retinotopic image. This means that the convolution kernel, or filter – in this case the Gaussian G – is convolved or folded over the image by moving it pixel-by-pixel over the image and then applying the full kernel relative to the current image position. Overall, the input image I is thus transformed into the output image H by means of $H = I * G$. This convolution essentially smoothes the input image, that is, it reduces noisy artifacts in the input. The smoothed output H then enables the better extraction of edges and other important, locally distributed information.

8.5.3 Edge detection

After the smoothing operation, which appears to take place mainly in LGN, the *primary visual cortex* (V1) extracts those structures that are typically present in retinotopic images. In natural images, these are mostly edges, that is, local, slightly extended nonlinearities in light intensities. It turns out that such edges are typically not only present, but also highly informative about the outside environment.

As shown in Figure 8.11, edges can be categorized into particular types, all of which result in similar edge-like properties. First and foremost, *depth discontinuities* occur at the border of objects that are closer to the observer, blocking the view of further distant entities. Second, *surface discontinuities* lead to the reflection of light from a different subset of light sources and thus produces a visual edge. These discontinuities can be concave or convex. Third, the material or color on the surface may yield additional discontinuities, reflecting a different spectrum of the incoming light. Finally, shadows of objects can produce further edges.

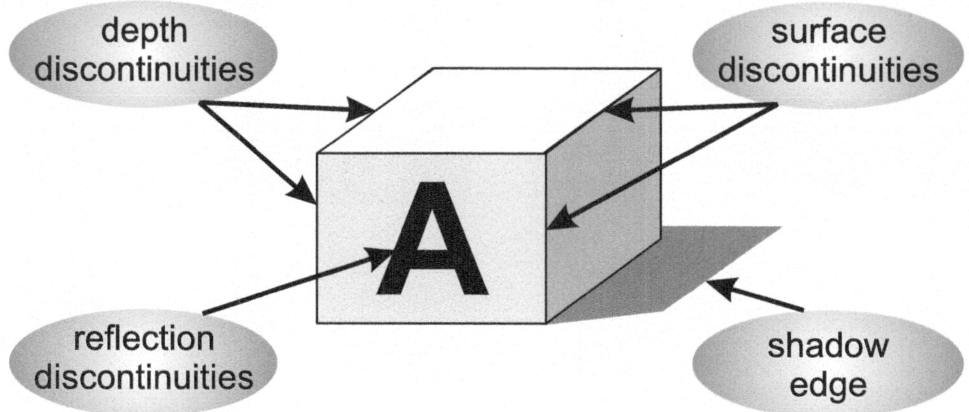

Figure 8.11: Visually perceivable edges are caused in four fundamental ways. The contours of an object cause visual edges due to surface discontinuities (and consequent differences in the distribution of incoming light that hits the surfaces) and depth discontinuities. Also reflection discontinuities due to differences in the coloring of a surface as well as shadows produce additional visual edge types.

V1 may be viewed as analyzing the visual image, searching for edges in the incoming, pre-processed sensory information. Computationally, a classical edge-detection filter is the **Canny Edge Detection**, which identifies local image changes by means of the partial derivatives in horizontal and vertical directions of the smoothed input image I:

$$H_{hor}(x,y) = I(x,y) * \frac{\partial G}{\partial x} \cdot G_\sigma \ \ \text{and} \ \ H_{vert}(x,y) = I(x,y) * \frac{\partial G}{\partial y} \cdot G_\sigma \qquad (8.5)$$

where the operator $*$ denotes the convolution of the image I at point (x, y) with the respective local partial derivatives of the Gaussian.

An even simpler approach in computer vision is the application of a **Sobel-filter**, which is defined by a horizontal and vertical 3x3-matrix:

$$S_x = \begin{bmatrix} -1 & 0 & 1 \\ -2 & 0 & 2 \\ -1 & 0 & 1 \end{bmatrix} \ \ \text{and} \ \ S_y = \begin{bmatrix} 1 & 2 & 1 \\ 0 & 0 & 0 \\ -1 & -2 & -1 \end{bmatrix}, \qquad (8.6)$$

which is then convolved with the image. In this manner, the Sobel-filter detects vertical and horizontal edges by seeking activities that are highly nonlinear. The matrix makes it rather obvious: a uniform input image with similar activity values will yield a result that is close-to zero because the sum of the matrix values yields zero. This is also the case for the Gaussian derivatives of the Canny edge detection. On the other hand, an image gradient in horizontal or vertical direction will yield activities significantly different from zero in S_x or S_y, respectively. Dependent on the direction of the image gradient, the activity will be significantly below zero or above zero, thus being not-only edge- and edge-direction sensitive, but also edge-gradient-sensitive. Figure 8.12 illustrates the edge detection process, extracting dominantly vertically and horizontally-oriented edges, respectively.

Regardless of which filter is used, edge gradients in horizontal and vertical directions are detected and can then be further processed. A general edge map, independent of the

| (a) Original Image | (b) Sobel X | (c) Sobel Y |

Figure 8.12: To extract the edges of some image (a), the simple Sobel filter suffices to yield the main vertical (b), and horizontal (c) edges (shown in inverted gray scale). Note that diagonal edges show up with equal intensity in (b) and (c). Note also how the different edge types are particularly suitable to detect particular facial features.

individual edge directions and gradients can be computed by squaring and summing the detected edge signals:

$$H(x,y) = \sqrt{H^2_{hor}(x,y) + H^2_{vert}(x,y)} \tag{8.7}$$

If the resulting general edge detection value is greater than a particular threshold, $H(x,y) > \theta$, then one may speak of an actual edge detection. However, it is probably better to think about these values as edge indicators or edge estimates. With the horizontal and vertical edge gradients, it is possible to also compare the relative proportions of the edge signals. Mathematically, it is thus possible to estimate the actual orientation of the dominantly detected edge by:

$$\arctan 2\left(\frac{H_{vert}}{H_{hor}}\right) \tag{8.8}$$

This short mathematically-oriented overview thus shows that edge gradients contain much information about surfaces, the edges of surfaces, and the edges' orientations.

The filter type that appears to be most similar to the cell responses identified in V1 is the *Gabor filter*. Gabor filters are generated by multiplying a Gaussian with a cosine (focusing on local dark or light contour-like structures) or sine function (focusing on edges with one bright and one dark side), which is similar to the Canny edge detection described previously. Often, Gabors come in four directions, covering edge orientations of $0°, 45°, 90°$, and $135°$, where negative values essentially indicate mirrored edges. As a result, the whole edge orientation spectrum is covered. Moreover, several Gabor edge detectors that indicate a similar edge orientation in a local neighborhood along the edge orientation are likely to signal the edge of a common, larger source.

8.5.4 Motion detection

Besides edge sensitivities, however, Hubel and Wiesel detected motion selective neurons. In addition to the information content that is found in visual edges, motion also gives many clues about the outside world. When an object moves, it is typically the whole object that moves, thus yielding similar motion signals in a broader area; when we move ourselves, the whole visual field is shifted and/or expanded. Thus, motion detection provides additional information about objects, their motion relative to us as the observer, as well as our own motion relative to the surrounding.

To detect visual motion, the *optic flow* offers itself as the crucial source of information: the faster we move forward through the world, the faster the world passes by. However, this passing by is very nonlinear: the further ahead the next visual stimulus, the slower it will move; the closer and the more to the side, the faster the stimulus will move. For example, driving on the highway yields hardly any visual flow in the fovea when focusing on a straight road ahead. On the other hand, the edge of the road immediately to the side of the car will move fast. This optical flow due to self-motion, shown in Figure 8.13, thus gives a lot of information about how we are currently moving through the environment.

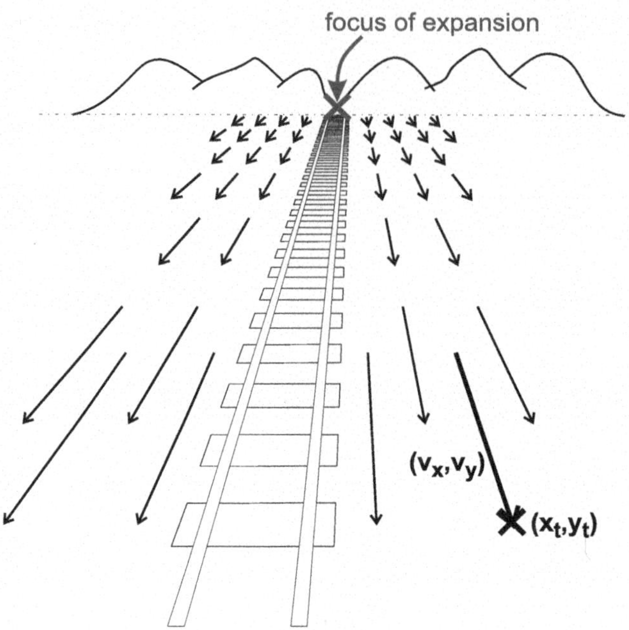

Figure 8.13: Optical flow (here generated by self-motion) are the velocity vectors $(v_x(x, y, t), v_y(x, y, t))^T$ in points (x, y) of an image between two temporally close points in time t and $t + \delta t$.

To determine this optical flow in successive images $I(t)$ and $I(t + \delta t)$, particular image points are compared to each other and reassigned, essentially establishing a mapping between successive points. To compare mappings that attempt to correlate points in an image of distance $\Delta x, \Delta y$ relative to each other, the difference between the two mappings is typically analyzed in one of the following two manners:

- Sum of squared differences:

$$SSD(\Delta x, \Delta y, \delta t) = \sum_{(x,y) \in (X,Y)} [I(x, y, t) - I(x + \Delta x, y + \Delta y, t + \delta t)]^2, \qquad (8.9)$$

  or

- Cross-correlation:

$$CC(\Delta x, \Delta y, \delta t) = \sum_{(x,y) \in (X,Y)} I(x, y, t) \cdot I(x + \Delta x, y + \Delta y, t + \delta t) \qquad (8.10)$$

where (X, Y) refers to all the pixels in the considered image or a sub-area in an image. Note how such information can be extracted best when the image is full of textures. Uniform surfaces or large free spaces do not typically allow a proper computation of optical flow in this manner. Thus, other knowledge sources may need to be used.

Similar to the neural wiring map for detecting edges proposed by Hubel and Wiesel and many others shown in Figure 8.9, a motion detection wiring mechanism was proposed by the German biologist and physicist Werner E. Reichardt (1924–1992) in the 1960s. Figure 8.14 shows the principle of this neural motion detector, which in this case prefers motion from right to left. In contrast to the neural edge detector wiring above, a crucial difference is that the Reichardt detector compares neural activities locally over time and space, rather than only in space. The figure shows excitatory and inhibitory connections, where the inhibitory connections are assumed to persist slightly longer over time. Given a sensory stimulus on the left, which moves to the right, it will successively activate the stimulus-sensitive neurons in the top row from left to right. This activation is projected onto the motion-sensitive layer in the middle row. Because an activation in this row inhibits the activity of the next right neuron, the projection of a signal that moves to the right in the top layer will be inhibited in the middle layer. If the stimulus is coming from the right in the top layer and moves left, however, no inhibition takes place in the middle layer. Consequently, the bottom layer integrates the activity in the middle layer, yielding higher activity when a stimulus moves from right to left. Directional optical flow is encoded by directionally selectively inhibiting and exciting activities in neighboring cortical columns, and thus neighboring receptive fields. Integrating the resulting activity over a mid-size receptive fields yields motion sensitivity, as observed by Hubel and Wiesel in some of their *complex cells*.

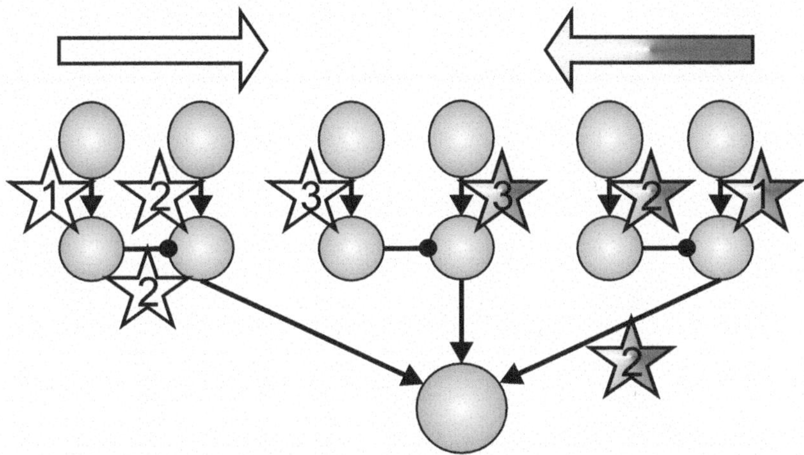

Figure 8.14: Basic principle of the Reichardt detector, which here detects motion from right to left. Numbers indicate an imaginary progression over time, at which point cells may fire. Arrow heads indicate excitatory connections, while circles indicate inhibitory connections.

The following computation nicely illustrates how the optical flow relative to the observer allows the selective activation of particular environmental interactions. Let us denote that a certain point of some surface, object, or similar, which is perceived at location (x, y) on the retina, has a distance of $Z(x, y)$ from the observer. When the observer now moves in direction (T_x, T_y, T_z), then the point on the retina will be translated as follows:

$$v_x(x, y) = \frac{T_x + x \cdot T_z}{Z(x, y)} \quad \text{and} \quad v_y(x, y) = \frac{T_y + y \cdot T_z}{Z(x, y)}, \tag{8.11}$$

where v_x and v_y denotes the velocity – or visual translations – of the point, which was previously perceived in (x, y) on the retina. Thus, after the movement (T_x, T_y, T_z), the point previously located at (x, y) will be located at $(x + v_x(x, y), y + v_y(x, y)$. When executing such a movement, the *focus of expansion* is the point from which the optical flow field expands. Due to the notations used, the focus of expansion is located at:

$$e_x =_{\text{def}} \left(\frac{-T_x}{T_z} \right), \quad \text{and} \quad e_y =_{\text{def}} \left(\frac{-T_y}{T_z} \right). \tag{8.12}$$

The focus expansion point is unique and does not depend on the distance of the observer to the object.

Flies and other flying insects exploit these physical principles to coordinate their landing behavior (essentially employing a Braitenberg vehicle principle, cf. Section 3.6.2), as well as when avoiding obstacles. To coordinate the landing, the most important information is contained in the time when the point of the focus of expansion is so close that the legs of the insects should be projected toward the approaching surface. The most valuable information thus is inherent in the time-to-landing, which is equal to Z/T_z.

The expansion around the focus of expansion is directly related to the time-to-land. For convenience, let us define (d_x, d_y) as the distance of a point (x, y) on the retina from the focus of expansion:

$$d_x \quad =_{\text{def}} x - e_x \quad = x + T_x/T_z \tag{8.13}$$

$$d_y \quad =_{\text{def}} y - e_y \quad = y + T_y/T_z \tag{8.14}$$

At this retinotopic distance, the velocity will depend on the perceived distance of the point in space perceived relative to the observer as follows:

$$v_x(d_x, d_y) = \frac{d_x \cdot T_z}{Z(x, y)} \quad \text{and} \quad v_y(d_x, d_y) = \frac{d_y \cdot T_z}{Z(x, y)} \tag{8.15}$$

Thus, when integrating over the whole area around the focus of expansion, the signal can be used to estimate the time-to-landing: the faster the motion toward the point and the closer the point, the larger its lateral translation. Once a particular expansion speed around the focus of expansion is reached, it is clear that impact must be imminent, enabling the fly to decrease speed and prepare for landing at the right moment.

Bees and other insects have also been shown to maintain a certain height during flight by monitoring the optical flow below them: the faster the flow, the closer to the ground. The same principle works for avoiding impact with objects to the side, such as trees or walls.

Interestingly, pigeons also appear to make use of this principle. In this case, though, a more active information gathering process is observable. By moving their head back and forth, they actively create an optical flow field while walking around on the ground, searching for food, for example. The flow field in this case provides them with depth information. This information is particularly important for pigeons and other kinds of animals whose eyes are turned too far to the sides of the head to use stereo vision for inferring depth.

Note also how the visual flow may be influenced by other movements in the environment, fooling the inferences made by the brain based on optical flow. Bees could be made to fly higher or lower by artificially moving a carpet-like band on the floor in a forward or backward direction, respectively. However, we do not need to consider other animals actually. Human brains consider optical flow for estimating the current speed, as for example when in the car the speed appears faster when trees are nearby in comparison to on a totally open road. Moreover, our brain can be fooled by other movements in the environment. For example, when sitting in a train it typically feels rather weird when the train on the next platform suddenly starts moving – the large optical flow field created by the train generates the expectation of self-motion and because our vestibular system meanwhile does not signal any motion change, the brain notices that something unusual is going on, producing a slightly dizzy feeling.

8.5.5 Integrating edge information

With sufficient information about visual edges in hand, brains appear to attempt to integrate these edges for deducing common causes. That is, brains appear to bind sets of edges, assigning them to an object with a consistent, three-dimensional shape and size. This is why we do not perceive a bunch of disconnected edges, but rather a set of connected wholes, that is, connected surfaces and objects. For us, it seems trivial to solve this problem, as in most cases we see objects and object surfaces – such as the table surface, floor, wall, ceiling, etc. – seemingly without any effort. It comes with some surprise that such a visual, edge-based binding problem is actually rather difficult, when viewed from an abstract, computational point of view.

Let us consider the objects in Figure 8.15. It seems easy to bind the present edges into one or several cube objects. With hardly any effort, we bind the perceived lines and indicated

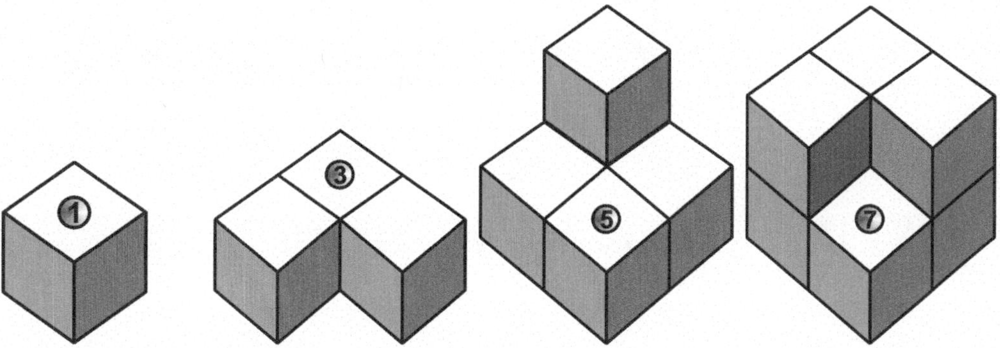

Figure 8.15: Huffman and Clowes (Clowes, 1971; Huffman, 1971) independently suggested analyzing polyhedrons at the center corner of which three surfaces meet forming a trihedral corner.

surfaces into one cube, or a particular collection of cubes. In 1971, Huffman and Clowes formalized this problem and could show that the formalized problem without additional information is actually a computationally highly challenging problem. The problem, now known as *Huffman–Clowes labeling*, starts with assigning each edge a particular type, which can be either concave or convex – due to surface nonlinearities – or it can be an occluding edge, in which case a closer surface occludes the view of parts of a surface that lie further to the back and the closer surface may either lie to the left or to the right of the edge. All types of edges are shown in Figure 8.16.

In consequence, edges that meet each other at particular corners need to be consistent with each other. Huffman and Clowes have shown that all edges and edge intersections can be illustrated by means of cube combinations, which are shown in Figure 8.15. The focus was how the edges of the cube meet at the central point, thus identifying all possible edge junctions. Views of the edge junction from anywhere within a particular surface away from the edge junction yield the same type of junction. Figure 8.17 shows all possible junctions of the three major edge types (convex "+", concave "-", and occluding "→", where the surface left of the arrow direction is the one further to the back). It is worthwhile to verify some of the junction types and to imagine how the surfaces that meet need to be oriented relative to each other.

Edge types	
+	convex edge
-	concave edge
→	occluding edge (surface to the front on the right side)
↛	shadow edge (arrow head points into the shadow area)

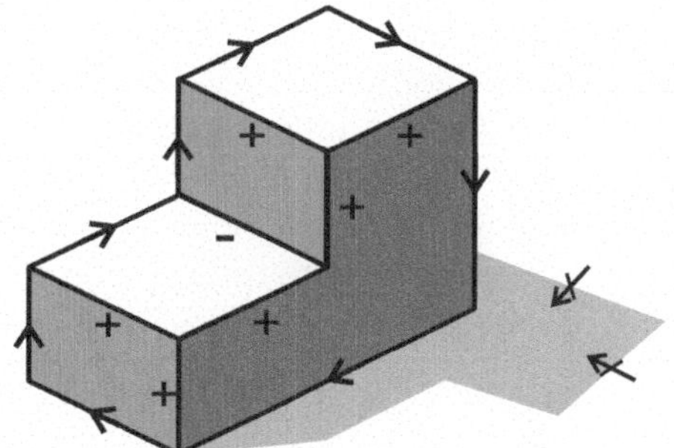

Figure 8.16: Huffman–Clowes edge labeling example

Starting with the Huffman–Clowes labeling, Waltz then formulated one of the first constraint satisfaction problems in artificial intelligence, which formalizes the challenge to find a consistent type assignment for all edges in an image, such that the overall image can be realized by three dimensional polyhedral shapes. The problem is defined as follows:

- Junctions are defined as variables.

- Each possible junction type (cf. Figure 8.17) is identified by one integer value.

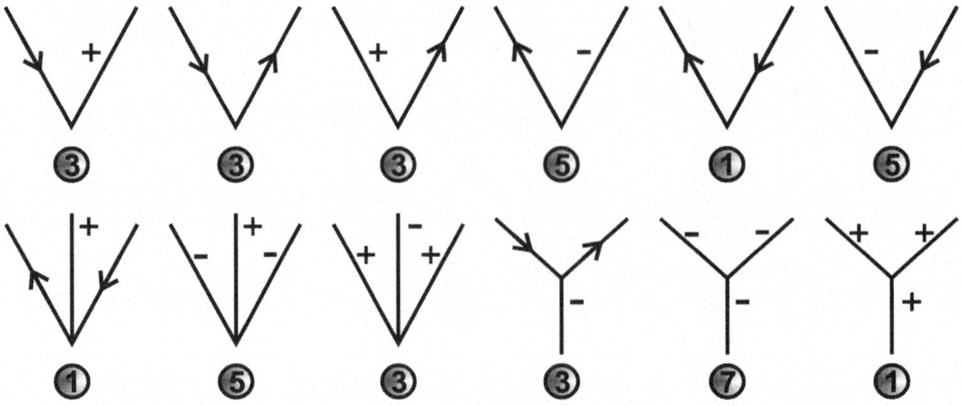

Figure 8.17: Junction types in Huffman–Clowes labeling in relation to trihedral corners. [Redrawn from Huffman, D. A. (1971). Impossible objects as nonsense sentences. In Meltzer, B. & Michie, D. (Eds.) *Machine Intelligence*, Edinburgh University Press. 6, 295–324.]

- Each edge, which connects to junctions, must thus have the same type, given junction type assignments for each junction.

This problem is now known as an *edge-assignment* or *junction-assignment problem*, and essentially belongs to the class of NP-hard constraint satisfaction problems. Although it still has not been proven, it is assumed that these types of problems can only be solved by an algorithm in non-polynomial time, such that the time to solve a problem grows faster than any polynomial of the size of the problem. Essentially, this means that when viewing a scene, the algorithm may in certain instances – given a large number of edges – take a very long time to find a proper edge assignment. From a cognitive point of view, the result also implies that the brain cannot solve the edge assignment problem in exactly this way. Rather, heuristics need to be applied and other, complementary sources of information need to be taken into account.

8.5.6   Further sources of visual information

Knowing the difficulty of the edge assignment problem, one must ask how the brain seems to solve this problem apparently nearly instantly, even when only provided with line drawings without further depth information. One may also ask how the brain solves this problem in the real world. What further information may be available?

What other sources of information may be used to reconstruct a three dimensional visual scene, given retinotopic visual information only? Besides edges and motion information, which allow the deduction of depth information by processing optical flow, there are actually quite a number of additional purely visual sources, which can give further information:

- Color and color gradients suggest that they belong to the same or neighboring surfaces and give additional information about the shape of a surface (for example, concave or convex). In this way, for example, surfaces common to one object can more easily be identified.

- Occluding edges can also be further identified by perceiving objects further back as smaller as well as in a different brightness, dependent on the light source distribution.

- Lines that are converging in depth are perceived as parallel lines.

- Texture information about object surfaces, which become smaller and the patterns of which converge in depth, often give further information about surface orientations.

- Shadows can be identified by their distinct darkening with little color changing property, which can be used to estimate the sizes and shapes of the objects that cast the shadows.

- Individual surfaces typically have distinct reflection properties, which reflect light in a distinct manner dependent on the distribution of light sources.

- Stereo-vision information gives additional clues about the depth of surfaces and objects (with declining information content in distance).

While we are not going into any further detail about these other sources of information at this point, it soon becomes clear that the problem is alleviated when taking into consideration all these additional sources of information. In fact, when considering cartoon-like, black-and-white drawings, artists often use drawing "techniques" to simplify the perception of the actual scene, and these drawing techniques typically provide some of the listed types of information. Visual illusions are deceptive in that they manipulate these types of information in such a manner that they give the wrong clues about the actual three dimensional scene, because of, for example, an atypical irregularity in the used non-uniform textures.

8.6   Summary and conclusions

Although there are many redundant and complementary primarily bottom-up sources of visual information available in a retinotopic image, some of the properties listed work best only when *top-down expectations* are included. For example, we have stated that reflection properties depend on the distribution of light sources. Thus, the brain needs to use internal estimates about light source distributions in order to infer surface orientations given the perceived visual surface reflections. Inferences based on shadows, texture, occlusions, and typical object properties also depend on prior assumptions about the respective typical visual perceptions of these visual phenomena. In Chapter 9, we focus on how the brain may use and learn top-down visual expectations about particular types of objects, surfaces, reflections, and motions. Moreover, we will also see that particularly prior assumptions about objects, including their sizes and surface properties, can simplify the edge assignment problem as well as the more general, object recognition problem.

In conclusion, it should be emphasized once more that visual perception is not only relevant for object recognition and scene modeling, but evolutionary initially primarily for facilitating environmental interactions. To improve environmental interactions it is useful to be able to determine where things are currently situated in a scene relative to the observer. Moreover, it is useful to identify the objects, entities, and other humans that we see. It has been shown, for example, that we can recognize other people not only by their individual faces, but also by their hand writing, by their motion dynamics, or by their voice. Similarly, objects seem to be classified not only dependent on their physical properties, but also dependent on their behavioral properties and on the types of interactions they offer to us. To do so, the brain appears to form hypotheses about what is currently in the scene and to verify these hypotheses given the currently available information. If this information is not sufficient, the brain sometimes issues *epistemic actions* (that is, information-seeking actions) to render the object or scene less ambiguous, such as when we use our hands to search for a particular tool in a drawer, or when we rotate an object to see its other side. In Chapter 9, we introduce how such top-down expectations and distinctions can be made and can develop, aiding both, the basic edge assignment and the scene understanding problem, as well as the problem of issuing proper and versatile environmental interactions as fast as possible.

8.7   Exercises

1. Why is the diffuse, Lambertian light reflection property of surfaces essential to be able to visually perceive objects?

2. Show that the visual flow around the focus of expansion progressively increases in speed when approaching the focus of expansion with a constant speed.

3. Show that the information about an object in the visual field (with fixed location X_0, Y_0) moves progressively closer to the center, the further distant the object is located from the lens of the eye. Show also that the width of the image covered by object decreases with increasing distance.

4. We are typically not aware of the blind spot on each of our two retinas. Why might this be the case?

5. The systematical wiring from the retina via LGN to V1 ensure that the visual information that enters V1 has a retinotopic topology. Why might this be useful?

6. Two main visual processing pathways have been contrasted in the cerebral cortex. Name and characterize them. Why might such a partitioning of visual information be useful considering where objects typically can be found in our environment?

7. Choose a particular object and list five different aspects in the visual information that provide redundant and complementary information about the object.

8. Why do some animals with eyes far to the side tend to actively move their heads back and forth? Which visual information do they generate and exploit? In which manner is this related to the reafference principle?

9. Why is it useful to smooth, and typically normalize, visual information before proceeding with further image analysis?

10. Cortical columns in the visual cortex do not only exhibit retinotopy, but also a systematic edge orientation analysis within. How may deeper visual areas benefit from such a systematic encoding?

11. Which types of edges greatly help in the perception of the shape of an object?

12. How come our brain typically solves the edge assignment problem at ease?

Chapter 9

Top-Down Predictions Determine Perceptions

9.1   Introduction

Even though we are typically not aware of it, our brain continuously activates expectations about current and successive perceptions. This anticipatory principle was presented in Chapter 6, where we focused on motor behavior. Insights from cognitive neuroscience, neurobiology, and related disciplines suggest, however, that this anticipatory principle can be generalized also to purely perceptual mechanisms. Anticipations are not restricted to temporal predictions, but they also apply to static, top-down, perceptual predictions. As in the previous chapter, we focus here on visual perception, although there are many indications, which suggest that similar information processing principles also apply to other sensory modalities.

From neuroanatomy analyses of V1, it is well-known that most of the axons that project information from other areas of the brain to V1 do not originate in LGN, but rather in extrastriate cortical areas, that is, mainly from "higher" visual cortical areas in the occipital lobe. Neural activities in V1 are thus not only influenced by incoming visual information, but are also strongly influenced by *top-down* projections of neural activities. What is the purpose of such connections? Why may such top-down projections actually be more numerous than the actual, bottom-up incoming sensory information?

It is now generally believed that the top-down connections signal expectations about the incoming sensory information. The expectations may be viewed as predictions stemming from a generative model, which attempts to reconstruct important properties of the scene. Given a sufficiently successful scene reconstruction, the sensory information verifies the currently activated reconstructions. From a top-down point of view, it can be said that top-down predictions "explain away" the incoming sensory information by subtracting the top-down predictions from them. Left with hardly any *residual*, that is, with hardly any difference between top-down predictions and bottom-up sensory information, which is sometimes also called "evidence" in this relation, the system will become progressively more confident that the current internal generative model activities reflect the state of the environment.

Bayesian information processing is able to combine bottom-up sensory information with top-down, generative predictions, yielding probability estimates about the actual causes for the sensory perceptions. This can be understood when considering a partially occluded object. For example, a car may be parked at a corner, so that only the front of the car is visible. Do we perceive only half a car? Certainly not! Our perceptual system clearly assumes that there is a whole car, which is partially occluded, parked around the corner. Note, however, that, if there was no indicator for an actual occlusion of the back of the car, our curiosity would be aroused, leading to speculations of unusual explanations. For example, our brain may consider the possibility that, for some reason, there really is only half-a car standing there, or, alternatively, there may be an invisible occluder, which may block the view of the full car.

This chapter explains how such top-down expectations may be combined with bottom-up, incoming sensory evidence to enable highly robust perceptions about the outside environment. A basic understanding of predictive, generative models is given first. Next, Bayesian information processing as the fundamental principle that controls the interaction between top-down predictions and bottom-up information is introduced in detail. Probability distributions, mixture models, and neural predictive models are also introduced. With these techniques in hand, it will be possible to understand how information combinations can work optimally in principle. After that, several models of visual processing are briefly discussed, showing that different types of predictions may come from higher level areas of visual processing, particularly contrasting spatial expectations with object identity expectations.

At the end of the chapter, we illustratively discuss various visual illusions in light of such top-down-bottom-up interactions. The illusions essentially suggest that our brain continuously attempts to integrate all available visual information to form an internal representation of a consistent whole. Even very high-level scene information and scene aspects are considered during this process. We also show that if there is not enough evidence to fully settle on one scene interpretation, the brain tends to stay in a bistable or even multi-stable state, where the visual system randomly swaps from alternative to alternative. Finally, we take a short look at the advanced topics of predictive encodings and free-energy-based inference mechanisms, which are now believed by a growing community to be the key ingredients that foster the development of the known visual and neural encodings and their interactions.

9.2 Top-down predictive, generative models

In contrast to the classical feed-forward "sense→think→act" cycle of cognitive processing, generative models are part of the inverse, top-down direction (cf. Section 6.3). Feed-forward information processing essentially results in classifications and compressions of the sensory information, enabling decision making. However, as we discussed in Section 2.2.2, this results in the *homunculus problem*. The main reasons for the homunculus problem are that it remains unclear where classification error information comes from and that there is no general reason to believe that processed information, analyzed by a "homunculus", is better suited for decision making than the raw sensory information.

Top-down predictive, generative models solve the homunculus problem. Generative models possess an additional directional link from "think" to "sense", that is, "think→sense". As a result, "thinking" not only determines current motor activities, but also predicts current sensory information. *Perception*, that is, the actual recognition of the state of the world, does not occur on the sensory level at all in generative models, but rather on deeper, "cognitive" levels. In these deeper levels, the encodings essentially generate top-down expectations of "lower-level" neural activities toward sensory modalities.

From a developmental perspective, generative models may first learn to compactly encode sensory information. On a level deeper, another generative model may then encode systematicities in this compactly encoded sensory information. The progressive addition of even deeper levels then leads to progressive sensory information abstractions, looking for general systematicities in the data. The deeper the generative model, the more abstract is the encoded information. As a result, a hierarchy of generative models develops.

Figure 9.1 shows a generative model, where the model generates predictions about lower level neural activities and ultimately, sensory activities. In contrast to the traditional view, in such an interactive model it is currently assumed that the information fed forward to "higher" levels does not carry compressed – or abstracted – sensory information, but rather the residual. Generative models thus activate expectations about sensory information, which are then verified or falsified, by propagating the resulting residual error back up. Consequently, depending on the error information, the internal generative models will adapt their activities, thus modifying their top-down predictions, that is, their sensory expectations. This internal adaption results in a new residual and the updating loop continues.

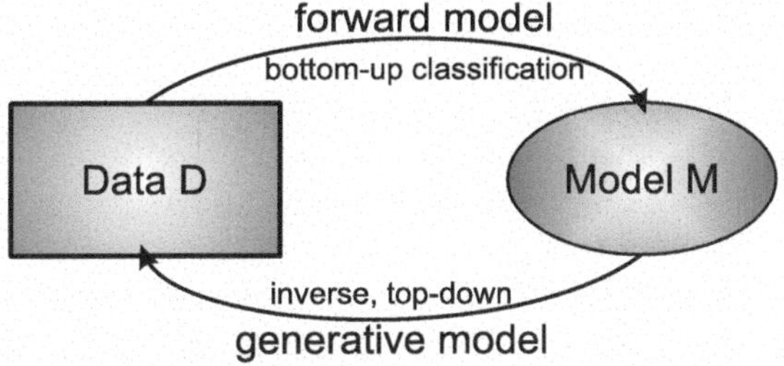

Figure 9.1: Traditionally, forward models were assumed to process data yielding classification or recognition-like activities without considering, inverse, top-down projections. Inverse, top-down predictions can be understood as generative models, which generate expectations about the current or next incoming sensory information. When top-down predictions are applied, typically it may suffice to project the residual back up.

Traditional, forward models can be viewed as *discriminative models*: given some data, for example, in the form of sensory information, a model is searched for that recognizes typical differences in the data in a most effective manner, that is, by minimizing a classification error given particular classes. This introduces the problem of the lack of supervised signals for identifying "relevant" differences in the data. In a cognitive system, an actual supervision signal, which may provide abstract, category-like information about the type of object, is generally not directly available. If we assumed the existence of such a signal, we would essentially be back at the homunculus problem, because we cannot assume the a priori existence of discriminative signals. Actual discriminative supervision signals can only come from the same or other sensory information, including sensory information about ones own body. For example, when babies put objects and other entities into their mouth, they get feedback in the form of taste, which can be interpreted as a discriminative signal. Additionally, from a behavioristic perspective, distinct reward signals can be interpreted as discriminative signals. Thus, the discriminative signal is based on sensory information sources, so that feed-forward, discriminative models turn into generative models.

In some of these cases, when, for example, sensory information is predicted given other sensory information, one can speak of a *self-supervised learning* process. Self-supervised learning essentially binds different sources of information together because of their reliable co-occurrence during particular environmental interactions. Redundant sources of information are particularly useful for learning in a self-supervised manner. In fact, the ideomotor principle (cf. Section 6.3) postulated such a self-supervised learning process, predicting sensory changes given motor activities.

In the case of vision, top-down generative models may predict particular visual information given other (possibly multiple complementary or redundant) visual information sources. The result is a purely visual, self-supervised learning process. When focusing on only one type of visual information – such as edge encodings – top-down predictions may come in the form of clustered edges and combinations of edges, predicting the individual edges that determine the actual generative clusters. In such modal generative models thus the feedback comes in the form of compact, typically occurring clusters of structured sensory information.

In contrast to discriminative models, generative models are top-down, predictive models, which expect particular data – or sensory information – given their current internal state. Generative models do not start with data, but with an internal state that predicts data. If the data corresponds to the internal state, the state is confirmed and remains active. If the data differs from the internal state, the internal state is adapted taking the difference into account.

The following example illustrates a well-known generative model: the model of a traffic light (cf. Figure 9.2). Most of us have a solid generative model of a traffic light in our head. A green traffic light, for example, has the lower of the usual three circles of a traffic light lit-up in a green color. This is a very obvious, top-down visual expectation on a rather abstract level. We also associated other predictions about the green traffic light: for example, we know that we may drive through the traffic light while it is green. Moreover, we may expect that the traffic light may soon turn to orange, which would correspond to the middle circle lighting up in orange color. Similarly, orange may turn to red and red may turn back to green, or, at least in Germany and some other countries, to red plus orange and then to green. Figure 9.2 shows the two major components of a generative traffic light model: given an internal, traffic light state estimate, particular visual perceptions are expected; moreover,

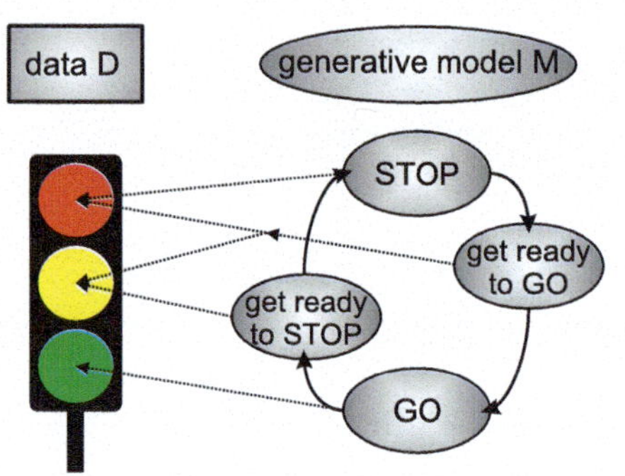

generative predictive state model

	STOP	grt GO	GO	grt STOP
STOP	0.9	0.1	0	0
grt GO	0	0.5	0.5	0
GO	0.05	0	0.9	0.05
grt STOP	0.5	0	0	0.5

generative sensor model

	red=1	yellow=1	green=1
STOP	1	0	0
grt GO	1	1	0
GO	0	0	1
grt STOP	0	1	0

Figure 9.2: The generative model of a traffic light highlights the importance of internal predictive states, which generate sensory expectations and state transition probabilities.

With this generative model in mind, we as the driver can drive happily on when a traffic light shows the green light. If we approach it from a distance, however, we may anticipate the switch to orange, thus preparing to use the brakes if necessary. We would be surprised if the traffic light behaved unexpectedly, such as suddenly switching to blinking orange, or turning all three lights on or off. Thus, the traffic light model nicely illustrates generative models, containing two important components: first, the current internal state of the model predicts corresponding visual perceptions; second, the current internal state also predicts the possible next internal state, and the possible next visual perception, enabling a sensory preparation in anticipation of possible next states over time. In the following, we formalize internal models and the resulting possible processing mechanisms by means of Bayesian methods.

9.3 Bayesian information processing

To formalize the concept of a generative model, let us define the following sets, which may also be viewed as current neural states or state estimates:

- D - the set of possible data, that is, sensory perceptions;

- C - the set of possible internal classes or state estimations;

The data set D can be viewed as any possible visual perception imaginable, or rather, perceivable by the visual sensory system. A concrete visual stimulation $d \in D$ thus specifies the bottom-up sensory input. The set of possible classes C can be viewed as all possible explanations of the data. A particular $c \in C$ can be viewed as a particular class distribution estimate, that is, a particular state estimate.

As sensory information is noisy and internal state estimations are based on these noisy information sources, it helps to formalize the actual current states by probability distributions. Note how uncertainty in the states not only comes from sensory noise, but also from uncertainty in how to interpret particular sensory information. Uncertainty may essentially come from:

- *Sensors*, including inaccurate and faulty measurements.

- *Ignorance*, that is, the incapacity to interpret the sensory information, which may even be fully accurate, about the world in the right manner.

- *Unobservability*, that is, aspects of the environment that are simply not perceivable by the sensors, but which nonetheless influence the behavior of the environment and as a consequence the sensory perceptions.

The probability of a particular state $c \in C$ may be denoted by $p(c)$ and, similarly, the probability of a particular sensory perception $d \in D$ by $p(d)$. As the data and the internal states depend on each other, the conditional probabilities are more important than particular state or sensory estimations without being given further evidence. In a *discriminative model*, the focus lies on $p(C|D)$, which denotes the general probability of a class in C given data D. In a *generative model*, on the other hand, the focus lies on the joint probability $p(C, D)$, that is, the probability of a match between class and data, or, in other words, the probability that particular class and sensory states co-occur. With this look ahead to the actual formalizations of discriminative and generative models, we now introduce the necessary basics of probability theory, before we get back to a complete, rigorous formalization.

9.3.1 Probability theory: a short introduction

Probability theory is based on the following axioms, which were proposed by the Russian mathematician Andrey N. Kolmogorov (1903–1987):

- $p(X = x) \in \mathbb{R}$ and $p(X = x) \geq 0$ specifies the *(unconditional) probability* that a random variable X will have a certain value x, that is, the probability that any state x in X is always greater or equal to zero. Often, the notation is shortened to $p(x)$.

- All possible events are contained in X, such that the probability that one event in X occurs is one, that is, $p(X) = 1$, or, in other words, $p(\mathtt{true}) = 1$.

- Finally, summing up the probabilities of all possible, mutually exclusive states of X yields a probability of one, that is, $\sum_{x \in X} p(x) = 1 = p(\cup_{x \in X} x)$.

Importantly, several consequences can be directly derived from the formulated axioms, including:

$$p(\emptyset) = 0, \tag{9.1}$$

$$p(A) \leq P(B) \ \mathtt{given\ that}\ A \subseteq B \subseteq X, \tag{9.2}$$

$$p(X = x) \in [0, 1], \tag{9.3}$$

in which the last property may be considered the most important one, specifying that any possible state x has a probability that lies between zero and one. A *probability mass* refers to the probability distribution over all possible values for a variable X, which adds up to one according to the axiomatic definition.

As pertaining to discriminative and generative models, more important than the probability of a particular state, is its probability given further information. In this case, one talks about *conditional probabilities* and denotes these by $p(x|y)$, which specifies the probability that $X = x$ given the fact that $Y = y$. Moreover, the joint probability of $X = x$ and $Y = y$ is important, especially when considering generative models. It is denoted by $p(x \wedge y)$, which specifies the probability that both states are true, that is, it is true that $X = x$ and that $Y = y$. In the case that the two states are mutually independent of each other, that is, they do not influence each other directly or indirectly, then the individual probabilities can simply be multiplied. If this is not the case, though, the conditional probability needs to be considered:

- *Joint probability*:

  independence of x and y:

$$p(x \wedge y) = p(x) \cdot p(y) \tag{9.4}$$

  dependence (more general):

$$p(x \wedge y) = p(x|y) \cdot p(y) = p(y|x) \cdot p(x) \tag{9.5}$$

- *Conditional probability*:

  independence of x and y:

$$p(x|y) = p(x) \tag{9.6}$$

  dependence (more general):

$$p(x|y) = \frac{p(x \wedge y)}{p(y)} \tag{9.7}$$

Note the interdependence between these relationships. In a more general case, dependencies have to be assumed between any two variables. However, if independence is known, the simpler joint probability equation directly derives from the independence equation for the conditional probability, and vice versa.

From these observations the *Bayesian rule* for conditional probability, which can be traced back to the English mathematician and priest Thomas Bayes (1701-1761), can be derived:

$$p(y|x) = \frac{p(x|y) \cdot p(y)}{p(x)}, \tag{9.8}$$

$$p(y|x, e) = \frac{p(x|y, e) \cdot p(y|e)}{p(x|e)}, \tag{9.9}$$

where the additional e in the second equation denotes additional evidence, which is simply carried along in the reformulation. Depending on the context, conditional probabilities are often called *likelihoods* for particular states y, for example, sensory measurements, given particular evidence x and further evidence e. This is especially the case when $p(y|x, e)$ is approximated by $p(x|y, e)$. Unconditional probabilities are also referred to as *a priori* probabilities, because they are assumed without, or prior to, the consideration of further evidence. Given additional evidence, the resulting conditional probability is often referred to as the a posteriori probability, that is, the probability given new informative evidence.

A further important concept is the *marginalization* over all possible states y of a particular set Y. Given all possible states y and their respective a priori probabilities $p(y)$, the probability for $p(x)$ can be derived by marginalizing over all possible states y:

$$p(x) = \sum_{z} p(x|z) \cdot p(z) = \sum_{z} p(x \wedge z), \tag{9.10}$$

$$p(x|y) = \sum_{z} p(x|y, z) \cdot p(z|y) = \sum_{z} p(x \wedge z|y), \tag{9.11}$$

$$\tag{9.12}$$

denoting essentially the fact that the a priori probability of a state x without any additional knowledge can be derived from the conditional probabilities of x given y when considering all possible states y. This equation thus essentially computes the mean conditional probability of x over all possible states y, weighted by the a priori likelihoods of y.

With these few equations in hand, it is now possible to compute many interesting probabilistic relations, derive dependencies, and particularly useful probability estimates, given other probabilities. For example, a particular conditional probability is rather easy to determine, such as $p(x|y)$, whereas the inverse case, that is, $p(y|x)$, may be rather hard to estimate. This is particularly often the case when data x is available and the generative model y needs to be estimated, that is, $p(y|x)$. On the other hand, given a particular generative model y, the data that may correspond to y can often be estimated more easily, that is, $p(x|y)$. In this case, the above specified Bayesian rule (Eq. 9.8) will be very helpful, allowing the inference of $p(y|x)$ based on estimates of $p(x|y)$ and a priori estimations for $p(x)$ and $p(y)$.

Before moving on to an illustrative example, it should be noted that probability theory can be easily extended to continuous, real-valued spaces. In this case, the set of possible states is infinite (because in a real-valued, continuous space there are infinite concrete states) and state estimations need to integrate information over a particular area around a particular location. For example, an n-dimensional state may be denoted by $\boldsymbol{X}_{\mathbb{R}} = \mathbb{R}^n$. The probability for a particular state value can then only be expressed by value ranges, such that in a one-dimensional space ($n = 1$):

$$p(x_l \leq x < x_h) \in [0, 1].\qquad(9.13)$$

With such a probability concept in real-valued space, the probability distribution over the space is called a *probability density*, where the integral over all possible event values inevitably integrates to one, that is:

$$\int_{x=-\infty}^{\infty} p(X = x)dx = 1.\qquad(9.14)$$

Similarly, marginalization in a real-valued space needs to consider the whole space:

$$p(x) = \int_y p(x|y) \cdot p(y)dy = \int_y p(x \wedge y)dy,\qquad(9.15)$$

such that the state x is marginalized over the continuous space y. Figure 9.3 shows exemplary conditional probabilities for different combinations of conditional discrete and continuous variables.

We can assume that there are no continuous probability density estimates in the brain. Approximations of such density estimates by means of neural activities, however, seem likely to be present in one form or the other. For now, however, we do not consider how the brain may actually represent probability distributions.

9.3.2 A simple example

Let us consider a simple example with which we can understand the basic principle of conditional probabilities in the context of a visual task. Let us assume that we go to a zoo and are looking forward to seeing some zebras. To recognize a zebra, various visual information is available, such as an entity that has four legs, a zebra-like head, black-and-white stripes on the body, a tail, etc. To simplify things, we here focus on the stripes and assume that we are equipped with a visual stripe-detector, which signals the detection of stripes when feeding in an image of an animal enclosure (cf. Figure 9.4). Since we like zebras, we are looking forward to seeing them, but we are uncertain when we will. Given we are looking into an appropriate animal enclosure and the stripe detector goes off, how likely is it that we are looking at a zebra? Formally, the question is what is the probability of $p(\mathbf{zebra}|\mathbf{stripes})$?

To answer this question, let us assume that the following information is available:

X_1	X_2	$p(Y = 1|X_1, X_2)$
A	1	0.42
A	2	0.17
B	1	0.66
B	2	0.30
C	1	0.99
C	2	0.02

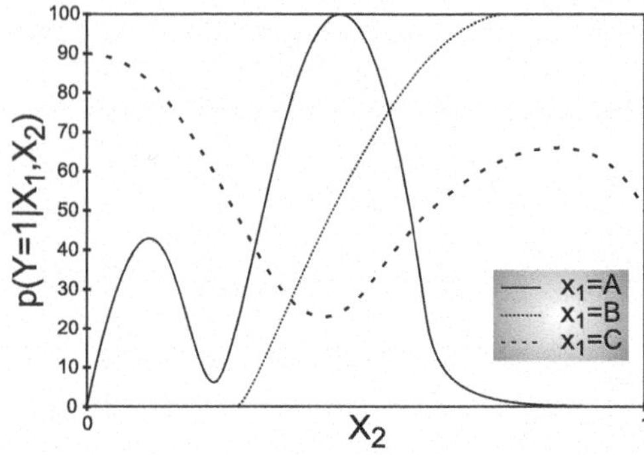

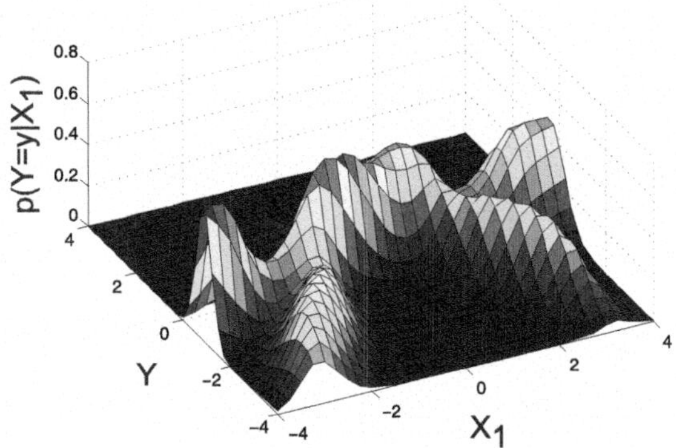

Figure 9.3: Conditional probability masses and probability densities can be represented in various forms. If the variable in question Y as well as the conditional variables, which Y is conditioned on, are discrete, the probability mass can be represented by a table. If Y is discrete, but the conditioning variable is continuous, *probability functions* need to be specified. In the example, the functions specify $p(Y = 1|X_1, X_2)$, given the discrete variable $X_1 \in \{A, B, C, \}$ and a second, continuous variable $X_2 \in [0, 1]$. On the other hand, when Y is continuous, but all input variables X are discrete, a set of probability density functions (one for each input value combination) is needed. Finally, when both values are continuous, a probability density function is needed, with additional input values X.

- $p(\texttt{zebra}) = 0.05$, which may be for example derived a priori given the knowledge that there are 20 different types of animals in the zoo that can be found in zebra-suitable enclosure.

- $p(\texttt{stripes}|\texttt{zebra}) = 0.9$, which states that our stripe detector is somewhat reliable, signaling stripes when there is a zebra in an image with a probability of 0.9.

- $p(\texttt{stripes}|\neg\texttt{zebra}) = 0.2$, which indicates that our stripe detector also signals stripes in other circumstances. As stripes are not only found on zebras, this probability seems somewhat plausible.

With these probabilities. it is now possible to compute the a priori probability that the stripe-detector will signal stripes $p(\texttt{stripes})$, when watching a particular zebra-suitable enclosure (assuming at least one zebra is always visible when looking into the actual zebra enclosure) by means of marginalization:

$$p(\texttt{stripes}) = p(\texttt{stripes}|\texttt{zebra}) \cdot p(\texttt{zebra}) + p(\texttt{stripes}|\neg\texttt{zebra}) \cdot p(\neg\texttt{zebra})$$
$$= 0.9 \cdot 0.05 + 0.2 \cdot 0.95 = 0.235 \tag{9.16}$$

Figure 9.4: While a stripe detector may indicate the presence of a zebra, certainly also other animals and image properties, such as the Okapi here, may activate the stripe detector.

With this marginalization, we can also derive the probability that we are looking at a zebra given the stripe detector signals `stripes`:

$$p(\text{zebra}|\text{stripes}) = \frac{p(\text{stripes}|\text{zebra}) \cdot p(\text{zebra})}{p(\text{stripes})} = \frac{0.9 \cdot 0.05}{0.235} = 0.1915 \qquad (9.17)$$

Note how we have used marginalization to estimate $p(\text{stripes})$, which we needed to apply the Bayesian rule.

The example shows that our probability of seeing a zebra in a zebra-suitable enclosure has risen from the a priori probability of 0.05 to 0.1915, due to the stripe detector. Clearly, though, the stripe detector is not enough to see zebras with absolute certainty. Other sensory signals, which give information about zebras, would need to be considered as well to corroborate enough evidence to reach approximate a posteriori certainty about looking at zebras.

9.3.3 Bayesian networks

The computations of conditional probabilities, which allow the inclusion of many sources of evidence as well as possible interactions between the evidences, are best accomplished by means of Bayesian networks. These Bayesian networks are essentially graphical models that specify conditional dependencies and interdependencies.

Bayesian networks consist of:

- *Nodes*, where each node specifies a certain state or event in the form of discrete or continuous probability distribution.

- *Vertices*, that is, directed edges, which connect a parental node with a child node.

- *Probability distributions*, which specify the a priori probability mass or probability density of a discrete or continuous root note, which has no incoming vertices, or the conditional probability mass or probability density of a discrete or continuous child node, given parental node values.

A complete Bayesian networks essentially allows the derivation of all possible joint probabilities. It does not have cycles, that is, there is not a path along the vertices in the network that results in a closed loop.

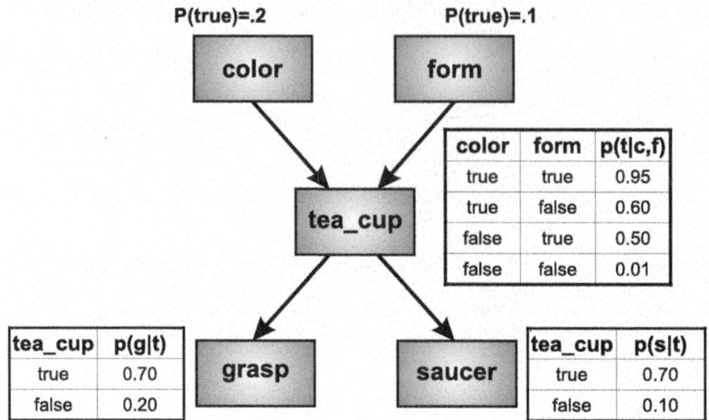

Figure 9.5: A simple Bayesian network, which sketches out conditional probability for detecting a teacup.

Figure 9.5 shows a simple Bayesian network, with which we show how to identify a teacup, given evidence about color and shape and the existence of a saucer. Moreover, a grasping decision is included, which adds an additional twist to the network's potential interpretation. We hypothesize that the color detector may indicate the brownish color of black tea, such that the presence of the particular color adds evidence of the presence of a teacup. Additionally, the form detector may add further evidence. We may assume that the form detector fires particularly often, when a teacup is currently in view.

We may furthermore assume that the Bayesian network has learned that over all images without other assumptions, the color detectors goes off in say 20% of the cases, while the form detector does so in say 10% of the cases. These are essentially two a priori, unconditional probabilities in the network. The actual detection of a teacup now depends on these two sources of evidence. Thus, to specify all conditional probabilities $p(\texttt{tea\_cup}|\texttt{color}, \texttt{shape})$, all value combinations for $\texttt{tea\_cup}$ and $\texttt{shape}$ need to be considered. Since

$$p(\texttt{tea\_cup} = \texttt{false}|\texttt{color} = c, \texttt{shape} = s) = 1 - p(\texttt{tea\_cup} = \texttt{true}|\texttt{color} = c, \texttt{shape} = s),$$

and both, $\texttt{color}$ and $\texttt{shape}$ can take on two values each (that is, $\texttt{true}$ or $\texttt{false}$), four conditional probability values suffice to specify all conditional probabilities. Figure 9.5 gives some (made up) values for these conditional probabilities, essentially suggesting that the presence of a teacup is most likely when both detectors are on (95%) and very unlikely when both detectors are $\texttt{false}$ (1%). If only one is $\texttt{true}$ (or "on"), then we expect a 60 or 50% chance given that the color or shape detector is the one that is $\texttt{true}$, respectively.

In addition the graph specifies consequences of the presence of the teacup. Here we distinguish between an action consequence and a perceptual consequence. On the one hand, the Bayesian network specifies that it is more likely to see a saucer when a cup is present (70%) versus when it is not present (10%). On the other hand, the Bayesian network specifies the probability of executing a grasping action given there is a cup present. While such a grasping decision would also depend on various other factors, the network emphasizes that evidences may not only be used for perceptual, discriminative tasks, but also for action decision making.

The network now essentially specifies particular independence assumptions. The main independence can be formulated: two nodes in the network are independent of all its non-descendants in the network given values for all its parents. This implies that the joint probability of N nodes in the network can be computed by:

$$P(X_1, X_2, ..., X_N) = \prod_{i=1}^{N} P\left(X_i | \mathtt{parents}(X_i)\right) \tag{9.18}$$

With respect to the Bayesian network in Figure 9.5, it can thus be stated that `color` and `shape` are independent of each other because they do not have any parents (thus all parent values are given) and they are non-descendants of each other. Similarly, given `tea_cup`, `grasp`, and `saucer` are independent of each other as are `saucer` and `color`, `saucer` and `shape`, `grasping` and `color`, and `grasping` and `shape`. Unfortunately, there is one confusing case, which is the one where nodes become dependent on each other when there is evidence about common descendants. This is the case when `tea_cup` or `grasping` or `saucer` is given, in which case the previously independent nodes `color` and `shape` become dependent on each other, that is, $p(\mathtt{color}, \mathtt{shape}|\mathtt{saucer}) \neq p(\mathtt{color}|\mathtt{saucer}) \cdot p(\mathtt{shape}|\mathtt{saucer})$!

The algorithm called *d-separation* specifies all conditional independences in a Bayesian network: two nodes X and Y are d-separated by a set of evidence variables $\boldsymbol{E}$ if and only if all undirected paths from X to Y are "blocked". A path is blocked in the following cases:

- There exists a node $V \in \boldsymbol{E}$ on the path where the vertices that connect V are "tail-to-tail".

- There exists a node $V \in \boldsymbol{E}$ on the path where the vertices that connect V are "tail-to-head".

- There exists NO node $V \in \boldsymbol{E}$ on the path OR in the set of descendants of nodes on the path for which the vertices that connect V are "head-to-head".

As a result, if the set of evidence nodes $\boldsymbol{E}$ d-separates X and Y, then X and Y are independent of each other given $\boldsymbol{E}$. Figure 9.6 shows the three cases of blocks and connected paths, dependent on the additional available evidence. A simple depth-first graph search algorithm can compute d-separation in linear time. However, given particular probability values, nodes may sometimes be independent of each other even if they cannot be d-separated.

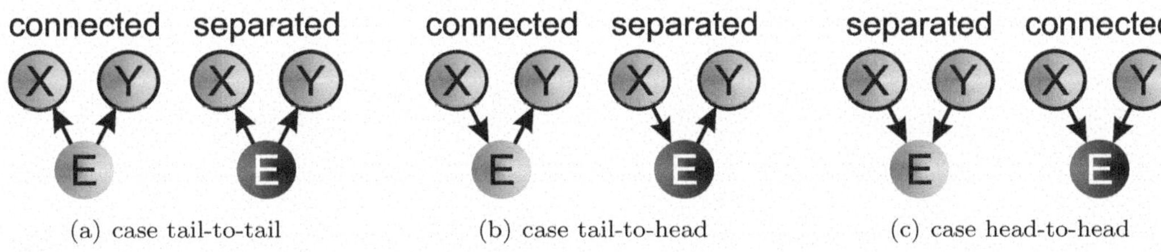

Figure 9.6: Nodes X and Y are connected (that is dependent on each other) or separated (independent) dependent on if the connecting evidence node E is unknown (light colored sphere) or known (dark colored sphere).

With the principles of conditional independences in hand, it is now rather easy to infer probability estimates given particular probabilities and conditional probabilities of interest. It is also easy to compute concrete joint probabilities by means of Eq.(9.18). However, to avoid the computation of all necessary joint probabilities, principles of *deduction* allow the exact inference of other conditional probabilities in Bayesian networks. Deduction is the foundation for computing inferences under uncertainties. It is thus very relevant for top-down influences on (visual) perception and allows the derivation of various machine learning techniques.

Four types of deduction can be distinguished, which are contrasted in Figure 9.7. In the following list, we exemplarily compute each form of deduction with respect to the exemplar Bayesian network discussed previously (cf. Figure 9.5). For reasons of brevity, we write

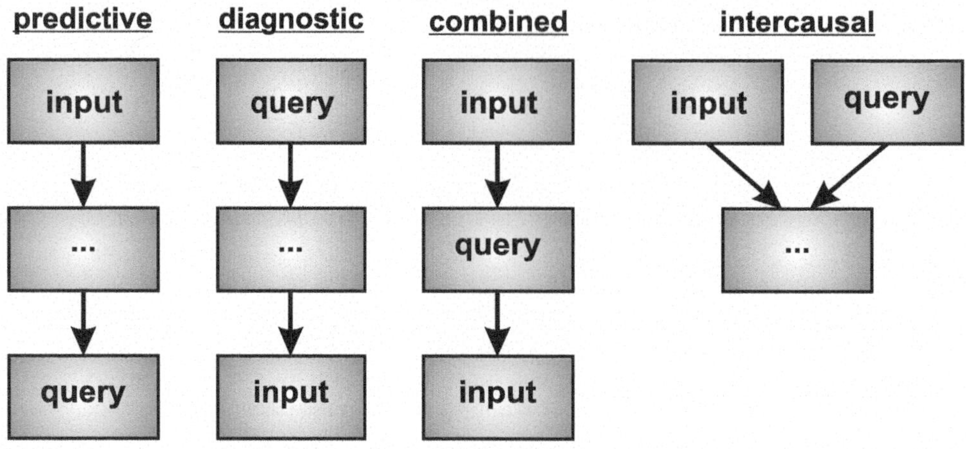

Figure 9.7: Depending on which information is available (marked as "input"), different types of deductions can be contrasted in Bayesian networks.

the probabilities using only the first letter of the names of the nodes, that is, for example, $p(\texttt{tea\_cup} = \texttt{true})$ or $p(\texttt{grasp} = \texttt{false}|\texttt{color} = \texttt{true})$ simply by $p(t)$ or $p(\neg g|c)$.

Predictive deduction is accomplished by means of marginalization, in order to generate predictions given evidences:

without evidence:

$$p(t) = \sum_{c,f} p(t|c, f)p(c, f) = \sum_{c,f} p(t|c, f)p(c)p(f)$$

$$= (0.95 \cdot 0.2 \cdot 0.1) + (0.6 \cdot 0.2 \cdot 0.9) + (0.5 \cdot 0.8 \cdot 0.1) + (0.01 \cdot 0.8 \cdot 0.9) = 0.1742$$

with evidence:

$$p(t|c) = \sum_{f} p(t|c, f)p(f)$$

$$= (0.95 \cdot 0.1) + (0.6 \cdot 0.9) = 0.635$$

Diagnostic deduction is computed by means of the Bayesian rule, given evidence further down the tree:

$$p(t|s) = \frac{p(s|t)p(t)}{p(s)} = \frac{p(s|t)p(t)}{p(s|t)p(t) + p(s|\neg t)p(\neg t)}$$

$$= \frac{0.7 \cdot 0.1742}{0.7 \cdot 0.1742 + 0.2 \cdot 0.8258} = 0.4247$$

Combined deduction is necessary when evidence for a node is available from above and below in the Bayesian network:

$$p(t|s, f) = ?$$

1. step: predictive deduction:

$$p(t|f) = \sum_{c} p(t|c, f)p(c)$$

$$= 0.2 \cdot 0.95 + 0.8 \cdot 0.5 = 0.59$$

2. step: diagnostic deduction:

$$p(t|s, f) = \frac{p(s|t, f)p(s|f)}{p(s|t, f)p(s|f) + p(s|\neg t, f)p(\neg t|f)}$$

$$= \frac{p(s|t)p(t|f)}{p(s|t)p(t|f) + p(s|\neg t)p(\neg t|f)} = \frac{0.7 \cdot 0.59}{0.7 \cdot 0.59 + 0.2 \cdot 0.41} = 0.8343$$

Intercausal deduction is given when evidence along a non-directed chain is available:

$$p(c|t, f) = \frac{p(t|c, f)p(c|f)}{p(t|f)} = \frac{p(t|c, f)p(c, f)}{p(t|f)p(f)} =$$

$$= \frac{p(t|c, f)p(c)p(f)}{p(t|c, f)p(c)p(f) + p(t|\neg c, f)p(\neg c)p(f)} =$$

$$= \frac{0.95 \cdot 0.2 \cdot 0.1}{0.95 \cdot 0.2 \cdot 0.1 + 0.5 \cdot 0.8 \cdot 0.1} = \frac{0.95 \cdot 0.02}{0.95 \cdot 0.02 + 0.5 \cdot 0.08} = \frac{0.019}{0.059} = 0.322$$

Note how intercausal deduction needs to consider the presence of the `form` indicator, despite the presence of the `tea_cup`. If we do not consider `form`, the computation yields a different result, which is due to the interdependence of `color` and `form` given `tea_cup`:

$$p(c|t) = \frac{p(t|c) \cdot p(c)}{p(t)} = \frac{0.635 \cdot 0.2}{0.174} = 0.718$$

This intercausal deduction thus predicts a lower probability for the `color`-detector being `true` when not only the `tea_cup` is given, but also the evidence `form`. The Bayesian network predicts that the probability of the color detector increases when a teacup is present. However, when the form detector already provides evidence for the teacup, the coactivation of the color detector is less likely. Note how different probabilities in the conditional probability table of the `tea_cup` node could also have produced the reverse effect, increasing the probability for `color` further given additional `form` evidence.

At this point we are not going into further detail about the maximally effective algorithmic realization of these deductions. It is well known, however, that generally the problem of calculating $P(X|Y)$ for some nodes or sets of nodes in a Bayesian network is #P-hard, which is more difficult than NP-hardness, and thus typically computationally intractable for large problems. Nonetheless, for non-extreme probability values, fast polynomial time algorithms exist that can give good approximations.

Indeed, from a cognitive perspective, approximations rather than exact estimations are probably the choice made by evolution. Especially seeing that sensory information is noisy anyways, internal state estimations are also prone to noise, so that perfect conditional probability estimates remain elusive. Moreover, state estimates will inevitably be in flux, being continuously adapted to the available sensory information.

When reconsidering the teacup Bayesian network from a cognitive perspective, it may be imagined that the teacup estimate may be represented by two neurons, which gather evidence for and against the presence of a teacup. This (highly simplified) neural representation of a probability mass will then be continuously updated by the incoming evidences, such as the evidence from the hypothesized `color` and `form` detectors. In this form, the Bayesian network is a *dynamic Bayesian network*, for which the probability flow over time needs to be specified. For example, the teacup presence estimate may stay stable over time while remaining in the same location, but it may decrease when, for example, leaving the kitchen, or increase when entering a café. Various researchers indeed consider the brain to approximate a dynamic, highly distributed, hierarchical, and modularized Bayesian network in various respects. However, this view is not sufficient to explain the goal-directedness of our brains – an aspect that we will re-consider in later chapters.

9.3.4 Probability distributions and densities

When considering the availability of bottom-up, topologically structured sensory information, such as the retinotopically organized structure in V1, it is useful to consider how particular sensory information may be predicted over these topological structures. Approximations of such predictions can be formalized by means of Gaussian probability density functions as well as by mixtures of these functions.

A *Gaussian* or *normal probability density* can be defined for one real-valued dimension $\mathbb{R}$ or for a number of real-valued dimensions $\mathbb{R}^n$ by:

- One-dimensional:

$$p(x) = \frac{1}{\sigma\sqrt{2\pi}} \exp^{\frac{-(x-\mu)^2}{2\sigma^2}}, \tag{9.19}$$

  where μ specifies the mean and σ the standard deviation (σ^2 is consequently the variance) of this Gaussian probability density.

- Multi-dimensional:

$$p(\boldsymbol{x}) = \frac{1}{(2\pi)^{n/2}|\Sigma|^{1/2}} \exp^{-\frac{1}{2}(\boldsymbol{x}-\boldsymbol{\mu})^T\Sigma^{-1}(\boldsymbol{x}-\boldsymbol{\mu})}, \tag{9.20}$$

  where n specifies the number of dimensions, $\boldsymbol{\mu}$ the mean column vector, Σ the covariance matrix (with dimensions $n \times n$), $|\Sigma|$ the determinant, ^{T} the transpose of a vector, and $^{-1}$ the inverse of a matrix.

Figure 9.8 shows several Gaussian probability densities in one and two dimensions.

To estimate Gaussian densities, given a number of samples N, simple statistics can be used:

$$\mu = \frac{1}{N}\sum_{i=1}^{N} x_i \tag{9.21}$$

$$\sigma^2 = \frac{1}{N-1}\sum_{i=1}^{N}(x_i - \mu)^2 \tag{9.22}$$

In higher dimensional spaces, the same principle holds:

$$\boldsymbol{\mu} = \frac{1}{N}\sum_{i=1}^{N} \boldsymbol{x_i} \tag{9.23}$$

$$\Sigma(\boldsymbol{x}) = \frac{1}{N-1}\sum_{i=1}^{N}(\boldsymbol{x_i} - \boldsymbol{\mu})(\boldsymbol{x_i} - \boldsymbol{\mu})^T =$$

$$= \begin{bmatrix} var(x^{1,1}) & cov(x^{1,2}) & \cdots & cov(x^{1,n}) \\ cov(x^{2,1}) & var(x^{2,2}) & \cdots & cov(x^{2,n}) \\ \cdots & \cdots & \cdots & \cdots \\ cov(x^{n,1}) & cov(x^{n,2}) & \cdots & var(x^{n,n}). \end{bmatrix} \tag{9.24}$$

The covariance matrix is a $n \times n$, symmetrical matrix, which essentially specifies the relative interdependencies between the dimensions. The diagonal of the matrix specifies the independent, axis-specific variances. The non-diagonal entries, on the other hand, specify the covariances between different axes, where values close to zero indicate approximate independence, while larger values indicate stronger dependencies. Figure 9.8 shows the effect of different covariance matrices on the resulting Gaussian distributions.

Gaussian distributions have many convenient properties. Most notably, the *central limit theorem* shows that the distribution of the arithmetic mean of any independently sampled variables will become a Gaussian distribution with increasingly more samples. Nonetheless, the expressiveness of Gaussian distributions is limited because they focus the distribution around one most probable value (the mean $\boldsymbol{\mu}$). Sometimes, though, probability densities have several peaks, yielding bi-modal or even multimodal distributions. Take, for example, a probability density for possible locations of your bicycle or your car without any given

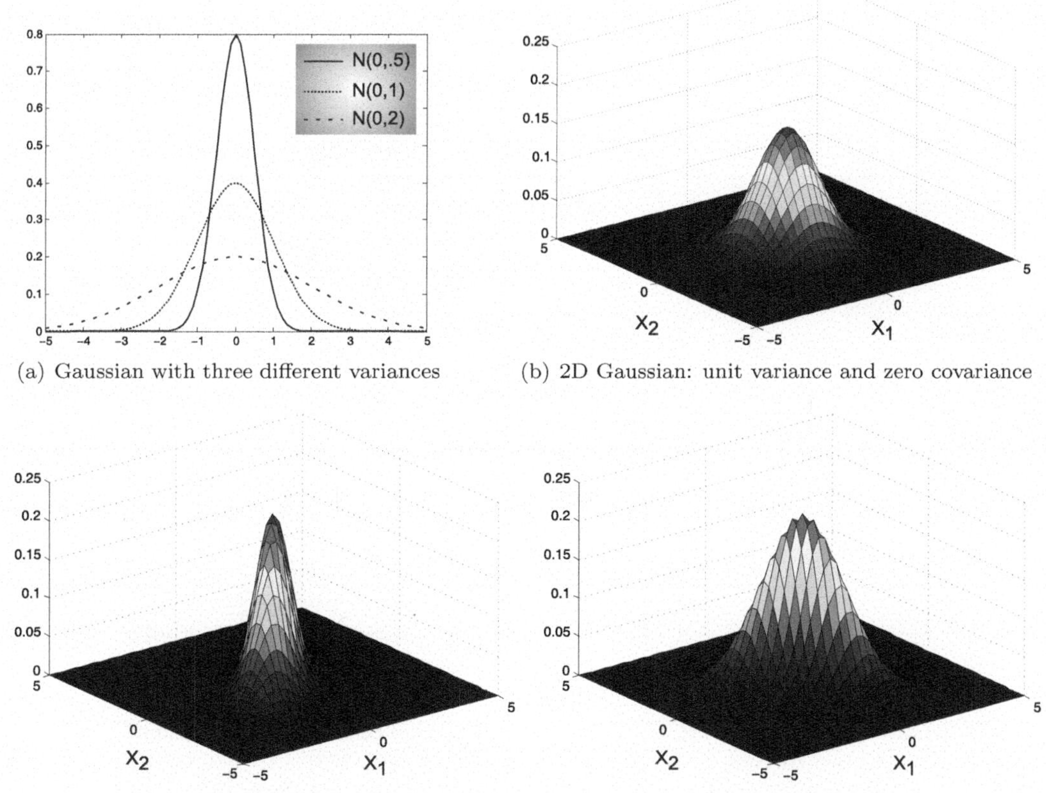

(a) Gaussian with three different variances

(b) 2D Gaussian: unit variance and zero covariance

(c) 2D Gaussian: unit variance and 0.7 covariance

(d) 2D Gaussian: unit variance and -0.7 covariance

Figure 9.8: Examples of several one dimensional and two dimensional Gaussian probability densities. In two dimensions, the covariance matrix allows the encoding of oblique, ellipsoidal densities. The diagonal variance values in the matrix stretch or compress the two-dimensional Gaussian distribution along the main axes (not shown). The non-diagonal values result in an oblique orientation.

information. Most of the time it will typically be either at work (or at university, school, or wherever you may spend most of your work days) or at home. A Gaussian probability cannot model such a distribution and indeed will make the mistake of estimating the mean to be somewhere between work and home, which certainly will not reflect the actual probability density.

Gaussian mixture models (GMMs) are well-suited to approximate more complex, bi- and multimodal probability densities. Figure 9.9 shows several Gaussian mixture models. As the name suggests, GMMs mix several Gaussian distributions by means of a mixture coefficient $\boldsymbol{\pi}$ of length m, where $\sum_{i=1}^{m} \pi_i = 1$ to ensure that the resulting GMM is a proper probability density function. The GMM distribution is thus defined by:

$$p(x) = \sum_{i=1}^{m} \frac{\pi_i}{(2\pi)|\Sigma_i|^{1/2}} \exp^{-\frac{1}{2}(\boldsymbol{x}-\boldsymbol{\mu}_i)^T \Sigma_i^{-1}(\boldsymbol{x}-\boldsymbol{\mu}_i)}, \tag{9.25}$$

where a Gaussian probability density is needed for each mixture coefficient, assuming a mixture of m Gaussian densities. Generally, mixture models exist in various forms and shapes, where the individual densities do not necessarily need to be Gaussian.

To illustrate the use of GMMs, let us assume we want to determine if we see an apple, an orange, or a banana, when looking at a collection of fruit (Figure 9.10). To do so, let us assume that we are shown images of apples, oranges, or bananas, where the images encode

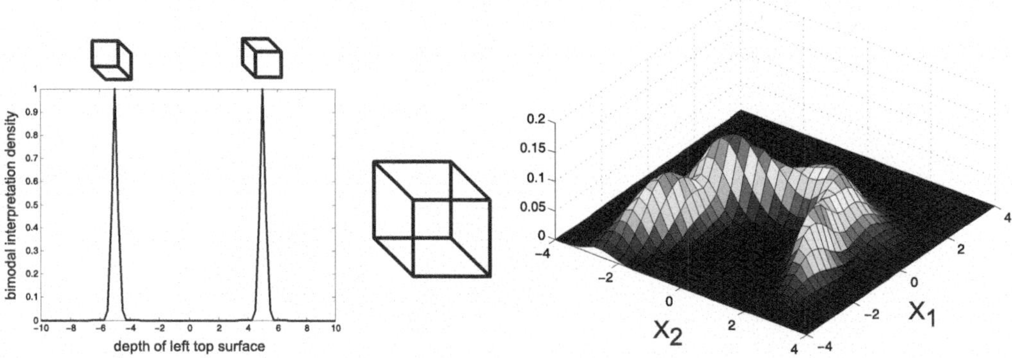

(a) Bimodal distribution as an interpretation of the Necker cube

(b) Complex Gaussian mixture density with four individual multivariate Gaussians

Figure 9.9: Gaussian mixture models can express bimodal distributions, such as when predicting multiple locations or also alternative feature expectation. Also continuous densities can be generated in N-D space, possibly by means of multiple multivariate Gaussians with individual means and covariance matrices.

the color value highly simplistically by means of one value, which specifies the dominant color wavelength. Considering for now individual pixels, we may then estimate if the pixel belongs to a particular type of fruit or not. We are mainly interested in classification given visual image data, that is, we are interested in $p(C|D)$ where $C = \{\texttt{apple}, \texttt{orange}, \texttt{banana}, \texttt{none}\}$ and D is an image. To compute the conditional probabilities, we need

- A priori probabilities for a specific class. For example, we may assume equal a priori probabilities for apples, oranges, and bananas, but we may also assume a 55% chance of seeing none of them; thus, $p(\texttt{apple}) = p(\texttt{orange}) = p(\texttt{banana}) = 0.15$ and $p(\texttt{none}) = 0.55$.

- Moreover, we need conditional probabilities for particular data signals given a particular class, assuming that the generative model approximations $p(D|C)$ will be easier to approximate than the discriminative conditional probabilities $p(C|D)$. Figure 9.10 shows exemplary, plausible probability densities, where apples are expected to be green or red, oranges are orange, and bananas tend to be yellow. Moreover, the probability for none of the fruits, given an image color, is modeled by a uniform distribution over the color space.

Given these probabilities, it is possible to compute the joint probability $p(D, C)$ by means of $p(D, C) = p(D|C) \cdot p(C)$ and classification can be accomplished by choosing the maximum joint probability, that is:

$$\arg\max_c p(C = c|D) = \arg\max_c p(C = c, D), \tag{9.26}$$

assuming that $p(D)$ is equal for all classification cases.

When assuming that D is an image of size $N \times N$, where the object in question is fixated centrally, the arg max operation may be applied over the evidence integrated over the whole image, possibly weighing evidence from the center of the image more strongly than evidence stemming from the periphery, that is,

$$\arg\max_c p(C = c|D) = \arg\max_c \int_{x,y} p(C = c, D(x,y))G([N/2, N/2]^T, \Sigma)(x,y), \tag{9.27}$$

with a suitable covariance matrix, for example, which has diagonal values of $N/4$ and zero values in the non-diagonal entries. With these specifications, we have formulated a generative model, which assumes that the object is located in the center and that the color distribution around the center is most informative about the object's identity. Figure 9.10 shows an illustration of the generative model. As we will see later, a more general formulation of this model, which has been proposed as a model for human vision, is able to learn and generate object-specific top-down location and feature expectations.

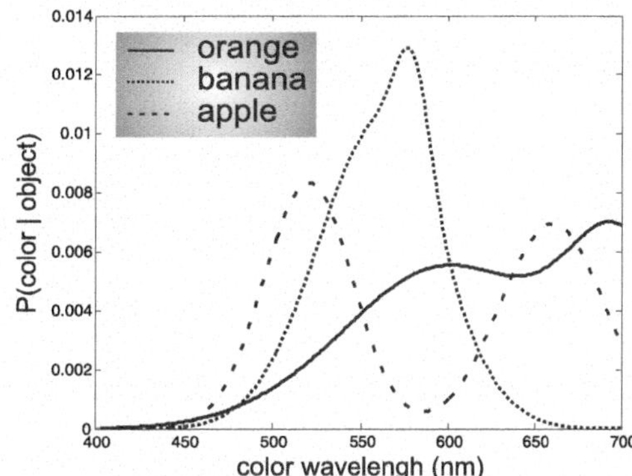

Figure 9.10: When expecting to see either an orange, a banana, or an apple, color helps to recognize a fruit correctly. The graph shows a Gaussian mixture model as a color estimation density model, given a particular fruit. By means of deduction, likelihoods for each fruit can be inferred when integrating over the color space. Additional priors for the a priori likelihood of each fruit would improve the likelihood estimates. However, most importantly other information sources, such as the shape of the fruit (as shown on the left), certainly help to disambiguate the identity of the fruit in question.

9.4 A Bayesian model of visual processing

As an example of a recent model of visual processing, we now give a short overview of a model developed by Serre, Wolf, Bileschi, Riesenhuber, and Poggio (2007) and Chikkerur, Serre, Tan, and Poggio (2010). The authors proposed an interactive visual processing model, which combines bottom-up with top-down interactive information processing mechanisms based on Bayesian techniques. It provides preprocessed visual information akin to a cortical columnar structure using a hierarchy of Gabor filters as edge detectors. These filters are used as input to the next visual processing layer, which combines the bottom-up information from the Gabor filters with top-down feature expectations. The expectations are combinations of ventral-stream-like feature expectations and dorsal-stream-like location expectations, both of which are modeled by means of Bayesian a priori densities.

Figure 9.11 shows the principled structure of the architecture. The model is a generative model, which is in principle able to generate image imaginations I given an internal scene description S. That is, the model can generate joint probabilities $p(S, I) = p(I|S)p(S)$. The scene description consists of two independent components:

- *Location estimates* about the scene L, which generate top-down location priors.

- *Object-specific* estimates O, which generate top-down feature distribution priors.

With no further evidence, L is assumed to be independent of O so that the joint probability model can be denoted by:

$$p(I, L, O) = p(I|L, O)p(L)p(O), \qquad (9.28)$$

where the scene description S was replaced by its two components L and O.

While the location component is not further differentiated in the model, the object component is assumed to contain N feature complexes F_i, whose activities directly depend on

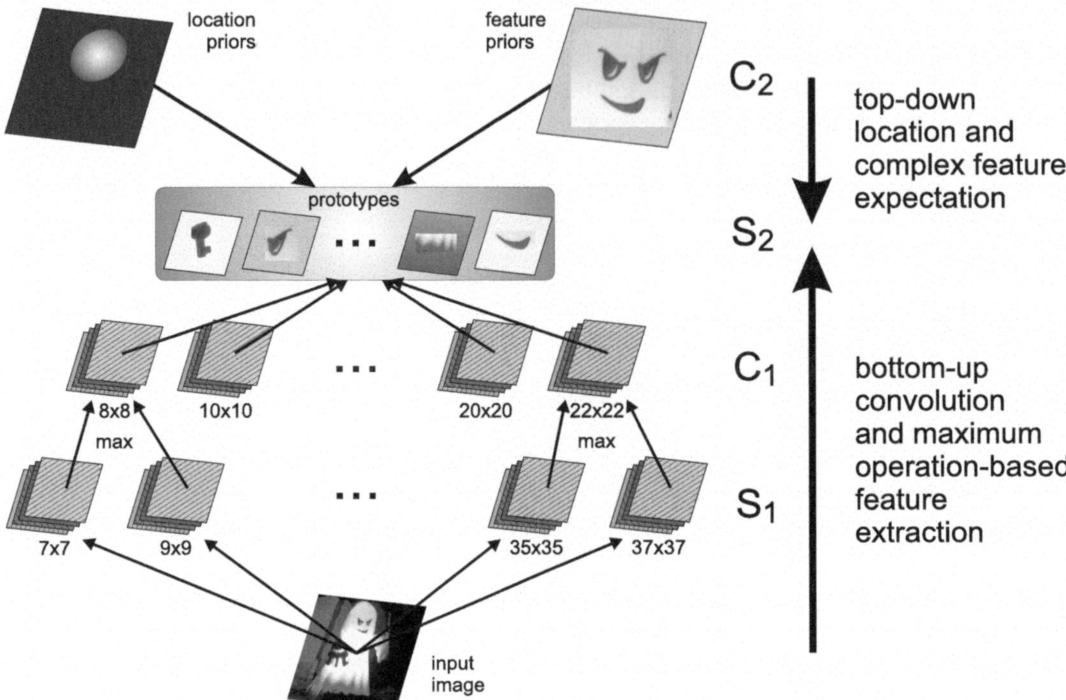

Figure 9.11: Given an image, first edges in various sizes and orientations are extracted by Gabor filters in S_1. These edges are then combined by a maximum operation into the first "complex" layer C_1, yielding a more coarse-grained feature distribution. Up to this stage, information is processed fully bottom-up without any top-down influence. Next, in the "prototypes" layer, combinations of complex features from C_1 are encoded, whose spatial resolution is even more coarse-grained than the one in C_1. In this layer, feature and location prior expectations are fused with the bottom-up evidence by means of Bayesian information processing.

the top-down expectations about which object is to be seen. Given a particular object prior O, a distribution of feature complexes is activated accordingly. This feature complex may be seen as a simplified object Gestalt, which essentially specifies object-particular critical visual features loosely binding them together given their prior, top-down activation (cf. Section 3.4.3). The feature complexes are then mapped onto feature maps X_i. The feature maps are location specific, such that the location priors L influence the induced feature distributions, enhancing and inhibiting the activated feature complexes selectively in particular image sub-areas. With this further differentiation, the model can be made even more concrete as follows:

$$p(I, L, O, X_1, \ldots, X_n) = p(I|X_1, \ldots, X_n) \left(\prod_{i=1}^{N} p(X_i|L, F_i)p(F_i|O) \right) p(L)p(O), \quad (9.29)$$

such that the joint probability depends on the location and object priors, which influence the resulting conditional feature activities $p(F_i|O)$, which, together with the priors, determine the spatial feature map distributions $p(X_i|L, F_i)$; this, in turn, determines the conditional probability of seeing the actual image, given all N feature map distributions. Note how this model is essentially a factorized Bayesian network, where the individual, factorized computation is possible due to the assumed conditional independence of the feature maps X_i given location L and feature priors F_i, which, in turn, are independent of each other given the object prior O.

The corresponding Bayesian network is shown in Figure 9.12. On the right side of the figure, the presumed relation of the model to cortical structures, which was proposed by the authors, is shown. The edge-detector structures, which constitute the image input I in the model, are assigned to the primary visual processing areas V1 and V2. The feature maps X_i are related to V4. The feature priors F_i are related to the ventral stream (inferior temporal area IT), whereas the location prior L is related to the dorsal stream (lateral intraparietal area LIP and possibly the frontal eye field FEF). Finally, actual object recognition is assigned to prefrontal cortical areas (PFC). While these cortical relationships should be taken with a grain of salt because the areas referred to are also involved in other neural computations and most likely they do not compute probability estimates exactly in the described manner, the relation is striking and very illustrative.

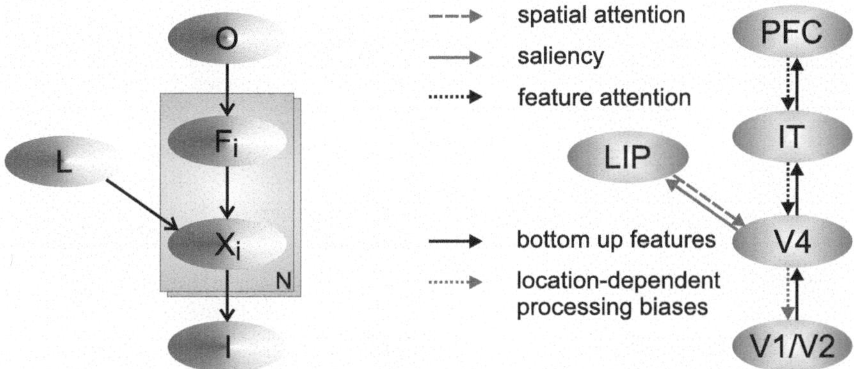

Figure 9.12: The Bayesian model of visual processing (left) has been related to the dorsal and ventral visual processing pathways (right). The lateral intraparietal sulcus (LIP) is involved in spatial attention, while the prefrontal cortex and inferior temporal areas (IT) have been related to object recognition. Visual area V4 serves as the interface in which top-down and bottom-up information is integrated. [Adapted with modifications from *Vision Research, 50*, Chikkerur, S., Serre, T., Tan, C. & Poggio, T., What and where: A Bayesian inference theory of attention, 2233–2247, Copyright (2010), with permission from Elsevier.]

The model separates dorsal and ventral streams in a manner that is rigorously Bayesian and that yields several visual processing properties, which are comparable to human-like visual processing. Simply by manipulating the priors $p(F_i)$ and $p(L)$, the following interesting attentional phenomena can be generated. These are illustrated in Figure 9.13:

(a) *Spatial and feature type invariance:* The a posteriori probabilities signal where and which singular stimulus is present in the image. The location of the stimulus and the type of stimulus are reflected in the a posteriori probabilities $p(L|I)$ and $p(F|I)$, respectively. Note how the two a posteriori probabilities generalize over the stimulus feature and the location of the stimulus, respectively.

(b) *Spatial attention:* By modulating the a priori probability $p(L)$ resulting in a spatial, attentional, top-down expectation, the a posteriori feature probability $p(F|I)$ highlights the feature type at the attended location.

(c) *Feature-oriented attention:* By focusing the attention on a particular feature type by means of the a priori probability $p(F)$, the a posteriori location probability $p(L|I)$ yields a higher probability mass where the attended feature can be found.

(d) *Feature popout:* This approach also highlights the typical pop-out of unique feature properties from many common feature properties. The a posteriori feature probabilities $p(F|I)$ signal the presence of vertical and horizontal features in all three cases shown in Figure 9.13d. The a posteriori location probability map $p(L|I)$ on the other hand, which in this case can be interpreted as an interactive saliency map, increases in value at the position of the unique vertical Gabor-like edge when the competing horizontal edges increase in number (cf. also 11.3.1). Intuitively, this effect emerges because the top-down influences of the feature probabilities (a priori and more so a posteriori) spread their probabilities over all locations where the feature is detected, thus focusing this top-down influence on a unique feature while distributing it among individual lower values when the features are found in multiple locations.

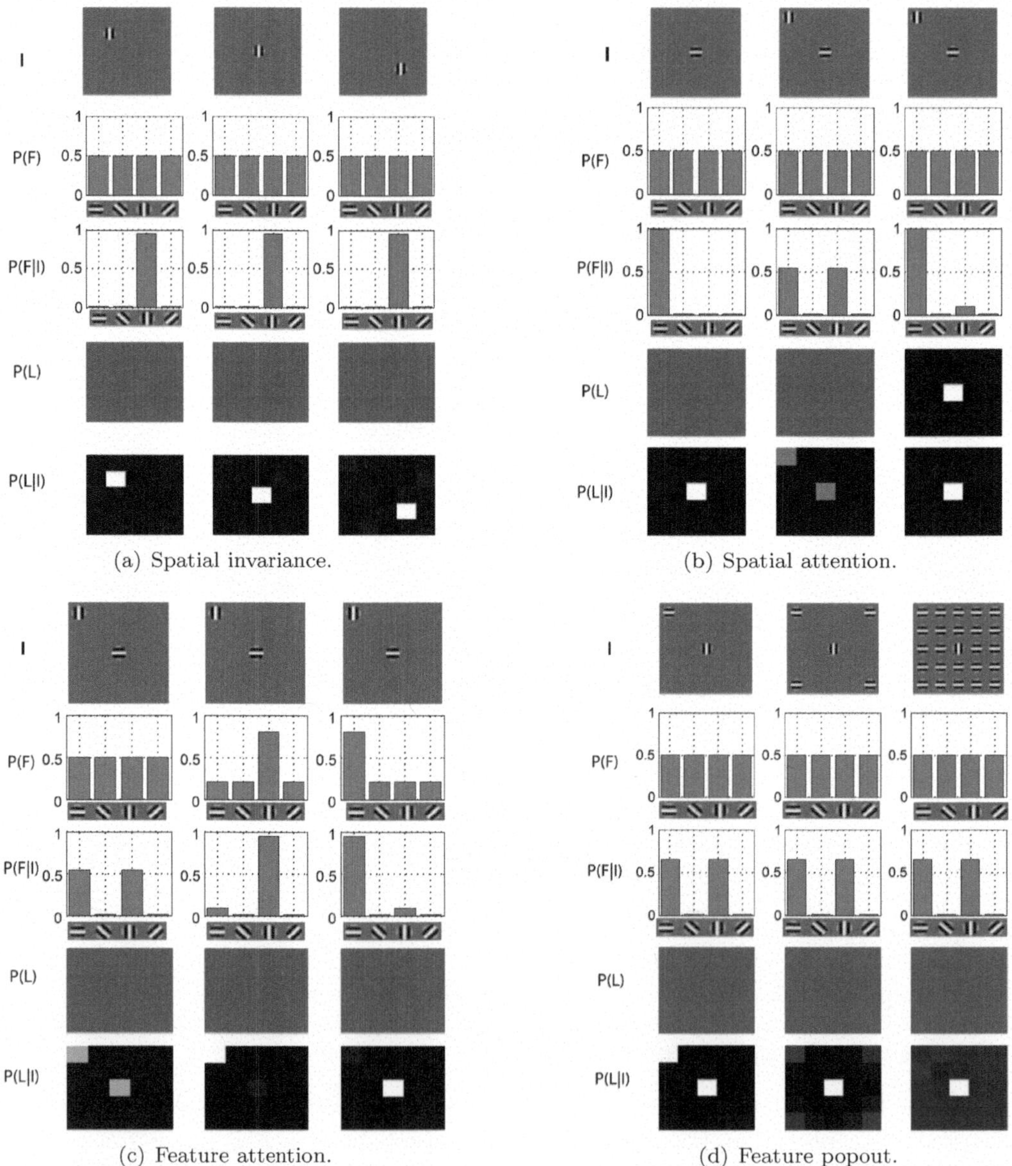

(a) Spatial invariance.

(b) Spatial attention.

(c) Feature attention.

(d) Feature popout.

Figure 9.13: The Bayesian model of visual processing enables the simulation of several typical results found in the visual processing and attention literature. [Reprinted with permission from *Vision Research, 50*, Chikkerur, S., Serre, T., Tan, C. & Poggio, T., What and where: A Bayesian inference theory of attention, 2233–2247, Copyright (2010), with permission from Elsevier.]

This model was not only applied to toy problems, but o real-world scenes as well. In fact, after being trained on natural images forming features F_i, it could be shown that the model's posterior on the image was closest to that of eye saccade behavior (in terms of fixation distributions) of human participants when both feature and location priors were included.

The model could also be trained to focus its attention on cars versus pedestrians, and this distinction also reflected the heat maps produced by eye fixations of human participants.

In conclusion, it has been shown that this model can mimic several typical psychological findings, which can be interpreted as evidence for the validity of the model. Moreover, the model adds additional evidence to its validity due to the relation to primary and deeper visual and cognitive processing areas. First, by processing actual image information with a columnar structure of Gabor filters of different scales, neural response properties of V1 and V2 are mimicked. Second, top-down location and object-oriented priors can be induced by the model based on a rigorous Bayesian formulation, which is believed to be approximated by many brain areas and cognitive functions. Finally, the separation of dorsal and ventral-like visual pathways and their bidirectional interactions via the integrated feature maps X_i is believed to be neuro-anatomically plausible and can explain how spatial and feature-based top-down priors can be induced.

Clearly though, the model does not tell the whole story and leaves open important aspects. For example, various parameter dependencies, such as the sizes of the Gabor columns or the number of features N, were hard-coded. Additionally, learning concentrated only on the problems at hand. Moreover, the dynamic attentional processes unfold on static images only. Extensions to a dynamic Bayesian network, which may generate temporal predictions and may self-adapt its feature encodings, are pending.

Recent developments on learning by means of predictive encodings based on the work of Rao and Ballard (1998) and many extensions and modifications thereof, have shown that Gabor-like visual detectors can be learned when feeding-in natural images. Abstractions into more complex feature detectors – such as corner detectors – have also been successfully implemented. Additionally, the *free energy principle* has been proposed as a general principle of cognitive processing (Friston, 2010, cf. Section 7.6). It can be used to derive the learning principles that underlie predictive encoding approaches from a rigorous mathematical approach, which subsumes Bayesian formalizations. While a detailed treatise of these techniques goes beyond the aim of this book, interested readers are encouraged to monitor the recent developments in these directions.

9.5 Visual illusions

Top-down, neural visual processing mechanisms can be nicely illustrated by various "visual illusions" as well as by bi- and multi-stable visual stimuli. Interestingly, the term "visual illusion" is often misleading – especially when analyzed from a top-down, cognitive perspective of visual processing. From this top-down perspective, one may talk about accurate top-down influenced inferences, where these inferences are based on the assumption that the presented image is a real-world image and that the objects in the image adhere to the typically applicable physical laws and consequent statistical regularities (Geisler, 2007). Although artists have shown that also in real-world scenes objects, walls, or textures can be arranged in such a way that particular objects are perceived much larger or smaller than they actually are, these are situations that occur rarely. Our brains generalize over these coincidental (or intentionally misleading) situations, use the visual bottom-up cues available, and integrate them in an interactive top-down-bottom-up manner, yielding maximally plausible estimates in the form of joint probability approximations. The following are examples, which illustrate how perception can be influenced by top-down expectations:

- The circles in Figure 9.14(a) are typically perceived as concave or convex, depending on whether the lighter parts are below or above the center of the circle, respectively. The explanation is that our brains assume a priori that most light comes from above, rather than from below. Moreover, the brain "knows" that surfaces that face a light source will reflect more light than those that face away from a light source. Consequently, the shades of gray suggest an accordingly tilted surface. These assumptions are totally reasonable in the real world and typically helpful for estimating surface orientations.

- The famous chess illusion suggests to us that the indicated fields A and B (Figure 9.14(b)) are dark and white fields, respectively. However, they have the exact same gray scale values. The top-down computation of the shadows and the grid regularity results in the illusion. Even more astounding are the chess pieces shown in Figure 9.14(e) ((Anderson & Winawer, 2005, p. 80)): the corresponding white and black pieces have the exact same gray scale values.

- Illusionary contours can be seen in various situations – especially when incidental corners seem to be systematically connected (cf. Figure 9.14(c)). Such a situation, which typically does not occur in nature, is thus interpreted as the invisible contour of an object, generating the illusionary object contour. Of course, the more well-known and regular the object – such as a simple geometrical triangle – the stronger the illusion. The illusion is false, but behaviorally rather helpful. If I were to grasp the illusionary object in 3D, I would know where to place my fingers.

- In the *Ponzo illusion*, the same sized bar looks larger when it is positioned in a location in space that appears to lie further away in depth (Figure 9.14(d)). Top-down expectations infer that the bar is likely positioned at a further distance, where the actual floor would be. As a consequence, our brain enlarges the perceived size.

- A particularly puzzling image is presented by the terrace illusion in Figure 9.15, where semantic cues such as the sky, handrails, stairs, strings, and people's orientations confuse our brain. Depending on which part we fixate, the top corner either extends into the image or out of the image. In the attempt to make everything consistent, our brain seems to partially bend the central tiled, squared area inward. Thus, a very confusing impression is created, which reminds somewhat of M.C. Escher drawings.

In addition to these visual illusions and their somewhat faulty, top-down driven interpretations, bi- or multi-stable images show how our brain constantly attempts to fall into local joint probability maxima to interpret a scene in a maximally plausible manner.

- The Necker cube is probably the most well-known example of a bistable stimulus. The cube can be interpreted in two manners, where the left larger square can be seen as either in the front or in the back. The combination of four Necker cubes in fact allows for 16 interpretations (Figure 9.16). Staring at this Necker cube chaos, the brain tends to switch its interpretation again and again. The maintenance of one of the interpretations over an extended period of time is actually very hard. Again, this behavior of our brains can be interpreted as helpful because it avoids getting stuck in a locally optimal interpretation, especially when other equally plausible interpretations are available. The interpretations may be thought of as constituting a Gaussian mixture model where the two interpretations of all four cubes are independently equally likely (cf. also Figure 9.9(a)).

- Figure 9.16 also shows a combination of the Necker cube with virtual contours and intersecting black circles. Note how the black circles seem to be unstable in depth and adapt with the current interpretation of the Necker cube.

- Bistable foreground/background illusions (Figure 9.17) are another example where the perception focuses on two alternative interpretations, which essentially constitute the two possible local minima. Note how it is virtually impossible to maintain multiple interpretations concurrently.

- Figure 9.17 shows how one attempts to interpret the depth of a combination of multiple figures so that the lower central circle seems to be behind the upper central circle, and the square in turn appears to be located behind the lower central circle. Consequently, the outside circles appear to be slightly further back in the image. Also, the triangle appears slightly tilted to account for the fact that on the one hand it seems to originate in the square, but on the other hand it appears to be above the lower central circle.

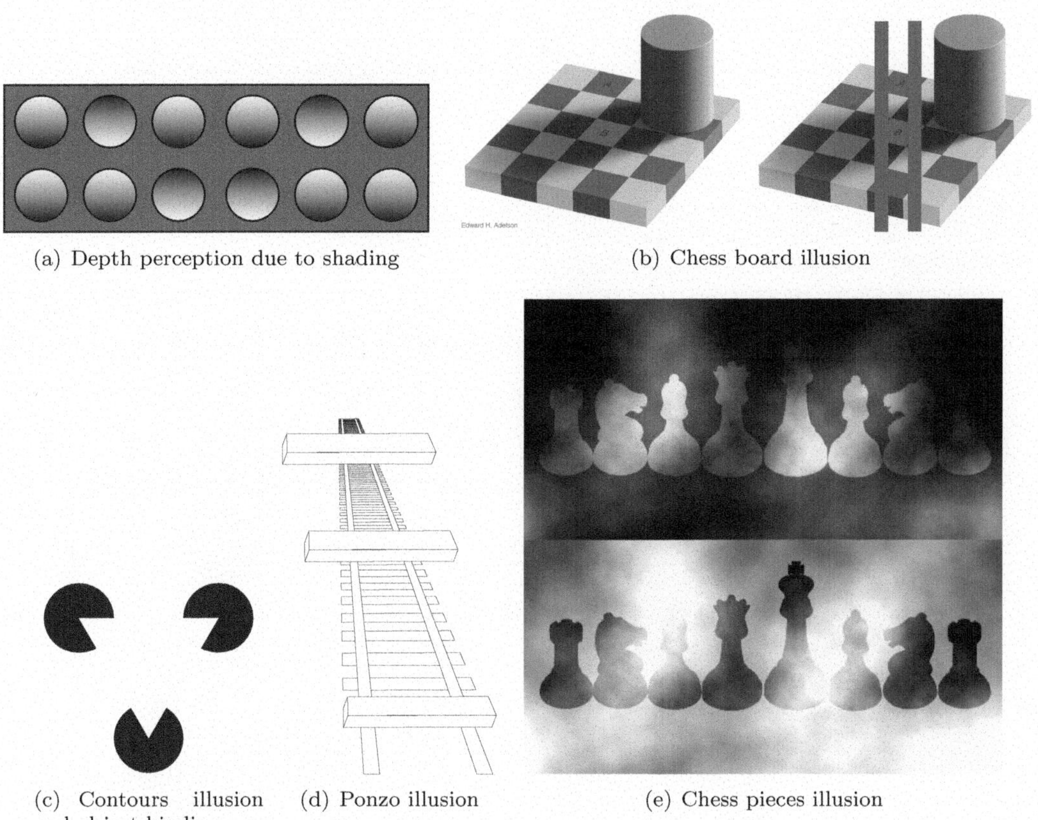

(a) Depth perception due to shading

(b) Chess board illusion

(c) Contours illusion and object binding

(d) Ponzo illusion

(e) Chess pieces illusion

Figure 9.14: Top-down expectations about our environment with its physical properties lead to several "optical illusions", which are actually rather optical phenomena, because the perceptual interpretation, which our brain constructs, closely corresponds to what would be typically the case in the real world. [Chess pieces illusion: Reprinted by permission from Macmillan Publishers Ltd: *Nature, 434*, 79–83, Image segmentation and lightness perception, Anderson, B. L. & Winawer, J., Copyright (2005).]

These illusions essentially highlight that our visual perceptual system is not a passive, feed-forward, observer-like system of the incoming visual information. Rather, it is a highly active, generative system that attempts to use its internal predictive knowledge about typical structures and structural relations in the world to interpret all visual scenes. The brain attempts to integrate all visual information and all available top-down prior knowledge into a consistent whole scene interpretation. When alternative interpretations seem equally plausible, the brain tends to spontaneously switch between them. All the results can be interpreted by means of a dynamic Bayesian information processing network.

9.6 Summary

This chapter has emphasized that our brain does not work in a purely, feed-forward, passive information processing manner. Besides the inevitable homunculus problem, which is the consequence of a feed-forward, passive information processing view, various observations from the cognitive vision literature and from visual illusions and bistable visual stimuli show that visual perception is actually accomplished by a highly interactive, interpretive process. This process attempts to integrate bottom up sensory information with top-down, interactive prior assumptions, where these assumptions reflect typical statistical properties

Figure 9.15:   The terrace illusion by David Macdonald is a great example of how much the brain attempts to generate overall consistent interpretations of the perceived environment, taking all available cues into
consideration.   [Reproduced with permission.   Copyright © David Macdonald.
http://users.skynet.be/fa414202/Cambiguities/Illusion_Site_David_Macdonald_Illusions
___Image___Terrace_Illusion_files/Cambiguities/Terrace.jpg]

found in the world (Geisler, 2007). As a result, it attempts to generate a consistent scene
interpretation by taking all available information and knowledge into account.

Bayesian models were shown to be able to interpret bottom-up visual information by
means of generative models. They allow combinations of continuous and discrete variables
and may be viewed as the fundamental information processing principle that is pursued by
our brain – at least in approximation. Various results suggest that interpretive, a posteriori probability densities are estimated, reflecting the internally constructed scene given
the available visual information. Probability densities can, for example, be represented by
Gaussian mixture models, and similar density approximations can be generated by neural
activities. An overview of an implementation of an artificial interactive vision model showed

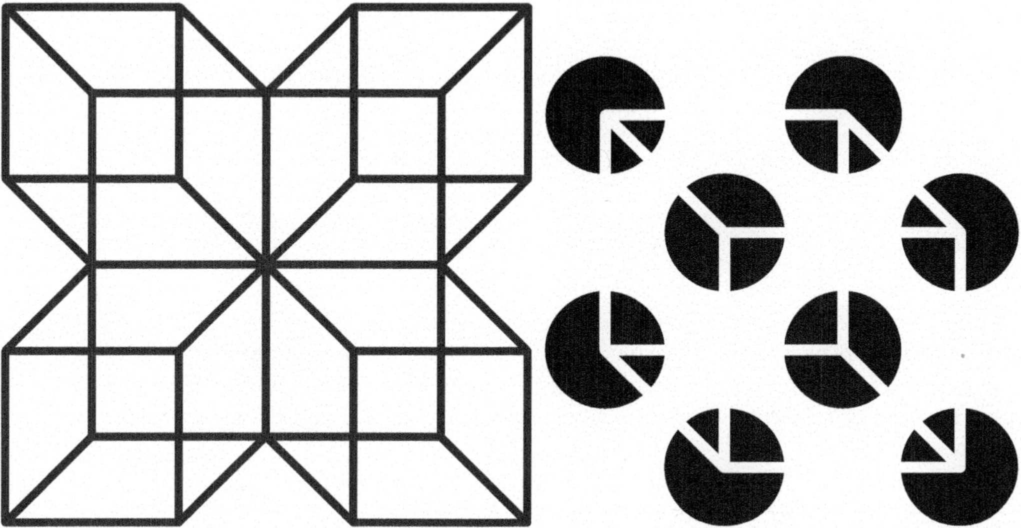

Figure 9.16: 16 interpretations are possible when viewing the quadruple Necker cube. Combined with illusionary contours, several other temporarily stable interpretations are possible.

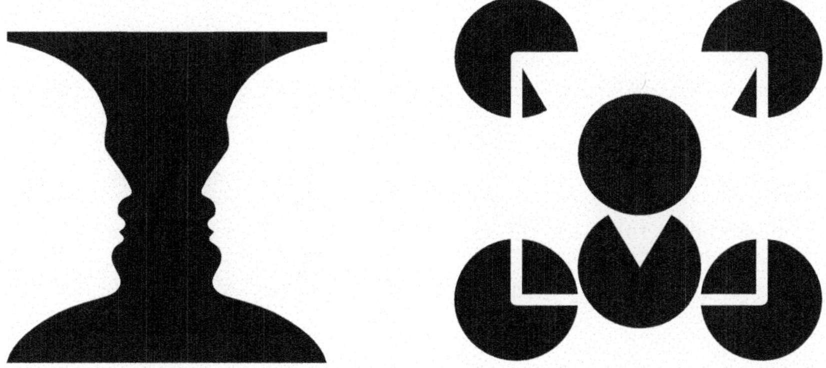

Figure 9.17: left: The Rubin vase is a great example of a bistable foreground/background segregation, which tends to switch spontaneously. right: Due to the different object shapes, their overlaps, and suggested interceptions, the brain struggles to find a consistent interpretation.

that spatial as well as object priors can selectively influence visual perception, even without direct connections between them.

To summarize, this chapter has shown how Bayesian information processing can combine multiple sources of information easily, where the sources may include both, bottom-up visual information as well as top-down expectations about visual features and their locations. The brain can be understood as a generative model, which combines top-down expectations and the knowledge about objects and scenes, which is inherent in top-down expectations, to generate maximally consistent scene interpretations. The scene interpretations continuously and dynamically adapt over time, identifying entities in a scene and arranging them in a maximally consistent manner. As a result, the perception of an object, another entity, or a scene comes in the form of internal generative model activities, which attempt to maximize the joint probability of internally generated, top-down expectations and bottom-up sensory evidences.

9.7   Exercises

1. Why do we not perceive half an object if we only visually see half of it?

2. Why are generative models very suitable to generate imaginations while forward models are not?

3. Consider the traffic light model shown in Figure 9.2. Start with a generative model that is on STOP, that is, set the prior state of STOP to 1 and all other generative model states to zero. Then iteratively compute the next state estimates of the generative model by iteratively computing updates via the generative predictive state model and the generative sensor model. Assume that the generative model states are independent of each other during the generative sensor model update, but normalize them to one after the update. During the temporal update, distinguish prior from posterior generative model states.

4. Give examples in the real world that illustrate how uncertainty about the state of the world can arise due to sensory noise, ignorance about the environment, and unobservability.

5. Reconsider the zebra example. Determine probabilities for the stripe detector such that the detector does not provide any additional information about the likelihood of looking at a zebra.

6. Construct a simple, three-node Bayesian network and show exemplary that the intercausal case can make two variables dependent on each other given their common descendant.

7. Show all conditional independences on a simple three node Bayesian network A→B→C.

8. Compute the joint probability table for the Bayesian network shown in Figure 9.5.

9. Consider the ball detection Bayesian network shown in Figure 9.18.

    (a) Compute the probability of perceiving a ball without further evidence.

    (b) Given the surface detector goes off, how does the probability of perceiving a ball change?

    (c) Does the probability of perceiving a circle change, when a ball is present/when the surface detector goes off?

    (d) Compute the joint probability table and verify your answers given to the questions 9(a-c).

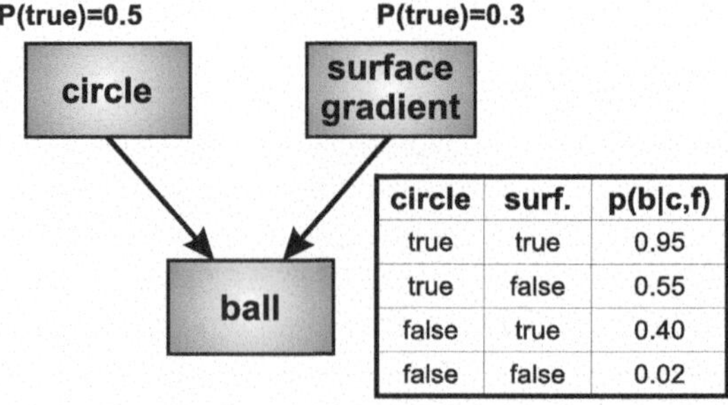

circle	surf.	p(b|c,f)
true	true	0.95
true	false	0.55
false	true	0.40
false	false	0.02

Figure 9.18: A simple Bayesian network that considers shape and surface property evidence to compute the probability of perceiving a ball.

10. Explain the difference between a probability mass function, a probability function, and a probability density function. What is the result when summing/integrating over the output value space of each of these functions?

11. Determine the covariance matrix given the input vectors $x_1 = (1, -2, 1)^T$; $x_2 = (0, -1, -2)^T$; and $x_3 = (-4, 3, -2)^T$.

12. Which types of distributions can be represented with Gaussian mixture models that cannot be represented with individual (multivariate) Gaussians?

13. Is it possible to independently activate particular feature and locations priors with the introduced Bayesian model of visual processing? Is this model a generative model?

14. Explain the contours illusion and the Ponzo illusion computationally with the help of the introduced Bayesian model of visual processing (without actual mathematical computations).

15. In which way do several of the illusions indicate top-down expectations of perceiving a particular "Gestalt"?

16. When the perception switches from the vase to the faces and back while looking at a Rubin Vase image, which two kinds of internal, generative, bistable states must be switching synchronously from the one attractor to the other and back?

Chapter 10

Multisensory Interactions

10.1   Introduction and overview

While we focused on vision in the last two chapters, in this chapter we will expand what we know about information processing to other sensory modalities, including the motor modality. The focus lies now on interactions between the modalities and the potential for emerging structures as a result of these interactions.

As we discussed in Section 6.2.3, multiple sources of information about a particular entity in the environment can be very helpful to disambiguate what is actually out there. Due to the redundant nature of different sensory (that is, modal) information sources about a particular stimulus, one source can supplement the other, when one is temporarily not available. Similarly, multiple sensory information sources may complement each other, particularly when providing different information aspects about the same entity. For example, an opaque milk bottle may be identified visually, but the amount of milk left in the bottle may be determined manually by lifting the bottle perceiving its weight by means of proprioceptive feedback.

As is the case for redundant and complementary visual information, other information sources may also be viewed as redundant and complementary in themselves. For example, auditory information includes pitch, volume, auditory dynamics, and spatial information, such that we are able to imagine the size of a piece of paper that we hear being torn apart, for example, or the type of object that just fell to the ground – including its material as well as if the object broke when hitting the ground – and where it hit the ground. Similarly, the tactile modality can give us information about different surface properties and relative distances, enabling the identification of objects given pressure-based and texture-based information, for example, on the skin of our fingers. Not surprisingly thus, dorsal and ventral processing pathways have also been identified for other modalities by neuroscientific techniques, revealing multiple resemblances with the two visual processing pathways (Dijkerman & de Haan, 2007, cf. Section 8.4).

Especially for somatosensory information, but for other modalities as well, an important additional component comes in when considering motor information in interaction with sensory information. Systematic, motor-driven somatosensory dynamics enable us to identify an object simply by exploring it with our hands. In doing so, brain areas are activated – such as the inferior temporal (IT) cortex – an area, which was previously believed to be a unimodal, visual area. In the case of tactile exploration with the hands, the parts of the object that carry critical information are explored more, such as probing the tip of an object when manually searching for a fork in a silverware drawer. Note how in this case motor behavior, that is, manual behavior, needs to be integrated with tactile feedback to be able to construct an image of the whole object from the touched parts. The motor behavior essentially provides additional, relative spatial information about the object's parts, which are identified by means of tactile interactions, supporting the mental creation of the whole object. Interestingly, saccades can be interpreted in the same way, where they explore (particularly larger) objects providing in this case visually-grounded relative distance information between object parts.

In all of these cases, recognition can be achieved through a Bayesian process, which integrates the available information approximately optimally (Ernst & Banks, 2002). In contrast to the unimodal case, in the multisensory case multiple modal sensory sources of information are integrated for the creation of a consistent whole. Nonetheless, the principle of optimal information integration is also valid in this case, where estimates about the reliabilities of the respective information sources, and estimates about their respective information contributions for recognizing a particular entity, need to be considered (Fetsch, Pouget, DeAngelis, & Angelaki, 2012).

Because the different sources of information are grounded in different, sensory-specific frames of reference, proper information fusion is not straightforward, however. To be successful, the brain needs to integrate different sources of information about entity identities by projecting or mapping the different sources of information into a common frame of reference. As these projections need to take the current posture of the body into account, our brain needs to learn and selectively activate those mappings between particular spatial frames of reference that currently apply. In this way, respective sensory-grounded information can be flexibly, adaptively, and continuously integrated.

Consider, for example, our hands: when placed on the keyboard, our eyes provide information about the approximate positions of the hands and fingers relative to the keyboard as well as relative to our body. The tactile feedback from the fingers furthermore gives information about the correct positioning with respect to the individual keys on the keyboard. As another example, think about washing your hands. In this case, the hands are perceived rather differently from a visual perspective. Visual information helps, for example, to position the hands under the water. Meanwhile, the hands themselves perceive each other via the tactile modality, confirming the interaction with the water as well as with each other. In both examples, visual information and tactile information are complementary. Moreover, the posture of arms and hands, that is, proprioceptive information, needs to be taken into account when integrating the visual information with the tactile information. In fact, in virtual reality setups an uncomfortable feeling typically arises when the hands' postures do not visually correspond with the proprioceptively perceived posture.

Thus, to integrate multisensory information from multiple modalities, appropriate *spatial mappings* between the different sensory-grounded modalities need to be active. Additionally, when interacting with an object, the object itself also needs to be mapped into the different, modality-specific frames of reference. In this way, it becomes possible to recognize the same object using different modalities, such as touch or vision.

Note also the effects of such multisensory information integration processes when considering surprise. Typically, we are not surprised about the onset of touch sensations when the hand begins to touch an object because the eyes or other sources of information have informed our brain that the hand will very soon touch the object, so the touch is anticipated and thus not surprising. This can be most easily verified when walking in the dark a couple of steps and suddenly touching a wall with the extended hand earlier than expected, which then yields a feeling of surprise. In this case, the prediction was slightly off and surprise became apparent.

The learning of spatial representations and mappings between modality-grounded frames of reference is another challenge that needs to be considered when investigating multisensory information integration and interactions. Various researchers have suggested that only through manual and locomotive interactions with space is it possible to learn spatial representations and mappings. Matches between sensory signals across multiple sensory modalities may serve as a crucial learning signal in this case. However, evolutionary predispositions probably also play an important role.

Many multisensory spatial encodings have been termed *peripersonal spaces*, because they appear to encode the space surrounding the body. Peripersonal space refers to any topological encoding that exhibits spatial sensitivities relative to the body or a particular body part. Peripersonal spaces are sensitive to multiple sensory information sources, so several redundant or complementary sensory modalities may invoke neural activities in these spaces. Topographic, sensory-grounded encodings as input modalities may be very suitably for developing such body-centered, spatial encodings, and interactions between them. How exactly peripersonal spatial encodings are learned, however, remains an open question.

In the following sections, we first focus on peripersonal spaces, the involved topological neural encodings, multisensory information fusion, and the development of such spatial encodings and spatial mappings given sensorimotor interactions of our body with the outside environment. Next, we focus on object codes, which seem to exist somewhat independent of spatial codes, because objects can be perceived anywhere in the space surrounding our body. We will show that such object codes also extend to dynamic, interaction codes, enabling, for example, the recognition of biological entities purely by seeing dynamic motion patterns in the form of point-light motions. Finally, we consider how *external space* may develop, which is often referred to as a *cognitive map*. External spatial representations most likely develop to enable planning and navigation to particular goal locations, that is, particular, desired locations in external, *allocentric*, space.

10.2 Body-relative spaces

When facing the task of interacting in a goal-directed manner with the world by means of our extremities – including arms, legs, mouth, etc. – we need to be able to coordinate our motor behavior with the incoming sensory information. As we saw in Chapter 6 on anticipation, to act goal-directedly, forms of encodings need to be available that allow the activation of a reachable goal, which in turn activates those motor commands that are believed to reach the goal. As interactions with the environment can typically be accomplished only when a part of our body – or a tool that is controlled by our body – is in contact with the object that is to be manipulated, spatial frames of reference that relate the controlled body part with the object need to be learned and available. Thus, during cognitive development, spatial representations and mappings between them need to be learned very early on to enable the execution of goal-directed object interactions.

An illustrative example of spatial representations and spatial mappings between body parts is the *rubber hand illusion* (Botvinick & Cohen, 1998). To invoke the illusion, traditionally a rubber hand is placed in front of the subject in a position and orientation that could also be produced by one hand of the subject. The subject's hand, however, is hidden behind an occluder. When concurrently stroking both, the rubber hand and the real hand with a brush in a spatiotemporal congruent manner, typically subjects report afterwards that it appeared as if the brush strokes came from the rubber hand and that it felt as if the rubber hand was the subject's own hand. More indirect measures, such as estimates about the position of the subject's hand or even about the angle of the subject's elbow, once the illusion is invoked and the rubber hand is occluded, show that the rubber hand illusion is a robust phenomenon (Butz, Kutter, & Lorenz, 2014; Rohde, Di Luca, & Ernst, 2011). Figure 10.1 illustrates the set-up in a typical rubber hand illusion experiment. Neurological investigations have suggested that the premotor and parietal cortex as well as the cerebellum are critically involved when the feeling of ownership of the rubber hand is elicited (Ehrsson, Spence, & Passingham, 2004; Ehrsson, Holmes, & Passingham, 2005).

From a computational perspective, sensory information from distinct frames of reference – including visual, proprioceptive, and body posture information – need to be considered to elicit the illusion (Ehrenfeld, Herbort, & Butz, 2013b). The brain attempts to integrate these multisensory information sources into a consistent whole. The more synchronous and compatible the brush strokes are in time and in hand space, the more the brain starts to "believe" that the brush stroke that is visually observed on the rubber hand must be on one's own hand. The brain attempts to integrate the rubber hand into the postural body image and adjusts the estimates about how the body is situated in space accordingly.

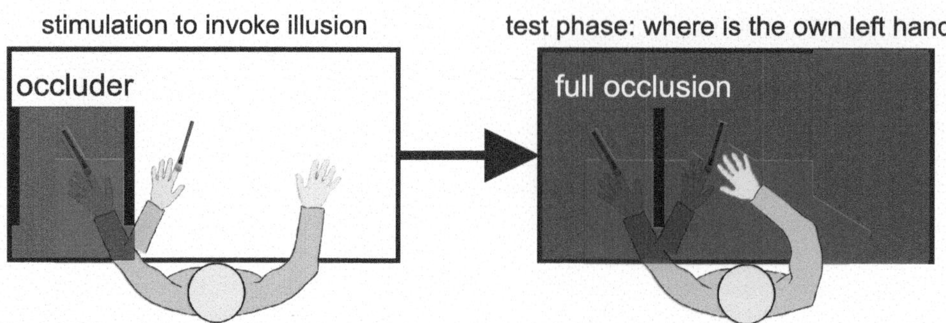

Figure 10.1: In the typical rubber hand illusion experiment, first, the rubber hand and the participant's hand, which is not visible to the subject, are stimulated with a tool, such as a brush. After a duration of a few minutes, the rubber hand is covered, and participants give an estimate of their left hand's location, either verbally or by pointing with the right hand to the location. After the trial, the participants are asked to complete a questionnaire about their phenomenal experiences. [Re-sketched from Butz, M. V., Kutter, E. F., & Lorenz, C. (2014). Rubber hand illusion affects joint angle perception. *PLOS ONE, 9*(3), e92854. © 2014 Butz et al.]

This "belief" can be modeled once again by means of generative Bayesian information processing principles (cf. Section 9.2). The more synchronous the strokes are, the greater the certainty that the observed strokes came from one's own arm, and in consequence, the stronger are the attempts of the brain to integrate the stimulus into current body state estimates. As even the elbow estimate can be affected by the illusion (Butz et al., 2014), a full *postural body schema* needs to be at work, which relates the individual arms with each other (Maravita, Spence, & Driver, 2003; Holmes & Spence, 2004). Only such a schema can enable the translation of the false visual hand posture information from the rubber hand into joint angle estimations of the arm, attempting to maintain a consistent body schema. A postural body schema essentially needs to maintain various limb-relative frames of reference as well as flexible mappings between them to maintain an overall and consistent body image. Indeed, it has been shown that various body-relative frames of reference can be found in our brain (Holmes & Spence, 2004; Maravita et al., 2003; Soto-Faraco, Ronald, & Spence, 2004).

In addition to sensory-grounded topographic visual feature maps, many other cortical areas have shown body-relative, topographically encoded maps. The simplest and most obvious codes of this kind were covered in Chapter 7: the somatosensory and motor homunculi (cf. also Figure 7.12) represent the body in a skin-relative, somatosensory (including pressure, heat, but also joint and muscle-tension), as well as a muscle-relative topology (for issuing motor control commands), respectively, where the topologies also reflect the density of receptors or types of muscles in the respective body areas. Multisensory integrative topographies, such as peripersonal spaces, can be found in "deeper" brain areas, whose neural encodings are not dominated by one sensory or motor modality. Particularly in the parietal and premotor areas, neurons exhibit selective spatial sensitivities, and the spatial frame of reference can often be characterized as surrounding the body (centered on the torso) or a particular body part (such as the face or an arm).

10.2.1 Redundant, body-relative, multisensory spaces

Each of us has probably had the experience of feeling somewhat trapped in a constricted space – such as an overfull subway train – where the space for ones body is extremely limited and one feels pressure from all sides. Why do we feel so uncomfortable in such a situation – especially when we are not very used to it? The anthropologist and social psychologist Edward T. Hall (1914-2009) explains the phenomenon by proposing the concept of a *personal space*. With personal space, the *comfort zone* characterizes the space surrounding the body that we prefer to have for ourselves. If another person comes too close, the reaction is to retreat because the other person has invaded our comfort zone. Because the size of the comfort zone differs between cultures, cultural conflicts during communication can emerge when one person continuously invades another person's comfort zone. The first person (the "invader") may get upset in this situation as the other person continues to retreat, which may be interpreted as a lack of interest or a personal dislike. Meanwhile, the latter person may also get upset feeling offended because the invader continues to invade her comfort zone.

Neuropsychologists have characterized the personal space as directly relative to one's own body. Figure 10.2 shows these spatial characterizations and their distinctions. Three main types of spaces can be distinguished:

- *Peripersonal space*: refers to the space surrounding a particular body part or the whole body. It integrates not only visual and tactile, but also auditory information and even taste and odor. Peripersonal spaces were shown to exist for our face, hands, and arms, although virtually every part of our body seems to be encoded with a peripersonal space – albeit with differing amounts of detail.

- *Reachable space*: refers to the space that is reachable by our limbs without locomotion – that is, the space our hands or also feet can reach in a particular posture.

- *Extrapersonal space*: refers to the space that is not currently directly reachable with our hands or feet.

In each of these particular spatial representations, the brain integrates sensory as well as motor information to estimate body-relative distances and orientations. Social spaces – such as the comfort zone mentioned previously or the *flight zone* of an animal – seem to be encoded in a similar manner.

As in the primary sensory and motor areas, a battery of overlapping receptive fields, which are often referred to as *population codes*, cover the relevant space. Particular neurons in these population codes selectively fire when a stimulus – such as an object or a flash of light – is perceived at a particular body-relative position. For example, a neuron may indicate the presence of a stimulus close to ones right cheek, or close to the left forearm, regardless how the respective body part is currently positioned with respect to the rest of the body. Moreover, the neuron will fire regardless if the stimulus is only seen or only felt, although both sensory sources of information enhance the firing rate. It will even continue firing when no stimulus information is temporarily available, but the stimulus source, that is, an object or other kind of entity, is believed to still be present at the relevant body-relative location. Such neural activities most likely are also the reason why we seem to feel a feather that is moved above our skin, for example, even if the feather does not touch the skin. The fact that we cannot tickle ourselves seems to be due to this integrative nature of peripersonal space (Blakemore, Wolpert, & Frith, 2000), successfully inhibiting the tickling sensation caused by our own fingers or the self-applied feather, due to the concurrent tactile and pressure feedback stemming from our own hands.

While we have so far focused on the fact that multisensory information is integrated, another aspect is the relevance of integrating multisensory information sources for motor control. As shown in Figure 10.2, peripersonal spaces are not only found in parietal areas, but also in premotor areas. In their review of studies on peripersonal spaces, Holmes and Spence (2004) conclude that

> [...] 'body schema' and 'peripersonal space' are emergent properties of a network of interacting cortical and subcortical centers. Each center processes multisensory information in a reference frame appropriate to the body part concerning which it receives information, and with which responses are to be made. (Holmes & Spence, 2004, p. 104.)

The authors emphasize the close link to motor encodings and the likelihood for behavior-oriented codes that lead to the development of peripersonal spaces. Note how this view fits

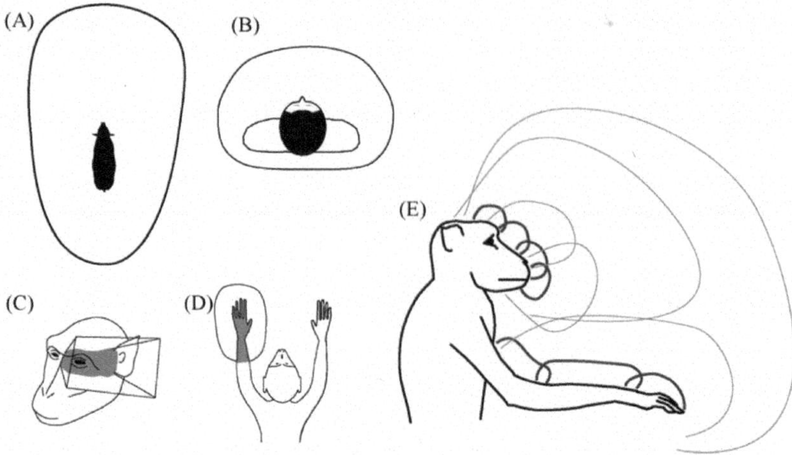

Figure 10.2: A peripersonal space of an animal (A) or a human (B) also may encode the zone of flight: if some somewhat aversive or neutral entity enters this area, we tend to retract or animals may attempt to flee. To encode this space, information from overlapping visual (C) and tactile (D) receptive fields is integrated. Within a peripersonal encoding, the density of receptive fields typically decreases with increasing distance to the relevant body part (E). Neurons encoding the peripersonal space of an arm, for example, have been identified in the ventral premotor area (F4) and the anterior intraparietal area (AIP), while face-relative spatially responsive neurons were identified in the ventral intraparietal area (VIP) and the polysensory zone (PZ). [Reprinted from *Neuropsychologia, 44*, Graziano, M. S. A. & Cooke, D. F., Parieto-frontal interactions, personal space, and defensive behavior., 845–859. Copyright (2006), with permission from Elsevier.]

very well with the principle of anticipatory behavior and the need to represent interaction goals: peripersonal spaces encode the body in such a way that interactions with and by means of the body are facilitated. Peripersonal spaces also solve the homunculus problem (cf. Section 2.2.2) to a certain degree: the brain does not represent space for its own sake, but rather the internal representations develop to be able to convert sensory information in such a way that motor behavior can be executed effectively. Thus, peripersonal spaces focus on those subspaces that are maximally behaviorally relevant. These subspaces are those that are surrounding hand, arm, and face, where the latter is particularly important for coordinating social interactions. Timely stimulus interactions are also supported by neural signals from peripersonal encodings: neurons were shown to actually fire in anticipation of a tactile impact on the arm from a flying object. That is, neurons were shown to respond to an approaching stimulus (such as a fly or a projectile) and they did so as if they estimated the time until impact: faster approaching entities invoked earlier neural activities than slower approaching stimuli, apparently in anticipation of movement onset or tactile contact (Fogassi, Gallese, Fadiga, Luppino, Matelli, & Rizzolatti, 1996). More recently, these encodings were directly related to defensive behavior, proposing that peripersonal neural activities can trigger actions to avoid the encoded, approaching stimulus (Graziano & Cooke, 2006). Furthermore, the encodings were show to re-map peripersonal space in anticipation of future finger locations, just before a grasping action is executed (Brozzoli, Ehrsson, & Farnè, 2014; Farnè, 2015).

The discovery of peripersonal spatial encodings suggests that the brain does not represent the surrounding space once, but redundantly and interactively in various, partially multisensory, often overlapping frames of reference. The reference frames orient themselves relative to particular body parts – such as arm, hand, or face – but also relative to particular bodily axes, such as the body mix axis or the head. These relative encodings most likely do not develop purely as a result of genetic predispositions, but rather for developing encodings that are maximally suitable to execute effective, goal-directed interactions with the environment. The space surrounding the body is thus encoded in redundant, limb-relative topologies because these encodings are best suited to selectively issue manipulation-oriented or defensive behavior.

10.2.2 Simple population encoding in a locust

Interestingly, spatial, behavior-relevant encodings – albeit still very much sensory grounded – can be found even in animals with rather small brains. In Section 3.6.1 we described several examples how sensory information is mapped by animals, such as frogs or flying insects, onto motor behavior. Here we consider neurobiological insights from locusts, where individual neurons have been identified that encode the surrounding space by means of a simple population code.

Pouget, Dayan, and Zemel (2003) investigated single cell recordings from locusts, as illustrated in Figure 10.3. Let us consider four neurons c_1 through c_4, which focus their receptive fields on certain locust-relative directions. The neurons focus on disturbances in the preferred direction and the close surroundings, reacting, for example, to air puffs, which may have been generated by an approaching predator. Their maximum firing rate $r_i^{\max}$ is generated when the air is blown directly from their respective preferred direction, that is, $s_1 : 45°$, $s_2 : 135°$, $s_3 : 225°$, and $s_4 : 315°$. The firing rate decreases with increasing deviation from the preferred direction. To generate directional receptive fields, a directional tuning curve is necessary. A rectified cosine function turns out to yield an approximation that correlates rather well with actual experimentally recorded firing rates:

$$f_i(s) = r_i^{\max}[cos(s - s_i)]^+ \quad \texttt{where} \quad [x]^+ = \begin{cases} x, & \text{if } x > 0 \\ 0, & \text{otherwise} \end{cases}, \tag{10.1}$$

where the $[\ ]^+$ operator generates the rectification and s encodes the current stimulus direction.

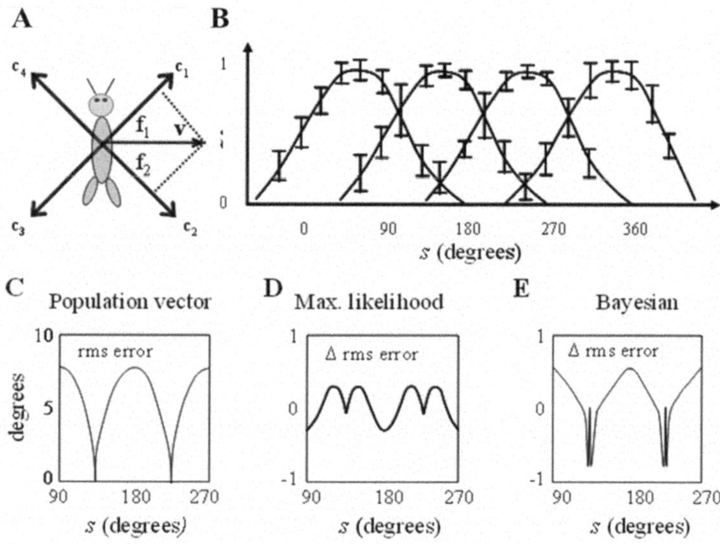

Figure 10.3: Simplified illustration of a peripersonal space identified in a locust. The space is covered by four neurons, with complementary preferred directions of maximum sensitivity (A,B). [Reproduced with permission of Annual Review from Pouget, A., Dayan, P., & Zemel, R. S. (2003). Inference and computation with population codes. *Annual Review of Neuroscience, 26,* 381–410. Copyright © by Annual Reviews, http://www.annualreviews.org.]

To avoid the potential predator, the activity patterns need to be mapped onto behavior. For example, when a stimulus from about 160° is applied, three neurons will fire. The differing firing intensities may be used to compute an actual direction estimate. Pouget et al. (2003) have investigated various methods of decoding the neural patterns, yielding various reconstruction errors. From an embodied perspective and when reconsidering the functionality of Braitenberg vehicles (cf. Section 3.6.2), however, it may be the case that at least the brain of the locust is not very interested in exactly from where the stimulus came from. Rather, it is interested in wiring the directional stimulation onto flight behavior, which should be directed sufficiently away from the stimulus. As the stimulus direction can be reconstructed to a certain extent, the signal is certainly crisp enough to wire it to an appropriate directional behavioral code, which may cause the locust to jump away from the stimulus source.

10.2.3 Learning peripersonal spaces

The claim that peripersonal spaces are optimally suited to control behavior has not yet been proven. Models that develop such spatial encodings from scratch while focusing on optimal behavioral control only exist in rudimentary forms. However, it appears that artificial neural networks (ANNs) generally offer the best tool to be able to develop such spatial encodings, including emergent spatial distributions for optimizing behavioral control. Here, we give a short overview of neural population encodings, which can be learned by standard ANN techniques using Hebbian-like, associative learning. Advanced techniques have been developed, but they cannot be covered in detail in this book (Ehrenfeld et al., 2013b; Ma & Pouget, 2008; Pouget & Snyder, 2000).

The overarching goal of such an ANN may be characterized as striving to develop *neural fields*, where each neuron has a particular receptive field and the neural field as a whole fully covers a particular space in question. Moreover, the ANN should develop into a stable neural field representation, such that receptive fields do not keep drifting in the encoded space. The ANN should also be able to integrate multiple sources of information. To integrate the information, it essentially needs to be able to communicate, that is, to exchange information with other, possibly single modality-grounded, neural fields. *Self-organizing artificial neural networks* (SOMs) have been developed, which are capable of accomplishing the task at hand (cf. also the ANN introduction in Section 3.5). SOMs essentially are able to iteratively learn a self-organized network structure, which covers a particular spatial distribution defined by the sensory signals that are processed by the SOM.

Prominent examples of such ANNs are:

- *Kohonen networks* (Kohonen, 2001), which distribute a pre-wired n-dimensional lattice structure (in two dimensions a simple grid of neurons). Kohonen networks essentially first determine a winning neuron in each iteration, which is the neuron whose weight vector is most similar to the current input stimulus. Next, the neighborhood around this winning neuron inside the lattice structure is activated – the closer to the winning neuron, the stronger the activation. Finally, all activated neurons' receptive fields are moved toward the input stimulus – the higher the neural activation, the stronger the displacement. In this manner, the complete input space is soon covered by the lattice structure and the lattice reflects the input dimensionality and distribution.

- *Neural gas* (Martinetz, Berkovitsch, & Schulten, 1993) does not pre-wire a particular lattice, but starts with a randomly distributed number of m neurons, which are not connected to each other in any way. In contrast to the Kohonen principle, neural activations are rank-based, that is, the neurons whose receptive fields lie closest to the input stimulus are ordered and the activation of the respective neurons depends directly on the order. As a result, the receptive fields are moved toward the input stimulus depending on their rank-based activation strength. The resulting distribution thus typically nicely covers the sampling distribution. The topology itself can only be deduced indirectly by considering the neurons' local neighborhoods.

- *Growing neural gas* (Fritzke, 1995) also obeys the rank-based update principle. However, it starts with $m = 1$ neurons and grows new neurons on demand when the current input is not covered sufficiently by any available neuron. Moreover, it also grows connections between the two best matching neurons given a particular input. As a result, GNG develops an explicit topology, which typically reflects the input topology including the underlying dimensionality of the input.

Figure 10.4 illustrates the principle behind these three algorithms when uniformly randomly sampling input values from the shaded subspace. Kohonen networks are too restricted in this case: the two-dimensional grid does not fit particularly well in the complexly shaped and partitioned subspace. In contrast, Neural gas and GNG can distribute their neurons well. The GNG approach additionally allows the execution of path-planning methods (such as model-based RL) within the developing lattice structure.

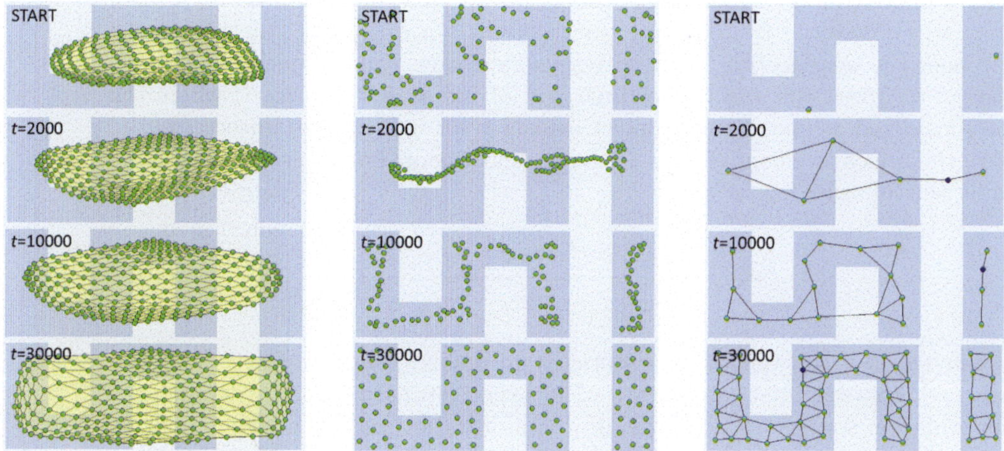

Figure 10.4: Kohonen networks (left), Neural gas (center), and Growing neural gas (right) are examples of self-organizing neural networks that are able, to a certain extent, to deduce the structure of the underlying data. The two-dimensional input is uniformly randomly sampled from the shaded area during learning. [Generated with DemoGNG, Loos & Fritzke, 1998.]

The lattice structures developed by a GNG-like algorithm can easily be related to the concept of peripersonal spaces; the shaded area may be viewed as corresponding to the peripersonal space to be represented. Given, for example, retinotopic and somatotopic stimulus information, a GNG-based neural field may develop that maps the respective input spaces onto each other. The whole space is thus represented by partially overlapping, local receptive fields, which may predict visual information given corresponding tactile information and vice versa. Several additions are necessary to accomplish such bidirectional mappings, but GNG principles still find their way into neurocognitive models.

Such a lattice structure, regardless of how it is actually learned, enables the representation of spatial activities simply by activating the neurons that overlap with the location in question. Probability densities can be activated to induce, for example, spatial priors as done in Chikkerur et al.'s architecture (cf. Section 9.4). When stimulating a particular location, the receptive fields surrounding this location will be maximally active, and the neural activity will decrease with increasing distance from the stimulus. Similarly, bimodal spatial distributions can be represented when, for example, considering two relevant stimuli concurrently. Such bimodal and multimodal distributions, once again, can be viewed as probabilistic mixture models (cf. Section 9.3.4), where the mixture in this case is not combining Gaussians, but neural receptive fields.

10.2.4 Optimal information fusion

Neurons that encode spatial locations are fed by information from sensory and typically also from motor systems. In peripersonal space, this multisensory information is integrated – or fused – in a certain manner. It has been shown in various studies that the brain often integrates the available information in an approximately optimal, Bayesian fashion. This optimality can be explained best when considering Gaussian probability density estimates in a particular space.

Generally, we can assume that a particular sensory system S_i provides information about a particular spatial location by mapping the sensory information onto the spatial space by a mapping function f_i. Given the current sensory reading $s_i(t)$, the function f_i thus yields a sensor-specific location estimate $l_i(t) = f_i(s_i(t))$. However, to fuse this location estimate with another location estimate $l_j(t)$, which possibly stems from another sensory information source S_j (with $j \neq i$), another aspect of the two information sources needs to be available. Imagine that one source is much more precise in the estimate than the other, then the former should be weighted more heavily than the latter when fusing the two estimates. This information can be deduced from sensory source *precision* estimates: the higher the precision, the stronger the weight of the respective sensory source. The simplest measure to estimate precision is to invert the variance estimate σ_i^2 of the sensory source i, where the variance may be considered as one of the simplest measures for information uncertainty.

Equipped with these measurements it can be shown that the optimal sensory integration is accomplished by the following equation:

$$\hat{L}(t) \quad = \quad \frac{\sum_{i \in \mathcal{I}} f_i(s_i(t)) \frac{1}{f_i(\sigma_i^2(t))}}{\sum_{j \in \mathcal{I}} \frac{1}{f_j(\sigma_j^2(t))}}, \tag{10.2}$$

where the denominator is a constant that essentially normalizes the estimate, yielding a properly relative precision-weighted integration of location estimates, and $\mathcal{I}$ denotes the set of sensory information sources that contribute to the location estimate. The resulting estimate corresponds to the *maximum likelihood estimate* of information theory. That is, $\hat{L}(t)$ is the maximum likely location when assuming that all information sources about the location are independent and the respective uncertainties σ_i^2 can be projected without biases into the location space. While these assumptions are not totally valid in most cases, the estimate typically serves as a good approximation. Note how this estimate is strongly related to Gaussian distributions: Eq.(10.2) is exact when all individual distributions and their respective projections into the location space are Gaussian, with means and variances in location space specified by $f_i(s_i(t))$ and $f_i(\sigma_i^2(t))$, respectively.

As we usually do not integrate singular sensory information about a stimulus, but rather integrate this information over time, another component needs to be added that can project the location estimate forward in time. When thinking about peripersonal spaces, this forward prediction will usually depend on behavior: when we move our face, arm, or hand, for example, the objects surrounding the respective body part will systematically shift their location depending on the motion. We may denote this projection of a location estimate forward in time by a function $g(m(t))$, which is a temporal prediction of how things change due to motor activities. This projection converts the currently executed motor command $m(t)$ into an anticipation of a shift in location space, that is, $g(x) : \mathcal{M} \to \Delta L$, where $\mathcal{M}$ is used to denote the motor command space. Given that a motor command was executed, we then encounter an actual spatial transition, which yields an estimate about the resulting location:

$$\hat{L}'(t+1) \quad = \quad \hat{L}(t) + g(m(t)). \tag{10.3}$$

Note how this estimate is related to two concepts, which were introduced in previous chapters: first, we have formalized the *reafference principle* (cf. Section 6.4.1), which anticipates the sensory consequences – in this case actually the location consequences – given a motor command; second, we have generated an *a priori estimate* of a location, given information from the past, according to Bayesian information processing principles (cf. Section 9.3).

To expand the location anticipation to a full probability density estimate, we require some sort of uncertainty estimate in order to yield a location-distribution, rather than one location estimate. Again, let us keep things simple and assume that a variance estimate $\sigma_L^2(t)$ is carried along. How should this estimate change over time? Assuming that we have an a priori location estimate in the form $[\hat{L}'(t), \sigma_L'^2(t)]$ available, we may then consider the incoming sensory information. Assuming further that the location estimate itself is independent of all sensory information, the location estimate can be considered as a fully independent estimate, similar to the sensor-based location estimates. Thus, information fusion simply extends to:

$$\hat{L}(t) \quad = \quad \frac{\left(\sum_{i\in\mathcal{I}} f_i(s_i(t))\frac{1}{f_i(\sigma_i^2(t))}\right) + \hat{L}'(t)\frac{1}{\sigma_L'^2(t)}}{\sum_{j\in\mathcal{I}}\frac{1}{f_j(\sigma_j^2(t))} + \frac{1}{\sigma_L'^2(t)}}, \tag{10.4}$$

$$\sigma_L^2(t) \quad = \quad \frac{\left(\prod_{i\in\mathcal{I}} f_i(\sigma_i^2(t))\right)\sigma_L'^2(t)}{\sum_{j\in\mathcal{I}}\frac{1}{f_j(\sigma_j^2(t))} + \frac{1}{\sigma_L'^2(t)}}, \tag{10.5}$$

yielding the a posteriori location estimate, where the a posteriori uncertainty mixes the a priori uncertainty with other independent information sources, yielding information gain, that is, a decrease in uncertainty.

When then projecting the location estimate into the future by means of the motor-dependent projection function $g(m(t))$, the location will be shifted and the uncertainty should again increase to a certain extent. This extent may depend on the motor function, but it may also add by default some uncertainty, such that, for example:

$$\hat{L}'(t+1) \quad = \quad \hat{L}(t) + g(m(t)), \tag{10.6}$$

$$\sigma_L'^2(t+1) \quad = \quad \sigma_L^2(t) + g(\sigma_m^2(t)) + \sigma_c^2, \tag{10.7}$$

where σ_c^2 adds uncertainty, which may account, for example, for neural processing noise. As a result, the processing loop is closed and the system can continuously maintain an internal estimate $[\hat{L}'(t), \sigma_L'^2(t)]$.

The formalized loop certainly simplifies the actual neurocognitive processing that is going on in several respects. Moreover, it is not known to what extent and exactly how the put-forward formalization is implemented by the brain. However, from a computational perspective, some sort of processing, which mimics this optimal information processing sketch, needs to be realized in order to be able to maintain internal spatial estimates about locations in the environment, as well as about one's own body posture. Various behavioral studies have confirmed that a process similar to this one is at work (Butz et al., 2014; Ehrenfeld et al., 2013b; Ernst & Banks, 2002).

Advanced formalizations of these equations can be derived from free-energy-based minimization principles, thus providing an even more general formalization (Friston, 2009; Kneissler et al., 2015). Additional information processing steps appear to be at work. In particular, it appears that different sensory information sources are compared with each other, fusing only those sensory information sources with the a priori location estimates that provide *plausible* information. Moreover, the resulting a posteriori spatial estimates may be further compared with other information sources – such as estimates about other objects – further fostering the consistency between these estimates given knowledge about the body and the environment. For example, an object may not be exactly located where another object is already located. Similarly, the limbs of the body can only be arranged in certain ways, given limb lengths and joint angle flexibilities. In fact, experimental and modeling results suggest that our brain attempts to maintain a consistent postural body schema estimate over time (Butz et al., 2014; Butz, 2016; Ehrenfeld et al., 2013b).

10.2.5 Spatial, topology-preserving transformations

When fusing multiple sources of sensory information, which may stem from different sensory modalities and which are thus grounded in different frames of reference, it is inevitable that mappings between these different frames of reference need to be available. The different sensory modalities need to be mapped onto each other to enable sensor fusion, or they need to be mapped onto an integrative spatial representation – possibly one that is maximally suitable to support information fusion. Furthermore, as our body moves and thus our limbs change their locations and orientations relative to each other, different sensor-grounded frames of reference also change relative to each other. Thus, spatial, body-relative mappings need to be available that depend on the current body posture. Often such forms of encodings are termed *body images* or more precisely, a *postural body image* or a *postural body schema*. Sometimes a body image is also philosophically related to being consciously accessible, while a body schema is related to bodily encodings that influence motor behavior (Gallagher, 2005). As we focus on the functional, behavioral relevance of such mappings, we will use the term *postural body schema*.

When reconsidering the spatial mapping equation (10.2) in light of a postural body schema, it becomes apparent that the mapping function f_i, which was characterized as mapping sensory input S_i onto a different space L, needs to become more flexible. In particular, f_i needs to depend not only on the sensory information, but also on the current bodily posture. With such an advanced mapping function, it is then possible to flexibly map sensory information onto other frames of reference.

In the brain it appears that *gain-field* structures accomplish such mappings (Salinas & Sejnowski, 2001). Gain-field neurons are multiplicatively modified given, for example, a change in bodily posture. It appears that arm positions and body-relative locations are encoded, which appear to be used by the brain to generate reach plans or to execute directional movements. ANNs have been developed that show that such gain-field-like, multiplicative combinations of postural with directional or location-oriented information sources are well-suited to generate temporal forward model predictions about motor-induced changes – essentially implementing the reafference principle. By means of Hebbian, associative learning principles in combination with multiplicative units, as well as by means of generative model learning principles and backpropagation learning, ANNs with multiplicative units have been developed that can learn such gain-field structures (Kneissler & Butz, 2014; Kneissler, Stalph, Drugowitsch, & Butz, 2014; Schrodt & Butz, 2015). While we will not go into further detail here, it should be remembered that evidence is accumulating that gain-field-like structures are quite suitable for learning mappings between different frames of reference, and that the learning can be accomplished by focusing on learning motor-dependent, temporal forward models.

Meanwhile, it has been shown that such temporal forward models, when suitably inverted, are well-suited to issue goal-directed motor control commands (cf. also Section 6.4). When combining the learned gain-field structures with the principle of maintaining internal location estimates, including postural body estimates in limb orientation space, a postural body schema emerges, which adapts the mappings between different frames of reference on the fly and thus strives to maintain optimal location estimates in different frames of reference. Advanced techniques have shown that these estimates should most likely attempt to maintain consistency in the redundant, but overlapping spatial estimations (Butz et al., 2014; Butz, 2016; Ehrenfeld & Butz, 2013).

Overall, we have sketched-out a system that is able to learn and maintain a postural body schema as well as internal, body relative location estimates. Due to the available postural body schema, different sensory information sources can be mapped onto each other, thus enabling approximately optimal information fusion when precision estimates are available. On the other hand, due to the different spatial estimations, the postural body image can be continuously kept up to date and the mappings can be further adapted over time. Thus, a highly adaptive spatial system may support the spatial location and orientation of the body in space. Moreover, entities can be located relative to the body, facilitating bodily interactions with these entities in the surrounding space.

10.3 Multisensory recognition

Learning of spatial multisensory integrating forms of representations seems a prerequisite to being able to coordinate goal directed interactions with the environment. Almost as important, however, is the need to learn to recognize objects. Only when we are able to distinguish between different objects, are we able to interact with them in a selective manner. A frog shows selective behavior with regard to small moving objects, which leads to tongue protrusion and the quick movement to approach the source, while large shadow-like motion can cause flight behavior (cf. Section 3.6.1). While this frog behavior is certainly largely pre-wired akin to a Braitenberg vehicle, we do not always want to interact with particular objects in the same way. Therefore, it is useful to be able to distinctively recognize objects and thus to enable object-specific decision making and versatile interactions.

In addition to objects, other things need to be recognized as well, including animals, other humans, body parts, faces, surfaces, or the ground. Particularly with respect to other humans, it is useful to be able to identify them, but also to determine their current intentions. For example, it is useful to know if another person is friendly or hostile, or if the other person currently wants to hand you something or to throw something at you. Thus, multisensory recognition is concerned not only with object recognition, but also with behavior recognition.

10.3.1 Object recognition

Numerous developmental psychological studies indicate that infants, within a few months if not days, are already able to distinguish between animate entities, that is, humans and animals, and inanimate objects. Then, objects become progressively classified into categories, starting with vehicles, planes, manipulatable objects, and furniture. At the end of the first year, these distinctions become even more differentiated, forming more and more categories. This differentiation process continues intensely for the next few years of a toddler's life, but generally through a lifetime.

How do such object recognition and differentiation capabilities develop in infants? From very early on, the ability to recognize one's own body develops. Two month old infants have been shown to have a rudimentary body image (cf. also Section 4.2.2). Even at birth, a very crude postural body image seems to be available, which most likely developed before birth (Rochat & Striano, 2000). After learning about what the body and body parts, such as the hand, approximately look like, it becomes possible to ignore visual information about one's own body, once again applying the reafference principle (cf. Section 6.4.1). Thus, babies can more easily focus on other things in the environment by subtracting sensory information about their own body.

Luckily, objects have a couple of convenient properties, which can be exploited while learning to recognize them. Distinct visual properties have been discussed in the bottom-up vision chapter (Chapter 8), including surface properties, systematic edge distributions, or particular properties of occluding edges for background segregation. When babies and toddlers actively interact with objects – a behavior that is supported by the grasp reflex – objects behave in a systematic fashion. Most importantly, the object will systematically move with the hand that holds the object. In a robotics-oriented developmental article, termed the *Birth of the object* (Kraft et al., 2008), the development of an object concept was modeled with a humanoid robot, characterizing the birth of an object as follows:

> First, by making use of an object independent grasping mechanism, physical control over potential objects can be gained. Having evaluated the initial grasping mechanism as being successful, a second behavior extracts the object shape

by making use of prediction based on the motion induced by the robot. This also
leads to the concept of an "object" as a set of features that change predictably
[...]. (Kraft et al., 2008, p. 247.)

Thus, an object is particularly characterized by a *Gestalt*, that is, a set of features and how
this set of features changes when the object is being manipulated. In contrast to the visual
model in the previous chapter (cf. Section 9.4), though, this Gestalt not only specifies visual
features, but also other sensory features, movement properties, and even behavioral aspects,
such as the affordance of the object, that is, in what manner the object may be typically
interacted with.

Of course, for babies and toddlers changes are not restricted to the visual system. A
rattle, for example, makes a particular sound. Objects make distinct sounds when being hit
against some other object. Objects produce particular taste impressions when probed by the
mouth. Objects also typically provide distinct tactile feedback. Finally, when considering
food, taste and consistency associations (how does it feel when being chewed on?) will be
formed once the diet of the infant goes beyond milk. Thus, objects will soon be associated
with various types of sensory and sensorimotor encodings, enabling their distinction.

Taking the predictive, generative stance once again (cf. Section 9.2), assuming that
babies want to be able to reconstruct objects with all their properties, the most important
properties of objects are probably how they behave on their own and how they behave when
one interacts with them. In all cases the object changes in space, which is easily perceivable
(at least when sufficiently close and in sight), and which demands disambiguation. With
this in mind, it is not surprising that objects are first distinguished largely based on their
behavioral properties: animals and humans can be quickly separated from artifacts because
they move on their own in an intentional, biological manner; artifacts do not move on their
own, but movable artifacts (such as kitchen utensils or toys) can soon be differentiated
from unmovable (or difficult to move) objects (such as furniture). Finally, moving or flying
artifacts, such as cars or airplanes, can also be separated from other objects rather easily
due their distinct behavioral properties. Only after about twelve months become individual
objects in a category further differentiated (Hoffmann, 1986; Mandler, 2004, 2012).

How can an object thus be recognized? In general, any information about an object
including sound, taste, tactile impressions (including softness of the surface, hardness of the
material, squishiness, heaviness, etc.), smell, and visual impressions can lead to the recognition of an object. Some researchers speak of an *object file*, which may be equated with the
temporary activation of particular object characteristics in working memory (Kahneman,
Treisman, & Gibbs, 1992; Pylyshyn, 2009).   From a predictive encoding perspective (cf.
Section 9.4), such object files are not abstract forms of representations, but rather predictions about how the object is expected to behave and how it can be perceived given
particular object interactions. Recognizing an object thus means that a temporary, consistent, distributed encoding of the object is activated, which properly anticipates the behavior
of the object upon interactions.

As is the case for spatial, multisensory integrations, object recognition will depend on
the information that is available about the object. Various studies have suggested that
during object recognition the brain attempts to integrate the different sources of information
optimally to disambiguate different objects, once again following Bayesian principles. In the
case of object recognition, a disambiguation process is at work rather than a localization
process. This disambiguation process depends on object features and their distinctiveness
for identifying particular objects. Thus, disambiguation is only possible when the baby has
learned distinct properties and can differentiate them, suggesting that over-generalizations
of object categories is very likely early in life and is indeed often observed (cf., for example,
Mandler, 2004).

In contrast to the spatial frames of reference and the spatial mappings described previously, it appears that to a certain extent objects are encoded independent of space. It is a
simple fact that any object can be recognized regardless of where it is located as long as it
is close enough to actually be perceivable. On the other hand, object expectations activate

spatial priors, which characterize where an object should typically be detected (Lachmair, Dudschig, De Filippis, de la Vega, & Kaup, 2011). This once again points to a tendency to separate object property encodings from spatial encodings: the brain is flexible enough to recognize an object anywhere in space. While spatial prior properties of size and location are available, they can be flexibly adapted given the current sensory evidence.

Recent ANN models have attempted to implement this spatial versus property separation of objects, as described in the model of Chikkerur et al. in the previous chapter (cf. Section 9.4). These models, however, need to be expanded to models that include all perceivable object features, including typical spatial locations and the other types of features mentioned previously. Such models may then generate spatial, object-oriented attention in the form of suitable spatial mappings. The temporary activation of an object-specific spatial mapping can be assumed to map all sensory impressions, which are currently being gathered about an object, onto object-particular feature encodings, thus supporting or actually enabling the current object recognition process. This is for example the case when exploring an object with ones hands, eyes, or mouth, where the individual impressions need to be linked to the object that is perceived relative to the body. Thus, while an object is being recognized in the form of an object file, which includes various object-specific features, spatial mappings need to be activated and flexibly adjusted while exploring the object to enable the proper mapping of the impressions onto the object files.

Later, objects are further differentiated in various manners. Research has shown that even neurally, the posterior, inferior temporal cortex tends to cluster objects with particular properties locally. It is now generally believed that this clustering mechanism is not purely visually driven, but behavioral aspects also have a strong influence. For example, hand-held tools seem to be separated from food and these again from furniture. Meanwhile, furniture and tools, for example, seem to be further separated depending on how one interacts with them, so that reclining objects such as chairs, sofas, or beds are separated from closets and wardrobes and, similarly, tools are further differentiated such that hand-held tools, for example, are separated from movable objects, which do not directly imply a usage (Creem-Regehr & Lee, 2005; Mahon, Kumar, & Almeida, 2013; Martin, 2007; Rueschemeyer, Lindemann, van Rooij, van Dam, & Bekkering, 2010). Thus, these aspects indicate that the *affordance* of an object (cf. Section 3.4.1) also influences the structure of object files. Moreover, these and other behavioral studies suggests that imagining an object not only activates the anticipation of sensory impressions, but also the usual routines, which one executes when interacting with the object (Bub, Masson, & Cree, 2008; Masson, Bub, & Breuer, 2011).

10.3.2 Behavior recognition

Especially when socially interacting with other humans, but also when interacting with animals, it is very useful to anticipate the behavior of the other. Also in this case it appears that multisensory sources of information are combined. A very important information source is the dynamics of perceived biological motion. Indeed, infants only a few days old have been shown to be somewhat more interested in biological motion patterns than in random dot patterns that exhibit a similar amount of motion energy (Pavlova, 2012). Even other animals, such as little chicks, have been shown to be more interested in biological motion, suggesting a particular type of genetic predisposition for biological motion (Kerri & Shiffrar, 2013). While it remains unclear, which visual motion dynamics constitute the key biological motion signals, it is apparent that behavioral recognition is tied to our tendency to be interested in biological motion. Learning focuses on biological motion patterns and appears to map these biological motion patterns onto our motion capabilities (Kilner, Friston, & Frith, 2007; Pavlova, 2012; Perrett et al., 1985; Schrodt, Layher, Neumann, & Butz, 2015).

The discovery of *mirror neurons* supports this last point most directly. Mirror neurons have been discovered in Macaque monkeys. These neurons fire not only when the monkey is executing an action, but also when the monkey monitors another monkey or a human caretaker executing a similar action (Fogassi, Ferrari, Gesierich, Rozzi, Chersi, & Rizzolatti, 2005; Gallese, Fadiga, Fogassi, & Rizzolatti, 1996; Jellema & Perrett, 2006; Rizzolatti, Fadiga, Gallese, & Fogassi, 1996). Various differentiations of mirror neurons have been discovered. For example, mirror neurons seem to selectively co-encode the current apparent goal of an observed action, the reachability of the object that is manipulated by the action, as well as the viewing angle onto the observed action (Gallese & Goldman, 1998; Caggiano, Fogassi, Rizzolatti, Thier, & Casile, 2009; Caggiano et al., 2011). Even purely auditorily perceived object interactions, such as hearing how a paper is torn into two parts, seem to activate mirror neurons in premotor areas, which are also active when one executes the same action oneself – even when the resulting noise is actually masked (Umiltà et al., 2001; Rizzolatti & Craighero, 2004). However, how these mirror neurons develop is still being debated (Cook, Bird, Catmur, Press, & Heyes, 2014; Kilner et al., 2007; Kilner & Lemon, 2013). What is commonly accepted, however, is that human brains develop neural encodings that support the recognition of another person's action by interpreting those actions with the help of one's own behavioral repertoire (Pavlova, 2012; Turella et al., 2013).

Behavior recognition is possible via various sensory information channels. Even congenitally blind people have a good model of the current behavior of other people, by focusing more on information from the auditory channels. Visually, it has been shown that behavior can be recognized from motion cues only, without providing any form information. In these cases, *point-light displays* are used where the points move in a particular, systematic fashion. When the point-lights were attached to the human body, the general shape of the human body is soon inferred based solely on the visual motion cues (Garcia & Grossman, 2008; Johansson, 1973; Thurman & Grossman, 2008). In cartoons, the fast succession of static images of shapes gives a motion impression enabling behavioral recognition. Once again, the currently available knowledge and neural models suggest that all available information is approximately optimally fused in order to maximize recognition success. It thus appears that biological motion recognition is accomplished by considering both, static and dynamic movement information, fusing the respective information sources approximately optimally to generate an overall percept (Giese & Poggio, 2003; Layher, Giese, & Neumann, 2014; Schrodt et al., 2015; Thurman & Grossman, 2008; Vanrie, Dekeyser, & Verfaillie, 2004).

A final, very illustrative example of such an information fusion process for behavior recognition is known as the *McGurk Effect* (McGurk & MacDonald, 1976). The authors paired the repeated utterance of the syllable /ba/ with a video showing the face of a person who concurrently utters the syllable /ga/. In about 98% of the cases, the participants reported to actually hearing the syllable /da/. The visual information modulated the auditory information, leading to a fused perception that integrated the two modal information sources. In fact, the same information fusion process that was quantified for fusing location estimations (cf. Eq. 10.2) can also be used in this case. By projecting the visual information about the mouth and lip dynamics onto the auditory sound space, where the optimal integration of these two somewhat contradictory, but usually complementary information sources leads to the auditory perception of the syllable /da/, because it lies in terms of both, lip dynamics and sound dynamics in between /ba/ and /ga/. Thus, multisensory recognition of behavior – in this case expanding behavior to speech recognition – once again exhibits approximately optimal information fusion.

10.4　Cognitive maps

After the integration of multisensory information for inferring the current constellations of body-relative spaces and objects within these spaces and for inferring the identities of the objects, the last major challenge when integrating multisensory information is the inference of allocentric spatial properties. Clearly our brain is typically able to localize us in the building, in which we are currently located in, as well as where we are currently located in the city, county, state, and country. Moreover, we are able to imagine other places, and, depending on our knowledge about these places, we are able to sketch approximate maps about these spaces, including the location and orientation of buildings, trees, rivers, and other significant entities within those spaces. Often, such spatial knowledge is referred to as a *cognitive map*, such as a cognitive map of a building, a city, or a country, but also of a natural environment, such as a national park, a forest, or a mountain range.

Imagining a particular location in such a cognitive map is typically referred to as a *spatial image* or *spatial imagery*. The spatial imagery itself, however, provides only crude sensory images at best, and rather focuses on relative spatial properties, such as the location and orientation of particular entities – such as buildings, roads, or rivers – relative to each other and possibly relative to a global coordinate system, such as to the north or to the south.

Once again, cognitive maps do not seem to be learned for their own sake. Rather, cognitive maps serve navigation purposes. Depending on our sense of orientation, we tend to remember the paths from a particular location to another rather than the actual map in which this path is integrated. In fact, various studies have shown that abstract map knowledge and navigational knowledge are two somewhat independent entities, both of which we can use to execute effective navigation-oriented path planning. Here, we first focus on the relevant forms of representations and the learning of cognitive maps given multisensory information sources. Then we explore how this knowledge can be used for effective decision making and spatial navigation.

It is generally believed that cognitive maps are formed in the hippocampus of the human brain, as they are in rats and probably many other animals. It is well-known that the hippocampus is crucial for the formation of new episodic memory, such that cognitive maps and episodic memory appear to be closely related to each other. Thus, we will survey the currently available knowledge about the hippocampus and its apparent multisensory, integrative nature on the one hand, and its temporal episodically predictive nature on the other. We then ask the question how cognitive maps are learned given multisensory information sources, expanding the possibility of learning topological representations about allocentric space and spatial constellations.

10.4.1   Hippocampus and episodic memory

While memory in general seems to be distributed throughout our brain, particular regions can be identified that process particular aspects of our experiences and thus contribute crucially to the formation of particular aspects of memory. The hippocampus is essential for the formation of novel episodic memories. Episodic memory refers to those memory capabilities that enable us to remember past events and interactions, occasions such birthdays, a vacation, a hike, a conversation, or a party. Humans, who have lost both of their hippocampi (most often when suffering from dementia, but also in rare cases where they were removed due to medical reasons), suffer from anterograde amnesia: they are not able to build new episodic memory. However, their old memory remains generally intact. Thus, the hippocampus is essential for forming new episodic memory, but not so much for the recall of already consolidated episodic memory entries.

Interestingly, the hippocampus is a very old brain structure found in many animals and is thus certainly highly important for other aspects of memory and brain functionality, besides the capability to explicitly remember particular episodic events in the past. It is part of the limbic system and located medially in the telencephalon (cf. Figures 7.7 and 10.5). Because of its particular shape (akin to a seahorse, the scientific name of which is hippocampus), which fosters the development of internal recurrent connections across several areas within the hippocampal complex, and because of its highly distributed connectivity from and to various other brain regions, the hippocampus seems to have evolved for remembering interactions with the world in an episodic manner. Insights from pathology further support this hypothesis. Due to its capability to various sources of cortical information over time, it seems crucial to consolidate this integrated information. It is also interesting that the hippocampus is one of the few brain structures where neurogenesis continues throughout a lifetime, probably for supporting the formation of new episodic memories.

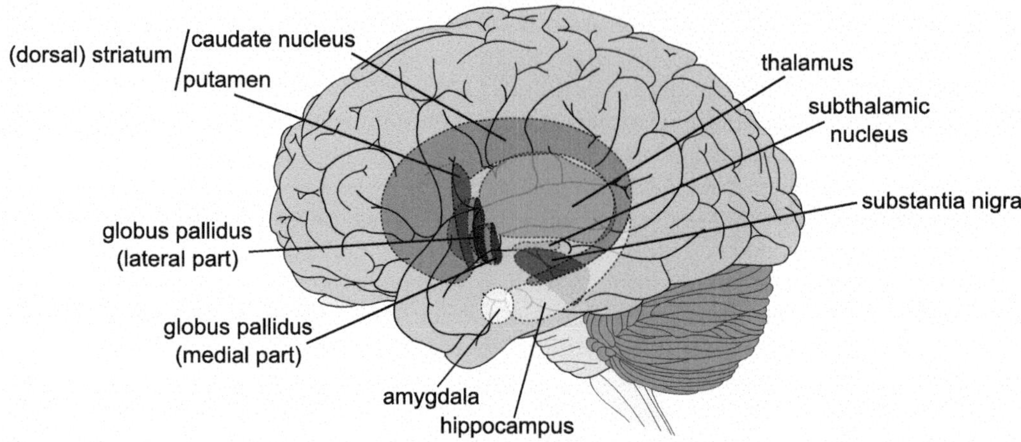

Figure 10.5: The hippocampus, which is an important structure of the limbic system, is crucial for building episodic memories. Additionally, single cell recordings indicate that particular hippocampal cells are selectively activated when at particular locations in space or when taking on particular external space-relative orientations. Additionally, the basal ganglia are shown, which include the caudate nucleus, putamen, and globus pallidus, which functionally closely interact with the subthalamic nucleus and the substantia nigra.

In animals (mainly in rats) individual cells in the hippocampus have been identified that appear to be crucially involved in spatial orientation and navigation tasks (Buzsaki & Moser, 2013). More recently, temporal selectivity has also been shown. Firing rates of particular cells can be correlated with external, allocentric space. Particular types of space-sensitive cells have been characterized as:

- *Place cells* have *firing fields* in a particular subspace of a room, a maze, or similar spatial structures, within which they show maximal neural activity (Moser, Kropff, & Moser, 2008).

- *Head direction cells* appear to be sensitive to the heading direction or head orientation of the animal in a globally-grounded coordinate system, such as the head orientation relative to a particular wall of a room (Taube, 2007).

- *View cells* are particularly responsive when the animal looks toward a certain location, such as a window or a door, regardless from which direction (Gaussier, Revel, Banquet, & Babeau, 2002).

- *Time cells* exhibit *temporal firing fields*, that is, they show selective firing patterns temporarily within an interaction episode, such as during a particular time window while walking on a treadmill (Rowland & Moser, 2013).

Interestingly, neighboring cells in the parts of the hippocampus where place cells can be found do not necessarily encode neighboring locations in space. Moreover, neighborhood relationships of cells change from space to space, such that cells that encode neighboring locations in one space do not necessarily encode neighboring locations in another space. Even more interestingly, the same cell may show place cell as well as time cell characteristics, by, for example, firing selectively during a particular time interval while walking on the treadmill as well as when walking through a particular (other) area in a maze. As implied by these facts, place cells do not necessarily only encode one location, but they may be active in several spatial locations. Thus, it appears that the hippocampus offers highly flexible, reusable encoding structures.

An important information source that is directly connected with the hippocampus is the entorhinal cortex, where *grid cells* can be found (Moser et al., 2008; Rolls, Stringer, &

Elliot, 2006). Individual cells in the entorhinal cortex exhibit grid-like activities of varying spatial distributions. These activities are believed to disambiguate space, such that individual spatial locations can be decoded by giving a sufficient number of grid cell activations. Unfortunately, it remains unknown what are the crucial information sources for these grid-cell encodings. However, it has been shown that landmarks – such as a window or a large stationary item that is visible from anywhere in a particular space – are used by the brain for self-localization and for determining place-cell and grid-cell activities.

In addition to these spatial localization properties, however, hippocampal neural activities suggest that the hippocampus is also involved in goal-directed planning and behavior:

- Forward directed *sharp waves* have been recorded, which suggest that the animal is considering a particular path through a maze. Moreover, similar forward-directed sharp waves have been recorded during sleep, suggesting the involvement in consolidation and a kind of dreaming (Diba & Buzsaki, 2007).

- Inverse-directed sharp waves have been recorded, for example when eating, as if the animal reflects on the path it has taken to reach the food (Diba & Buzsaki, 2007; Foster & Wilson, 2006). Interestingly, these inverse sharp waves have been related to dopamine activities, suggesting that the food-induced dopamine gradient may be back-projected onto the path to the food location, similar to eligibility traces in reinforcement learning (cf. Section 5.3.3).

- Theta-rhythm respective shifts in place field activities seem to co-encode if the place field is currently being approached or left behind.

- Finally, place cells have been shown to be somewhat active in anticipation of a potential goal location as well as being indicators for particular newly detected goal locations, in which function they have also been referred to as *goal cells* (Fyhn, Molden, Hollup, Moser, & Moser, 2002).

Thus, in addition to its relevance for spatial memory formation, the hippocampus also seems to be involved in planning, behavioral learning, and adaptation processes (Hirel, Gaussier, Quoy, Banquet, Save, & Poucet, 2013).

These neuroscientific insights suggest a rather diverse and crucial involvement of the hippocampus in the formation of new episodic memory as well as in the structuring of cognitive maps. The hippocampus helps to integrate various impressions about interactions with the environment, including where these interactions took place, what actually happened, and when particular interactions took place within an interaction episode. When abstracting over the temporal domain, focusing solely on spatiotemporal proximities, but generalizing over the temporal direction, it may be possible to abstract over these proximities and thus to form a behaviorally grounded cognitive map of the encountered space.

Most likely, however, "space" needs to be understood in a general, behavior-grounded sense in such cognitive maps. As the formation of any type of episodic memory in humans is disabled after the removal of both hippocampal structures, it appears that any type of environmental interaction is somewhat spatially grounded, but is also further differentiated depending on the types and durations of the interactions that actually unfolded within the encoded space. For example, the same room may be remembered quite differently when interacting in the room only with the computer or when interacting (on a different occasion) in the same room with people, such as when celebrating a particular event or when having a productive discussion. Similarly, a rat appears to encode the interaction of a treadmill in its hippocampus, not only that it has interacted with the treadmill, but also how long the interaction took and how laborious it was (Rowland & Moser, 2013).

10.4.2 Behavior-oriented cognitive map

How are cognitive maps learned from a behaviorally-grounded, computational perspective? Hippocampus-like structures apparently focus on clustering multisensory experiences in an episodic manner, leading to the remembrance of places and interaction episodes. Thereby various sources of information become integrated and encode allocentric, spatially relevant information as well as goal-relevant information.

As is the case in learning peripersonal spatial encodings, self-organizing ANNs offer themselves as one possible modeling approach. In the following example, we briefly introduce an exemplar ANN model, which enables the behavior-oriented buildup of a cognitive map (Butz, Shirinov, & Reif, 2010). To succeed, it uses a temporal extension of the GNG algorithm. Moreover, it links the cognitive map with somatic markers, that is, with markers about rewarding areas in the maze. In effect, inverse, goal-directed planning becomes possible, which implements a motivationally-driven, model-based reinforcement learning approach.

The cognitive map is built up by an algorithm that grows neurons on demand when apparently new spatial locations are reached. Moreover, the spatial locations are linked by means of edges when particular neurons fire in temporal succession. As a result, a cognitive map is learned where

- *Neural nodes* exhibit spatially local firing fields.

- *Neural edges* connect temporally neighboring nodes. Moreover, motor codes are associated with the edges, which specify which direction needs to be pursued to reach the successive node.

With the cognitive map at hand and given a current location in the cognitive map, the system can anticipate which neighboring destinations can be reached and how they may be reached.

Experiments were carried out with this approach by simulating a point-like system that is able to execute step-wise interactions with a particular maze (Butz et al., 2010). Distance sensors to the maze's border provided enough information to disambiguate every position in the maze. Moreover, global orientation information was provided akin to the head direction cell information found in the hippocampus. By exploring the maze with a random or a novelty-oriented behavioral selection strategy, the system then learns a cognitive map. Due to the distance sensor information, nodes develop that characterize particular spatial locations. Moreover, recurrent neural edges link these locations to each other. Depending on the exploration speed and the target-granularity of the network, a complete cognitive map soon develops (cf. Figure 10.6).

The learned cognitive map then allows the execution of goal-directed behavior by means of model-based reinforcement learning principles (cf. Section 5.3.3). By activating particular, desired locations in the map, that is, by activating those nodes where a particular reward or where a particularly interesting interaction was encountered previously, the activation is propagated inversely through the network by means of model-based RL. Once the propagated activity gradient reaches the node that encodes the current location of the artificial agent, it can choose to execute the behavior that is expected to lead to the maximally active neighboring node. By iteratively climbing the reward gradient, the goal is eventually reached.

The described model is essentially able to build a cognitive map "from scratch", developing place fields and temporal connections between these fields. Planning and behavioral control is possible by model-based RL and a closed-loop control process, which progressively climbs the reinforcement gradient. By associating particular nodes with particular rewarding events, self-motivated goal-directed behavior can be realized. Moreover, behavioral adaptations are possible when the maze changes or previously rewarding places become neutral. Even the combination with other priorities, such as avoiding open spaces, are possible when the system is enhanced with a motivational module (Butz et al., 2010, cf. also Section 6.5). This module can adaptively issue reward signals onto the nodes and the edges of the cognitive map depending on its current internal state, thus flexibly adapting the activity propagation process.

While the model thus shows how a cognitive map may be learned in principle by integrating sensory as well as motor information, the relationship with the hippocampus is very

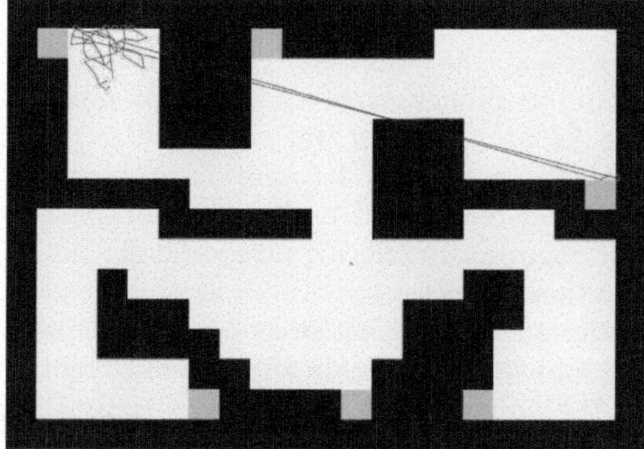

(a) Learning in a maze

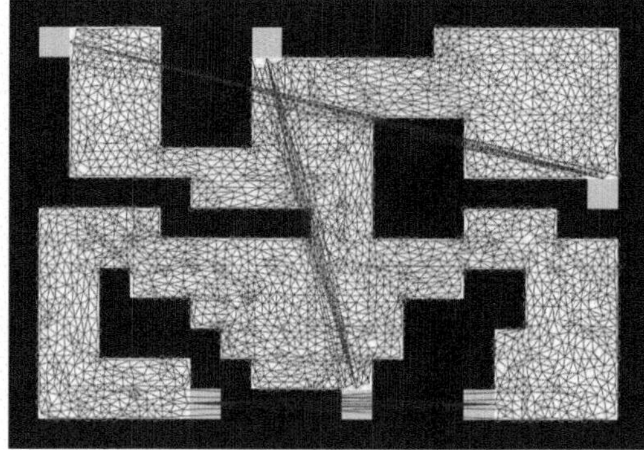

(b) Learned cognitive map

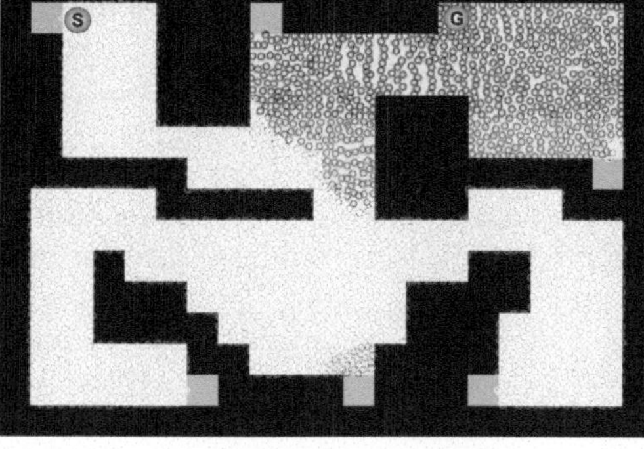

(c) Expanding reward gradient

Figure 10.6: The point-like system explores the maze forming in this case a rather fine-grained cognitive map. Grey squares are paired "teleportation" connections. When a goal is activated (circled "G"), reinforcement-based activity is propagated through the network (by means of model-based RL) throughout and until the starting position (circled "S"), leading to the goal-oriented execution of behavior along desired edge directions.

crude at best. The formation of episodic memory is not really supported, but the temporal information is immediately abstracted into step-wise spatiotemporal proximities. Moreover, the information is not processed in an integrative Bayesian fashion and at this point the sensory information is not selected for, for example, its spatial properties, but rather it is provided in the implementation. Multisensory Bayesian integration of additional sources of sensory information should be considered in future models. The combination with episodic memory-oriented learning should also be considered further. Nonetheless, the model offers a neural implementation of a behaviorally grounded cognitive map learning mechanism and it shows how such cognitive maps can be employed to plan and execute goal-directed and even self-motivated behavior.

10.5 Summary and conclusions

When multiple information sources interact bidirectionally in the attempt to mutually predict or generate each other, spatial forms of representations and spatial mappings between such topological representations can develop. Such spatial topologies and mappings can be found in various cortical brain areas, including parietal and premotor areas, where mainly peripersonal spatial encodings can be found, as well as in the hippocampal complex, where spatiotemporal encodings can be identified. While the former is mainly responsible for coordinating direct interactions of the body with the outside environment, the latter is very helpful when the task is to navigate through the world in a goal-directed manner.

In addition to spatial topologies and mappings, however, recognition-oriented multisensory integration processes appear to be at work. In this case, the recognition of static stimuli, such as objects, can be contrasted with other recognition processes that are based on dynamic motion information. In other, connected parts of the brain, these two types of recognition processes are combined, once again offering a redundant, fail-safe system for recognizing particular entities and motion patterns.

Neural population codes and Bayesian information processing principles are the "key players" in developing such encodings and structures. In the case of recognition processes, top-down Gestalt hypotheses about the object's structure with its typical features, which may be perceived by means of various sensory modalities, and consequently expectable sensory impressions, are fused with actual sensory information. In the case of spatial encodings, multisensory spatially-relevant information as well as temporal predictive information appears to be fused in an approximately optimal manner. In the case of external, allocentric spatial encodings, that is, cognitive maps, it appears that episodic, behavior-grounded encoding principles are crucial, which once again integrate predictive temporal with spatially-relevant information. Both, spatial encodings as well as entity recognition encodings can be used as goal signals, where their activation then leads to inverse, model-based RL-based planning and the resulting execution of goal-directed behavioral control (cf. sections 5.3.3 and 6.4.2). Moreover, both types of encodings can be coupled with a motivational module, which may adaptively modify the goal activations over time, depending on the current motivational state of the system (cf Section 6.5). Chapters 11 and 12 consider these adaptive processes in further detail, first considering attention and then decision making and motor control.

10.6   Exercises

1. Show that uncertainty decreases when computing information gain.

2. Name three information aspects each that can be extracted from the somatosensory / tactile / auditory / and visual modalities.

3. How is a postural body schema relevant when perceiving the own body and estimating its current posture?

4. We all know the annoying feeling when we hear a mosquito close to our ear or face. Explain how this feeling may come about with reference to the peripersonal space. How are we able to hit it sometimes successfully when we felt it sitting down on our skin (or actually feeling the bite) without having seen it?

5. Studies indicate that items in reach may be perceived as nonlinearly closer than items just beyond reach (Witt, Proffitt, & Epstein, 2005). How may this perceptual phenomenon come about?

6. Neural population codes have been shown to encode directional fields, peripersonal spaces, and even object-relative spaces. What is the advantage of such population codes when contrasted with approximate Gaussian encodings?

7. Name and shortly explain the main differences between Kohonen networks, Neural gas, and GNG.

8. Given three sensory sources about the one-dimensional location of the same object (for example, visual, tactile, and auditory) in the form of Gaussian densities with means and variances;

   (a) Compute pair-wise interactions applying sensor fusion.

(b) Given these pair-wise comparisons, how could the plausibility of one sensory source be determined relative to the others?

(c) Given furthermore an internal temporal prediction in location that predicts no location change, but an increase in variance of 50%, compute one information processing cycle.

9. Multisensory spatial information was contrasted with multisensory information for recognizing objects and other entities. Which information do these two generalize over respectively? How do the two types of encodings complement each other?

10. How can behavior recognition be understood as the recognition of a spatiotemporal Gestalt.

11. How can behavior recognition and Gestalt recognition abilities complement each other?

12. Traditionally, animated cartoons were created by showing still images in fast succession. How come we perceive an actual motion of the animated characters?

13. Relate cognitive map learning to model-based RL. In which manner can a cognitive map be used in model-based RL?

14. Neural activities in the hippocampus indicate that cells are not only involved in memorizing episodes, but also in planning, model learning, and RL. List the gathered evidences and explain them briefly in your own words.

15. Imagine an implementation of two motivational modules, one that generates negative reward while passing through open spaces, an another one that strives to reach goals. Consider an artificial agent that is equipped with these modules and that needs to pass diagonally through a room to reach a goal. Sketch likely trajectories of the agent with the following relative strengths of the two modules: $(0, 1)$; $(0.1, 0.9)$; $(0.5, 0.5)$; $(0.9, 0.1)$; $(1, 0)$.

Chapter 11

Attention

11.1 Introduction and overview

The discussions of multisensory and motor interactions in the last chapter always assumed that all information available is processed when estimating egocentric or allocentric spatial locations and orientations, as well as when recognizing entities and behaviors. However, the amount of sensor information available is often too large and too complex to be considered in its entirety. Consider, for example, your desk and think about how you are not confused about where the monitor is positioned, where pen, pencil, and paper can be found, which sensory signals give information about the phone, the keyboard, and so forth. Seeing the diversity of stimuli, it needs to be acknowledged that the brain is continuously solving challenging *binding problems*, needing to correctly bind together those subsets of stimuli that belong to individual objects.

However, even if the binding problem is solved entirely, the desk example highlights another problem, which is the *frame problem* (cf. Section 3.4). Seeing that there are typically a large number of objects and other entities around in our world – regardless if in an artificial world, such as an office, or in a natural world, such as a forest – and seeing furthermore that one cannot interact with all of these object and other entities in parallel, it is in the interest of behavioral decision making and control to focus on those aspects that are currently most relevant. Given the task at hand, some information sources may be more important than others, and some of those other information sources may even be disruptive. Mechanisms of *attention* enable our brain to focus on particular sensory information sources, on particular internal encodings, and on the execution of particular motor behavior.

Another motivation for the need for attentional processes comes from the apparent fact that our brain is limited in its capacity to concurrently make multiple, distinct, behavioral, or cognitive decisions. This can be seen when realizing that it is usually impossible to control multiple cognitive processes in parallel without any interaction effects. For example, when walking down a sidewalk or in a building with a group of friends and conversing, the conversation is typically interrupted when navigation becomes challenging – such as when a door needs to be opened, a street needs to be crossed, or another group of people needs to be passed. Similar effects occur when talking on a cell phone, which is certainly the reason why it is illegal in most countries to talk on a cell phone while driving: the cognitive demands of driving and talking on a cell phone – thus thinking about the unfolding conversation – partially overlap. Because of this overlap, cognitive processing interference occurs, which may unfortunately result in detrimental consequences in the case of steering a car. In general, given tasks with high *cognitive load*, which are particularly those where many challenging decisions need to be made in parallel, processing interferences can be registered.

Another consideration that stresses the need for attention is the fact that our sensory capabilities are embodied, that is, they are grounded in bodily frames of reference. To gather information about the world with a particular sensory organ, regardless of which one it is, it needs to be directed in the right direction, the distance to the stimulus may need to be adjusted, and other potentially disruptive sensory information needs to be avoided. In order to gather particular sensory information, *active, epistemic behavior* is often necessary. For example, to identify an object in the dark with the hands, the hands and fingers need to actively explore its shape and surface properties. To perceive an object visually, the eyes need to look into the right direction, the line of sight must not be occluded, and the object needs to be sufficiently, but not overly close. To hear an auditory stimulus sufficiently well, other auditory stimuli need to be avoided, and again one needs to be close enough to the sound source. Sensory information is thus determined and actively selected for enabling particular bodily interactions with the environment, requiring to focus on anticipated information-relevant aspects.

In the following, we focus on visual attention, but emphasize the general applicability of the put-forward principles. We first contrast bottom-up with top-down attention and discuss their interactions. Next, we give an overview of several highly revealing attentional phenomena and interpret them in an information-oriented fashion. We then explain attentional mechanisms in further detail in light of several models of attention and attentional processing. At the end of the chapter, we expand the view of attention to cognitive processing in general.

11.2   Top-down and bottom-up attention

Attention is often considered to be driven by two main components: *bottom-up* attention refers to sensory-driven attention, which focuses attention on *salient features*. Bottom-up attention is thus considered to be *exogenously* driven by salient sensory signals from the environment. Particularly unique and unexpected stimuli tend to capture our attention. The more unique and unexpected a feature, the stronger the *attentional capture*, focusing attention on this feature possibly even against our will. *Top-down* attention on the other hand, refers to *endogenously* driven attention, that is, attention that is determined by internal motivations, current behavioral and cognitive tasks, and task-respective goals.

Top-down visual attention has often been compared metaphorically with a "spotlight", which determines, which information is currently processed most intensively. Several features are included in this *spotlight metaphor*:

- We seem to be able to direct our attention toward a focused area, within which the stimuli are processed in detail whereas the surrounding stimuli are attended to much less intensely.

- The width of the spotlight focus can be adjusted, directing most of our cognitive information processing resources into the spotlight's focus.

- It seems very hard to split attention. Typically it is assumed that the attentional spotlight either switches between multiple sources of independent information or it expands to cover all information aspects.

- The spotlight can be directed *overtly*, that is, visibly to an external observer, for example, when we actively direct our eyes by means of saccades toward particular stimuli.

- Attention can also be directed *covertly*. We look at items "from the corner of our eyes" without actually changing the posture toward the attended stimulus.

- The spotlight can be directed toward any modality, toward particular stimulus aspects, and even toward abstract thoughts.

The spotlight metaphor provides a good general intuitive understanding of the essential effect of top-down attention: cognitive processing resources can be directed toward particular aspects or subspaces of information. When focusing on visual attention and saccades, a *premotor theory of attention* (Rizzolatti, Riggio, Dascola, & Umiltá, 1987) was proposed, which closely relates the overt reorientation of the visual spotlight to the oculomotor program necessary to execute this reorientation. The more complex the oculomotor reorientation is (the further and the more into the other hemifield), the longer it will take, even when it is only executed covertly.

As the characterizations suggest, however, the spotlight metaphor should not be taken too literally. Originally formulated exclusively for visual attention, various studies have shown that the "spotlight" of attention is much more flexible than an actual spotlight. For example, using a feature identification task, Duncan (1984) has shown that the spotlight can not only be directed toward particular aspects or subspaces of sensory information, but also toward particular objects. In the study participants were shown two overlapping objects, such as a line and a rectangle (cf. Figure 11.1). In both objects two features were critical. The line could be tilted slightly to the left or to the right and it could be either dashed or dotted. The rectangle was either small or large and had a gap on the left or on the right side. When the two overlapping objects were displayed sufficiently briefly and masked afterwards, it was observed that it was much easier for the participants to report two properties of one object rather than one property from one object and another property from the other object. For example, it was much easier for them to report if the box was small or large and where the gap was located, versus if the box was small or large and if the line was dotted or dashed. As this effect was stable for all paired combinations, the authors concluded that top-down attention was supported by an object-oriented spotlight, which bound the individual object features together and thus enabled better recognition of two features that belonged to the same object. Many other studies have confirmed these results, implying that the spotlight metaphor should be interpreted rather loosely, suggesting that top-down attention is possible toward any cognitive imaginable set of "bindable" features.

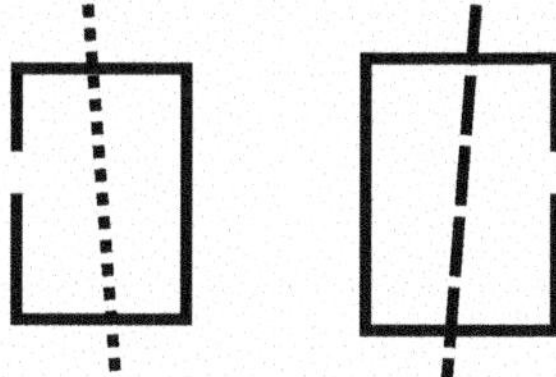

Figure 11.1: When subjects are asked to remember two visual properties, it is easier to remember two properties of one object than one property of each object, even though the actual features are completely overlapping.

While top-down, spotlight-like attention is at work, however, bottom-up salient stimuli also continuously influence our overall attention. Salient stimuli that result in attentional capture are typically hard to ignore, especially when they occur unexpectedly and in an irregular pattern. It has been shown that the more unexpected or unique particular sensory stimuli are, the stronger the bottom-up influence on attentional processing.

When interacting with the real world, bottom-up and top-down attention are continuously interacting. For example, when viewing a scene, our eyes tend to focus on unique stimuli, but also on stimuli that "interest" us or that are task-relevant in one way or another. When manually interacting with the world, top-down attention is focused on the task. It can, however, be distracted by bottom-up salient stimuli, such as a sudden disturbing visual change. On the other hand, it may also be partially complemented by salient bottom-up stimuli, especially because task-relevant stimuli are often partially characterized by bottom-up salient features (Belardinelli, Herbort, & Butz, 2015). Thus, during natural interactions with the environment, our attention is typically co-determined by both, endogenous and exogenous influences.

11.3   Phenomena of attention

With the knowledge of the interactive nature of top-down and bottom-up driven attention, it is worthwhile to consider particular attentional phenomena in further detail. We start with one of the most fundamental paradigms to investigate visual processing, which is *visual search*. This paradigm reveals our capability to distinguish stimuli as well as to focus our attention on particular stimulus aspects. Next, we consider the *attentional blink* and *repetition blindness*, both of which reveal temporal aspects during attentional processing. We then consider *change blindness* and *inattentional blindness*, where the former reveals that we encode only a few details about our world at a time, and the latter shows how well we can direct our top-down attention toward critical stimulus aspects. Finally, we emphasize that similar attentional phenomena can also be observed in non-visual modalities.

11.3.1   Visual search

Visual search is an important paradigm for investigating the basic mechanisms that underly visual attention. Images are shown to participants on which a target stimulus is present amidst a number of distractor stimuli. The task for the participant is to identify the goal stimulus as fast as possible. The investigated question is typically how strong the distractors disrupt the identification of the target stimulus, where the disruption may depend on the feature properties of target and distractor stimuli, the number of distractor stimuli, and other task-relevant aspects.

Two main types of visual search settings have been contrasted (cf. Figure 11.2):

- *Disjunctive search:* the target stimulus can be identified uniquely by focusing on only one feature dimension – such as color, orientation, or shape – because all distractor stimuli differ from the target stimulus along this feature dimension.

- *Conjunctive search:* the target stimulus does not differ uniquely in one stimulus dimension, but can only be identified by a feature conjunction, that is, at least two feature dimensions are relevant for identifying the target.

The resulting reaction times show that visual search hardly depends on the number of distractor stimuli when facing a disjunctive search task, while they somewhat linearly depend on the number of distractors when facing a conjunctive search task. In the former case, one speaks of a *popout effect*, because the target stimulus "pops out" of the distractor stimuli and is detected nearly immediately. In essence, the unique feature is apparently directly identifiable and thus attention can immediately be directed toward the stimulus. In the conjunctive case, however, serial search appears necessary because every stimulus needs to be probed.

Several other observations have been made while considering the stimulus material and variations thereof. For example, it has been shown that the absence of a feature is harder to detect than the presence of a feature. Evolutionary influences also seem to play a role, such that an angry face is detectable more easily amongst happy faces than a happy face amidst angry faces. Moreover, the popout effect can be trained to some extent, such that an initially conjunctive search becomes progressively disjunctive after extensive training. Finally, visual search can be hierarchically structured. Participants have been shown to be able to search for the conjunctive stimulus faster by considering only those objects with one feature conforming to the target, thus on average cutting the serial search time in half (Pashler, 1998).

11.3.2   Attention over time

In addition to the fact that top-down attention can be directed toward spatial regions as well as toward particular features and object-bounded sets of features, temporal, top-down attention is possible. Many researchers have experimented with *stimulus onset asynchronies*

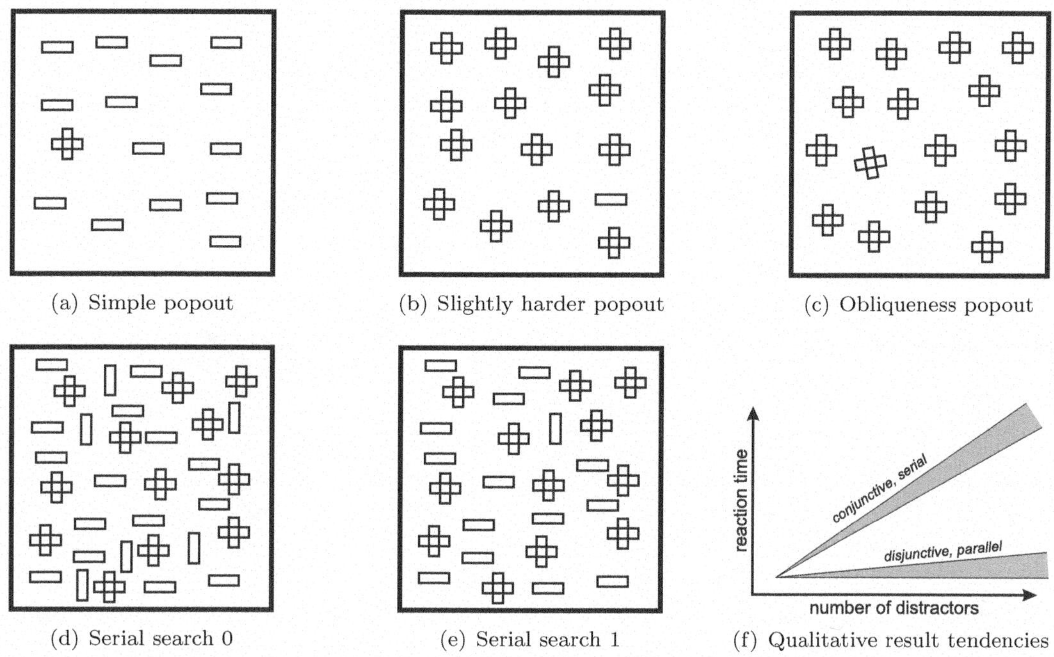

(a) Simple popout   (b) Slightly harder popout   (c) Obliqueness popout

(d) Serial search 0   (e) Serial search 1   (f) Qualitative result tendencies

Figure 11.2: Depending on the uniqueness of the target stimulus, disjunctive search tasks can be contrasted with conjunctive search tasks. Slightly idealized, it is typically observed that the target stimulus "pops out" in a disjunctive search task, while serial search takes place when the target stimulus can only be identified by considering a conjunction of features.

(SOAs), where a target stimulus onset varies temporally with respect to a stimulus prime. When the target onset occurs always after a certain number of milliseconds, say 400ms, then attention appears to focus on this point in time. When the target is displayed somewhat later or earlier, say after 450 or 350ms, then the reaction time to the target (regardless of which target aspect a participant needs to react to) is typically slower. When the target occurrence is blocked, occurring for example in one block after 400ms and in another block after 800ms, then the reaction time is slower for the block of 800ms – indicating that temporal attention is more precise for shorter time intervals. However, when several SOAs are equally likely, such as 400ms and 800ms within a block of trials, then the reaction time becomes progressively faster. This effect is explained by the additional stimulus onset asynchrony. The target stimulus may occur with only 50% chance after 400ms, but, if it was not presented after 400ms, it will definitely be presented after 800ms, yielding a 100% chance and the possibility to maximally prepare attention (Niemi & Näätänen, 1981; Rolke & Hofmann, 2007). These results show that attention can be directed toward particular points in time, expecting and preparing for the processing of a particular event at that time.

Another interesting aspect with respect to time is the fact that the processing of a stimulus appears to temporarily block further stimulus processing. The effects are best characterized by the phenomena of the *attentional blink* (Raymond, Shapiro, & Arnell, 1992; Shapiro, Raymond, & Arnell, 1994) and of *repetition blindness* (Kanwisher, 1987).

The attentional blink characterizes a covert blink effect while processing a visual stimulus. When using a rapid serial visual presentation (RSVP) paradigm, images are presented in rapid succession (< 100ms). For example, a series of letters may be presented among which two numbers of different color are to be identified. Figure 11.3 shows an illustrative trial. The attentional blink characterizes the fact that the first to be identified stimulus somewhat masks the stimuli that occur in a time window of about 100–450ms *after* the first stimulus. That is, when the second target stimulus is presented during this time window, it is much more often overlooked than when it is presented immediately after the first target or more than about 500ms after the first target. It is as if visual processing is temporarily inhibited shortly after processing of the first stimulus commences. As the attentional blink remains present even when only the presence and not the identity of the target stimuli needs to be indicated, it appears that the attentional blink can be characterized as a very early, temporal attentional selection process.

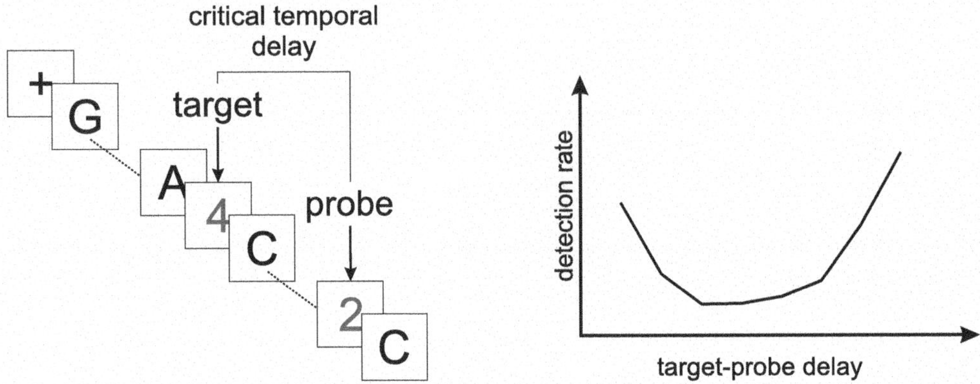

Figure 11.3: In the RSVP paradigm, participants have to identify two target (or target and probe) stimuli in a rapidly presented sequence of distractor stimuli – for example, two gray numbers are to be identified amongst black distractor letters. In most of the trials, the first target stimulus is identified correctly. However, when the second target is displayed in a time window of about 100ms to 450ms after the first target, then the second target is often overlooked. The plot on the right shows the idealized, but typically observed result pattern of detecting the probe stimulus.

In contrast to the attentional blink, *repetition blindness* occurs on a slightly higher processing stage. In this case, the RSVP paradigm is again used, but words or images are presented. In the original work (Kanwisher, 1987), for example, pairs of words had to be detected or a sentence had to be recalled. Often it happened that a duplicate word was overlooked – especially when the words were presented in a sufficiently fast succession (< 250ms per word). In the case of sentences, the effect is most surprising: a sentence such as "The brown couch and black couch were stolen" are often perceived as "The brown couch and black were stolen." On the other hand, when the first word "couch" was replace by "sofa", then the second "couch" word was typically not overlooked. Thus, the second, identical word was overlooked even if the sentence consistency could not be maintained. It is as if once a word has just been detected, that word is temporarily not available cognitively. Later experiments have duplicated *repetition blindness* effects with pictures and even with combinations of words and pictures (Bavelier, 1994).

In both, the attentional blink and repetition blindness, a temporary stimulus-specific processing inhibition is taking place. Most likely the inhibition is not an active process, though, but it is a correlate of the recognition and memorization processes that unfold when a critical item was presented and detected. Even if the exact identity of the target item does not need to be memorized, its actual occurrence does. In the attentional blink, the inhibition takes place very early making the subjects blind to particular visual features of a target, probably because the recognition and memorization processes focus on the visual features. In repetition blindness, the inhibition occurs on a higher somewhat semantic level disallowing the processing of an identical semantic stimulus while still or shortly after processing the first one, probably because the semantic identity of the target item is still being recognized and memorized.

11.3.3 Change blindness and inattentional blindness

At least two further observations need to be contrasted when considering attentional phenomena: *change blindness* and *inattentional blindness*. Both have in common that particular, seemingly obvious elements in a scene are not recognized or noticed. In change blindness experiments, rather obvious changes in successively presented images are overlooked. In inattentional blindness, the viewer's attention focuses on particular aspects of a scene – typically presented as a video – such that other rather salient and interesting aspects are overlooked.

Change Blindness In the classical change blindness paradigm, participants are viewing two images in succession. The images are identical except for one possibly rather large and highly visible element, which does not, however, change the general gist of the scene. If the two images are shown in succession without any intermediate mask, bottom-up attentional capture immediately reveals the difference between the two images. Bottom-up change or motion detectors signal the particular visual change, directing attention toward that change, and thus enabling immediate detection.

However, in change blindness experiments this *motion transient* is masked by one of multiple ways. The simplest mask is a short gray screen (that is, a *blank*) shown between the picture presentations. For example, in the original study (Rensink, O'Regan, & Clark, 1997) an image was presented for 240ms followed by a blank of 80ms, and then followed by the next image, and so forth until change detection. It was shown that the participants struggled to detect the difference in the successively presented images. However, when verbal cues about the change or a verbal image descriptions, which included the critical image component, were provided before the trial, change detection significantly improved.

Figure 11.4 shows two exemplar picture pairs from the original paper. In the dinner scene the handrail behind the couple changes height. In the helicopter scene, the other helicopter that is visible through the cockpit changes position. Without any prior knowledge, it typically takes quite a while to detect the aspect or item that changes in such images. The detection rate depends on various factors, where the factors dominantly depend on the significance of the changed item or entity in the scene. In the original study, the other helicopter's position was perceived as a rather significant object, so the change of position was detected rather quickly (after about 4 alternations on average). On the other hand, the handrail change, which is equally significant visually speaking, was only detected after slightly more than 16 alternations on average.

The authors concluded that top-down attention is necessary to detect the specific change in an image as long as the bottom-up motion transient is masked. Interestingly, later studies have shown that this mask does not need to be a blank screen. "Mud splashes", which are random small masks or blinks in the image that co-occur with the actual change and result in additional distracting motion transients, can perfectly mask the actual change as well. Even when the change is perfectly timed with the blink of our eyes, the bottom-up motion transient does not apply and we need top-down attention to identify the change. Finally, very slow gradual changes, for example in color, typically remained undetected (cf., for example, Rensink, 2002).

In conclusion, the results show that we perceive our environment in much less detail than we assume. When interacting with our world, we typically have the feeling that everything surrounding us is perceived. And this is indeed so, but only because we can look at it. However, when we are suddenly asked to close our eyes, we soon realize that most of us actually only have a general, abstract image in our brain about what exactly is out there. These observations thus suggest that our brain does not re-represent everything in the world in detail – which would only lead to the homunculus problem – but it rather follows the principle that *the world is its own best model*. As long as the world is accessible, we can look at it at any time and thus verify details, such as the existence of particular objects and their particular properties. Sudden changes are typically detected because we notice them due to the typically occurring motion transients when an item is removed (things hardly

Figure 11.4: In the change blindness paradigm, subjects are successively and repeatedly shown two nearly identical images. A short blank or other kind of distractor between the two images masks the motion transient, which would otherwise lead to immediate change detection. Prior knowledge, or the scene-specific importance of the change, can strongly influence change detection performance. [Reprinted with permission from Rensink, R. A., O'Regan, J. K. & Clark, J. (1997). To see or not to see: the need for attention to perceive changes in scenes. *Psychological Science, 8*(5), 368–373. Copyright © 1997, © SAGE Publications.]

ever just disappear without any signal of their disappearance). Thus, we typically feel quite comfortable in our environment and have the feeling that we are aware of the things that are there. Change blindness experiments illustratively show that we really know neither exactly what is there nor exactly where it is.

In essence, we appear to direct our attention mainly to those items that are somewhat relevant, for example, to comprehend what is going on in a particular scene or to facilitate interaction with the scene. Thus, we typically scan a scene by fixating items in a very information-driven manner – striving to disambiguate those aspects of a scene that seem relevant. We will discuss what may actually mean "relevant" when we consider behavioral control, conceptualizations, and language in Chapters 12 and 13.

Inattentional blindness   As in change blindness experiments, also in inattentional blindness experiments significant things in a scene are overlooked. In contrast to change blindness, however, the things that are overlooked are co-occurring with other things that are actually monitored with full concentration. Top-down attention focuses on many aspects of the scene, but not on the thing that changes during change blindness. In contrast, during inattentional blindness top-down attention continues to focus on one particular aspect, thus overlooking another interesting aspect. In other words, during inattentional blindness participants do not actively search for a particular thing, but they concentrate on one particular thing and thus overlook something else, even if it is temporarily fully fovealized (Mack & Rock, 1998; Neisser & Becklen, 1975; Simons & Chabris, 1999).

In the inattentional blindness paradigm, participants are typically asked to focus on a particular aspect in a video – such as the number of basketball passes by a particular team.

While the participants intently focus on the teammates and the ball, another event takes place. For example, a person in a gorilla costume or a person with an umbrella walks through the scene. Surprisingly, participants often overlook this additional person, even if the person walked right through the center of the action and even temporarily overlapped with the ball.

Figure 11.5 shows some of the stimulus material and sketches-out the results reported in Simons and Chabris (1999). Because the focus is on the task, the gorilla is often overlooked. It was also shown that the gorilla is more often overlooked when concentrating on the white team. When attending to the white team, all darker persons are filtered out, including the gorilla. However, even when the basketball passes of the black team were counted, the gorilla was still overlooked frequently. This indicates that top-down attention in this case focused not only on color, but it kept track of the individual team members over time, filtering out other individuals (that is, the white teammates as well as the additional gorilla in black). When the task was made more difficult – such as having to count both the aerial and bounce passes of a team – the detection rates dropped further, indicating an even stronger top-down attentional filtering process.

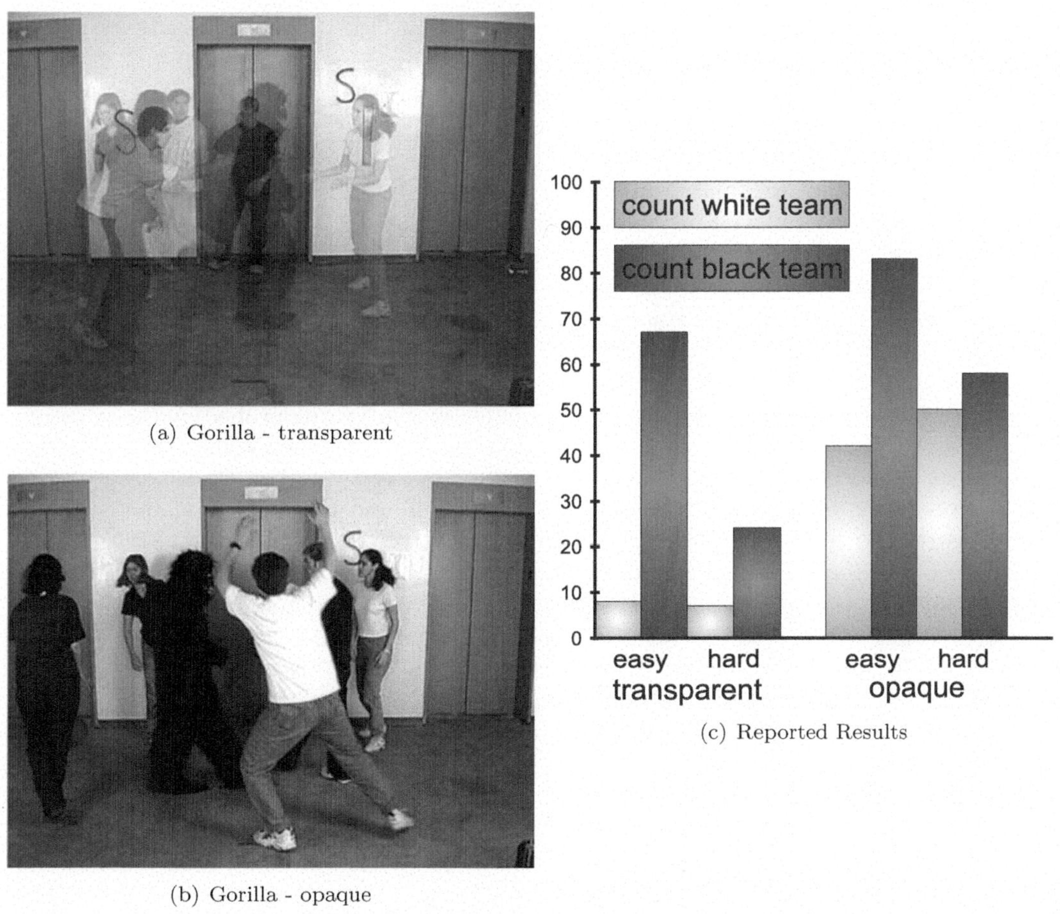

(a) Gorilla - transparent

(b) Gorilla - opaque

(c) Reported Results

Figure 11.5: In the inattentional blindness paradigm, participants watch a movie of two teams each passing a basketball to each other. While counting the basketball passes, the person in a gorilla costume or with an umbrella is often overlooked – and more so when focusing on the white team than when focusing on the black team. When the video is shown without any task, on the other hand, the additional unusual person is noticed nearly every time. Images reprinted with permission from Simons, D. J. & Chabris, C. F. (1999). Gorillas in our midst: Sustained inattentional blindness for dynamic events. *Perception, 28*, 1059–1074. Copyright © 1999, © SAGE Publications.]

These and other experiments highlight how well top-down attention actually works to accomplish a particular task – sometimes, however, too well such that we ignore or overlook other significant events. Top-down attention can be directed toward particular aspects of a scene very intently and in a highly focused manner, filtering and thus ignoring other aspects of the scene. This capability is usually very helpful, because it enables us to, for example, solve a task, study, play a game, listen to a talk, or participate in a conversation with full concentration. However, it can be also disadvantageous, for example, when we overlook a relevant aspect of the environment because we were fully focusing our attention on another aspect.

11.3.4   Other attentional capabilities

Similar attentional capabilities can be detected with respect to other sensory modalities and even with respect to abstract thought. However, in these cases it is typically much harder to conduct proper experiments that can reveal the properties of attention at work.

With respect to the auditory domain, auditory filtering is known to be reasonably effective. The most well-known phenomenon is the *cocktail party effect* (Cherry, 1953). Our auditory system is capable of focusing on a particular conversation while filtering out other conversations, music, etc. Interestingly, though, it appears that we do partially listen to other conversations as well. For example, it was shown that when one's own name is uttered in the another conversation, our auditory attention tends to temporarily shift toward the other conversation because the auditory pattern is highly relevant.

Cherry originally investigated such effects focusing on dichotic hearing. Participants were instructed to listen to a story that was played in one ear, while they were to ignore a second story that was concurrently being played in the other ear. While the participants typically barely remembered the semantic contents of the "ignored" story, and often did not even notice a switch from English to German in that story, other aspects such as the sex of the speaker as well as a shift in the tone of the voice were noticed. These and many other more recent studies have shown that while auditory attention can be directed toward a particular auditory source, other auditory signals can be ignored to a certain extent. A certain amount of stimulus processing, however, seems to occur in all circumstances regardless if attended or unattended.

Many other studies on attention and attentional processing have been conducted also for the tactile and other sensory domains. In all cases, it is typically found that attention can be directed toward particular information aspects. On higher cognitive levels, attention can be directed to whole objects and even groups of objects. When considering language and abstract thought, attention seems to be similarly selective. However, to be able to execute such higher-level, abstract attentional processes, suitably abstracted forms of encodings need to be available. In subsequent chapters, we consider attention in a more general context, where it seems to be relevant not only for stimulus selection and filtering, but also for selective, behavior-oriented information processing and for the selective, goal-oriented direction and focus of thought.

11.4   Models of attention

In the last sections we have gained various insights on different aspects of attention and their functionalities:

- Bottom-up attentional capture occurs when a particular feature is uniquely present or, to a lesser degree, when it is absent. Such features include not only color, edge orientation, size, edge types, and other static features, but also dynamic motion cues including motion direction and motion dynamics.

- Bottom-up salient stimuli need not to be completely unique in a scene, but unique in the local surrounding within which they occur.

- Very early in sensory processing, sensory stimuli are selected and filtered so that less salient, common feature-based stimuli are hardly recognized as long as top-down attention does not specifically focus on them.

- While a particular stimulus or entity is temporarily attended to, other competing, similarly relevant, or identical stimuli or entities may be overlooked. Attentional processing requires resources, which may disrupt the processing of other stimuli.

- Attention can be flexibly and intentionally directed toward specific locations, sensory features, objects, entities, and even collections of such entities. While top-down attention focuses information processing resources somewhat like a spotlight, the spotlight can also be divided to a certain extent, especially when integrative, higher-level grouping is possible.

In this section, we introduce several cognitive models that can explain aspects of the introduced properties of attention on several levels of understanding.

11.4.1 Qualitative models of attention

An important qualitative model of attention is the bottleneck theory of attention, which has been proposed in various forms and was modified in various ways. The bottleneck theory proposes that due to limited capacity information processes, a bottleneck of information processing exists, by means of which information can only be processed serially. Over the last century, this theory has been enhanced, questioned, and differentiated in various ways. Essentially, it could be shown that top-down attentional, control- and selection-oriented filtering can occur on various levels.

Early selection was recognized by a model of attention in Broadbent (1958), which was based on experiments on dichotic hearing during which two distinct auditory signals are presented to the right and to the left ear. The model characterized processes that filter early sensory information before any deeper, semantic analysis takes place. The model was also supported by the insights from Cherry about dichotic hearing, as mentioned above. Treisman (1964) modified the model in 1964 showing that early filtering processes can indeed be influenced by top-down, semantic processing mechanisms. In relation to the cocktail party effect, it was suggested that humans are able to listen to the person they are currently talking to, focusing on the sounds originating from the spatial location of the speaker while filtering most of the information from other sound sources.

These early selection processes were often contrasted with *late selection* processes, which allow to filter sensory information on a semantic level. Late selection processes were revealed in various studies and have led to formulations of several related theories of attention (Deutsch & Deutsch, 1963; Norman, 1968). Generally it can be stated that the bottleneck of attention processing can be observed on different levels of information and cognitive processing, where in all cases the involved attention can be characterized as selecting or deciding between alternative stimuli and their potential interpretations. Thus, attention filters what is transferred to some form of output modality – regardless if this modality is an actual motor action or a cognitive interpretation. A main problem of the characterization of attention as a filtering process is that it assumes purely sequential processing. Seeing, however, that various results show that parallel stimulus processing can take place to certain extents and that attentional resources can focus selection processes onto anticipated task-critical aspects, a pure filter theory of attention is inadequate.

Controlled parallel scheme

A particularly well-known enhanced model of attention is the *controlled parallel scheme* of information processing, put forward by Pashler (Pashler, 1998). Crudely speaking, the controlled parallel scheme proposes that to a large extent parallel information processing occurs in early stimulus processing. During this parallel processing stage, hardly any reduction in efficiency is detected. Depending on the task and current goals on "higher" levels, next, a selective analysis takes place, which is capacity limited. On this level, cognitive processing focuses on the attended stimulus aspects, which are expected to be behaviorally or cognitively relevant. Other aspects, on the other hand, are largely ignored. The attended stimuli are thus semantically analyzed, resulting in decision making and response selection. Many studies, which employed a *psychological refractory period* (PRP) paradigm (Welford, 1952), support the controlled parallel scheme. PRP assumes that parallel sensory processing is followed by a capacity-limited serial processing stage, during which decision making and response selection is assumed to take place, and which ends in a motor execution phase, which can again be executed in parallel with further sensory processing and serial processing.

Recently, however, it has been shown that depending on the task and dual task interferences investigated by means of PRP, the purely serial processing stage is not always as serial and capacity-limited as originally thought. Also, the motor execution and sensory processing stages do not always unfold fully independently in parallel. The gathered results suggest that the capacity bottleneck can occur at many levels of information processing, including visual feature levels, motor control components, spatial encodings, and even fully abstract, linguistic forms of encodings. When several tasks can be separated well, interference decreases despite the hypothesized capacity limitations. Moreover, motor execution components can be influenced by parallel visual processing and by further decision-making processes, and vice versa.

Attention may thus be viewed as a highly dynamic process, which flexibly focuses cognitive processing on particular information aspects and inhibits others (Freeman, Dale, & Farmer, 2011). When the currently activated selection, stimulus interpretation, and decision-making processes interact with each other – regardless of on which level – interference can occur (cf., for example, Segal & Fusella, 1970).

Feature integration theory

While the controlled parallel scheme mainly focused on capacity limitations, it remained rather fuzzy on where these limitations occur. Moreover, capacity limitations and thus filtering was typically assumed to occur locally at a particular cognitive information processing stage. In contrast to such a single stage interference view, the *feature integration theory* (FIT) (cf. Figure 11.6) proposes that attentional filtering can occur at multiple stages (Treisman & Gelade, 1980). At a so-called *preattentive stage*, FIT assumes that multiple, redundant visual features are processed in parallel, such as edges, color, size, texture, and motion, generally similar to the parallel processing in the controlled parallel scheme. This analysis takes place automatically and in parallel in disjunct feature spaces. Focal attention is needed in the integrative stage, during which the features from the individual feature spaces are compared with each other and bound into whole object percepts. These feature integration processes are assumed to require cognitive, attentional resources, such that it is impossible to integrate multiple independent objects in parallel.

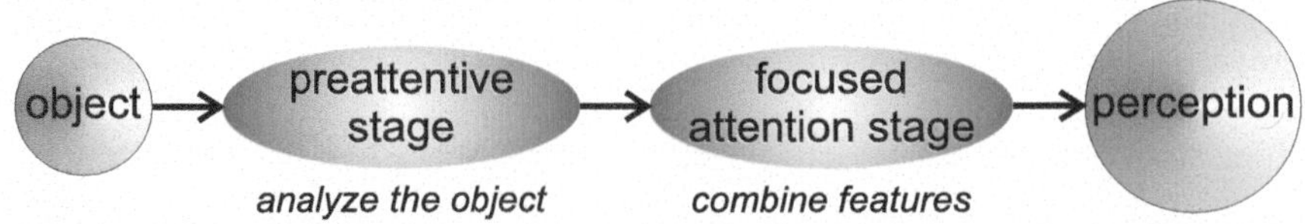

Figure 11.6: Focusing on object-oriented attentional processes, FIT proposes that vision extracts single features in parallel during a preattentive stage, while serial processes then bind the individual features together to enable object perception.

The advantage of FIT is that it is able to explain the popout effect and the linear increase in conjunctive visual search directly. The popout effect occurs because the individual feature spaces can be analyzed independently in parallel.   Serial search becomes necessary when feature conjunctions need to be analyzed such that no obvious feature or spatial separation is possible. Additionally, the preselection of a subset of features, which enables a directed search through a particular feature subspace, or of subareas, which are expected to contain a particular feature, is possible.

Integrated competition hypothesis

The very visually oriented FIT postulates that cognitive resources are necessary to focus attention on a particular object, binding the individual features together into a whole object percept. It does, however, not specify, how the focus of attention is directed toward particular bottom-up features. In subsequent work, FIT was enhanced further by projecting processes of selecting and binding individual features not only to object-oriented recognition processes, but to any integrative process that may lead to recognition or decision making. The *integrated competition hypothesis* (ICH) views attention as a competition for cognitive resources wherever an information integration process – such as a decision making or recognition process – currently needs to take place (Desimone & Duncan, 1995; Duncan, Humphreys, & Ward, 1997). This competition perspective is closely related to the *controlled parallel scheme* discussed previously.

ICH further details how cognitive resources may actually be distributed, given current tasks and goals. *Capacity sharing* distributes the limited resources among these tasks and goals, enabling simple parallel processing, such as executing routine motor actions and solving simple perception tasks in parallel. On the other hand, *time sharing* takes place when features need to be integrated for deciding between alternatives or accomplishing recognition. In this case, parallel processing is not possible, but time sharing is necessary. That is, the individual competitive processes share the available processing time between them, yielding the serial processing bottleneck of attention.

Interestingly, these two processes are assumed to possibly take place anywhere in cortical and possibly even subcortical areas (Duncan et al., 1997). In any such brain area, the authors assume that a competitive process between alternative "object" representations may take place, regardless of what kinds of individualized object-like encodings may be found in each of the hypothesized modules. Competitive selection, recognition, or decision making can thus take place in any cognitive encoding space available. Additionally, it is assumed that the modules in which competitions unfold are mutually interactive, such that a gain in activation of some object representation in one module also fosters a corresponding gain in related modules. Task-dependent neural priming is assumed to further influence this competitive, integrative process.

As a result, exogenous aspects and endogenous, task-oriented processes can influence the competitive processing stages. The stronger and more unexpected an exogenous stimulus, the more it will influence the serial processing stages in ICH. Similarly, the stronger the endogenous, top-down attention, the stronger the unfolding integration processes will be influenced by it. From this perspective, attention becomes part of the actual processes that are unfolding during perception and recognition, and even decision making and motor control. An object is perceived and recognized because relevant cognitive encodings are integrated into an overall percept, which is the internal encoding of the perceived object. Similarly, a behavioral or cognitive decision is made by integrating all decision-relevant aspects and converging to one decision. After a decision is made, the resulting behavior is executed, controlling one motor process, or the resulting train of thought is followed, controlling the thought along the lines of the decision.

11.4.2 Bundesen's theory of visual attention

The *theory of visual attention* (TVA) quantifies the qualitatively characterized competitive processes during feature integration. TVA determines which object in a scene may gather enough attentional resources to be recognized (Bundesen, 1990). It quantifies the current significance of a stimulus, integrating task-relevance with bottom-up salience in a probabilistic model. Recognition is influenced by the attention-dependent categorization of particular visual features, where the categorization depends on bottom-up evidence for, as well as top-down relevance of, particular features, feature categories, and subspaces. In addition, TVA assumes that categorized elements are available in short term or *working memory*, which has a limited capacity.   If this capacity is reached, categorized elements enter working memory by means of a stochastic process. The categorization itself unfolds in parallel, but competitively.

Under the assumption that an item $x \in \mathcal{S}$ of the set of displayed items $\mathcal{S}$ has not been categorized so far, the following equation calculates the probability $\nu(x, i)$ that this item is categorized into category $i \in \mathcal{C}$:

$$\nu(x, i) = \eta(x, i) \ \beta_i \ \frac{w_x}{\sum_{y \in \mathcal{S}} w_y} \quad \text{with} \quad w_x = \sum_{j \in \mathcal{C}} \eta(x, j) \ \pi_j \tag{11.1}$$

The probability is determined by the sensory evidence $\eta(x, i)$ for the item x to belong to category i, the sensory (bottom-up) bias β_i to categorize any item as belonging to category i, and the bottom-up salience weight w_x of item x relative to the sum of all salience weights of all items in the display $\mathcal{S}$. The salience weight is thereby determined by the sum over all categories $\mathcal{C}$ of products over the category-respective sensory evidences $\eta(x, j)$ with $j \in \mathcal{C}$ and the respective, task-specific pertinence (or relevance) π_j of category j.

Note how TVA enables the quantitative modeling of top-down, endogenous attention toward particular categories via π_j: the higher the prior bias for a particular category $j \in \mathcal{C}$, the more likely it is that a particular item x will be categorized into this category. Similarly, the more likely it is that first that item will be categorized into category j that has the highest sensory evidence for the particular category. Bottom-up, exogenous attention is modeled by category biases β_i, such that particular colors or shapes may be particularly distinctive, leading to attentional capture.   Moreover, attention is influenced by each particular sensory evidence $\eta(x, i)$ for each item $x \in \mathcal{S}$ with respect to each possible category $i \in \mathcal{C}$. For example, fuzzy displays or particularly exaggerated visual features may bias the categorization tendencies negatively or positively, respectively.

TVA not only enables the modeling of categorization processes, but it also models quantitatively the concept of attention as a resource to be competed for. This is particularly realized by the salience weights w_x, which depend both on top-down category biases and bottom-up category evidences. As a result, TVA enables the modeling of performance in experiments that are based on the visual search paradigm.   Even the modeling of iconic memory is possible by means of TVA (Bundesen, 1990) and a neural implementation of TVA has been proposed (Bundesen, Habekost, & Kyllingsbaek, 2005), which closely relates the computational mechanisms in TVA to neural processing pathways and areas in the brain.

Despite these advantages and its great promise, the theory is focused only on how items become temporarily encoded in short term memory, offering a quantitative theory of when an item actually enters short term memory. Actual dynamics, such as forgetting processes and other possible neural dynamics within short-term memory are not considered. Moreover, the storage space is assumed to be constant, although it is well known that the storage space is item- and set-specific. For example, our verbal working memory is characterized by a *phonological loop*, which emphasizes that pronunciation speed correlates with the number of items that can be maintained in verbal working memory. Lastly, the theory does not attempt to ground the parameters in actual sensory stimuli or to learn the parameters. In contrast, the following neural models of visual attention ground the parameters and evidences in actual sensory stimuli by means of information-theoretic approaches.

11.4.3   Saliency maps and eye saccades

A very well-known model of visual attention is the model by Laurent Itti and Christof Koch (Itti & Koch, 2001). It combines *saliency maps* of redundant bottom-up sensory information, focusing on early, bottom-up driven attentional processes. In their review article about the *Computational modeling of visual attention*, the authors stress the importance of these early selection processes, which nonetheless can be strongly modulated by endogenous, top-down attention, stating that:

> [...] although attention does not seem to be mandatory for early vision, it has recently become clear that attention can vigorously modulate, in a top-down manner, early visual processing, both in a spatially-defined and in a non-spatial but feature-specific manner. (Itti & Koch, 2001, p. 4.)

Reviewing the results and models of the available data about visual attention at the time, Itti and Koch emphasized that visual attention is predominantly influenced by the following four aspects:

- The *perceptual salience* of a stimulus mostly depends on the stimuli found in the close surrounding. These saliency computations result in specific *feature maps*, in which the features are encoded in a sensor-specific topology (for example, retinotopic). The feature maps essentially encode feature-specific saliences of locations with respect to their local surrounding.

- The combination of these feature maps is influenced by top-down biases and results in an overall, single *saliency map*, which encodes overall stimulus salience.

- *Inhibition of return* processes result in the local inhibition of locations that were recently attended to, such that saccades to recently fixated positions become highly unlikely.

- *Scene understanding* as well as *object recognition* processes additionally influence the selection of the subsequently fixated positions.

Apart from the capability to include top-down attentional influences, one major contribution of the model is the flexible combination of local, feature-based saliences. Figure 11.7 shows the main aspects of the attentional model:

- The input image is used to extract *low-level features* such as the orientation, intensity, color, motion, and other purely sensory-driven aspects. These features are topologically encoded.

- By considering center-surround differences in feature intensities encoded in the feature-specific topologies, and by introducing spatial competition, the low level features are converted into individual feature maps.

- The feature maps are then combined into one *saliency map*. The combination depends on training, such as knowledge about typical feature importances, and current endogenous, top-down influences about feature importances.

- The resulting saliency map is used to determine the next winning location (the one with maximal saliency) toward which the next saccade is directed. By means of the *winner takes all* principle, the selection process results in the actual fixation and realizes a resource competition process.

- In addition, the selection process is modulated by *inhibition of return*, which prevents saccades to recently fixated positions, as well as a distance-specific influence, such that shorter saccades are slightly preferred.

This model is related to several of the principles and models introduced earlier: first, the resource-based competition for visual attention is made specific by modeling saccades and fixations. The winner takes all mechanism essentially plays out the competitive process.

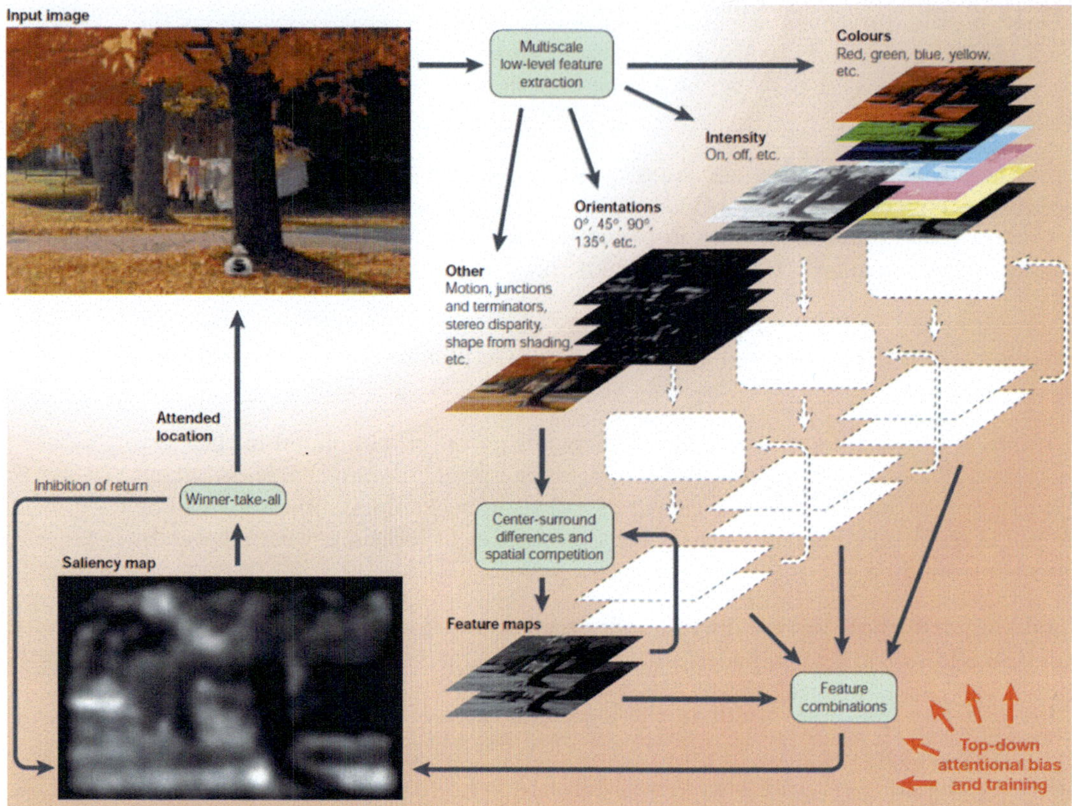

Figure 11.7: The model of visual attention based on Koch and Ullman. [Reprinted by permission from Macmillan Publishers Ltd: *Nature Reviews Neuroscience, 2,* 194–203, Computational Modeling of Visual Attention, Itti, L. & Koch, C. Copyright (c) 2001.]

Early feature selection processes are modeled by the top-down attentional biases and training effects, where top-down selection does not extend to the feature maps themselves. A connection to TVA must also be acknowledged, as the top-down biases may be related to the category specific top-down biases, which may differ with respect to the task. The saliency map then integrates the evidences and may thus be used to deduce item-specific evidences at the item-specific locations in the image. Finally, there is also a close connection with the Bayesian vision processing model, which we introduced in Section 9.4. In the Bayesian model, however, the computations were motivated by probabilistic information processing and focused on edge perception. Additionally, spatial biases were modeled more explicitly. Nonetheless, many of the features of both models are closely related.

11.4.4   Dynamic neural fields of attention

While the model by Itti and Koch focuses mainly on the construction of a saliency map and its control of saccades, another important aspect is the sustenance and dynamic redirection of attention onto particular subspaces. The *Dynamic Field Theory (DFT)* (Erlhagen & Schöner, 2002; Wilimzig, Schneider, & Schöner, 2006) offers a neural model of such dynamics. It implements several interactive neural fields, which are influenced by bottom-up saliences as well as by top-down, target-related activations and recurrent inhibitions, thus generating interactive dynamic fields.

These dynamic fields can represent feature- or object-spaces as well as size, direction, or location spaces, where the latter can be grounded in various frames of reference. The fields are represented by means of neural lattices, that is, neural fields or *neural population codes* (cf. also Chapter 10). The dynamics within in the model are encoded by differential equations, which denote how the activity $u(x)$ of a particular neuron in the population code changes dynamically over time t:

$$\tau \dot{u}(x,t) = -u(x,t) + h + S(x,t) + \int_x' \omega(x - x')\sigma(u(x',t))dx', \qquad (11.2)$$

where the change is influenced by the adaptation factor τ and the current activity $u(x,t)$, leading to self-inhibition, a resting activity h, the stimulus input $S(x,t)$, and the state of the surrounding, which is accomplished by the integral. The integral essentially integrates over all neurons in the field, weighing the respective activities according to their distance to the neuron x: $\omega(x - x')$ typically yields positive values in the close vicinity of x and negative values further away. Moreover, the sigmoid function $\sigma(u(x',t))$ yields positive values between zero and one – the closer to one, the larger the activity $u(x',t)$.

The result of this differential equation is that when there is no input at all from neighboring neurons, then the neural activity tends toward a resting activity h. Sensory input activity to x increases the activity. In the local neighborhood, neurons are reinforcing each other leading to local peaks in activations. However, larger neighborhoods mutually inhibit activities, leading to a dynamic competition of sufficiently distant, local activities. Overall then, the dynamic equation, depending on the exact implementation of ω and σ as well as on the parameterizations and the strength of the input, yields dynamic neural peaks of activations. These peaks can be interpreted as target selections, determining, for example, the next saccade or hand movement, but also as a converged interpretation of a stimulus, such as the recognition of an object or the localization of a particular item at a certain location. Once a peak has established itself, it can also be interpreted as a temporary bottleneck of attentional processing because all other activations in the particular neural field are temporarily inhibited. Due to the self-inhibiting neural dynamics, even bistable behavior can be modeled by neural fields, such as the bistable perception of the Necker cube (cf. Figure 3.1(a)). Figure 11.8 shows typical DFT dynamics, in this case associating retinocentric positions and gaze directions with a body-centered position code. Attention on two of these aspects yields to the derivation of the third aspect. Also an ambiguous or a bimodal state can be disambiguated by activating other associated information in other modalities.

DFT can be considered as a very general neural architecture of attention, working memory, sensory interpretation, perception, and even coordinate transformation, information exchange, and decision making (Sandamirskaya, Zibner, Schneegans, & Schöner, 2013). By associating various frames of reference with each other, sequential as well as parallel processing, efficient information exchange, and various forms of stimulus abstractions can be modeled. Accordingly, DFT has been used to successfully model various perceptual, motor, sensorimotor, and higher-level cognitive processes. The current drawbacks, which are drawbacks of most architectures considered herein, are that no learning or neural connectivity adaptation takes place. The neural fields and the connections within and between the neural fields are typically hard-coded. Self-regulating learning processes have not been implemented so far and pose a grand challenge to all the models of attention.

11.5 Summary and outlook

Processes of attention are essential for cognition. Without attention it is impossible to ignore irrelevant aspects of information, to actively focus on the relevant aspects, or to selectively bind relevant information sources together. Thus, attention is essential to solve the *frame problem* as well as the *binding problem*. While we have mainly focused on the sensory processing aspects of attention, several times we have also hinted at the fact that attention is not only relevant for perception, but it is probably even more relevant for decision making and for the control of motor behavior.

Generally speaking, attention is a cognitive processes that bundles cognitive resources to come to a form of decision, whether it is *recognition*, that is, deciding that an item is

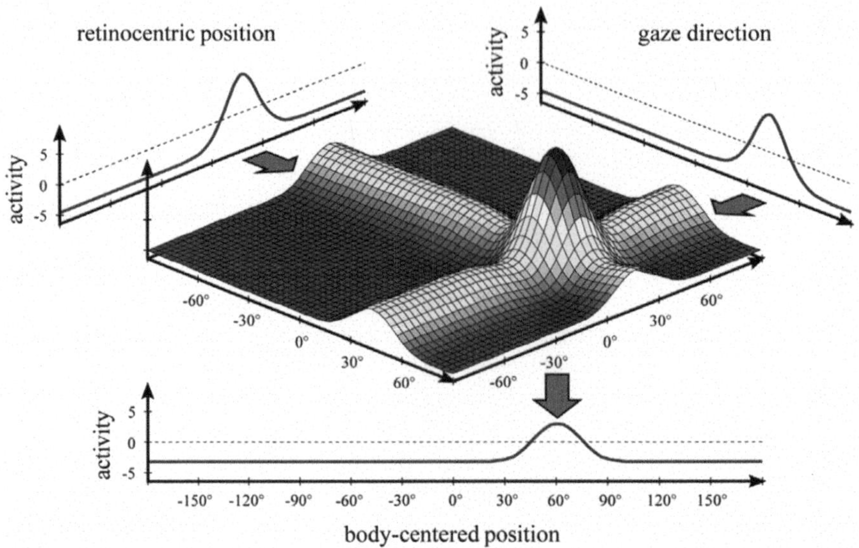

Figure 11.8: The activation of this illustrative, motion direction-sensitive dynamic neural field develops over time: a preparatory signal activates two potential motion directions. Once the response signal is perceived, the preparatory, ambiguous prime is quickly disambiguated leading to the actual motion response. [Reprinted from *New Ideas in Psychology, 31*, Sandamirskaya, Y., Zibner, S. K., Schneegans, S. & Schöner, G., Using Dynamic Field Theory to extend the embodiment stance toward higher cognition, 322–339. Copyright (2013), with permission from Elsevier.]

of a particular kind, a *behavioral decision*, that is, deciding to execute a particular action or interaction due to the presence of particular stimuli, objects, or entities, or a *mental decision*, that is, deciding to pursue a particular train of thought.

We have focused on phenomena and insights gathered from various research studies and models on visual attention. Although visual attention has been metaphorically compared to a *visual spotlight*, this spotlight should not be taken too literally. In fact, it appears that attention can be focused not only on sensory features and sensory subspaces, but also on more abstract encodings, such as whole objects.

The *bottleneck of attention* characterizes the fact that it is very hard or nearly impossible to recognize several particular entities or to decide on several particular actions in parallel. The bottleneck was originally closely related to the PRP paradigm, which postulates that recognition, decision making, and response selection processes can only be accomplished in a serial manner. However, this perspective has been challenged multiple times. In particular, it was shown that additional interactions between serial and parallel processes can occur and that skilled parallelizations of typically serial processes can be accomplished (as, for example, in skilled typing). In more general terms, it may be stated that attention is a process that selectively integrates features into a percept, a decision, or an action, while it filters-out other features. During the integration process, the involved cognitive resources prevent the integration of the currently considered features into other percepts or decisions.

The put-forward computational models of attention show how attention is influenced by top-down, endogenous, task- and goal-specific biases well as by bottom-up, exogenous, feature specific, salient environmental aspects. Bundesen's TVA model focuses on the competitive encoding of items into working memory. The saliency map model of Itti and Koch illustrates how bottom-up features can result in a competitive process for visual attention. Due to the top-down influenced integration of individual feature maps, top-down influences are accounted for as are inhibition of return processes. As a result, overt visual attention could be modeled, which predicted successive eye fixation locations. Finally, Dynamic Field Theory focuses on the dynamic neural interaction processes that unfold during cognitive processing, offering a model that has been applied not only to visual attention problems, but also to cognition and motor control.

Overall, the characteristics of attention show that our cognitive apparatus can be and usually is very selective in what information about the world is processed in detail. "The world is its own best model" seems to be used as a general principle by our brain. As a consequence, the brain continuously attempts to focus its cognitive processing resources on those aspects of the environment that currently seem most behaviorally and cognitively relevant. Thus, attention continuously strives to solve the *frame problem*, focusing on those aspects of the environment that are believed to be most relevant for making good cognitive and behavioral decisions.

11.6 Exercises

1. Discuss why the frame problem seems unsolvable without mechanisms of attention.

2. Seeing the highly nonlinear distribution of rods and cones on our retina (cf. Section 8.3), why may it be correct to say that visual attention is to a certain extent – or even necessarily – "morphologically grounded"?

3. Relate the spotlight metaphor to the Bayesian model of visual processing, which was introduced in Section 9.4). How can the results of Duncan (1984) be explained with the help of this model? What does it imply for the nature of the "spotlight"?

4. What do studies on visual search reveal about the nature of bottom-up attentional processes?

5. The attentional blink and repetition blindness indicate that the perception of a particular stimulus may temporarily block the processing of another related stimulus. In which sense are the two phenomena related? In which sense do they differ?

6. Change blindness indicates that our top-down generative visual expectations are not as crisp as one may think introspectively. Change detection occurs only when sufficient attention was put on the item or property of an item that changes. In which manner do these observations support the statement that "the world is its own best model"? Why is change blindness usually not a problem in the real world?

7. Name cases in the real-world where inattentional blindness may lead to negative consequences.

8. In which manner do the reported results on inattentional blindness indicate that attention can be more or less focused. Moreover, to which extent do the results indicate that the top-down attention of the participants did focus on both, visual features, such as colors, as well as on the individual persons and the ball in the video, tracking them over time?

9. Contrast early and late selection processes in visual attention.

10. In which manner characterizes the feature integration theory early and late selection processes in further detail?

11. Recall the bistable perception of the Necker cube and of the Rubin's vase (cf. Figure 3.1(a) and3.1(b)).

12. Shortly explain the parameters in Eq.(11.1) of TVA to calculate categorization probabilities and relate them to bottom-up and top-down mechanisms of attention.

13. Koch and Ullman's model of visual attention integrates top-down and bottom-up mechanisms into saliency maps. Can these top-down biases in principle be either spatial or feature-oriented? In which way does the model compute saliencies in the individual feature maps?

14. Explain the neural dynamics unfolding in dynamic neural fields over time by means of the differential equation (Eq 11.2). How can DFT be used to convert a retinotopic position into a body-centered position?

Chapter 12

Decision Making, Control, and Concept Formation

12.1 Introduction and overview

In the last four chapters, we have seen that sensory processing mechanisms start with a large set of partially redundant and complementary sensory sources of information. The redundancy and complementarity comes from the available, multiple sensory modalities as well as from the different kinds of information, which can be extracted from the same modality. Given these redundant and complementary information sources, the brain then abstracts over these sources forming and integrating more complex stimulus encodings.

Information is integrated within a modality and between modalities, generating integrated wholes. Along the *ventral pathway*, multisensory areas encode entity identities with all their perceptual properties, enabling entity recognition. Along the *dorsal pathway*, the focus lies on spatial environmental properties, including body- and object-relative locations and orientations. During multisensory integration, sensory information and motor information sources and the consequentially expected perceptual changes (that is, predicted *reafferences*) are considered, further optimizing spatial estimations and entity encodings. Attention focuses sensory and sensorimotor processing on those aspects that are currently most behaviorally relevant, binding and fusing them selectively task- and goal-oriented.

When interacting with our environment, our complex body with all its muscles needs to be controlled – once again typically task- and goal-oriented. Even eye saccades, which we had considered in the previous chapter on attention, may be considered as a particular form of motor control. In fact, attention and motor control are closely related; however, when controlling the whole body – instead of just the eyes when executing a saccade – matters become more complex. The body is influenced by inertia such that the brain needs to be able to cope with, control, and stabilize the body's dynamics. Moreover, our body is full of redundant degrees of freedom, which enable us, for example, to reach a certain point in space with different extremities as well as with different postures. In this way, *redundant behavioral alternatives* offer themselves while interacting with the world.

Planning and decision-making processes need to resolve these alternatives, choosing, if possible, the alternative that is optimal given certain criteria of optimality. When more complex sequences of actions are needed to reach a certain goal, progressively more abstract reasoning processes need to be involved. Additionally, subgoals need to be maintained in working memory and need to be striven for sequentially.

To enable effective reasoning and decision making, however, suitably abstracted encodings need to be available. Similar to abstract forms of sensory encodings, which lead to object-specific and spatial encodings, our brain has developed abstract, *interaction-specific encodings*, which enable the flexible execution of complex interaction sequences.

In the chapter on reward-oriented behavior (Chapter 5), we saw that *dynamic movement primitives* (DMPs) (cf. Section 5.4.4) are suitable for encoding such abstract, interaction-specific encodings. Once DMPs and combinations of DMPs, or generally speaking *motor primitives* and *motor complexes*, are available, higher-levels of planning and decision making do not need to worry about the actual implementation of an interaction. Rather, it suffices to focus on *conditional encodings*, which specify under which circumstances a particular DMP is typically successful, and on *effect encodings*, which specify the typical final consequences when executing a particular DMP.

Note how such schematic forms of interaction encodings are useful in applying *hierarchical, model-based reinforcement learning* and higher-level planning in general (Konidaris, Kaelbling, & Lozano-Perez, 2014, 2015). Moreover, when the conditional encodings only specify the *relevant* circumstances and the effect encodings only specify the *actual final* action effects, *factorized representation* become available. That is, representations are learned that specify (i) *conditional encodings* of those factors that are relevant to execute a certain action and, (ii) *effect encodings* of those factors that are affected by the action. In Section 5.3.3 we showed that such factorized state representations are suitable for propagating reward by means of *factored reinforcement learning* and thus for planning self-motivated and goal-oriented (cf. Section 6.5).

From a cognitive perspective, these conditional and effect encodings can be considered as leading to behavior-oriented *conceptualizations* of the environment. For example, the concept of something being "graspable" can be equated with an encoding that specifies that an object is in arm range and is openly accessible. Even simpler, the concept of a "thing" develops as something that can block a path or that can be moved around. As a last example, the concept of a "container" can develop as an entity that can be manipulated in certain ways to extract another entity or other entities within it or out of it (for a baby, the first container experience may be the own mouth and, later, drinking from a baby bottle). Thus, the *frame problem* (Section 3.4.2) can be solved to a certain extent, because behavioral considerations focus on those aspects of the environment that are actually influenced by behavior, ignoring irrelevant aspects.

To learn suitable schematic encodings, another principle becomes relevant: *event encodings* (Zacks & Tversky, 2001). Behaviorally speaking, an event may be characterized by an unfolding behavior, which has a beginning and an end. The beginning typically coincides with behavioral movement onset and is characterized further by conditional encodings. Similarly, the end coincides with the behavioral offset and can be characterized further by the achievement of schematic effect encodings. For example, when grasping a mug, behavior starts when the hand moves toward the mug and conditional encodings specify, for example, that the mug is located in a reachable distance and that it is graspable (not blocked by other objects or entities). The grasping event then unfolds until the hand closes around the mug and establishes contact and object control. The final effect is that the mug is held by the hand, that is, the hand feels the grip by means of the pressure feedback on the fingers and palm and, when starting to lift the object, the object's weight is perceived by "heavier" sensorimotor feedback (Roy, 2005b, 2005a). Given that the mug is a transportable object, "transportability", that is, the prediction that the object will move when pushed, pulled, or lifted, and "manipulability", that is, the prediction that the object will change in particular manners when interacted with by particular motor complexes, may be specified as final effects.

Note, however, that in the general sense behavior can be comprehended as any type of system behavior. For example, a behavior of a physical system is "raining", which can be interpreted as a behavior that is generated by the clouds. Similar, a behavior of a cognitive system, such as our brain, is "thinking", which can be interpreted as a behavior that is generated by attention or, generally speaking, by cognitive mechanisms.

To detail this behavior-oriented abstraction-toward-conceptualization perspective, we will first give an overview of the current knowledge about how the brain accomplishes motor behavior. We then consider how decision making unfolds and how the brain appears to abstract over actual behavioral control mechanisms, which are invoked selectively once a particular behavior is executed. To form suitable abstractions, schematic, *behavior-oriented event encodings* need to develop, which specify (i) when a behavior can be executed, (ii) which online changes and (iii) which final changes the behavior causes, and (iv) when the behavior typically ends. Given such behavior-, control-, and decision-making oriented event encodings, we take a look at how planning, decision making, and reasoning processes can unfold in a self-motivated, goal-oriented manner. In fact, we will see that such schematic encodings offer themselves for both, intricate planning of behavioral sequences in the real world, as well as abstract reasoning in hypothetical worlds. We also further detail the conceptualization aspects inherent in the development of event encodings. Finally, we relate these conceptualizations to the symbol grounding problem and to language.

12.2 Compositional motor control in the brain

Motor control in our brain is accomplished by a cascade of control loops and control processes, starting with the muscles and going up to deeper, less modality-influenced cognitive levels in the cerebral cortex. On a rather low level, redundant motor encodings are activated, essentially resulting in the directional behavior of particular limbs or other body parts (cf. Figure 12.1). These motor primitives are then combined into motor complexes – such as a reaching, grasping, typing, or walking. Motor complexes are flexibly invocable and adaptable to the current context. Thus, similar to hierarchical sensory processing, motor processing interactively unfolds modularly and hierarchically.

Looking back at the basic principles of bottom-up perceptual processing (cf. Chapter 8), several fundamental principles should be reconsidered:

- Most basic sensory processing is normalizing and smoothing the raw sensory input.

- Neural activations are determined by a weighted sum of sensory input activations, for example, during feature extraction processes.

- Stimuli are encoded redundantly by feature maps with local receptive fields, where the feature maps cover a particular sensor-grounded topology.

- Sensory processing is hierarchically structured beginning with simple sensory signals, which become progressively abstracted and more complex feature or spatial detectors.

Many lines of research suggest that similar principles also take effect in motor control (cf., for example, Herbort, Butz, & Hoffmann, 2005; Poggio & Bizzi, 2004):

- Motor activity, which is invoked by motor neurons, is self-stabilizing and contains basic, behavior-smoothing feedback loops.

- Bodily motion is accomplished by means of a weighted linear combination of directional or postural neural activities.

- Motor control structures are encoded redundantly and in a distributed manner, enabling the flexible invocation of behavioral alternatives.

- More complex behavior is encoded hierarchically, where motor complexes are encoded in an abstract manner, leading to the unfolding of a complex behavioral routine over time; this routine can be flexibly spatially modified to the current circumstances – such as to the surface orientation on which a manipulation unfolds, or to the particular location and orientation of an object.

Critical for the effectiveness of these modular and hierarchical structures seems to be an important additional point, which is often referred to as *compositionality*. As Poggio and Bizzi put it:

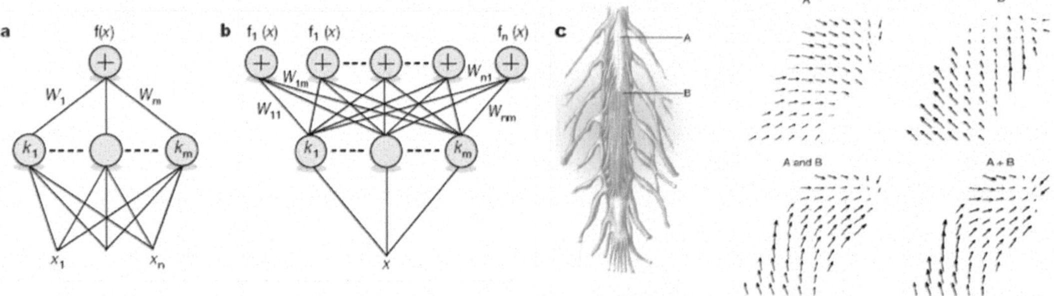

Figure 12.1: In visual perception (a,b) a visual feature is often considered to be computed by a weighted sum of sensory input activities. In motor control (c), the output signal can be considered to be generated by a linear combination of directional encodings, stemming from the spinal cord. The result is a combination of directional motion fields (A,B), where the concurrent stimulation of both fields results in an additively combined motion field (A+B). [Reprinted by permission from Macmillan Publishers Ltd: *Nature*, *431*, 768–774, Generalization in Vision and Motor Control, Poggio, T. & Bizzi, E. Copyright (c) 2004.]

> [...] roughly speaking, the issue is about compositionality [...]: neuroscience suggests that what humans can learn – in vision and motor control – can be represented by hierarchies that are locally simple. Thus, our ability to learn from just a few examples, and its limitations, might be related to the hierarchical architecture of cortex. (Poggio & Bizzi, 2004, p. 772.)

In other words, compositionality refers to structures that are locally simple, but that can be combined in a partially modular, partially hierarchical manner. For example, as we have seen, the brain partitions the visual recognition of an object, crudely speaking, into (i) a *spatial encoding* of the current object's position and orientation in space relative to the observer and (ii) an *identity encoding* of the object's typical perceptual properties. Thus, location and identity encodings are individually rather simple, but flexibly, compositionally combinable.

Motor control exhibits similar properties. For example, we can grasp the handle of a mug under various circumstances and coming from various directions. A grasp template may specify a general grasp motor complex, which may specify specific hand postures for mug handles. However, the orientation of the handle relative to oneself as well as the size and precise shape of the handle may lead to adjustments in the approaching arm and hand before and while reaching, and in the shape of the hand and fingers before and while grasping (Belardinelli et al., 2015; Belardinelli, Stepper, & Butz, 2016). According to the visual-motor analogy, object identity templates may be related to grasp templates, which the object affords (Cisek, 2007; Gibson, 1979), and object locations, orientations, and sizes may determine the actual parametrization of the considered or selected grasp template.

To further explore this analogy, we now consider actual basic motor control structures and their interactive encodings in the brain. We then also consider several modeling approaches, which shed further light on how motor control may unfold, and which exhibit the challenges that need to be solved to ensure successful action decision making and execution.

12.2.1 Muscles and the spinal cord

Starting at the bottom, any interaction with our environment (except for perspiration) is the result of the contraction of particular muscles and groups of muscles. These include, for example, the movement of a limb, such as an arm or a hand during object manipulation or the legs during locomotion, as well as the contraction of the diaphragm while breathing, and even the interplay of diaphragm, tongue, larynx, and facial muscles while speaking.

Many sensory systems are directly coupled with particular groups of muscles. For example, tiny muscles in the inner ear and middle ear determine the orientation of the auditory ossicles, by means of which we are able to filter and modulate sound perceptions. Similarly and more obviously, our eyes are controlled by a set of muscles, including outer and inner eye muscles, which are responsible for microsaccades, normal saccades, eye fixations, as well as the accommodation and regulation of the pupil.

Despite their different sizes and general functionality, all these muscles function based on the same general, evolutionary-determined principles: muscles consist of bundles of muscle fibers, which are muscle cells that in turn consist of many chains of myofibrils (Figure 12.2). When we contract a muscle, the thick and thin myofilaments in the myofibrils slide along each other, leading to the shortening of the muscle's length.

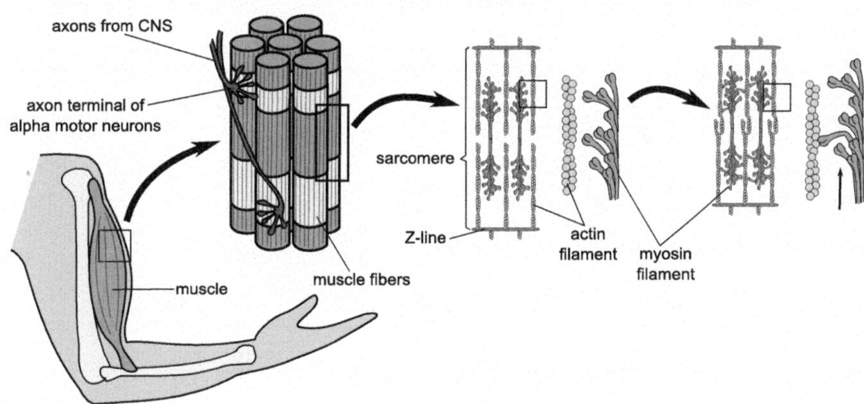

Figure 12.2: The spinal cord is the crucial hub between muscles and the brain. Sensory afferences from the muscles as well as motor efferences to the muscles are systematically wired. Moreover, local muscle control mechanisms are realized by means of low level neural feedback loops. [Adapted with permission from Mark F. Bear, Barry W. Connors, Michael A. Paradiso, Exploring the Brain, 3rd Edition, (c) Lippincott Williams and Wilkins, 2007.]

Often groups of muscles interact in a self-stabilizing manner, forming *muscle synergies*. Particular muscle pairs, which are often referred to as *antagonist* and *agonist*, stabilize each other. For example, biceps (*musculus biceps brachii*) and triceps (*musculus triceps brachii*) control the flection and extension of the elbow joint. While the biceps contracts, for example, the triceps typically relaxes and vice versa – but only to the extent that the flection does not become overly strong or even uncontrolled. Moreover, by contracting both muscles, the stiffness of the controlled joint can be flexibly increased. Thus, a joint can be positioned in an arbitrary pose and with a variety of stiffnesses, simply by contracting and relaxing a synergistic group of muscles.

In addition to morphologically intelligent muscle arrangements, simple self-stabilization mechanisms are first handled by neural self-stabilization mechanisms within the peripheral nervous system, the simplest of which is realized by *muscle spindles*, which are sensory receptors in the heart of a muscle that provide information about its current length. These muscle spindles are positioned between the extrafusal and intrafusal muscle fibers. α-motor neurons innervate the extrafusal fibers, while γ-motor neurons focus on the intrafusal ones. When the α-motor neurons are activated, the extrafusal fibers are contracted, which leads to a relaxation of the muscle spindle so that it cannot send any length information. However, when γ-motor neurons are activated, the intrafusal fibers contract so that the muscle spindles remain sensitive. As a consequence, α- and γ-motor neurons are typically coactivated to maintain sufficiently accurate muscular stretch and length signals. The coactivation of the two motor neurons essentially leads to self-stabilization and the maintenance of particular muscle stretches – enabling the continuous control of force and position.

While there are several more types of neurons in muscles and their interaction is more intricate than described, it is important to note that individual muscles contain mechanisms that are self-regulatory. Particular reflexes and rhythmic motions, such as movement of the legs during locomotion, can be directly generated via the spinal cord structures. Thus, muscles, their morphology, and the arrangement of muscle groups, as well as the spinal cord, and self-regulating neural loops within it, offer a motor system to the rest of the brain, which is well structured and much easier to control than individual muscle fibers would be. As a result, the control challenges, which the (rest of the) brain and the neocortex in particular have to solve, become simpler because basic, fast, self-stabilization mechanisms, as well as length and force maintenance mechanisms are generated by the morphology of and the local neural wirings within the muscles and the peripheral nervous system.

However, the muscles also continuously communicate with the rest of the brain via the spinal cord sending *proprioceptive feedback* about, for example, their current length and muscle tension. The sensorimotor pathways are very well-organized and have a compartmentalized structure (cf. Figure 12.3). Ascending, afferent pathways signal sensory feedback about the state of the muscles and joints. Descending, efferent pathways yield muscle activity invocations, that is, directional fields of motion (cf. also Figure 12.1). The control challenge faced by the central nervous system is thus not only simplified, but also modularized in that motor neurons of particular muscle groups can be activated by local neural activations, facilitating the activation of muscle groups and consequent synergistic motor behavior (Latash, 2008).

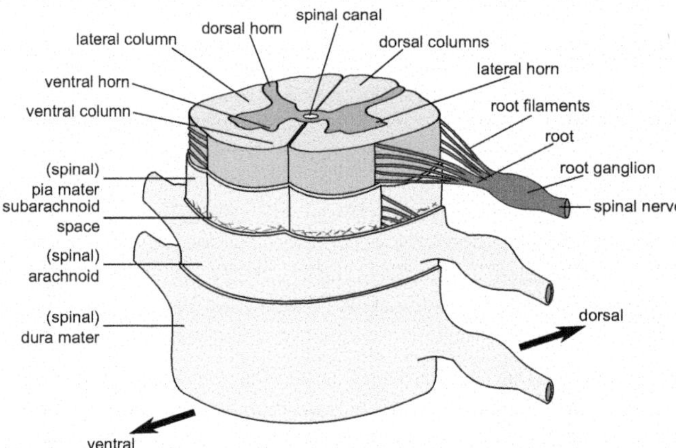

Figure 12.3: The spinal cord is the crucial hub between muscles and the brain. Sensory afferences from the muscles as well as motor efferences to the muscles are systematically wired. Moreover, local muscle control mechanisms are realized by means of low level neural feedback loops.

12.2.2 Motorcortex and beyond

From the spinal cord, afferent feedback from the muscle control loop and efferent control signals to the muscle control loop interact with the cortex – typically via particular thalamic nuclei. In the cerebral cortex, the main motor control areas is the motor cortex. It is responsible for the control of voluntary behavior, which can be a simple directional movement or a highly complex execution of a motor complex.

Functionally and histologically it can be partitioned into the:

- *Primary motor cortex*, which has a systematic, muscle complex-oriented organization (cf. the illustration of the motor homunculus in Figure 7.12) and sends motor signals via the efferent connections of the spinal cord to individual muscles and groups of muscles.

- *Secondary motor cortex*, which contains various additional compartments. It is believed to be mainly involved in goal-directed action encoding. Two of the main compartments are:

  - The *premotor cortex*, which is found anterior to the primary motor cortex. It is typically further partitioned into four subareas. These distinctly encode reaching motions, grasping motions, guided reaching, and other functionalities. Even slightly more anterior, the frontal eye field (FEF) can be found, which control saccades.

  - The *supplementary motor area (SMA)*, which is the most dorsal, medial part of the secondary motor cortex. It is believed to be mainly involved in the control of behavioral sequences and complex movement patterns.

Figure 12.4 illustrates this partitioning for the motor cortex of macaque monkeys as well as a relation of areas involved in motor control and particularly grasping in the human brain.

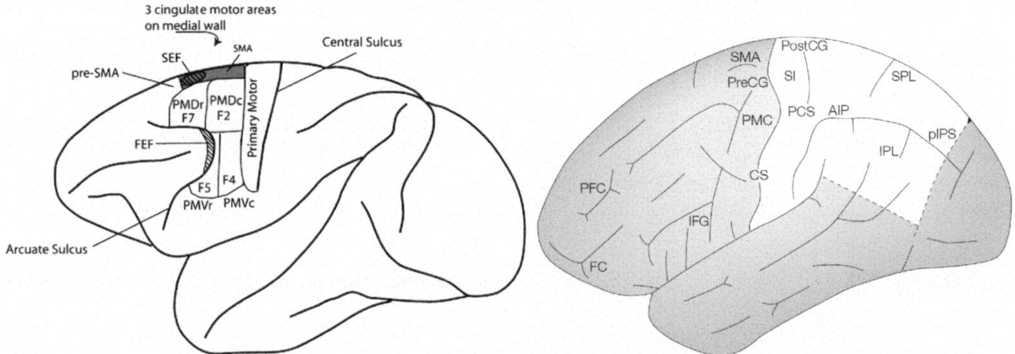

Figure 12.4: Investigations of the motor cortex of monkeys suggest that systematic, behavior-oriented partitionings can be identified. While the primary motor cortex shows a muscle synergy-oriented topology, the secondary motor cortex is believed to be strongly involved in the initiation and control of more complex, goal-oriented actions. [Reprinted from *Neuron*, 56, Graziano, M. S. A. & Aflalo, T. N., Mapping Behavioral Repertoire onto the Cortex, 239–251. Copyright (2007), with permission from Elsevier.] Similarly structuring are believed to be found in the human motor cortex (right). The human brain image shows additional critical areas, which are believed to be crucially involved in motor decision making and control. [Reprinted by permission from Macmillan Publishers Ltd: *Nature Reviews Neuroscience, 6*, 726–736, The neuroscience of grasping, Castiello, U. Copyright (c) 2005.]

Actual decision making is believed to be realized in the prefrontal cortex, which mediates between the motor cortex and other cortical areas. In particular, it appears to integrate information from the posterior parietal cortex, posterior temporal cortex, and motor cortex.

We already related the posterior parietal cortex to multisensory, spatial encodings with respect to peripersonal spaces (cf. Section 10.2). The dorsolateral prefrontal cortex is involved in evaluating entities in the environment for their desirability and chooses the particular entity with which a current interaction is initiated. Figure 12.5 shows these interactions. Table 12.1 lists the most important components, which are involved in the coordination of action decision making, initiation, and control. The sketch and table are certainly highly simplified and are intended to give a first, very crude and abstract idea of the actual modularity and complexity when considering neural behavioral decision making and control.

Areas in the premotor and parietal cortex of macaque monkeys have been mapped onto particular action complexes (Graziano, 2006). When moving from superior to inferior areas, action complexes for climbing, reaching for an object, moving the hand to the mouth, manipulating the space in front of the body with the hand, and the execution of defensive

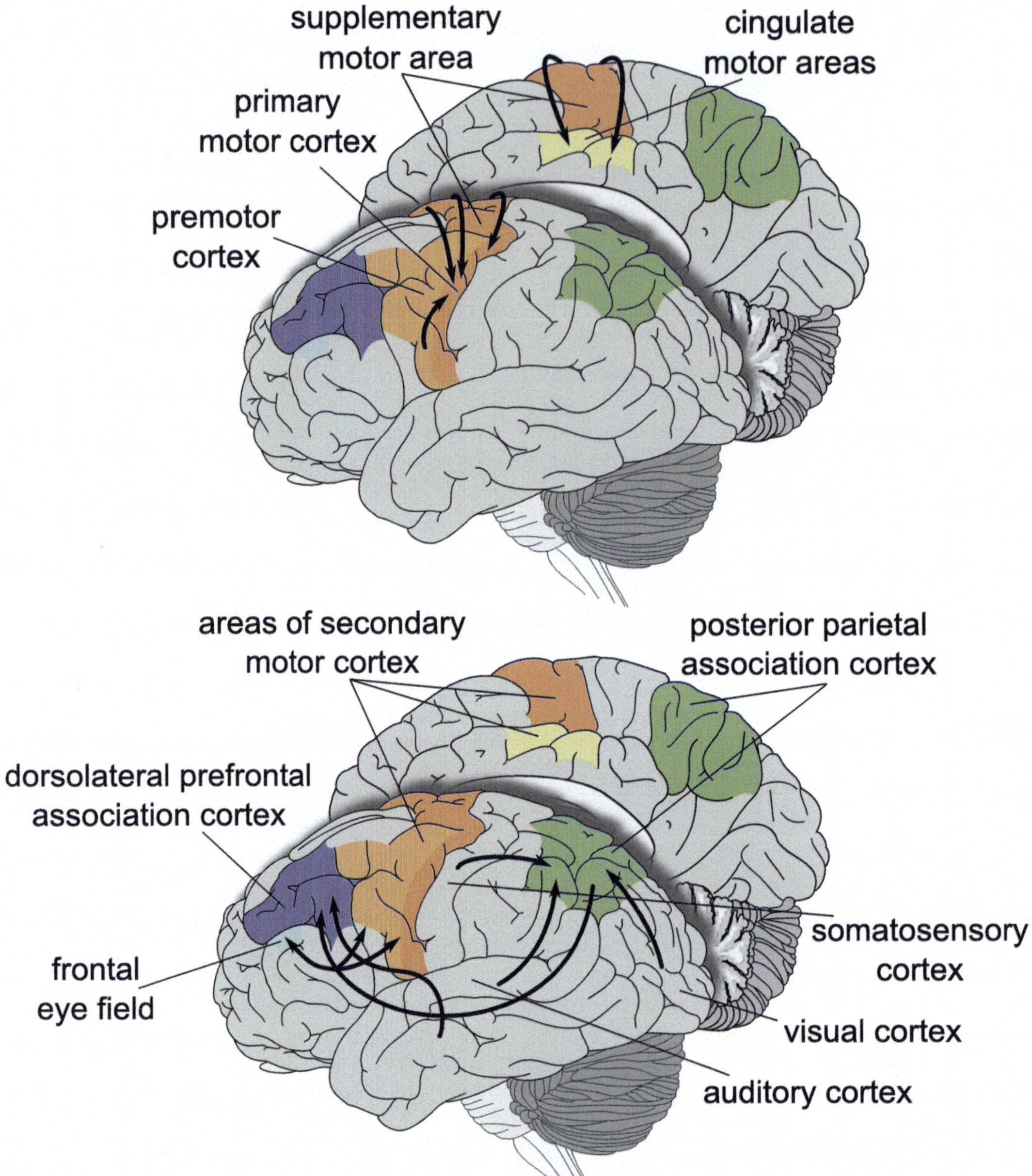

Figure 12.5: The dorsolateral prefrontal cortex interfaces motor cortex with other cortical areas, including parietal and temporal areas. It is involved in motor decision-making processes. As shown, a whole network of interactive areas is involved in motor decision making and control.

motions have been identified (Aflalo & Graziano, 2006). Not surprisingly, these areas overlap with the motor cortex homunculus to a certain extent in that the body parts involved in a particular action complex are neurally close to that action complex encodings. These insights come mostly from the microstimulation of neurons in motor cortex and parietal cortex (Aflalo & Graziano, 2006; Graziano, 2006). However, also TMS studies with human subjects do exist (cf., for example, Gentner & Classen, 2006), where particular motor cortex stimulations led to the generation of particular arm and hand postures. According to a study with single-cell electrode stimulations in human patients (Desmurget, Reilly, Richard, Szathmari, Mottolese, & Sirigu, 2009), premotor cortical stimulations invoked felt or actual

Table 12.1: Important neural wirings from and to motor- and prefrontal cortical areas (highly simplified).

Areal	Input	Output
Primary motor cortex	PMA, SMA, feedback from output areas	thalamus, spinal cord, basal ganglia, cerebellum, sensory input areas
Secondary motor cortex	Prefrontal cortex, primary motor cortex	same as primary motor cortex
Posterior-parietal cortex	visual cortex, auditory cortex, somatosensory cortex, dorsolateral prefrontal cortex	frontal eye field, secondary motor cortex, dorsolateral prefrontal cortex
Dorsolateral prefrontal cortex	posterior-parietal cortex, inferotemporal vortex	motor cortex, frontal eye field, posterior parietal cortex

behavioral executions – such as lifting the arm – while parietal stimulations led to reports of a current behavioral intention or even the belief that a particular action had just been executed.

Similar mappings in the parietal and premotor cortex have been identified with respect to eye saccades. The frontal eye field is involved when saccades are controlled and when a particular stimulus is fixated. The lateral intraparietal area (LIP), which we discussed in Section 7.4.2, is believed to be involved in planning the successive fixation of locations as well as in maintaining focus on the currently fixated stimulus (Patel et al., 2014).

The cerebellum and the basal ganglia are also decisively involved in motor control (Shadmehr & Krakauer, 2008) (cf. also Figure 10.5). The cerebellum is believed to be mainly involved in the generation and activation of sensorimotor forward models of particular motor behavior. The forward predictions are used for direct behavioral control, for the fluent, automatized execution of dynamic behavioral complexes, and for sending predictive information to the neocortex, where it is integrated with sensory feedback. The cerebellum thus appears to provide crucial online forward predictions of the immediate sensory reafferences caused by motor actions (cf. Chapter 6). While the reafferences are well predicted, execution can unfold smoothly because no significant error signals are registered. For automatized control, the cerebellum may temporarily substitute actual sensory feedback, thus enabling a very fast, open-loop control of highly trained behavioral complexes, such as walking, riding a bicycle, shooting a ball, writing, or typing.

The basal ganglia modulate the motor output and appear to coordinate the execution of succinct behavioral complexes. They were also shown to be involved in issuing motivational reward signals. Accordingly, they were related to being involved in reinforcement learning in concert with the hippocampus, by, for example, providing reward gradients during an inverse sharp-wave ripple for remembering *eligibility traces* (cf. Section 5.3.3 and Section 10.4.1). In addition to the learning aspect, the basal ganglia seem to also be crucially involved during motor executions: in patients that suffer from Parkinson's disease neurons in the basal ganglia that produce dopamine are dying off, leading to progressively weaker neural activities. As a result, patients suffer from progressively severe muscular rigidity, muscle tremor, and postural instability, apparently lacking the necessary dopamine rewards to maintain full body control.

In sum, similar to sensory processing areas, motor decision making, and control areas in the brain can be separated into various, interactive modules. Particular behavioral complexes, such as reaching, climbing, or eating are mainly controlled by local neural clusters in the premotor and motor cortex as well as in the parietal cortex. Parietal areas additionally appear to encode the world spatially – probably because relative spatial frames of reference are highly suitable for planning and controlling particular behavioral environmental interactions. Besides these interactive cortical encodings, subcortical areas are also involved, forming an interactive information processing hierarchy.

While it had been believed until recently that the cortex sends motor commands to the spinal cord and thus the muscles in a feed-forward manner, this belief has been replaced by a much more interactive, modularized, and hierarchical architecture. Higher levels in this architecture control lower levels, but also take into account the feedback from these lower levels to, for example, change the currently applicable spatial mappings and other higher-order motor complex parameters (Graziano, 2006). As a result, motor complexes can be invoked and flexibly adjusted in a compositional manner to the current circumstances, such as the one's own current body posture or the actual position and orientation of the targeted object. While grasping an object, for example, interactive and partially overlapping modules control the unfolding hand postures, the arm trajectory, the speed of the movements, the grip force, and other movement aspects. In the following sections, we consider how these modules and hierarchies may actually work computationally, considering also particular models of motor control where applicable.

12.3   Computational motor control

Over the last two centuries, progressively more refined models of motor control have been developed. Two model types can be contrasted: one type progressively focused on the control of executing a particular motor control routine; the other type considered how particular motor parameters, such as the final posture of an action, may be selected.

In the following, we focus on goal-directed object manipulations and, to keep things simple, we first consider motions to particular locations. In this respect, we address the following questions:

- How is a goal selected?

- How can a goal activate alternative behaviors given the circumstances?

- How is a goal-directed behavior controlled?

- How do such capabilities develop ontogenetically?

Movements toward a goal, such as a reaching motion, are well suited to address these questions. The behaviors are rather short and can thus be systematically investigated. Goal selection has been shown to depend on various factors, including task and circumstances. Once a goal is determined, the motor behavior is often flexibly adjusted given the currently relevant environmental circumstances. Moreover, various studies have investigated how behavior is controlled and adapted to unexpected disturbances. Finally, developmental studies have shown how behavioral competencies develop ontogenetically.

In fact, it appears that goal-directed motor control develops very early in childhood. Rudimentary forms of goal-directed behavior have been shown to be present from birth. Figure 12.6 shows how manual reaching behavior develops over the first two years of age. Interestingly, while the behavior is goal-directed from the beginning (Konczak, Borutta, Topka, & Dichgans, 1995; Rochat, 2010; von Hofsten, 2004), the hand's trajectory is only slowly optimized to an apparently approximately optimal stereotypic trajectory. These considerations will be relevant in the following models, where behavior is optimized based on optimal control principles.

12.3.1   Models of online motor control

The first computational models of online decision making and motor control were proposed in the 19th century. We already considered the *ideomotor principle* in relation to constructivism and development in Section 2.4.2, as well as in relation to sensorimotor learning and

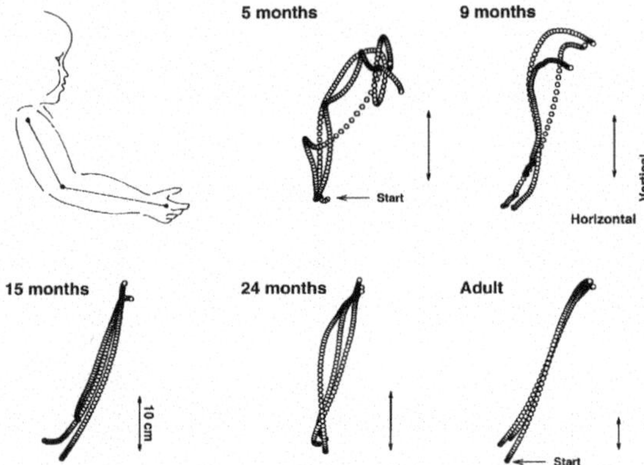

Figure 12.6: Reaching movements toward goal locations become progressively more stereo-typic over the first years of life. While even infants show goal-directed arm motions, these motions are initially clumsy. [*Experimental Brain Research, 117*, The development toward stereo-typic arm kinematics during reaching in the first 3 years of life, 1997, 346–354, Konczak, J. & Dichgans, J. Copyright © 1997, Springer-Verlag Berlin Heidelberg. With permission of Springer.]

adaptation in Section 6.3.   Postulated by Friedrich Herbart (Herbart, 1825), it essen-tially suggests that the anticipation of the desired sensory effects of an action triggers the best-suited motor program.

In 1852, the English physiologist William Benjamin Carpenter (1813-1885) confirmed this hypothesis for the first time: he showed that seeing a particular motion and, to a lesser extent, even the thought about a particular motion, can result in muscular activities that mimic this motion. He also coined the term idemotor, referring to the two components: *ideo* as an internal representation or idea (of a desired future state for that matter) and *motor* for the corresponding motor representation. Later, William James integrated Herbart and Carpenter in his groundbreaking work *Principles of Psychology* (James, 1890). Although originally related to occult practices – such as the attempt to mentally influence a pendulum – in its modern form, the ideomotor principle solely refers to the concept that ideas about the future are realized by suitable bodily means when possible, that is, a desired and achievable goal state is reached by executing suitable action complexes.

While in its essence probably true, the ideomotor principle does not tell the whole story. Over the last century, multiple challenges to the idemotor principle have been considered and solutions have been proposed over the last century. The Russian physiologist and biomechanic Nikolai Alexandrowitsch Bernstein (1896-1966) is considered to be one of the founders of modern motor control research. Bernstein introduced the *Bernstein Problem*, which is also called the *redundancy problem*: seeing that the body has redundant degrees of freedom, how does the brain manage to effectively, and usually successfully, reach certain locations in space?

Consider, for example, a human arm. Each of our two arms has seven degrees of freedom: three in the shoulder, two in the elbow, and two in the wrist.[1] If we want to place our hand at a position in space with a particular orientation, however, only six coordinates are fixed: the location and the orientation of the hand, each with three degrees of freedom. As a consequence, one degree of freedom remains. In fact one can explore this degree of freedom when doing a self-experiment: place your hand on the table in front of you and neither move the shoulder nor the hand; in this case it is only possible to rotate the elbow to a certain extent (remember not to move the shoulder). Thus, the remaining degree of freedom forms a one-dimensional complex axis in seven dimensional joint angle space. As a consequence, when placing ones hand on a table, the motor system has to choose a posture from a one-dimensional manifold of possible postures. When only touching a point in space, regardless

---

[1]Inner rotation degrees of freedom in the upper and lower arm are for simplicity reasons assigned to the respective upper joints.

with which orientation, the choice needs to be made from a four-dimensional manifold of possibilities.

Besides the redundancy in the final posture, an infinite number of trajectories can be executed to reach this posture. Although we tend to choose a particular, stereotypic path (cf. Figure 12.6), generally any trajectory of the hand to the target, and even of the whole arm and body, may be chosen. Even worse, the actual acceleration and deacceleration commands can be varied. Thus, the challenge is to choose an appropriate behavior among a huge set of alternatives.

In light of this challenge, Bernstein investigated the principles that allow us to effectively choose and execute appropriate actions on the fly. To find an explanation, he investigated how humans control their hands during reaching and manipulation tasks in controlled psychophysical experiments. His three main observations were that :

- The trajectories are chosen somewhat independently from where the start and end positions are actually located.

- They are typically rather straight.

- They exhibit a bell-shaped velocity profile.

In light of these observations, Bernstein proposed that the large number of degrees of freedom allows the flexible adaptation of actions to the actual situation. The goal of a motor control routine, then, is to optimize the smoothness of each behavior. The choice and adaptations, that is, the parametrization of the chosen behavior, then focuses on those degrees of freedom that can control the achievement of the desired effects in the most effective manner.

Later, this principle was formalized in terms of optimality. The *optimal control principle* (OCP) essentially postulates that, given suitably modifiable feedback controllers, usually that feedback controller is invoked that yields the best performance, such as the least energy consumption.   Meanwhile, feedback controllers only correct those motion errors that yield directional motion away from the goal. As a result, corrective control will focus its control effort on those degrees of freedom that are critical for successfully reaching the goal, which is often referred to as the *minimal intervention principle* (Todorov & Jordan, 2002).   Bernstein himself had already generated experiments that confirm these principles (Bernstein, 1967). More recently, others have confirmed Bernstein's predictions with respect to manual actions and even while producing speech (Todorov, 2004, and citations within).

The OCP offers the fundamental mathematics to optimize particular motion controllers and has been shown to be able to model particular biological motion very accurately. Derived from the cost-to-go equations in reinforcement learning (RL) by means of the Bellman equation (cf. Section 5.3.1), optimality is defined as the control that minimizes the cost-to-go. Although the brain does not always fully minimize this cost, it appears that we at least tend toward the optimum during learning, that is, our brain attempts to progressively optimize particular body-environment interactions, such as grasping an object.

These optimization principles are also closely related to the formalization of *dynamic movement primitives* (DMPs, cf. Section 5.4.4).   An individual DMP can be considered a particular encoding of a motor primitive, which can be executed in a closed-loop or open-loop fashion. Interestingly, DMPs also follow the principle of a hierarchy, which unfolds dynamically in space and time. On the lowest level, a simple temporal dynamic unfolds from the start to the end of a primitive. On the second level, this dynamic is translated into a dynamic trajectory, which can be optimized to execute, for example, an optimal tennis volley shot or an optimal object grasp. Because the second level encodes the dynamics relative in space and time, the third level can flexibly translate these dynamics into the current frame of reference, accounting, for example, for the speed and angle of the flying tennis ball or the size of the object and the distance and orientation of the object relative to oneself (Ijspeert et al., 2013).   The learning of the involved motor primitives is based on policy gradients (Kober & Peters, 2011; Stulp & Sigaud, 2015; Wierstra, Schaul, Glasmachers, Sun, Peters,

& Schmidhuber, 2014) (cf. also Section 5.4), which essentially implement the OCP principle by directly optimizing motor control parameters.

Formalizations of OCPs and their implementations have shown that biological motion is often optimized toward particular optimality criteria. When striving for a smooth and accurate action execution, even the behavior given forcefield disturbances was modeled successfully (Whitney, 1969; Todorov, 2004). However, making flexible behavioral choices – such as when using the elbow to open the door when the hand is currently holding a shopping bag – is a challenge that has hardly been addressed. Note how this essentially mirrors the same dilemma that was encountered in RL: *policy gradients* are good to optimize particular motor primitives, but they do not address the challenge of selectively invoking the best particular motor primitive under consideration of the current environmental circumstances; *temporal difference learning* in discrete spaces, on the other hand, seems to be too brittle to accomplish the whole control task via learning a full value function. Hierarchical and factorized representations are necessary in order to make hierarchical, model-based planning effective (cf. Section 5.3.3).

12.3.2   Models of decision making

While OCP-based optimization routines enable the execution of particular start-goal motions approximately optimally – also rhythmic motions, such as walking, can be realized by similar principles (Ijspeert, 2008) – OCP does not offer a computational technique that is able to adapt behavior on the fly to account for novel circumstances. That is, OCP does not allow the direct derivation of an effective redundancy resolution algorithm, which may be able to choose which action to execute in the first place. To achieve this, additional optimality criteria need to be considered and need to be flexibly integrated into the planning process.

For example, consider the observation of David Rosenbaum, which led to the establishment of the term *end state comfort*-oriented behavioral control (Rosenbaum, Slotta, Vaughan, & Plamondon, 1991; Rosenbaum, 2010). Rosenbaum had participants take a dowel, which was placed horizontally in front of them, and place it vertically onto a target disk (cf. Figure 12.7). The dowel was white on one side and black on the other (cf. Figure 12.7). To make things more interesting, the participants were instructed to place the dowel in one particular orientation (for example, black part down) onto the disk. The observation was that participants typically used a normal, overhand grasp when a 90° clockwise rotation yielded the correct orientation. However, when the stick was oriented the other way, typically an underhand grasp was used, ending up in the same final grasp posture by rotating the dowel 90° counterclockwise. It thus appeared as if the initial grasp was chosen in anticipation of the final, comfortable posture when placing the dowel vertically. Initial grasp redundancy – in this case two alternative grasps – was considered during planning, choosing the one that would lead to a more "comfortable" (or "optimal") final posture.

Figure 12.7: The end state comfort effect beautifully illustrates how our behavior is often anticipatory, that is, directed toward the final goal of an interaction. In the illustrated case, the stick has to be put with the light side down into the base station.

To achieve this task, multiple control models need to be available that can execute each particular grasp. Along these lines, Wolpert and Kawato proposed that multiple forward-inverse model pairs are suitable to achieve effective motor control (Wolpert & Kawato, 1998). In their MOdular Selection And Identification for Control (MOSAIC), the authors proposed to combine multiple model pairs, where the forward model is predicting the unfolding behavioral consequences, while the inverse model generates suitable motor control commands. In addition, a responsibility module for each forward-inverse model pair determines which pairs are currently best suited to solve the task under specific circumstances. In the first implementation of MOSAIC (Haruno, Wolpert, & Kawato, 2001), it was shown that it is indeed possible to selectively choose amongst the available control modules, given particular objects with individual dynamic properties. Two years later, the model was enhanced to a hierarchical version, where the upper layer could suitably bias the lower-layer control activities (Haruno, Wolpert, & Kawato, 2003). In all cases, though, MOSAIC did not consider explicit goal representations and it also did not develop predictive encodings that could anticipate the final effect a particular module may generate.

The fact that particular control models need to be chosen selectively in anticipation of a particular goal state was already proposed in "A schema theory of discrete motor skill learning" (Schmidt, 1975). The article essentially focused on the question how it is possible to learn particular motor skills in a discretized fashion, such as shooting a ball during soccer, driving a car, or tying one's shoe laces. Schmidt was inspired by observations made in 1932 by the British psychologist Frederic C. Bartlett (1886-1969), who had noticed that:

> How I make the [tennis] stroke depends on the relating of certain new experiences, most of them visual, to other immediately preceding visual experiences and to my posture, or balance of postures, at the moment. [...] When I make the stroke I do not, as a matter of fact, produce something absolutely new, and I never merely repeat something old. The stroke is literally manufactured out of the living visual and postural 'schemata' of the moment and their interrelations. (Bartlett, 1932, p. 201f.)

Accordingly, Schmidt proposed that the control of behavior must be influenced by the following three main factors:

- *Generalized motor programs* (GMP) represent the control component of the system.

- *A recall schema* feeds the GMP with particular parameter values, making the actual desired motion concrete.

- *A recognition schema* finally provides feedback expectations given the co-encoded behavior is actually executed.

The major difference from the MOSAIC model is that Schmidt's schema theory focused on the invocation of discrete actions by explicitly taking the goal in the form of the *desired outcome* into account. As a consequence, responsibility signals could be more explicitly learned in a goal-oriented manner. However, Schmidt does not offer principles as to how these mechanisms may actually be learned and how the schema system may focus on the critical aspects to achieve a particular goal. Moreover, online redundancy resolution mechanisms are not considered.

Nonetheless, both motor control models contain

- A *forward model* to predict and monitor the currently unfolding sensorimotor interactions.

- An *inverse model* (the recall schema in Schmidt), which determines the motor control.

- A *responsibility determinant*, which specifies which control schemata should be activated given the current circumstances.

However, the concrete functionality and implementation of the responsibility determinant remains obscured in Schmidt's schema theory, while MOSAIC considers neither the goal nor the current conditional circumstances explicitly.

The neural *Sensorimotor Unsupervised Redundancy Resolving motor Control arcHitecture* (SURE_REACH), offers a means to determine responsibilities and adjust behaviors on the fly, given the goal and current circumstances (Butz, Herbort, & Hoffmann, 2007). The SURE_REACH model is structured as follows (cf. Figure 12.8):

- In extrinsic *hand space*, which is two dimensional in the figure, the current hand location, the goal location, and possibly also current obstacles, are encoded.

- With the help of an inverse kinematic model, which is termed *posture memory*, the hand space neurons are associated with posture space neurons; these associations can be learned.

- In intrinsic *posture space*, which is three dimensional in the figure, the joint angles of the arm posture as well as the joint angle manifold of possible goal postures that are currently unreachable, because they are blocked by an obstacle, are encoded.

- By means of a sensorimotor model within posture space, distances between arm postures can be determined easily.

- The control component uses the neural activations to determine those control commands that currently seem most appropriate.

SURE_REACH is a *closed-loop* control system in that it decides every iteration anew which action to execute. Moreover, it implements model-based RL in its sensorimotor model, ensuring an approximate optimal execution of goal-oriented behavior.

Given a location goal, the corresponding neurons in hand space are activated. Next, this goal-activity is projected via the posture memory mapping into posture space – typically activating a manifold of goal-compatible postures. Depending on where the arm is currently located, the shortest trajectory to the goal manifold activities is chosen. This is accomplished by spreading the respective activities inversely via the sensorimotor model. When obstacles are present, these can be represented by inhibited neurons in hand space, whose inhibition multiplicatively influences the activities in posture space. In doing so, fully overlapping posture neurons will be fully inhibited, while partially overlapping ones will be inhibited only to a certain extent.

Figure 12.8 shows schematically a typical resulting trajectory adaptation. In contrast to the model above, SURE_REACH develops a redundant sensorimotor model and learns the sensorimotor transitions for all postures. Thus, SURE_REACH has approximated the model of its arm and is thus able to execute model-based reinforcement learning. In fact, SURE_REACH can produce globally optimal behavior given the current circumstances nearly instantly, especially when each neuron is processing information suitably in parallel. The system resolves motor redundancies on the fly by principles of optimality, where the optimum is approximated by model-based reinforcement learning. All other systems choose their action biases heuristically based on previous experiences, thus redundancy resolution is predetermined and not further optimized on the fly.

Nonetheless, note that similar to the MOSAIC system and Schmidt's schema theory, SURE_REACH contains the aforementioned three main components, but in a very intertwined manner: the forward and the inverse models are embedded in the architecture in the form of the *posture memory* and the *sensorimotor model*; goal activities are projected top-down through the model, generating a temporary reinforcement gradient throughout the sensorimotor model. Thus, action choices are made by means of the activity projections and RL-based planning within posture space.

SURE_REACH accomplishes motor control and redundancy resolution on the fly, solving the Bernstein problem by means of a highly adaptive architecture. It is generally neurally plausible as it implements neural fields by means of a neural population code. Moreover, it

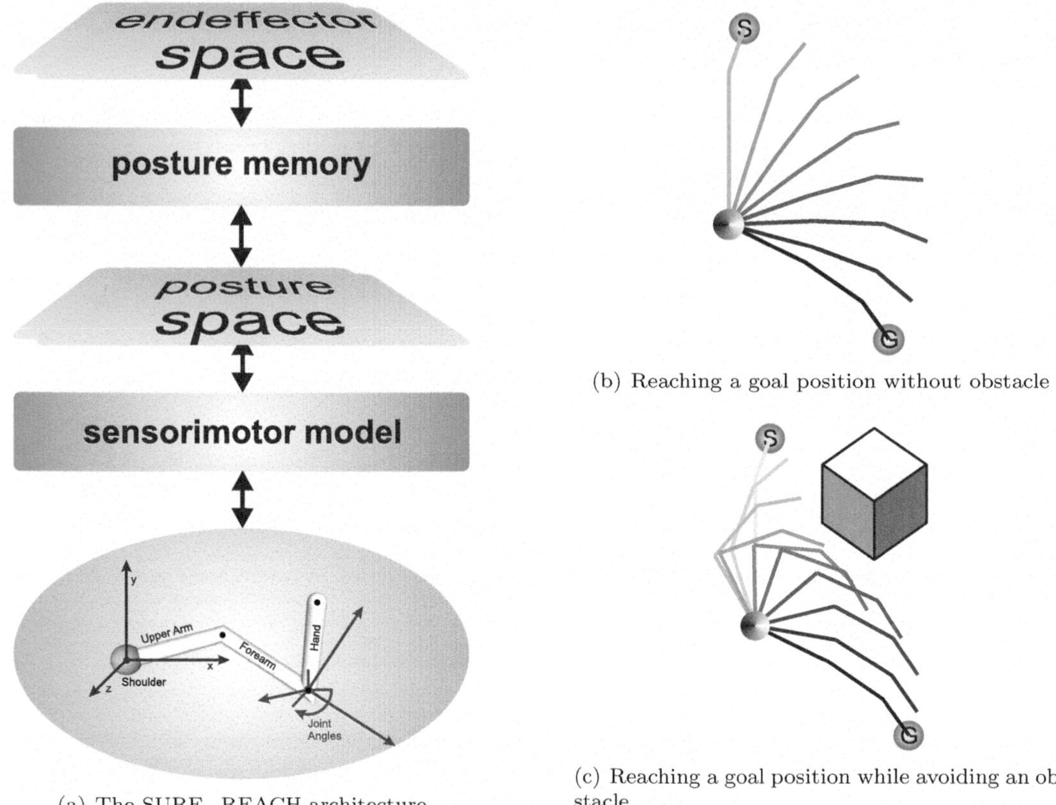

(a) The SURE_REACH architecture

(b) Reaching a goal position without obstacle

(c) Reaching a goal position while avoiding an obstacle

Figure 12.8: SURE_REACH is a modular, hierarchical, neural field architecture. Due to the redundancy encoding approach, the system is able to adapt its motor trajectories on the fly to the current circumstances, for example, bending its arm and then stretching it again to avoid an obstacle.

allows the adjustment of goal-directed trajectories flexibly and on the fly, so the system can indeed model the end state comfort effect (Herbort & Butz, 2007).

Clearly, though, SURE_REACH also has its down sides. The most severe one is that SURE_REACH does not scale without further modularization. That is, it is impossible to represent more than three to four dimensions with a neural field due to the exponential growth of the necessary number of neurons. Moreover, SURE_REACH focuses on the kinematics, that is, the arm's trajectory, but not the dynamics along the trajectory. Finally, model adaptations (not behavioral adaptations) are slow in terms of learning progress. Thus, while SURE_REACH is neurally implemented and has yielded interesting motor control capabilities, it is algorithmically not plausible when considering the exploding hardware requirements (exponential growth in the dimensions covered by a neural population code lattice). Modularization of the system is needed, which was pursued and was shown to yield very high noise robustness while maintaining an internal postural, probabilistic body schema over time (Ehrenfeld, Herbort, & Butz, 2013a). Unfortunately, the modularized system requires a more elaborate behavioral optimization mechanism, which cannot guarantee global optimality.

As a result of these modeling insights, it appears that the brain has evolved a compromise between full optimality and the effort that is involved when planning such fully optimal behaviors on the fly. For scalability reasons, both are not possible, but heuristics, approximations, and/or constraints need to be included, reaching a boundedly rational behavior. Recent psychological studies with human subjects have confirmed this suspicion. For example, in a series of experiments considering the *end state comfort effect*, it was shown that behavior is often neither fully optimal nor heading toward full optimality. In one study, people were asked to rotate everyday objects and to displace them slightly to the left or to the right. Surprisingly, when people had to displace the objects to the left they preferred to rotate the object counterclockwise, while they preferred a clockwise rotation when displacing it to the right (Herbort & Butz, 2011). Thus, the very slight directional motion necessary to accomplish the small displacement biased the preferred rotation direction and thus the consequent grasp orientation that was chosen to execute the object manipulation. Moreover, the initial orientation of the object influenced the grasp choice: when confronted with a cup to be rotated that was standing upright, only slightly more than half of the subjects showed the end state comfort oriented thumb-down grasp, while the others used the standard thumb-up grasp. On the other hand, when the cup was oriented up-side down, nearly all subjects showed the end state comfort effect using an anticipatory, thumb-down grasp (Herbort & Butz, 2011).

A simple model that accounts for these findings is the *weighted integration of multiple biases* (WIMB) model (Herbort & Butz, 2012). This model essentially suggests that, while preparing for an object interaction, behavioral choices are not fully optimized on the fly, that is, the brain neither plans the actual interaction completely nor considers all possible alternatives. Rather, the model and the modeled findings suggest that our brain is full of habitual behavioral tendencies, which are learned and activated depending on the current circumstances. For example, an oriented object, such as a cup or a glass, suggests to us a grasp according to the object's orientation, simply because this is the way we have usually grasped similar objects in previous successful interactions. Similarly, when intending to rotate an object clockwise, we use a pre-grasp, which may result in a thumb-down grasp given an elongated object (such as the dowel in the end state comfort experiments). Given a dial-like object, we still use a strong pre-grasp, even when only a very small rotation is needed. In sum, task- and circumstance-dependent as well as habitual biases co-determine decision making of concrete actions in a weighted manner.

12.3.3 Action and motor control components

To summarize, it appears that our motor control system neither plans fully ahead nor makes optimal decisions or generates fully optimal behavior in the light of the current circumstances. Fully optimal behavior requires training – as we know, for example, from sports or from playing an instrument. Similarly, behavioral decisions are continuously adapted given interaction experiences, thus also requiring training when these choices need to be optimized as much as possible. For example, to optimize behavioral decisions in sports, such as deciding if to bat or not in baseball, very fast heuristics need to be trained. Not only in sports, though, but also in our everyday life we tend to optimize our behavior and the involved decisions. Although typically not fully optimal in particular situations, our behavior with its biased decision-making processes and optimized control routines works typically very well because the decision biases and control routines develop for optimizing decision making in the longer run. When becoming an expert in a particular sport, but also in any profession, the brain is trained on a subset of situations in which only subsets of decisions and control routines need to be considered and optimized. As a result, the involved biases and control routines are progressively refined and optimized in the niche of the profession, developing effective behavioral control routines, such as highly effective reflex-like and intuitive behavior and cognition.

To make these reflections more concrete, it is useful to distinguish several main components when considering motor decision making and control. In accordance with the principles put forward by MOSAIC, Schmidt's schema theory, and SURE_REACH, an action may be thought of as an execution of a sequence of motor control processes, which typically lead to the achievement of a particular *goal* event or action effect.

Moreover, all models have considered the circumstances under which a particular control process may be applied. An action schema encoding may thus be viewed as consisting of four main components:

1. The *final action effect*, that is, the goal event, which is achieved once the action is finished.

2. *Temporal forward predictions* about the online unfolding movements of the body, and possibly also of other entities that are manipulated during the action.

3. *Inverse motor control processes*, which bring about the concurrently unfolding movements and ultimately the final action effect.

4. *Condition encodings*, which specify under what circumstances the action may be executed – possibly with estimates about the probability of achieving the final goal, given particular circumstances.

While MOSAIC and Schmidt's schema theory consider the final goal only indirectly by proposing a responsibility determination mechanism, SURE_REACH allows one to feed in the final goal directly. However, SURE_REACH does not distinguish behavioral alternatives beyond trajectory and end state redundancies. The WIMB model, on the other hand, completely focuses on the action choices and involved biases for achieving a particular goal.

In addition to these cognitive models, at this point we should also recall hierarchical RL (cf. Section 5.3.3). In hRL the goal is made explicit and determines the action choice, akin to SURE_REACH, but in a hierarchical manner (Botvinick & Weinstein, 2014; Sutton et al., 1999; Vigorito & Barto, 2010). Options in hRL can be viewed as motor primitives, which are a forward-inverse pair in MOSAIC terms or DMP in cognitive robotics. Thus, hRL chooses motor primitives conditioned on the current circumstances and the final goal, generating motor commands that yield goal-directed movements.

The downside of hRL is that, at the moment, there is no well-accepted mechanism available that robustly learns options starting from continuous, fine grained sensorimotor experiences. Various research labs in artificial intelligence and cognitive robotics tackled this challenge with various approaches. DMPs have been combined in various architectures. For example, a table tennis playing robot arm has been developed where a decision component chooses which DMP to execute in light of the current circumstances (Muelling, Kober, & Peters, 2010). Another approach has implemented a neural tree structure, which enabled the dynamic selection and sequential execution of particular object grasps, which were selected depending on the object's position and orientation relative to the robot arm (Luksch, Gienger, Mühlig, & Yoshiike, 2012). Even more recent work has introduced a more formal system that learned abstractions over sensorimotor interactions, enabling high-level, symbolic planning (Konidaris et al., 2014, 2015). However, the robustness and general applicability of these approaches is still under evaluation.

The anticipatory rule learning principle from the *anticipatory behavioral control* theory also considers the introduced four action components (cf. Section 6.4.2). Although the ABC theory never distinguished between final goals and immediate sensorimotor effects (Hoffmann, 1993, 2003), it formalized how conditions may be learned given experiences of various sensorimotor effects under different circumstances: the conditional focus should be on those environmental aspects that are necessary to ensure, or at least to increase, the likelihood of the sensory consequences when executing particular motor behaviors. The *anticipatory learning classifier system* (ACS), which was implemented based on the ABC theory, has shown robust and effective learning capabilities in a variety of behavioral learning tasks in discrete problem domains (Butz, 2002a). The behavior of rats has also been modeled successfully with ACS (Butz & Hoffmann, 2002). Interestingly, the system is closely related to factored RL approaches (cf. also Section 5.3.3), where ACS learns to focus its developing sensorimotor-effect-specific conditions on those factors in the environment that are maximally relevant for learning accurate effect predictions, rather than reward predictions

(Sigaud et al., 2009). The related XCS classifier system focuses on the latter (Wilson, 1995) and has been shown to be very well suited to learning to focus its conditional structures on those feature dimensions and even those oblique feature axes that are maximally relevant to accurately predict reward (Butz, 2006; Butz, Lanzi, & Wilson, 2008; Stalph, Rubinsztajn, Sigaud, & Butz, 2012).

To summarize, mechanisms and learning techniques are generally available that can learn the put-forward four components of an action. However, they have not been combined so far in a rigorous manner. In addition to action learning and control, though, the decision-making process needs to be considered in further detail.

12.3.4 Decision making components

Action encodings ideally allow for the flexible adaptation of behavioral choices and control commands to the current situation. Given the four component action structure described above, it seems necessary that action decision making is endowed with three further important components:

- The *probability of success* needs to be predictable given current circumstances, that is, given the action encoding conditions.

- The *action effort* needs to be estimated, also considering the current circumstances.

- The *expected payoff*, which may be reached when the action was executed, needs to be accounted for.

With this additional knowledge it becomes possible to execute goal-oriented Bayesian inferences within Bayesian networks (cf. Section 9.3.3). The SURE_REACH architecture, also allows for the inclusion of expected payoff and action effort, but not for the probability of success (Butz et al., 2007; Herbort et al., 2007; Herbort, Butz, & Pedersen, 2010). Given a particular goal, the action encodings can be used to propagate expected payoff inversely, that is, from final effects to conditions. Meanwhile, action efforts can be considered by means of model-based RL. On hierarchical, more abstract levels, efforts, and final effect distributions can be propagated inversely via condition-effect encodings, yielding a goal-directed planning process according to the options framework of hRL.

Let us look at an example. Consider the choice when confronted with grasping a glass, which you want to place into the dishwasher. The choice is to grasp the glass with either a standard, thumb-up grasp, with a thumb-down grasp, or with a top-grasp. First of all, current encodings about the spatial situation will determine that the glass in question is actually reachable. Moreover, such encodings will provide information about the actual distance and orientation of the glass relative to your own body. Situation and task essentially enable you to choose the actual grasp. Given the glass is standing upright, a top-grasp will allow you to transport the glass to the dishwasher, but the top-grasp is not suitable for placing the glass upside-down. Thus, the probability of success using this grasp under these circumstances is zero. On the other hand, a thumb-up grasp requires you to rotate your arm, ending in a thumb-down posture while placing the glass into the dishwasher. While this behavior can be assumed to be successful, so that the probability of success is one, the effort of the placement, which also depends on where the glass is intended to be placed, needs to be taken into account. When the goal is a location in the bottom shelf, the effort of placing the object with a thumb-down grasp may be higher than the effort associated with a thumb-up grasp. Thus, one will tend to prefer a thumb-down grasp when grasping the object in order to end up in a thumb-up grasp when placing the object. This essentially corresponds to the end state comfort effect described above, but the description here has algorithmically formalized it, thus explaining the computational mechanisms on a deeper level. Note once again that behavioral studies suggest that our brain does not execute the sketched-out combinations on the fly each time facing such a situation, rather, it optimizes the choices and behaviors over time by learning from gathered experiences (possibly including observations and demonstrations of others executing the task).

Once these considerations have come close to a decision, the associated motor control complexes will be initiated. As a result, temporal forward predictions of the intended motor control complexes will start to unfold, probing the likelihood and possibly adjusting the anticipated trajectory. While then actually executing the motor control, the forward predictions will unfold in tune with the motor commands according to the reafference principle, so as to be able to monitor behavioral success, to become aware of unexpected changes, and to detect action execution failure (cf. Section 6.4). Note again that thus action decision making and control processes are neither logical nor fully deterministic, nor fully optimal. Rather, they are highly dynamic and highly stochastic processes, which are computing likelihoods of success, behavioral efforts, intended trajectories, and anticipated consequences.

While the example has shown how the described cognitive architecture may bring about suitable action choices and motor control based on these choices, we have not yet considered how the actual goal may be chosen. Given we have many glasses to put into the dishwasher, with which one should we start? Given we have guests, should we even put those glasses in the dishwasher now – or do it later? Or should we rather go on a walk first because the weather is still so nice out?

To come to such decisions, the current goal itself needs to be selected and needs to become fully concrete. Reconsidering the dishwasher example, it becomes apparent that this needs to happen interactively at multiple hierarchical levels. On higher levels, a decision may consider the goal to fill the dishwasher. Once this goal is activated, consequent lower-level goals can be selected and activated. Lower-level decisions will thus fully unfold only once higher-level goals are set. For example, when it is decided that the dishwasher will be loaded, lower levels decide which glass to pick up next, and how to place the glass where in the dishwasher – whereby the placement consists of a "grasp" goal, a "placement" goal, and finally a "hand release glass" goal, which is the final goal. Given the "grasp" goal, on the next lower level, the exact type of grasp and the placement of the grasp on the object needs to be decided upon, attempting to maximize the success probability, but also the resulting, expected payoff, that is, avoiding sharp areas while grasping the glass at a maximally graspable position (Herbort & Butz, 2007; Trommershäuser et al., 2003b). Thus, a hierarchical decision system needs to be at play, which mutually biases goals, opens up specific goal considerations, considers behavioral effort, likelihoods of success, expected payoff, and thus biases action choices on multiple levels.

Note that the expected payoff may also depend on the current state of body and brain. That is, the motivational state of the system, which may be modeled by homeostatic reservoirs, may generate tendencies to activate currently desirable and achievable goals (cf. Section 6.5). Moreover, it may influence the payoff and effort estimates. This applies on all hierarchical levels, but may become more apparent on a higher level. For example, the higher-level decision if one should load the dishwasher or instead talk to the visitors, needs to consider social aspects, current energy levels, and the further context (for example, who are the visitors and how long they will stay still). In bidirectional interaction with the action decision and control mechanisms, then, those actions will be selected and executed that are expected to lead to a maximal satisfaction of the currently most pressing motivations – such as enjoying a nice conversation and the involved social interaction, or the prospect of a clean kitchen.

We have described computational mechanisms that are necessary to come to good action decisions and identified several decision making and motor control components, which facilitate this process. We have also covered generally how online control unfolds. DMPs (cf. also Section 5.4.4) and the related forward-inverse motor primitives are well-suited to accomplish the actual motor control. In the brain, the hierarchical cascade and the involved feedback control loops further facilitate the actual control process.

We also hinted at how conditions and final effects may be learned from a machine learning and AI perspective (cf. also Section 12.3.3), although there is no generally accepted learning mechanism available at this point. Regardless by which mechanisms the outlined action components are actually learned, however, from a cognitive science perspective the gathered insights hint at how the brain can accomplish the involved challenges. In particular, actions need to be associated with particular motor primitives, conditions, effect structures, as well as estimates of success probabilities, expected action effort, and expected final payoff.

When sequences of actions need to be executed, hierarchical implementations of such action structures are necessary. On higher levels, a motor complex specifies a set of actions that are necessary to achieve a particular final effect. Bernstein postulated that sequences of actions – such as an arm-extend, grasp, transport, release, retract sequence when relocating an object – can be viewed as a motor Gestalt, similar to a visual Gestalt postulated in Gestalt psychology (cf. Section 3.4.3). Similar to the way a visual Gestalt solves the sensory binding problem of perceiving a whole object, a motor Gestalt binds the necessary motor actions to generate a particular, semantic environmental interaction. Thus, a motor complex can be viewed as a motor-grounded Gestalt, which solves the problem of binding multiple motor primitives into a larger complex. Interestingly, such bounded, compactly encoded motor complexes offer a solution to aspects of the symbol grounding problem, where complex action verbs can refer to particular motor complex encodings, for example, "loading the dishwasher", "cleaning the kitchen", or "going on a walk".

12.4 Event-oriented conceptualizations

Although we have mentioned several approaches in artificial intelligence and robotics that tackle the problem of learning effective, factorized condition and effect structures for motor primitives and motor complexes, general means to learn these structures remain to be found. It should be noted that the learning of these encodings present further conceptualizations of the environment, and these conceptualizations offer themselves to reason on abstract, factorized and thus generalized levels.

For example, consider the case of cutting up a vegetable, such as a cucumber. A cutting action can usually be accomplished only with an appropriate knife. Thus, knife and cucumber need to be present. Moreover, to execute the cutting action the knife needs to be held by the hand in an appropriate orientation and the cucumber needs to be in reach. Furthermore, sufficient free space needs to surround the cucumber to be able to execute the actions without interferences. Several further considerations come to mind, but the point is to acknowledge that we have just described fundamental factors that are relevant to execute a particular behavior. Many other factors are largely irrelevant, such as the exact location in space, the temperature of the room, the time of day, or the presence of other objects. Thus, the characterization of action-relevant factors enables action-oriented conceptualizations and conceptual simulations of environmental interactions (Barsalou, 1999).

Two theories from cognitive science describe the nature of the condition and effect encodings, which the brain may learn in order to segment the continuous, sensorimotor stream into distinct action-encodings:

- The *theory of event coding* (TEC), which is closely related to the ideomotor principle, postulates that common event codes are learned during development, which integrate motor codes and their effects on the environment (Hommel, Müsseler, Aschersleben, & Prinz, 2001).

- The *event segmentation theory* (EST) focuses on how the sensorimotor stream may be segmented into particular events, which are separated by event transitions (Zacks & Tversky, 2001; Zacks, Speer, Swallow, Braver, & Reynolds, 2007).

Based on TEC and EST, conceptual understanding of events and event episodes can develop beyond action-oriented encodings, which have also been referred to as *event schemata* (Hard, Tversky, & Lang, 2006). From an ontological perspective, conceptual understandings of events can develop in the context of behavior and motor control. Cognitive development first focuses on the own sensorimotor experiences, thus segmenting the experienced environmental interactions caused by our own behavior. It thus becomes easier to segment observed environmental interactions of others, expecting that these interactions will also have a beginning, an end, which coincides with the final goal, and particular motor primitives and motor complexes, which control the unfolding event.

12.4.1 Events and event segmentations

Starting from theories on anticipatory behavior, TEC emphasizes that in order to choose and control an action in a goal-directed manner, the most suitable encodings are those that bind actions with their typically resulting effects. In accordance with the ideomotor principle and related theories, during decision making and actual behavioral control the focus lies on these desired effects, not on the control of the motor activity itself (Hoffmann, 1993; Prinz, 1990). This focus is believed to lead to *common event encodings* of motor actions and their sensory consequences (Hommel et al., 2001). As a consequence, interferences of anticipated motor effects on action decision making and action initiation can be registered. For example, it was shown that a strong muscle motion activation and thus expected fast visual motion acceleration signals are inherently associated with a high auditory volume, such that action decision making and initiating is delayed when, for example, an incompatible auditory effect is anticipated (Elsner & Hommel, 2001; Kunde, 2001; Prinz, 1997). TEC postulates that action control unfolds concurrently with the expected sensory consequences, and the anticipated consequences control the unfolding motor commands in the first place. TEC is closely related to the MOSAIC model, but focuses more on the cognitive, psychological encoding level. Moreover, TEC emphasizes that sensorimotor forward-inverse encodings integrate multiple sensory and motor modalities, producing generalized, possibly dynamic, multimodal common codings.

Derived not from behavioral experiments with humans, as TEC is, but rather from observations of how humans tend to systematically segment environmental interactions (Newtson, 1973), EST focuses on the characterization of events and event transitions. In particular, an event is characterized as

> [...] a segment of time at a given location that is conceived by an observer to have a beginning and an end. (Zacks & Tversky, 2001, p. 3.)

Based on such an event characterization, EST proposes how events may be segmented and also how event taxonomies are closely related to object taxonomies (Zacks & Tversky, 2001; Zacks et al., 2007). While objects can be viewed as distinct entities in space, which can be described or classified at various levels of abstraction, event can be viewed as distinct entities in space-time, which can similarly be described and classifiers at various levels.

The EST architecture essentially postulates that an event is represented by temporal forward models, which predict how the event typically unfolds. An activated event thus generates temporal predictions about how a situation will change over time. Given these predictions, error detection mechanisms can validate or falsify the predictions. Given that the predictions are validated, the current event encoding may be maintained and possibly further optimized. Given falsification, however, the current event encoding may be inhibited and the sensory information, as well as predictions about likely event transitions, may determine the next event model candidates.

EST is thus closely related to the four action components introduced earlier. Perceptual processing is determined by the currently active event models. These are, with respect to behavior, the currently unfolding motor primitives and the associated, motor-dependent forward models, which unfold while observing or while producing the event. Although the inverse models are not directly observable, it has been shown that when the observer is informed about the current intentions or goals of the actor, segmentations become intention-oriented, and thus more coarse grained. This observation suggests that goals co-determine the event interpretation, biasing the selection of currently active event encodings. Moreover, the perceived context has been shown to influence event perceptions (Zacks & Tversky, 2001), highlighting the conditional structure of event schemata. In accordance with Schmidt's schema theory, Land and Tatler (2009) have suggested that a schema system in the prefrontal cortex may be responsible for maintaining the current task activity and cause the planning of an overall action event sequence. As a result, attention, eye gaze, and manual control become goal-oriented, driven by the anticipatory parallel and sequential activation of event schemata.

Seeing this close relation to action, it may even be the case that the capacity to segment observed interactions and experienced events into meaningful units actually originates from the goal to control behavior in a goal-directed, flexible manner. EST proposes that significant visual motion changes are mainly responsible for event perceptions. However, visual changes seem to grasp only a fraction of the actual event transitions, as indicated from modeling the behavior of human subjects in event segmentation tasks (Zacks, Kumar, Abrams, & Mehta, 2009). Interpreting the available data in a broader manner, it seems that interactions are clustered into events and event transitions by focusing on the actually unfolding object manipulations. An object manipulation event, for example, starts when contact with the object is established. Next, the object manipulation unfolds maximally smoothly in accordance with the optimal control principle (OCP, cf. Section 12.3.1). Finally, the object manipulation ends by releasing the object, experiencing the object release, the tactile changes, and the consequences on the object (such as tipping over, falling, or remaining stable). Thus, EST can easily be embedded into theories of action decision making and control. Moreover, EST is closely related to hierarchical environmental models, which are needed for hierarchical, model-based reinforcement learning and hierarchical planning.

Recent work that focuses on longer chains of object interactions, such as making a peanut butter and jelly sandwich, has generated further evidence in favor of EST and its motor relevance. Hayhoe, Shrivastava, Mruczek, and Pelz (2003) tracked the hands and eye fixations of subjects whose task was to make, for example, a peanut butter and jelly sandwich. The results showed that the eyes indeed always anticipated the next action by several 100s of milliseconds, such that the scan-path of the eyes partially predicted the next hand motion. More recently, it was shown that individual objects are fixated with the task and final goal in mind; the eyes tended to fixate those parts of the object that were currently behaviorally relevant about 400ms before the hand actually executed the anticipated object manipulation (Belardinelli et al., 2015, 2016). These results suggest that planning and decision making precedes motor control, but it also strongly interacts with it. Moreover, the results show that final effect-oriented key parameters for acting successfully goal-directedly are determined. That is, the eyes precede our hands to determine the exact current environmental circumstances, and thus to prepare and parameterize the upcoming action execution to ensure that the final effect will be generated with high probability.

A computational model of the sandwich making task (Yi & Ballard, 2009) could show that a *dynamic Bayesian network* (cf. Section 9.3.3), which was fed with abstracted hand motion signals (reaching and manipulation indicators) and eye fixation signals (which object is fixated), allowed the derivation of the actual subtask event, which was being executed while making the sandwich. Subtasks were also related to each other by means of a Bayesian network. This latter network modeled, for example, the fact that knife, peanut butter, and bread need to be ready to be able to execute the behavior of spreading peanut butter onto the bread. Thus, a Bayesian model and Bayesian dynamics within the model allowed the inference of the actually observed interactions. The success of this model points out that not only observations, but also our own actions are most likely guided by the same action encodings, including conditional structures when an event commences, forward-inverse models, which control and predict the unfolding of the event, and final effects, which specify when the current event ends.

12.4.2 Event taxonomy

Let us take a look at the resulting taxonomy of events, when considering hierarchical event structures in space-time (Zacks & Tversky, 2001). Recollecting the dishwasher example from above, we considered three behavioral event levels: loading the dishwasher, putting one particular glass into the dishwasher, and actually reaching for the glass. Similar examples can be given in basically any everyday action and also with respect to the current situation you are facing as a reader. In fact, while reading this, you as the reader are experiencing and controlling the reading – on a rather low event level, scanning individual words and short text passages, and extracting meaning. On the next higher level, the situation you as the reader are in may be abstracted further. For example, you will probably read in a rather comfortable position to be able to focus on reading and comprehension – essentially being in a "reading event". On the next higher level, you are probably undergoing an event of studying, gaining knowledge, or similar characterizations, where this event guides your current behavioral and attentional focus and thus enables the lower event level choices and control. Possibly even one level higher, the event may reach over years, during which you are striving for knowledge, a higher education, a career, or similar. Another abstraction starting from the reading event may consider the mental side: while you are making sense of the words and sentences you are reading, you may be in the event of beginning to understand how behavior is related to conceptualizations, or, on a further abstract level, maybe even how the mind comes into being.

Note how such event abstractions form event taxonomies, which may be represented in a graph-like structure, where higher level events yield probabilistic priors on the activities of lower-level events, and lower-level events signal to the higher levels their current success, failure, or the end of the lower-level event (for example, reaching the end of a chapter or a section). Interestingly, as pointed out above (cf. also Zacks & Tversky, 2001), this taxonomy is structurally similar to object taxonomies, where objects may belong to particular categories, which can be divided into subcategories and so forth. Object taxonomies are, for example, an animal taxonomy – such as a particular dog that can be classified as a German Shepard, a dog, and a mammal – or a fruit taxonomy that distinguishes citrus fruits from berries on a higher level, whereas oranges and limes can be contrasted within the citrus fruit category. Similarly, event taxonomies can be formed, such as walking to school, which may consist of street crossing and side-walk walking events, where the street crossing event can be further partitioned into stepping down from the curb and walking across the actual street. Another example is a book reading event, which may be made more concrete as a science or fiction reading event, where the event can be made even more concrete by reading a particular chapter or page, or also reading a hardcover book or an e-book.

Even more obviously event-oriented is the conceptualization of a birthday party. Surely most of us have experienced many birthday parties in our lives, each of which was a particular instantiation of a "birthday party" event. Due to the large number of birthday parties, however, further distinctions may be possible such as kids' birthday parties, birthday parties at work, at a friend's place, or even in a restaurant or bar, similar to the distinction into citrus fruits, tropical fruits, or berries. Thus, similar taxonomic principles apply for events and for objects. Because events involve behavior, the four action and motor control components described above exhibit event-related hierarchical structures. However, the event concept enhances the action concept to other actors, social contexts, physical events, and even to mental events, such as when having an "idea" or when solving a mathematical equation in one's head.

12.4.3 Event conceptualizations and abstractions

As we described in Section 3.7, humans are readily able to abstract over actually perceived interaction events, interpreting even simple geometrical figures in a human-like, social context. This was first shown by social psychologists in 1944 (Heider & Simmel, 1944). The Belgian experimental psychologist Albert Michotte (1881-1965) was one of the first, who systematically experimented with our tendencies to interpret observed object interactions causally – even when we observe the interactions in abstracted form in a video or in a cartoon (Michotte, 1946). Take, for example, the interactions represented in Figure 12.9. In the first sketch, a circle moves to another circle, touches it, and thus appears to cause the other circle to move away. Essentially, a pushing event is perceived. In the second sketch, the ball attaches to the other ball and thus both balls move along – leading to the perception of an attachment event. Finally, in the third sketch, the ball simply moves over the other ball yielding no interactions and thus no interpretation of causality (but actually an interpretation of depth). The basics of such event perceptions were characterized by Michotte in 1946, and have since been elaborated upon and related to social cognition and language (Fleischer, Christensen, Caggiano, Thier, & Giese, 2012; Gibson, 1979; Jackendoff, 2002; Tomasello et al., 2005).

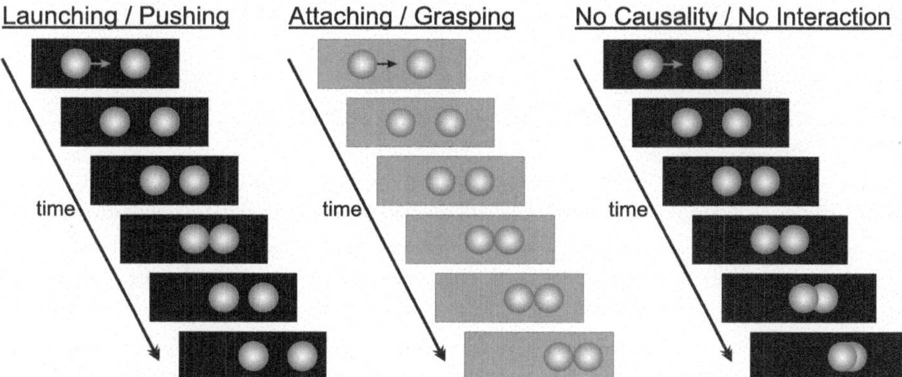

Figure 12.9: Illustration of the perception of causality when monitoring an object interaction. Note that causal events can be related to motor actions, where launching corresponds to pushing something with the hand while attaching corresponds to grasping something with the hand.

Event perceptions and anticipations in fact seem to be present from very early on in infants. For example, using the habituation paradigm (cf. Section 4.2.2), infants were shown two toys, one of which was grasped. After habituation, the same two toys were shown again with switched positions. Starting at about six months of age, infants tended to look longer at grasping events in which the other toy was grasped in comparison to grasping events in which the grasp trajectory differed, but the previously grasped toy was grasped again. Thus, infants appeared to have developed an understanding of the intention, that is, the final goal of the action, ignoring the differences in the executed trajectory (Woodward, 1998). More recent experiments have confirmed these insights numerous times, suggesting that at least by twelve months of age infants have action representations available that encode final goals and also take causal knowledge into account, such as when pulling on an object (for example, a cloth) to be able to reach the object (for example, a toy) on top of the pulled object (Sommerville & Woodward, 2005). These interpretations have also been confirmed in imitation studies, where infants re-enacted the intended action of a person, even if the observed person failed to execute the action successfully (Meltzoff, 1995). It thus appears that infants soon learn to abstract over the actual details of an action and to focus on the final goal, which corresponds to the intention behind an action.

Computationally, neurocognitive models have successfully simulated aspects of these interactions, offering a first neural model of understanding intentions. By generalizing over the actual observed hand and object identities and focusing fully on relative interactions between the two entities, a grasp event was classified by the neural model both when it was executed by a hand grasping an object and when it was executed by two artificial, ball-like stimuli (Fleischer et al., 2012, cf. also Figure 12.9). The model not only showed a certain degree of entity invariance, but action recognition was also achieved when being presented with the same interaction in different spatial areas as well as from different perspectives. The critical involvement of the goal object could be explained as well. In essence, the focus had to lie on the *relative* spatial interaction over time, centering the goal object and monitoring the motion of the other object relative to the goal object (Fleischer, Caggiano, Thier, & Giese, 2013). The model was also compared to *mirror neurons* (cf. Section 10.3.2), seeing that mirror neurons exhibit similar generalization properties (Caggiano et al., 2011; Fleischer et al., 2013; Gallese et al., 1996). Related models were also developed for the recognition of behavior, once again relying on relative spatial encodings (Giese & Rizzolatti, 2015; Schrodt et al., 2015).

When abstracting away from the actual objects that are being manipulated manually, or that interact with each other, an object interaction event ontology has recently been described. In all cases, the interaction events are structured in an approach, a manipulation, and a withdrawal phase. Only the presence or absence of objects, the changes in the object – such as an object destruction, modification, or the generation of smaller piece – and the changes in the relations between the objects – such as putting on top, connecting, or covering – were critical to generate the ontology. As a result, the authors could identify fewer than 30 fundamental manual object manipulations, which could be structured in a hierarchical taxonomy (Wörgötter, Aksoy, Krüger, Piater, Ude, & Tamosiunaite, 2013).

In sum, the models considered suggest that abstractions over object identities, locations, and viewing angles by means of relative, goal-centered encodings can result in very general event representations, including manual interaction events, locomotion events, physical events, and even social events. These representations can then be assumed to be generally applicable under various circumstances and can be flexibly adapted to the actual object that is to be manipulated or the actual event that is perceived. Thus, event encodings facilitate a view-invariant event recognition, regardless if one executes a behavior oneself, another person executes a behavior, or another biological system or physical system causes or exhibits a particular behavior.

12.5 Summary and conclusions

We have learned about the brain's motor control, action planning, and decision making capabilities. Essentially, body and brain maintain a cascade of motor control routines. The lower levels in the muscles and spinal cord, as well as the body morphology, contribute to maintaining and stabilizing basic limb postures and movement dynamics, yielding local morphological attractors, such as a comfortable seating posture (cf. also Section 3.6.1), and self-stabilizing dynamics. Even more importantly from a cognitive perspective, the cascade essentially eases the control burden and thus enables a deeper, interconnecting analysis of the experienced sensorimotor interactions on higher, cortical levels.

Moreover, we have shown that a well structured, associative network for decision making and high-level, goal-directed motor control develops in the cerebral cortex. This network with its typical, somewhat modularized, but highly connected structures develops and optimizes object and spatial action complexes for manipulating objects and the body itself. Following the ideomotor principle, motor control is triggered by internal motivations and resulting goals, bootstrapping the development of motor primitives and motor complexes. The cerebellum probably tends to optimize and smooth behavior by means of forward models, or even pairs of forward inverse models.

From a computational perspective, we then considered which control structures are maximally suitable for inducing behavior in a complex, but somewhat hierarchically arranged world. We have suggested that behavior, and even complex environmental interactions, seem to be clustered into events and event transitions. During an event, an action or a sequence of actions is executed. Thus, events and event transitions structure actions into bigger, and potentially social, conceptualizations and taxonomies.

The basis for developing cognitive event taxonomies appears to lie in action encodings, which consist of few, but distinct fundamental components. First, each action needs to be encoded by its final goal, that is, the final reward and/or the final effect, when the goal is reached. Second, particular forward-inverse sensorimotor control model structures, which are continuously optimized and partially diversified during development, enable local inverse control as well as local forward predictions – dragging motor control along the envisioned dynamic trajectory supporting the smooth and goal-oriented execution of behavior. Finally, conditions need to encode the relevant circumstances about when the action can be applied at all.

As a consequence, conditions of actions provide entry points to particular environmental interactions, resulting in a competition between different object affordances (Cisek, 2007). Given sufficiently developed action event structures, decision making essentially takes into account current behavioral options, that is, affordances (how can I interact with the world), the expected effort, the probability of success, and the expected reward. Interestingly, this same principle applies not only to actual motor control (as of moving the body), but also to the control of attention, and probably also to mental control, that is, thinking.

Abstractions, based on interaction experiences and event signals, such as sudden transitions, the establishment of a contact, the utterance of a speech sound, etc., help to conceptualize the environment. That is, the occurrences of event transitions are progressively predicted more accurately in that the condition structures of more actions and system behaviors in general precisely specify which sensory and dynamic motion encodings need to signal information about the environment in a particular manner. For example, parietal cortical areas may signal reachability and pre-activate suitable motor primitives, such as particular grasps, in premotor cortex. Meanwhile, temporal areas may activate the current object identities and associate reward expectations, anticipating potential interactions. An action decision takes these factors into account to decide with which object to interact.

Akin to an object or animal taxonomy, event encodings develop into hierarchical, spatial and temporal, event-grounded taxonomies. The overlap in the structure of these taxonomies is striking and may indeed enable analogical and metaphorical thinking across such taxonomies. For example, when stating that "he was as big as a tree," the concept of "human size" is associated with "plant size". Similarly, when stating that "this topic is so dry," the concept of a "topic" is associated with "dryness", such as dry powder or dust, which simply does not stick together, because the reader fails to integrate the topic into other conceptualizations (cf. Lakoff & Johnson, 1980 for many more examples along these lines).

Moreover, hierarchical event taxonomies enable the anticipation of final behavioral effects on multiple levels, enabling, for example, the experience of "Vorfreude", that is, pleasant anticipation, while preparing a nice dinner. In a social context, during action observations we attempt to infer the final goals of the observed actor, which is essentially the anticipated final effect of an observed event. Successful goal and intentional inferences consequently facilitate social interactions, because one's own actions can be attuned properly to the inferred, current goal-directed actions of others. Also, information transfer can be optimized by means of linguistic utterances, gestures, or even by the intentional exaggeration of particular behavioral components during an interaction (Pezzulo, Donnarumma, & Dindo, 2013; Sebanz, Bekkering, & Knoblich, 2006; Wagner, Malisz, & Kopp, 2014).

The gathered insights suggest that basic event encodings are grounded in the own motor behavioral system. Once some of these encodings have developed, it becomes easier to learn similar event-structured encodings about our physical environment, about other biological systems including plants, animals, and other humans, and even about abstract systems, such as a mathematical or a political system. For example, in a mathematical system an event may characterize the execution of a particular mathematical operation. As we can imagine an object as well as an object interaction event without its presence, we are able to imagine non-physical entities, such as an idea, a thought, or a political system, as well as non-physical entity interactions, such as "shooting down an idea", "dismissing a thought", or "destroying a political system". Thus, while being grounded in motor-controlled environmental interactions, event perception has generalized from motor behavior to any type of system behavior.

Event encodings, regardless if they encode own motor behavior or more abstract system behavior, also offer themselves as suitable environmental conceptualizations. Symbols can refer to particular action event encodings, which conceptualize, for example, particular object interactions. Similarly, other symbols can refer to system event encodings, which conceptualize, for example, particular behavior of the weather, a biological system, or even a political system. Thus, event encodings offer structures with which symbols can be associated, thus grounding the symbols in environmental conceptualizations. In the following chapter, we show how language structures, including word conceptualizations and grammatical compositions, are closely related to such event-oriented conceptualizations.

12.6 Exercises

1. In which manner do muscles and the spinal cord ease the control burden for the central nervous system?

2. Explain how eligibility traces in RL may be related to dopamine gradients and sharp-wave ripples in the hippocampus?

3. Relate the apparent compartmentalization of particular actions, such as defensive motions, manual manipulations, climbing, grooming, and scratching, to dynamic motion primitives and their potential optimization by means of policy gradients (cf. Section 5.4).

4. Reconsider the example of placing a glass into the dishwasher. However, consider now that the glass currently sits upside down on the kitchen counter next to the dishwasher. Which grasp will be applied most likely used given estimates of success and effort? What if the glass currently sits high up on some top-shelf tray?

5. The end state comfort can also be observed in social contexts, such as when handing over a knife to another person. Explain what it means to focus on a social end state comfort.

6. Determine how many degrees of freedom are available when intending to point your arm in a certain direction.

7. Explain how the SURE_REACH model is able to avoid obstacles while pursuing to reach goals.

8. Specify likely conditional, motor primitive, and effect encodings that characterize a baby bottle from the experience of a baby before being able to hold the bottle.

9. Specify the event of opening a bottle (as an adult) on two interactive levels of abstraction by means of suitable action event schemata. Also sketch-out the resulting simple bottle-opening taxonomy.

10. Give examples for hierarchical event taxonomies considering an event at work, at home, during vacation, and while pursuing a hobby.

11. Consider simple actions, such as kicking a ball, and sketch-out the sensory and motor aspects that may be integrated in a common event code. Particularly focus and contrast unimodal, sensorimotor, and multimodal dynamics, which may be bound together. Then do the same thing for an action complex, such as playing soccer.

12. Characterize simple action events by the seven properties introduced above - four to characterize the event itself and three to further specify the decision-making process.

13. In which manner to conditional encodings of an event schema conceptualize the environment and focus cognition on the behavioral- and goal-relevant aspects of the environment?

14. In which manner may conceptualizing conditional and final effect encodings help to properly reason about environmental interactions?

15. Notice that the task to navigate through an environment in order to reach a certain goal location is typically also accomplished by means of an event taxonomy. Navigation unfolds dependent on the current means of locomotion used (car, train, bike, bus, feet, etc.) and the knowledge about the environment. When considering everyday navigation tasks, a hierarchical taxonomy can be identified. Construct an event-related taxonomy of going to the movie theater on several levels of abstraction.

16. When considering navigation in a novel environment, we attempt to navigate either based on landmarks, using map knowledge, or by using directional knowledge, dependent on the knowledge we have available (Wiener, Büchner, & Hölscher, 2009). Discuss the relation to event goals and the unfolding inverse models in these respects.

17. Discuss how the perception of object affordances may ease the planning of particular action sequences. In which case, however, may it hinder planning and reasoning processes?

18. When working in the kitchen chopping up and frying up vegetables, characterize typical event transitions and available sensory signals, which signal these typical event transitions and which may thus be used to suitably segment the interaction experiences in the first place as well as to systematize and abstract the interactions in a goal-oriented fashion.

Chapter 13

Language, Concepts, and Abstract Thought

13.1   Introduction and overview

Human language is the most complex known form of natural communication. The understanding of the evolution of language has even been considered the hardest problem in science (Christiansen & Kirby, 2003b, 2003a). Language is amazingly interactive and the mechanisms that have led to the phylogenetic evolution of language as well as the mechanisms that lead to the ontogenetic development of language comprehension and production during childhood are still being hotly debated. On the one hand, statistical linguistics has uncovered statistical structures in large text corpi, implying that meaning and semantics may be extractable from statistics to a certain extent. On the other hand, theories of embodiment argue that meaning and semantics must be embodied and must thus be understood by means of sensorimotor, embodied structures and mechanisms (Bergen, 2012; Johnson, 1987; Lakoff & Johnson, 1980, 1999).

This chapter considers both aspects from a computational perspective and in light of the insights gained about encodings and mechanisms of sensory processing, attention, and motor control in the previous chapters. In particular, starting with basics from linguistics, we describe how language capabilities may be grounded in principles of embodied cognition (Barsalou, 1999, 2008; Beilock, 2009; Knott, 2012). To do so, we also consider the question how language may have evolved over the last two million years and how language development typically proceeds in young children. We will emphasize the importance of social interactions, and especially the great social benefit when becoming able to coordinate social cooperations and to share information in groups of individuals linguistically.

When considering how words are understood, the semiotic triangle is explained, which conceptualizes the fundamental relationship between a word, the thought that is expressed with the word, and the concept or entity, that is, the *referent*, the thought refers to. Seeing that the referents are existing in the world, being either a concrete object or entity or an abstract concept, such as a political system or the realm of mathematics, we establish the Greek philosophy concept of a universal idea, which is grounded and thus existing in the real world. We conclude that while the thought that is invoked by a word may individually differ, the referent the thought refers to is universal, because it exists in our world.

As referents are universals, Chomsky's Universal Grammar emphasizes that also universal grammatical structures can be found in all human languages. Derived from recent insights in cognitive systems and cognitive robotics research, we detail how the Universal Grammar may actually be shaped by event-oriented conceptualizations, which may have initially developed to support hierarchical planning and abstracted reasoning capabilities (cf. Section 12.4). With a grammatical language in hand, we then consider how conversations unfold in an imaginary *common ground*, which is shared among the conversation partners, and which conceptualizes the actual information, the ideas, and the larger conceptual understandings that unfold. During a conversation, the conversation partners can be viewed as attempting to shape this common ground based on their communicative intentions and their social assumptions about the knowledge and the unfolding understanding of their communication partners. Nonetheless, in the end all conversation partners understand a conversation in their individual *private grounds*, which overlap with the common ground the stronger, the more mutual understanding is reached. In conclusion, we propose an embodied, behavior-oriented, developmental perspective of language, whose structures resemble those of perception-, anticipation-, and behavior-oriented conceptualizations and compositions thereof.

In the following sections, we first provide an introduction to language from a linguistics perspective. Next, we give an overview of the fundamental brain structures that seem to support language generation and comprehension. We then consider language evolution and ontogenetic language development. In this respect, we sketch-out which conceptual structures are available in the brain before language comprehension and production commences, and how these conceptual structures can help to bootstrap language development. We also discuss how these structures appear to interact with the grammar of the particular language(s) encountered during childhood. Finally, we consider how language can support and enhance perspective taking, planning, spatial and feature-based reasoning, episode recollection, and even abstract thought.

13.2   Introduction to linguistics

To uncover the general functionality and possible origins of language, it is helpful to define language in general terms. Perhaps a good functional definition is that *language is a framework for communication that provides a shared system of signs and semiotic rules for the production, transmission, and comprehension of information.* The function of language is communication, that is, the goal-directed exchange of information between the conversing partners. Signs and symbols, that is, words, but also particular gestures in sign language and even pictures, are mutually associated with a certain *meaning.* Note that this association is essentially arbitrary, such that any arbitrary symbol can generally refer to any type of entity or thought. Meanwhile, though, during a conversation speakers and listeners assume that they share a common or at least an overlapping vocabulary, that is, the words of a language with their (arbitrary, but systematic) respective meaning associations. Semiotic rules govern how signs and symbols are combined into larger units of thought, which are expressed in sentences. These rules are *generative*, streaming the combination process, and contain considerations of *syntax, pragmatics,* and *semantics.*

To further differentiate these aspects of communication, meaning, syntax, semantics, and pragmatics, let us look at the differentiating characterizations set forth by the American linguist and anthropologist Charles F. Hockett (1916–2000), which highlight the diverse, but interactive characteristics of a language (Hockett, 1960; Hockett & Altmann, 1968):

- *Vocal-auditory channel*: language is accomplished through speaking (with mouth) and hearing (with ears). More recently, this characterization has been extended to also include tactile-visual (for example, in sign language) and chemical-olfactory (for example, pheromones in ant trails, cf. Section 3.6.1) communication channels.

- *Broadcast transmission and directional reception*: speakers transmit sounds in all directions; listeners perceive the direction from which the sounds are coming.

- *Rapid fading*: sounds fade rapidly (waves disappear once the speaker stops speaking), such that they exist only for a brief period of time, after which they cannot be perceived any longer.

- *Interchangeability*: speakers can broadcast and receive the same signal (so, anything that one can hear, one can also say, although there are exceptions to this rule).

- *Total feedback*: speakers hear their own speech and can monitor language performance while they produce language (note the close relation to sensorimotor forward models).

- *Specialization*: speech signals are intentional and solely serve communication.

- *Semanticity*: specific sound signals are directly tied to particular meanings.

- *Arbitrariness*: typically there is no intrinsic or logical connection between sound form and its meaning. Words only get their meaning via the objects, entities, or units of thought they represent. As a consequence, different words (for example, of different languages) can refer to the same object, entity, or unit of thought.

- *Discreteness*: speech can be broken down into small discrete units (*phonemes*), which are perceived categorically, even if physically different (though similar).

- *Displacement*: we can talk about things remote in place and in time.

- *Productivity*: we can say things never said before (create new words) or with other words (metaphors, analogies, poetry) without disturbing the understanding.

- *Cultural or traditional transmission*: although language capabilities are partly innate, language is learned in a social setting (by interacting with experienced language users). As a consequence, language and culture are woven together.

- *Duality of patterning*: meaningful messages consists of smaller meaningful units (*lexemes*), which in turn are made up of even smaller units (*morphemes*).

- *Prevarication*: we can lie or deceive, that is, we can make false, meaningless, or intentionally misleading statements.

- *Reflexiveness*: language can be used to talk about language.

- *Learnability*: language is teachable and learnable, such that we are not only able to learn our mother tongue, but also multiple mother tongues in parallel as well as other, foreign languages later on in life.

These characterizations have become reference points for contrasting animal with human communication systems, as well as for general considerations on the nature of human language. Some of the described features are certainly also present in animal communication systems: for example, honey bees communicate the location of food sources by a tail-waggling dance, which is a specialization of a tactile-dynamics channel. By means of the angular movement during the dance, the bee communicates the direction, and the dance speed indicates distances to the food source. Even the richness of the food source is encoded. Thus, clearly semantic meaning is communicated.

Monkeys have distinct alarm calls for different predators, such as a "leopard alarm", which causes monkeys to climb up trees, versus an "eagle alarm", which causes them to drop down from the tree tops. Thus, the alarm calls carry semantics, are specialized, and arbitrary. Moreover, they are interchangeable so that any monkey can produce and perceive the alarm call. The reaction to alarm calls is largely learned, so that the traditional transmission feature of language is also covered. Even prevarication can be exhibited to a certain extent, as a monkey may sometimes falsely utter an alarm call, for example, to distract the others away from a food source. However, several aspects seem to be only barely covered. Discreteness is only given in its simplest form. Displacement may be possible to a certain extent, but only in apes that have been trained to communicate via symbols. Productivity, however, has rarely been observed and the duality of patterning typically collapses to simple calls. Possibly the hardest trait, reflexiveness, has never been observed in animals.

These characteristics and the proposed definition of language should well serve us in taking a more detailed look at linguistics. In a certain sense, linguistics is the manifestation of Hockett's reflexiveness: human language enables the study of language itself. Several complementary approaches for studying language have emerged:

- *Theoretical linguistics* describes the principles underlying language as an abstract system.

- *Developmental linguistics* studies how infants and children begin to learn language in their individual ways.

- *Language evolution* considers the question how human language has evolved from animal communication in common ancestors. Evolutionary changes in modern languages over the last centuries are also considered.

- *The cognitive approach*, finally, understands language as a symbol-oriented conceptualization of the physical environment.

Naturally, we focus on the cognitive approach to language, but also take into account theoretical, developmental, and evolutionary considerations. Moreover, we relate the linguistics approaches to the computational perspective put forward in this book. As a result, we hope to show how language is generated and comprehended by the brain. To proceed, we first take a look at the historical development of linguistics as a separate discipline.

13.2.1   Historical sketch

Efforts to understand language have progressively intensified with the development of sophisticated cultures. Initial efforts focused on descriptive approaches to language. For example, in the 6th century BC, the grammar of Indian Sanskrit was described. Along similar lines, the Greek scholar Dionysios Thrax (2nd century BC) characterized old Greek grammar, starting to distinguish particular types of words, such as nouns and verbs, as well as fundamental morphosyntax categories, such as word cases and tenses. Similar works can be found for Chinese Mandarin and Arabic.

During the middle ages, Islam spread throughout Europe with Arabic as the lingua franca. This led to the further development of more detailed grammatical descriptions of Arabic for non-native speakers. At the end of the middle ages, interest intensified in linguistics, particularly for the purpose of producing accurate Bible translations, such as Martin Luther's (1483–1546) bible translation into German (1522–1534).

Already in 1244, the English philosopher and Franciscan friar Roger Bacon (~1214–~1292), also known as "Doctor Mirabilis" (the wonderful doctor), acknowledged that:

> [...] grammatica una et eadem est secundum substanciam in omnibus linguis, licet accidentaliter varietur [...] [In its essence, grammar is one and the same in all languages, even if there are somewhat accidental variations (own translation)] (Noland & Hirsch, 1902, p. 27.)

Starting with the renaissance in the 15th century, also questions about the origins of language and its development were considered. For example, in 1660 the *Port Royal Grammar* was published by Arnauld and Lancelot (cf. Section 2.4.3), which made particular universalities in grammatical structures more explicit. In 1869, the German linguist August Schleicher (1821–1868) introduced a *tree model* of the evolution of language according to which language has evolved based on the principle of natural evolution, similar to the origins of species (Darwin, 1859). Schleicher's tree model suggested that language is in permanent flux. Nonetheless, it also sketched-out roots in language evolution, tying different languages together and enabling the analysis of lexical and grammatical commonalities across different languages from an evolutionary perspective.

Today, studies in linguistics can be separated into five major branches:

- *Phonetics and phonology* – the anatomy and physiology of speech production across languages as well as the organization of speech sounds within a specific language.

- *Morphology* – the formation of words given phonetic and phonological principles.

- *Syntax* – the systematic formation of sentences from words.

- *Semantics* – the meaning behind words, sentences, and discourse.

- *Pragmatics* – language use in communicative contexts.

We will first detail these five topics and then proceed with how the brain learns to generate and comprehend language in light of these aspects.

13.2.2 Speech sounds: phonetics

Phonetics addresses the question how sounds are produced, considering the human speech production system. "Speaking" in terms of the deliberate production of specific sounds is one of the most fascinating motor processes: the respiratory system needs to be precisely controlled to guarantee a consistent subglottal pressure for a continuous outgoing stream of air. This outgoing stream of air is first modified at the glottis (a gap located in the larynx, which is formed by the vocal folds), which enables us to whisper, murmur, speak with a creaky voice, or speak normally.

Specific speech sounds are then produced in the vocal tract (that is, pharynx, nasal, and oral cavity) by forming constrictions in the resonating cavities or by deforming the cavities themselves, thereby causing turbulences in the stream of air, which are perceived as different sounds. In this manner, the distinctive sounds of consonants and vowels are produced. Different consonants are generated by altering the shape of constrictions in the resonating cavities. For example, both [t] and [s] are produced by touching the teeth ridge with the tip of the tongue (so called alveolar sounds).[1] Additionally, the manner of articulation, that is, how the constriction is formed further influences the outgoing sound. The sound [t] is produced by rapidly releasing a total oral closure (called stop or plosive), while for the [s] sound a turbulence in the air stream is formed producing a hissing sound (called fricative).

Vowels, on the other hand, are produced by bringing the lips and the tongue to specific positions, which change the shape of the oral cavity. As a result, vowels can be nasalized (by lowering the velum) and their length (for example, beat [biːt] vs. bit [bɪt]) can be varied, depending on the contraction of the muscles of the tongue during articulation.

These variations give only a glimpse of the major variations during consonant and vowel production. Many further variations as well as other sounds, such as click consonants in various African languages, can be distinguished, but are not further considered here. Figure 13.1 gives a crude overview over the involved organs and muscle groups.

A language learner has to figure out for him- or herself how to use this huge number of degrees of freedom provided by the speech organs to produce particular sounds. As in the motor control of other parts of the body with multiple redundant degrees of freedom, it is quite likely that *motor synergies* and evolutionary prestructured *motor primitives* (cf. Section 12.2) support the ontogenetic development of sound production capabilities. For example, every baby can produce a crying sound. Moreover, when suckling, for example, other typical shapes of the mouth are trained and probably assimilated and further modified for language production.

13.2.3 Words: phonology and morphology

As suggested by Hockett's language aspects, words in a language can be contrasted with mere vegetative or emotional sounds (like laughing or a shriek of fear/surprise). The utterance of a word conveys a specific meaning and the word is arbitrarily associated with the meaning.

---

[1]Note that written language often does not really reflect pronunciation, for example, the "oo" is pronounced differently in "boot" and "good". Vice versa, the vowel sounds in "good" and "would" are the same although they are spelled differently. Therefore, we use the international phonetic alphabet (IPA) to represent sounds of oral language.

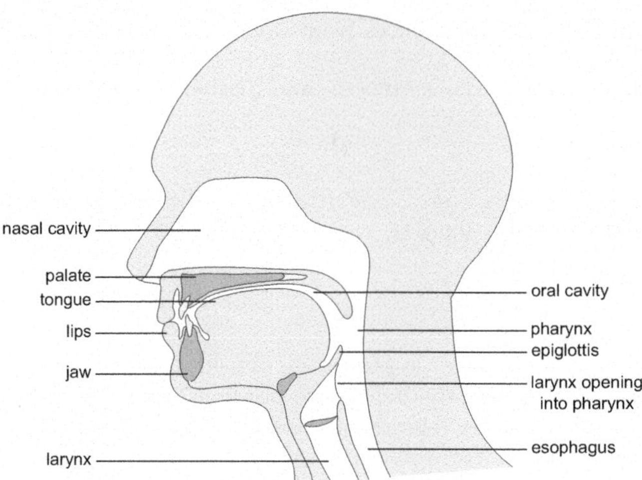

Figure 13.1: More than 100 muscles are involved in speech production. The image illustratively shows the most important cavities and organs involved.

Leaving the question of what meaning actually is to the side for now, here we focus on the units words are made of, which are phonemes and morphemes.

Even at this basic sound level, distinctions in meaning come into play. The smallest linguistic unit for conveying a distinction in meaning by means of sound alternations are called *phonemes*. For example, when changing the first consonant in the word "may" [meɪ] to "day" [deɪ], the meaning of the whole word is altered. Thus, /m/ and /d/ are different phonemes. However, it is not necessarily always the case that a different sound is a different phoneme. *Allophones* refer to phonemes that can be uttered with such variability that each is heard differently, but nonetheless they are all in the same phoneme category. For example, the standard [ʀ] phoneme may be uttered as a rolled, Bavarian [r] or a southern [ʁ] without any modification in meaning.

As there are inevitably always slight differences in the way we move our articulatory apparatus, technically speaking we never produce the same sound twice. Nonetheless, as is the case in vision and other forms of sensory perception, we are able to classify particular percepts into corresponding sound categories, which is a capability that develops very early in life (cf. Section 4.2).

While phonemes only alter meanings of more complex structures, *morphemes* are the smallest meaning bearing grammatical units. Morphemes can be single sounds, sound groups, or whole words. For example, the word "unhappiness" is made up of three morphemes: "un-" indicates the inversion of the meaning of the main word (the *root*); "happy", which is the root, conveys the core meaning; and "-ness" indicates a change of the root word into a noun, that is, into a state of being happy. Morphemes that only make sense in combination with a root – such as prefixes and suffixes, as well as inflection-indicating morphemes – are called *bound morphemes*. On the other hand, those parts that can function independently as a word (such as "happy" in our example) are called *free morphemes*.

At the word level, a similar distinction is made: while "word" refers to any word in any form, a *lexeme* refers to the basic unit of lexical meaning. *Synonyms* are different words that refer to the same lexeme. For example, "run", "ran", and "running" are all different words with slightly different meaning (differentiating the time in which the event takes place), but their general meaning, that is, their lexeme, moving quickly on one's own feet, is the same. Meanwhile, the same sound can have different meanings, such as the inflectional suffix "-s" in English, which can either denote the plural form in a noun or the third person form of a verb. In this case the meaning depends on the context in which the morpheme occurs. Similar to allophones, *allomorphs* exist that differ in pronunciation, but not in meaning.

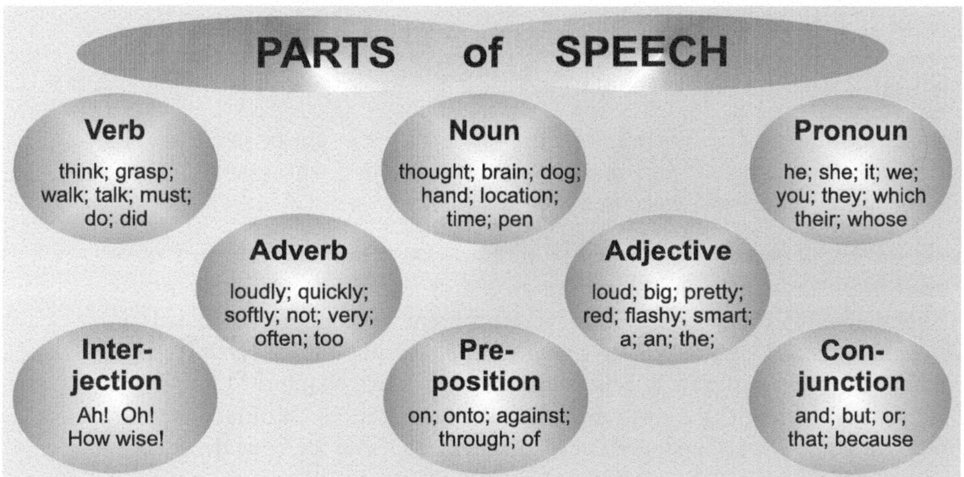

Figure 13.2: Words can be assigned different roles in a sentence. To assign such roles, different types of words play distinct roles. The shown fundamental word types can be considered building blocks in the creation of whole, meaningful sentences with the help of syntactic rules.

For example, the English standard plural making suffix "-s" is pronounced /-z/, /-s/ or /-ɪz/ without altering the meaning.

13.2.4 Sentences: syntax

Usually, it is not enough to utter single words in order to convey a message. Numerous words need to be put into context. However, to be able to identify the role of the individual word in such contexts and thus to be able to form an unambiguous mental composition of the words or, similarly, to produce an unambiguous utterance of a word-based composition, the combination of words needs to follow accessible rules. For example, rather different situations are described when uttering "the cat chases the woman" versus "the woman chases the cat," but how do we know who is chasing whom?

Because we know that English *word order* generally follows the rule subject–verb–object, whoever is mentioned first is the one who chases the other one. On the other hand, many other languages do not solely rely on word order and also English allows modifications. For example, the sentence "the woman was chased by the cat" is equally valid and has the reversed word order with respect to the meaning. Thus, word order as well as the form of the word and other accompanying words, such as the tense of the verb and the preposition "by" in the example, can influence the meaning of a sentence. Other languages, such as German, are not as strict with respect to the word order such that "die Frau jagt die Katze" (that is, "the woman chases the cat") can be interpreted in both ways, that is, either one chases the other one, so that the meaning has to be inferred by means of the context.

In its essence, then, syntax is about the rules, principles, and processes that tell us how to structure sentences in a given language to convey precise, unambiguous meaning. Thereby, syntax builds on types of words, as the building blocks to construct meaningful sentences. Figure 13.2 shows the main speech building blocks, or parts of speech and lists examples.

Most fascinating about syntax is the fact that a limited set of such (grammatical) rules is enough to generate an infinite number of utterances – a fact that ties closely to Hockett's productivity aspect: despite a limited set of grammatical rules, human languages can convey an infinite number of ideas. The German politician, author, and linguist Wilhelm von Humboldt (1767–1835) pointed to this stunning capability of language as follows:

> [...] Sprache [...] muß daher von endlichen Mitteln einen unendlichen Ge-
> brauch machen. ([...] language [...] makes infinite use of a finite number of tools.
> (own translation)) (von Humboldt, 1973, p. 477.)

As language is something that evolved naturally, the study of syntax is a rather descriptive process. Noam Chomsky (cf. Section 2.4.3) tried to model syntactic rule systems
mathematically by means of a *transformational generative grammar*:

> From now on I will consider a *language* to be a set (finite or infinite) of
> sentences, each finite in length and constructed out of a finite set of elements.
> All natural languages in their spoken and written form are languages in this
> sense, since each natural language has a finite number of phonemes (or letters
> in its alphabet) and each sentence is representable as a finite sequence of these
> phonemes (or letters), though there are infinitely many sentences. Similarly, the
> set of sentences of some formalized system of mathematics can be considered a
> language. (Chomsky, 2002, p. 13, author's emphasis.)

Chomsky thus formalized a generative grammar G as follows: $G = <N, T, S, R>$,
where: N is the set of *non-terminal* symbols, that is, of placeholders. Placeholders support
the generation of grammatical sentences, and are eventually replaced by other symbols in
N or in T. T specifies the set of *terminal symbols*. S is the set of *start symbols*, which are
particular non-terminal symbols from which the generative grammatical process commences.
Finally, R is the set of *production rules*, which define the possible replacements of strings
of start symbols and non-terminal symbols with strings of other non-terminal and terminal
symbols.

For example, let us consider the grammar for the formal language $\{a^n b^n | n \geq 1\}$. This
grammar includes essentially all words that begin with at least one a symbol followed by an
optional additional arbitrary number of a symbols (n in total), which is then followed by
the same number of b symbols. The corresponding grammar can be defined as follows:

$$G = (N = S, T = \{a, b\}, S = \{S\}, R = \{S \rightarrow aSb, S \rightarrow ab\}. \tag{13.1}$$

where R specifies two replacement rules. The first rule specifies a recursive rule, enabling the
generation of further a and b symbols. The second rule converts the non-terminal symbol
S into the minimum size terminal string, which is ab. For example, when applying rule one
twice and then applying rule two, the resulting grammatical word is $aaabbb$.

Depending on which rules are allowed, Chomsky distinguished four grammatical classes
(types 0–3). Type 3 grammars are also known as *regular grammars*. They are the most
restricted type of grammar, restricting the allowed set of replacement rules such that only
single non-terminal symbols may be replaced with one terminal symbol or with a combination
of one non-terminal and one terminal system in either right or left order. Type 2 grammars
are *context-free grammars*, such as the example given previously. The rules in context-
free grammars are restricted to replacements of one non-terminal symbol into an arbitrary,
possibly empty string of terminal and non-terminal symbols. They are context-free, because
the replacement of non-terminal symbols cannot consider the context of symbols surrounding
the single non-terminal symbol. Type 1 grammars are *context-sensitive grammars*, which
allow the rules in the grammar to specify a context (of terminal and non-terminal symbols)
within which replacements of one non-terminal to a string of non-terminal and terminal
systems can take place. Thus, type 1 grammars are more general than type 2 grammars,
because they can specify a larger variety of languages. For example, the formal language
$\{a^n b^n c^n | n \geq 1\}$ can only be specified by a context-sensitive grammar. Finally, the most
general type 0 grammars, which have been shown to be equivalent to a Turing machine and
thus a modern computer in their computational abilities, allow any type of production rule.

While grammars can be closely related to the automaton theory and different levels of
computability, Chomsky was mostly interested in their relevance for constructing sentences
in natural languages. For example, "The cat that the dog chased died" links "cat" with

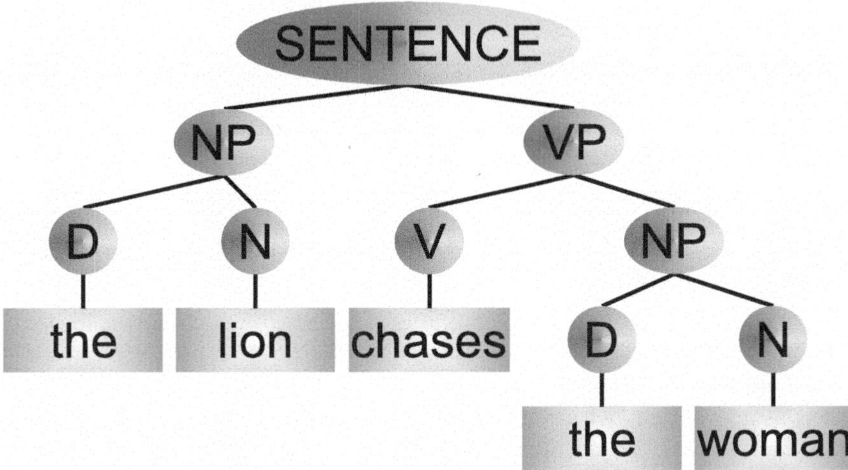

Figure 13.3: A context-free grammar allows to analyze sentences by means of tree structures, in which noun phrases and verb phrases can be flexibly combined.

"died" and "dog" with "chased". Thus, the sentence obeys the structure **abba** and expresses a *recursive* construction. Chomsky viewed recursion as a critical element of natural languages. It allows for a recursively more detailed description of a particular state of affairs. For example, the sentence version posited earlier could be enhanced by stating that "The cat(a) that the dog(b) that the rat(c) bit(c) chased(b) died(a)" and so forth. Today, it is assumed that most natural languages are context-free (Pullum & Gazdar, 1982), although some counter examples can be found (Shieber, 1987).

The advantage of a context-free grammar is that sentences of a natural language can be expressed by a tree structure of *constituents*, where constituents refer to a word or a group of words, which function as a unit in a sentence. Well-known units of a sentence are the sentence (S) itself, which can typically be broken down into noun phrases (NP) and a verb phrases (VP). For example, the sentence shown in Figure 13.3 – "the lion chases the woman" – can be broken down into an NP ("the lion") and a VP, which can be broken down further into a verb V ("chases") and another NP ("the woman"). Both NPs consist of a determiner (D) and an actual noun (N). Figure 13.3 shows the corresponding context-free grammatical tree.

In sum, syntax specifies the production rule with which grammatical sentences can be constructed in a natural language. Typically, the syntax of one language differs somewhat from the syntax of other languages. However, Chomsky postulated that all humans have the universal, possibly inborn capability to learn the grammar of a natural language and that natural languages offer particular systematicities that facilitate this learning process (Chomsky, 1965). While the universality of these systematicities has been questioned numerous times and natural languages have been found that somewhat contradict the universal grammar principle, certain regularities are indeed found across all natural languages (Jackendoff, 2002). From an embodied, computational perspective, this should not at all come as a surprise. After all, all languages speak about the world we live in and thus inevitably must reflect the typical structures that can be found in that world in one way or the other. Seeing that Chomsky with the term *Universal Grammar* referred mainly to the universal readiness of children to learn a natural, human language, we will reconsider Chomsky's take on it in more detail when addressing language development (Section 13.5).

13.2.5 Semantics and pragmatics

Historically, linguistics has focused most of its efforts on phonology and syntax, owing to the groundbreaking work of Chomsky with his generative grammar perspective. This generative characteristic was often considered to be the central mechanism for constructing sentences and thus for language production. Thus, a *syntactocentric* point of view (Jackendoff, 2002) was pursued by many researchers, focusing on getting the syntactic rules right and categorizing semantics as "something that also seems to be important," but that is too hard to understand – or even as "something that is produced by syntax," possibly in combination with a lexicon whose lexical entries may influence the applicability of syntactic rules. The main problem with this point of view is that the meaning of a sentence, which is considered by semantics, remains completely obscured. It is assumed to develop out of a production rule-based, generative, grammatical system.

Recall the Chinese Room problem, which attempts to highlight that most likely the room, which provides the rules, and the library to produce answers to questions posed in Chinese, is not a conscious or thinking being, but just a machine (cf. Section 3.7). From a syntactocentric perspective, we are to a certain extent facing the Chinese Room problem from a linguistic angle. If the perspective was correct, then we would indeed be language-producing machines and the *strong AI* perspective would be correct as well.

From the computational perspective put forward in this book, however, we have seen that motor behavior, attention, and even sensory processing are typically goal-directed and that any top-down process is essentially generative. Thus, we have seen that cognition is generative in itself, before even considering language.

The *parallel architecture* of Ray Jackendoff (Jackendoff, 2002) hypothesizes that semantics may be as generative as syntax and phonology are, such that all three components, i.e. phonology, syntax, and semantics, may contribute to language generation and comprehension interactively in parallel. As Jackendoff puts it:

> [...] *language comprises a number of independent combinatorial systems, which are aligned with each other by means of a collection of interface systems.* Syntax is among the combinatorial systems, but far from the only one. (Jackendoff, 2002, p. 111, author's emphasis.)

Interactivity, according to Jackendoff, leads to temporary alignments between these combinatorial systems. From a predictive, probabilistic point of view, one would rather speak of establishing temporary consistency by activating consistent subcomponents and mappings between these components. That is, similar to the challenge of maintaining a consistent internal postural body schema of one's arm during the rubber hand illusion, the brain attempts to process language by establishing parallel phonological, syntactic, and semantic interpretations, which need to stand in agreement with each other.

When attempting to comprehend a sentence, phonological input is interpreted respecting the applicable syntactic rules and the common ground that is currently assumed to be shared between you as the listener and the speaker. The common ground may be equated with a form of background semantics, which lead to expectations about the next intention, that is, the next aspect the speaker may verbalize. Meanwhile, the syntactic rules may anticipate how such a verbalization will typically unfold. Finally, the phonological input is interpreted accordingly, interpreting the perceived sound biased toward the parallel expectations. As a consequence, the auditory signals are interpreted by the listener in the attempt to extract the meaning behind them, taking all the listener's knowledge (including the assumed knowledge of the speaker) into account.

Such an interpretation of sentence comprehension and production is essentially part of pragmatics, the study of which can be traced back to the American philosopher and logician Charles Sanders Peirce (1839–1914) (Peirce, 1934), who proposed that a symbol essentially represents something to speaker and listener. While the symbol itself is the same for both, the interpretation of the symbol may differ. Essentially, Peirce proposed that the symbol invokes a kind of *idea* in the listener, which will typically overlap with the idea of the speaker, but which may yet significantly differ from the speaker's idea. Pragmatics, from a general perspective, is thus about conversations between people, the application of syntax, semantics, and phonology during these conversations, as well as the unfolding meaning-oriented interpretations of words and sentences.

The *semiotic triangle* (Ogden & Richards, 1923), which may be traced back even to Greek philosophy, explicitly states that a single word or symbol has three, correlated concepts to it. Each word by itself is an arbitrary symbol, which may exist in its own right. During a conversation, "thought units" of the speaker are mapped onto appropriate words. Meanwhile, the referenced thoughts typically are referring to particular referents in the environment, which may be a concrete object or also a category or even some abstract, general concept. Seeing that the symbol, that is, the uttered word itself stands for the referent and symbolizes the thought that is referenced, a triangular structure emerges. Note, however, that the link between the word, or symbol, and the referent is only indirectly realized via the mapped thought. Only in cases of iconic symbols or onomatopoeias (words whose sounds stands for the referent) can a direct link be drawn. Figure 13.4 shows this triangular relationship.

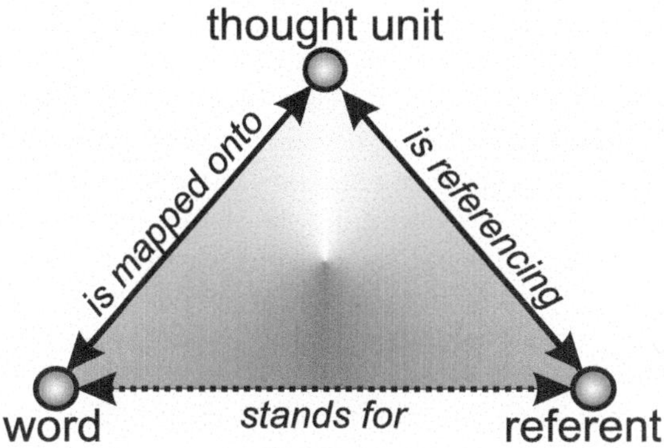

Figure 13.4: The semiotic triangle emphasizes that a symbol only indirectly stands for a referent via the actual idea or thought about the referent.

When again considering the perception of the listener, the semiotic triangle makes it obvious that it is only the symbol itself that speaker and listener have in common. The idea about the symbol and the embedding of the symbol in the current context, with all its interpretations and deduced meanings, are as personal to the listener as they are to the speaker. As the American linguist Leonard Bloomfield (1887–1949) put it:

> We have defined the *meaning* of a linguistic form as the situation in which the speaker utters it and the response which it calls forth in the hearer. [...] In order to give a scientifically accurate definition of meaning for every form of a language, we should have to have a scientifically accurate knowledge of everything in the speakers' world. The actual extent of human knowledge is very small, compared to this. (Bloomfield, 1933, p. 139.)

Thus, the interpretation and usage of a symbol as well as the interpretation of whole sentences and conversations depend on the respective mental worlds of speaker and listener. They are mediated over an assumed common ground, but are grounded and thus interpreted in the end by the listener by means of his or her *privileged* or *private ground*. The same holds true for the speaker, who utters words and sentences based on his or her privileged ground and the assumed common ground, where the latter does not necessarily fully overlap with the assumed common ground of the listener.

As a final important point when given an overview of semiotics and pragmatics, *speech act theory* needs to be considered (Austin, 1962; Searle, 1969). The theory highlights that utterances can be categorized in particular acts of speech, which can be analyzed on three levels: the *locutionary act*, the *perlocutionary act*, and the *illocutionary act* (Austin, 1962).

The locutionary act concerns the actual utterance with the implied meaning of words and sentences; the perlocutionary act concerns the effects of the utterance on the listener(s); and finally, the illocutionary act considers the intention of the speaker and thus the reason why the speaker actually generated a particular linguistic utterance.

John Searle, whom we introduced with respect to the Chinese room problem in Section 3.7, analyzed and categorized illocutionary acts in further detail, proposing a taxonomy (Searle, 1969, 1976). He proposed that five basic kinds of utterances can be contrasted: *representatives* (or assertives), *directives*, *commissives*, *expressives*, and *declarations*.

Representatives are utterances in which the speaker commits to something being the case to varying degrees. A speaker may believe something, know something, suggest something, insist on something, or have heard of something, which is expressed in a representative utterance.

In contrast, *directives* are statements in which the speaker asks the listener to do something. Again, directives can vary greatly. For example, the speaker may utter a question about, or a request for, a piece of information or an object, but the speaker may also give an order, make an invitation, or give an advice.

Commissives are closely related to directives, but differ in that it is the speaker who commits to a certain type of future course of action, without necessarily needing to execute the action himself. For example, the speaker may state that he or she will do something in the future, he or she may thus promise, guarantee, or commit to a future course of action.

Expressives are utterances in which the speaker expresses his or her own emotional or intentional state, such as when congratulating, welcoming, or thanking somebody, or when apologizing for something. In contrast to the other three types, expressives do not attempt to match words to the world, which is the case in representatives, or, vice versa, to match the world to the words, which is the case in commissives and expressives. Searle gives the example of "I apologize for having stepped on your toe" (Searle, 1976, p.12), where the speaker neither attempts to claim that your toe was stepped on, nor that the toe should be stepped on. Rather, the internal state of the speaker is expressed. Interestingly, English syntax contrasts such expressives by requiring a gerundive construction rather than a "that"-clause construction. That is, it is incorrect to state "I apologize that I stepped on your toe."

Finally, *declarations* are utterances that change the state of affairs in the world directly by the utterance, given the speaker has the power to do so. A most obvious example is the statement "I hereby pronounce you husband and wife," effectively declaring that the couple is married from now onwards. Another harsher example is the statement "You are fired."

13.3   Language in the brain

With the basic linguistic principles in mind, we now consider how the brain is able to comprehend and generate language. Neuroscientific evidence suggests that several specialized language areas exist. However, language capabilities are dependent not only on these areas, but rather seem to involve most of the cerebral cortex.

We already introduced Broca's speech production area in the left inferior frontal gyrus as well as Wernicke's speech comprehension area in the left superior temporal gyrus in Section 7.4. By means of lesion studies Broca and Wernicke independently discovered the involvement of these areas in language production and comprehension, respectively. More recent studies with transcranial magnetic stimulation (TMS) have confirmed the findings, selectively impairing speech production or comprehension when temporarily disrupting the functionality of the respective region. As a result, over the years not only have the respective brain areas been named after the persons, who had identified their relevance, but also the types of aphasia (impairment in speech production or comprehension) are sometimes referred to as *Broca's aphasia* (also called *expressive aphasia*) and *Wernicke's aphasia* (or sensory or *receptive aphasia*).

In addition to these core types of aphasia, however, other types have also been identified. For example, *conduction aphasia* is diagnosed when auditory comprehension is generally intact, speech production is fluent, but speech repetition is affected. That is, people suffering from conduction aphasia have difficulties repeating phrases and frequently produce typical, spontaneous speech errors, such as substituting or transposing sounds. Patients with *anomia*, on the other hand, have problems finding the words or names for things they want to talk about. A third example is *agrammatism*, in which patients are unable to speak in a grammatically correct fashion, falling back to a telegraphic kind of speech, omitting function words and mostly forming only two-word sentences as is the case during language acquisition. These examples suggest that language comprehension, processing, and production is somewhat compartmentalized in the brain. That is, different aspects of language are processed in different brain areas.

To a large extent, brain areas that are critically involved in language processing are found in one brain hemisphere, forming a *language network*. In both hemispheres, a *perisylvian network* was identified, which systematically connects the temporal, parietal, and frontal lobes.   Due to the lateralization, in most humans the network in the right hemisphere is dominantly involved in spatial processing and interactions, while the left hemisphere is dominantly involved in language processing (Damasio & Geschwind, 1984; Suchan & Karnath, 2011). In the following paragraphs, we detail a couple of critical modules, which are involved in or provide information for the language network.

The bundle of axons called the *arcuate fasciculus* bidirectionally connects Broca with Wernicke's area (among others). It is mandatory to align the processing of the two areas. Considering primary sensory areas, clearly the primary auditory cortex should be mentioned, which lies anterior of Wernicke's area in the superior temporal gyrus, and is mandatory for auditory speech comprehension. Similarly, the visual cortex is mandatory for processing written language. On the other hand, motor cortical areas transform output from Broca's area into concrete motor commands, including the vocal apparatus for auditory speech production as well as hand and fingers for writing. The *angular gyrus*, which lies between the middle and superior temporal gyrus and the inferior parietal cortex, as well as the *supramarginal gyrus* (SMG), which can be found anterior of the angular gyrus in the inferior parietal area, also need to be mentioned. The angular gyrus is known to be involved in number processing and spatial cognition, but also seems mandatory for integrating multimodal information into a phonetic code. It has also been related to the ability to understand metaphors and to form crossmodal abstractions (Hubbard & Ramachandran, 2003). The supramarginal gyrus lies even closer to primary somatosensory areas. It is involved in tool usage, but it is also known to be relevant for appropriate phonological processing, including auditory and visual processing, such as when asked about the number of syllables in a word (Hartwigsen, Baumgaertner, Price, Koehnke, Ulmer, & Siebner, 2010). Moreover, it has been shown that the SMG is involved in the causation of semantic phonological errors, implying its involvement in binding linguistic features (Corina et al., 1999).

Many more areas are known to contribute to the human language system, such as the fusiform gyrus and the inferior frontal lobe. This overview can only give a glimpse at the complexity of the language system in our brain; however, at least two further observations should be mentioned. First, when only Broca's and/or Wernicke's areas are affected by a lesion, the general intelligence of the patient typically suffers very little. Thus, while these areas are important for speech production and comprehension, it appears that they only play a minor role in general intelligence. Another interesting observation comes from a study with a deaf signer, who had to be tested with a cortical stimulation mapping procedure, activating neurons in Broca's area as well as in the SMG (Corina et al., 1999). The findings suggest that Broca's area was involved in the execution of sign language, which implies that Broca must have some general, evolutionarily determined predisposition for language production, regardless by which means. Indeed, Broca's area has also been shown to be involved in planning, recognizing, and organizing sequences of actions (Fadiga, Craighero, & D'Ausilio, 2009; Fazio et al., 2009; Fogassi et al., 2005; Graziano & Cooke, 2006). Figure 13.5 shows a sketch of the first neurolinguistic model of speech perception and production, which was proposed by Wernicke in 1874 and later on refined by the American neurologist and neuroscientist Norman Geschwind (1926–1984) in the 1960s.

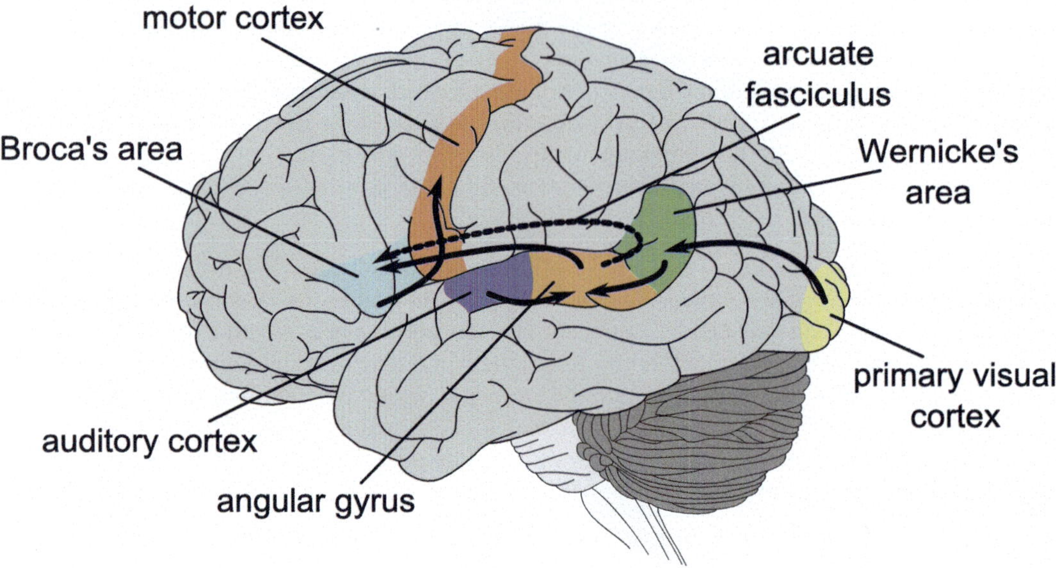

Figure 13.5: The Wernicke–Geschwind model considers the auditory perception of speech via the auditory cortex, reading via the visual cortex, and speech production via Broca's area to the motor cortex. [Adapted with permission from Mark F. Bear, Barry W. Connors, Michael A. Paradiso, *Exploring the Brain*, 3rd Edition, (c) Lippincott Williams and Wilkins, 2007]

Since the invention of writing, language is not a purely vocal-auditory matter any longer; reading and writing also come into play. Interestingly, reading and writing can also be affected by particular language-related deficits. *Alexia* refers to the disruption of reading, and letter and word recognition capabilities. On the other hand, *agraphia* refers to the disruption of writing abilities. Thus, both, reading and writing seem to be modularized in the brain in that dedicated subregions specialize in the respectively relevant sensory and motor processing areas. Finally, it should be mentioned that it appears as if foreign languages that are learned later in life (say after the age of seven), are processed in different brain areas than the mother tongue(s) (Perani et al., 1996). This finding may be viewed as another indicator for an evolutionary predetermined readiness of particular areas to learn a language during the first years of ontogenetic development.

13.4   Language evolution

When considering the question what makes us humans unique when compared with other animals, language is certainly high up on the list. However, when comparing the human brain with that of our closest non-human relatives, the great apes, it becomes apparent that the structures do not differ in any fundamental manner, although particular areas are more pronounced. In fact, it appears that while apes do not have Broca and Wernicke areas, the same brain regions are involved when apes interpret or produce vocalizations that are common in their species, such as distinct alarm calls (Gil-da Costa et al., 2006; Petrides et al., 2005). Thus, although the step from the common ancestors to humans and apes was not a very big one, it was nonetheless a very significant one.

Several evolutionary developments appear to have led to the fundamental cognitive differences in apes and humans:

- The human body has evolved in such a way that speech production is fully voluntarily controllable and is much more differentiable than the one in monkeys. In particular, the larynx, for example, is shaped differently and has moved deeper into the throat.

- The brain has expanded in size, most pronounced in particular areas, and it has increased its brain surface to further accommodate more neurons within the same volume (Deacon, 1997).

- Social abilities of cooperation, sharing, and fairness have evolved further, enabling more intricate human interactions, making linguistic communication even more necessary and valuable.

- While human ancestors initially also likely used manual gestures for communication, the gesture system was eventually replaced by an auditory language system.

- The grammar system has evolved, making human babies language ready during cognitive development.

These five aspects only give a glimpse at the likely complexity involved in language evolution. Moreover, these aspects have not evolved sequentially, but they have most likely co-evolved and are mutually interactive. Furthermore, at least at later stages during language evolution, the co-evolution of culture probably played a significant role (Corballis, 2009; Deacon, 1997).

To proceed, we first consider insights from the development of social skills that distinguish humans from the great apes and probably our common ancestors. Next, we consider the co-evolution of grammatical structures and a gestural communicative system. Finally, we integrate these and highlight the apparent importance of a particular gene code, which may have significantly supported the phenotypic evolution of language capabilities.

13.4.1 Shared intentionality, cooperation, and communication

Possibly a precursor, but most likely co-evolved with linguistic communication is the human ability to cooperate in unprecedented ways (Tomasello, 2014). Indeed, the human capability to cooperate appears to go far beyond the cooperative capabilities found in other species. While the great apes and various other species, including dolphins, elephants, and some bird species such as scrub-jays, show forms of explicit cooperation and communication, a fundamental and distinct difference seems to lie in our ability to trust in the cooperative minds of others, especially when in the same peer group or "tribe". This trust enables cooperation in an explicit, coordinated fashion.

The key to developing such a capability may be *shared intentionality*, that is, the capability to explicitly share current intentions and involved high level goals (Tomasello et al., 2005). The shared intentionality must essentially be built on the belief that all group members have the same high level intentions. Moreover, the shared intentionality extends to the expected final reward, which is expected to be achieved when engaging in a cooperative activity, in that the individuals trust in their cooperative partners' intent to share this reward. In fact, it has been shown that children have this sense of *fairness*, which manifests itself in the fact that they tend to share reward among all group or team members that have contributed to gaining the reward. In contrast, if a reward was gained without the help of others, it is shared significantly less often (Tomasello, 2014).

Given shared intentionality and a sense of fairness, the individualization and differentiation of other humans and especially of the people with whom one interacts is another important factor for enabling effective cooperation. When individualization is not sufficiently evolved, game-theoretic aspects suggest that the best strategy for interaction is *tit-for-tat* – that is, interact with your game opponent or partner in the way he or she recently interacted with you; if you were always good, your partner could cheat on you by always being bad – typically benefiting from your goodness. Thus, one has to punish bad behavior by means of a tit-for-tat, or a *win-stay, lose-shift*, strategy (Axelrod, 1984; Nowak & Sigmund, 1993).

Because we are able to individualize other people, we can selectively interact with each of them according to such a strategy. In particular, we remember if another person was typically good or bad to us, if they previously communicated or cooperated well or not. If cooperation worked out well in the past, one can expect further mutually beneficial cooperations and thus one will value interactions accordingly. On the other hand, if cooperation was refused or reward was not previously shared, one will probably avoid further cooperations. As a result, a society of cooperative partners and teams can develop, where uncooperative people will soon be singled out and have little influence. These principles have culturally evolved into written laws in modern societies, which also specify the negative consequences when disobeying particular laws.

To agree on a particular cooperative actions – possibly including how the expected reward will be shared – requires communication, because the coordination of particular cooperative acts is much facilitated when means of communication are available. Thus, it can be expected that progressively more intricate cooperative acts co-evolved with progressively more versatile communicative capabilities. The evolutionary niche of cooperation thus probably resulted in a fitness pressure toward better means of communication, and, vice versa, better means of communication enabled even more intricate means of cooperation.

Language offers itself as the perfect means to coordinate not only physical, but also mental interactions by communication. Sharing information is a very effective means of warning others about upcoming danger, or, generally of informing them about aspects in the environment that may be relevant for them or that may at least interest them (such as, "I saw tigers by the water."). Moreover, group efforts can be coordinated in light of anticipated future events (for example,"Winter is coming, we need to store food.") or in light of a current situation (for example, "The mammoth is trapped in the gorge. I block its path, and you kill it from the top of the cliff."). Similarly, the coordination of labor division becomes possible, such as who is going to hunt, to gather, watch the kids, take care of the fire, and so forth. Coordination also allows the group to split up, where some may go hunting and others gathering elsewhere. Moreover, individuals could spend more time doing one particular thing for the group and thus perfecting the necessary skills. Such coordination thus can improve the situation of the group, increase the likelihood for survival and reproduction, and open up access to new food sources (Harari, 2011).

13.4.2　Gestural theory and verbal communication

In contrast to gesture based communication, verbal communication has the great advantage that it does not require the conversation partner to be visible and it does not occupy the hands and fingers – as during sign language – so that communication is possible concurrently with manual, other environmental, and social interactions. Accordingly, a *gestural theory of language evolution* has been put forward, which builds on insights from the mirror neuron system (Arbib, 2005, 2006; Graziano & Cooke, 2006; Rizzolatti & Craighero, 2004) (cf. also Section 10.3.2). As homologue areas of Broca's and Wernicke's areas have been found in monkeys (Petrides et al., 2005; Gil-da Costa et al., 2006) and as these areas can be found close to areas in which mirror neurons have been identified, it is suspected that these areas evolved from the mirror neuron system. Moreover, the following analogy can be drawn: while the mirror neuron system is involved in behavior recognition and probably also the inference of the current intentions of others, speech acts express verbal intentions and speech comprehension is essentially an auditorily driven behavior recognition. Thus, it seems likely that gesture-based communication evolved before verbal communication and enabled the evolution of verbal communication (Arbib, 2006; Corballis, 2009).

Interestingly, insights from genetics further support the gestural theory of language evolution. A study with over 150 adopted children whose biological parents exhibited language deficits showed that these children also developed similar language deficits, despite having grown up with parents that had no deficits. Vice versa, children who grew up with adoptive parents with language deficits, but whose actual parents did not have the deficit, also did not have the deficit (Fisher, Lai, & Monaco, 2003). In a particular family, 50% of the family members were affected by a severe speech and language disorder; the trait proved to be dominant as it was passed down through the family. A genetic analysis revealed that a mutation of one single nucleotide of the so-called *FOXP2*-gene (which consists of several hundreds of thousands of nucleotides) on the long arm of chromosome seven caused the disorder. Apparently, the nucleotide encodes a certain transcription factor, which in turn regulates thousands of other genes.

From these findings, one can make deductions about the evolution of verbal language capabilities: *FOXP2* is remarkably stable in mammals. In man and mice it differs only in three amino acids. However, genetic analyses suggest that two of these changes occurred in humans only after the separation from the last common ancestor of humans and chimpanzees. Thus, a small, human-specific genetic change, which probably occurred only about 200,000 years ago, most likely led to a profound improvement in vocal communication abilities (Fisher & Scharff, 2009). As a result, it seems likely that verbal communication overtook gestural communication not very long ago.

However, this does not mean that no vocal communication took place before that. The fact that many African languages still include up to 48 different click sounds for communication, plus the likelihood that even the anatomy of Neanderthals allowed the differentiated production of such sounds, suggests that click sound-based communication existed before fully developed verbal communication (Corballis, 2009). Nonetheless, the transfer to the versatile verbal speech production abilities in humans most likely facilitated the further abstraction of gestural and click sound-based communication to completely abstract, non-iconic, verbal communication. In such verbal communication, sounds are arbitrarily linked to meaning and sound combinations lead to meaning combinations. Thus, verbal communication enables a complete abstraction and detachment from the current state of the environment, and greater flexibility in verbal expressions enabled the construction of more complex, grammatically structured utterances. Cognitively speaking, verbal communication thus facilitates the generation of abstract thoughts.

13.4.3 Mind and grammar

It is still being debated, whether or not grammar developed before verbal communication (Corballis, 2009). In modern sign languages, full grammatical rules exist and thus sign languages are as expressive and grammatical as verbal languages are. What evolutionary pressures may have evolved grammatical structures?

It seems apparent that grammar supports communication about events that are currently not occurring, thus supporting communicative, social planning about the future (for example, how to hunt the group of mammoths) as well as reflections on the past (for example, "You jumped out of the bush too early so the mammoths ran away."). In essence, grammar enables *mental time traveling*, imaging the future and the past, and particularly cooperative events and social interactions (Corballis, 2009). When going on such mental time traveling journeys, the perspectives of the included individuals also need to be considered, that is, *perspective taking* becomes necessary. Once again considering the social realm, perspective taking can be further differentiated into *spatial perspective taking*, that is, what another person may (mostly visually) perceive at a certain moment in time, and *mental perspective taking*, that is, what another person may think or know at a certain point in time (Frith & Frith, 2005). *Mentalizing* situations where peers and oneself cooperate, allows for social planning and interactions far beyond individualized goal-oriented behavior (Frith & Frith, 2003). Words help to individualize each person by giving names (Butz, Lanzi, Llorà, & Loiacono, 2008) or by identifying them in the grammatical context. Grammar assigns subjects and objects, actors, and recipients, and thus supports getting the perspective right; grammar supports the separation of individual perspectives.

With a society of individuals in each of our minds, it becomes possible to mirror one's own thoughts and knowledge to peers, similar to the mirroring of actions and intentions by means of mirror neurons (Gallese & Goldman, 1998). It has even been theorized that empathy may be rooted in the mirror neuron system, as it allows one to simulate the perspective of others by means of one's own behavioral, and probably mental, repertoire (Gallese, 2001). Considering prehistoric tribes and clans as well as our modern societies, individuals must have developed assumptions about the knowledge of other people in their group. And this knowledge not only considers the current knowledge about a certain event, but also general knowledge and general behavioral capabilities, such as the knowledge that the other person will understand certain gestures (be they vocal or manual) and will know how to accomplish certain things, such as carrying something. Essentially, a social, conventional theory of mind develops about the group with its knowledge and its conventions.

Through communication then, the current knowledge and thoughts of other individuals can be probed, modified, enhanced, or questioned. To do so, a *theory of mind* is necessary that enables us to tune our communication to the conversational partner's knowledge and understanding (Frith & Frith, 2003; Frith & Wolpert, 2004; Frith & Frith, 2005). Even before verbal communication, knowledge about individual skills and about knowledge differences is helpful to effectively cooperate and divide the necessary subtasks and labor among themselves.

During communication it is mandatory that the speaker and the listener share an overlapping vocabulary such that nouns, verbs, adjectives, and so forth are understood in a similar manner. More generally speaking, speaker and listener must share a *common ground*, which may be described as a complex structure of mutually shared knowledge, assumptions, and current understandings. Communication then unfolds within and about this common ground. During a communication, the common ground develops, in that, for example, particular aspects are emphasized and new aspects are discovered (Brown-Schmidt, 2009; Clark & Marshall, 1981; Stalnaker, 1978). Only by means of mentalizing and the involved perspective taking it seems possible to accomplish such communicative abilities. The evolution of a generative grammar has certainly helped to get planning and the perspective right when planning social interactions, that is, when coordinating group efforts and cooperations.

In sum, language has probably evolved with progressively greater fitness benefits reaped from improving social cooperations and social coordinations. Planning and reasoning capabilities are also improved because events and people are individualized and flexibly put into specific perspectives. Mental time traveling is supported by language, but, vice versa, it also creates additional pressures toward the evolution of even more elaborate, grammatical language, supporting the explicit, linguistic recollection and reflection of events, and thus the learning from mistakes, and the better planning of future events. Finally, language, and even much more so fully developed grammatical language, supports the exchange of knowledge and the transmission of knowledge to the next generation, even without ever having experienced particular events or needing to face certain situations.

13.4.4   Further implications of language evolution

To a certain extent some animals also transfer information from generation to generation, implying that a grammatical language is not really necessary. Chimpanzees, for example, are known to use particular tools to reach honey in trees and show this knowledge to their young. Particular tool knowledge is transferred to the next generation and was indeed shown to differ between groups of chimps. By means of language, however, much more intricate knowledge can be transferred. Episodic stories and memories can be shared, so it is not necessary to experience them oneself (especially when they are dangerous situations) and so that one knows what to do when in a similar situation.

Tool usage can be compared to language usage in this respect. For example, once a good stone tool has been created, it can be transferred to another person. The need to create a new one only arises when it breaks; and how to use the tool can be shown to the offspring. Language helps in explaining the tool and in showing how to use it in different situations and particularly any situations that can be linguistically expressed. Language can thus be seen as a general tool to exchange information and to teach the offspring, far beyond what is possible by actual demonstrations.

Clearly, the invention of writing lifted the progress of cultural evolution to yet another level. Suddenly, communication does not need to be verbal, but it can also be communicated indirectly in written form. Writing essentially externalizes and conserves thoughts and knowledge in an abstract, conceptualized manner. Whoever can read can access the externalized knowledge, which enables learning and access to much more knowledge. New knowledge can then build on what is known, enhancing, modifying, or correcting it. New ideas, inventions, and insights, once written down and sufficiently often multiplied and read, become permanent knowledge. With the rise of the Internet and the near universal access to general knowledge, we are currently experiencing yet another hugely significant speed-up in knowledge accumulation and exchange.

Writing also enables a much better and regulated social coordination. Written laws, and consequences when laws are broken, prevent lengthy discussions about the law, or at least about how the law was remembered. As a result, it appears that even the very first complex cultures could evolve only because clear rules could be written down, which could not be changed at will, and administrative duties could be coordinated much better. As a result, humans have uncovered an evolutionary niche that offers unprecedented opportunities and allows the evolution of knowledge itself, including reflective knowledge about language, culture, and even consciousness and how the mind develops.

13.5   Language development

As we discussed in detail in Chapter 4, evolution not only determines particular bodily and cognitive capabilities, but it needs to embed these capabilities into ontogenetic bodily and cognitive development. According to Chomsky, the "innateness" of a Universal Grammar makes children ready to learn a grammatical language and expect a universal grammatical structure (Chomsky, 1965). Moving one step away from genetically-encoded innateness, from the reasoning side it has been proposed that it appears that the brain develops by means of a sophisticated inferential machinery (Pearl, 2000). This inferential machinery learns from the embodied, structured experiences from the world and the systematic, goal-oriented experimentations of these structures by means of one's own body, as well as from linguistic experiences and interactions, that is, from "linguistic advice" (Pearl, 2000, p. 253.).

When considering language development, it should be kept in mind that language co-develops with other cognitive and behavioral capabilities. Moreover, computational cognitive science suggests that large parts of the brain attempt to form predictive structures by means of inference, where the developing internal predictive structures strive to predict the external structures found in the environment (Friston, 2009); language is just one of these structures. Semantics, which reflects real-world sensorimotor and social structures (initially without language), yields many other structural components, which are learned in parallel and progressively in close interaction with language, as also implied by the *parallel architecture* (Jackendoff, 2002). Thus, during language development, structural semantics about the world are linked and augmented with language and linguistic advice, that is, words and sentences that index and characterize the world and social interactions within.

We will first consider which conceptual structures appear to be present prelinguistically. Next, we detail how phonological competence and grammatical knowledge can develop, given this prelinguistic, conceptual, semantic knowledge. We also detail how mutual development of semantic, phonological, and syntactical knowledge progresses.

13.5.1   Prelinguistic concept development

Developmental psychology has proposed innate concepts upon which further knowledge is built (Mandler, 2012). These concepts, which are described as *conceptual primitives* in Mandler's Perceptual Meaning Analysis (PMA) model, include spatial concepts of locations ("location", "start of path", and "end of a path"), visual considerations ("± visible", "move out of / into sight"), things ("thing", "container"), static spatial relations ("± contact", "link", "in", "out"), motion ("move", "path", "path to", "blocked motion"), and dynamic spatial relational changes ("[move] into", "out of", "behind"). Starting from these conceptual primitives, PMA develops more elaborate concepts including object characterizations, such as an "animal", an "inanimate thing", a "heavy thing", a "colorful thing", or a "flying thing". Additionally, goal concepts are developed, such as "establishing contact", "drinking", or "throwing something".

PMA states that the conceptual primitives are "innate". When acknowledging that structures in the brain typically develop behavior-oriented, and that genetic predispositions probably mainly encode bodily and cognitive developmental biases rather than exact structures, the question where these "innate" structures come from needs to be considered. Recollecting the behavior- and event-oriented conceptualizations of action conditions and effects, we suspect that infants learn these conceptual primitives very early in life and even before birth (cf. also Section 4.2).

It is well known that fetuses move a lot within the womb, suck on their thumb, touch their own body, and so forth. If we assume that behavior-grounded event structures develop before birth, several of the conceptual primitives may indeed be prestructured during that time of cognitive development. Consider as an example a fetus who has learned to suck his or her thumb. The hands and fingers register contact with the mouth, thus enabling thumb sucking. Thumb sucking can be viewed as an *event*, during which particular sensorimotor interactions unfold. Thus, "± contact", "link", "in", and "out" can be distinguished. Also "start of path" and "end of a path" can be segmented and related to "move", "path", "path to", and "blocked motion" (for example, when the infant's mouth is closed while he or she attempts to insert the thumb). While it is unclear how much visual considerations may be processed, certainly spatial relational changes are encountered when the infant's hands move "behind" each other or when the thumb "moves into" and "out of" the mouth. These observations suggest that primitive conceptual learning may indeed be possible even before birth.

Note that nearly all of Mandler's *conceptual primitives* are spatial or spatiotemporal in nature. The differentiation of "things" is hardly developed in infants, which is not too surprising because the number of objects that are encountered before birth is extremely limited. Possibly the mouth may be the reason for the presence of a rudimentary "container" concept, but this is highly speculative and hard to test. Nonetheless, when considering both, Mandler's conceptual primitives as well as the experiences that are possibly gathered before birth, space dominates over object types and object properties. The same argument can be made regarding behavioral development: before the rest of the environment can be explored, space, spatial relationships, and the body in space needs to be handled sufficiently well. This is necessary in order to be able to filter out spatial changes that are caused by one's own movements following the *reafference principle* (cf. Section 6.4.1), and to manipulate the outside world goal-directedly, thus being able to differentiate properties of the experienced entities.

After birth, the world changes radically in various respects. Nonetheless, similar spatial relations hold and one's own body with its hands is still present. Thus, similar spatial relationships can still be encountered and can be compared to new spatial situations, for example, similar relations that are more distant from the body. Interactions with the environment have been shown to be very important to foster cognitive development (Byrge et al., 2014), essentially actively shaping the developing, extensive brain-body network. Moreover, it has been acknowledged numerous times that these interactions are not passive, but goal-directed from very early on (von Hofsten, 2004). As detailed in the previous chapter, the event-oriented analysis of the environment enables further conceptualizations, including the deduction of relevancies. As a result, progressively more systematically, hierarchical, conceptualized, compositional structures develop for the sake of improving action decision making, motor control, and social interactions by these means.

As Mandler emphasizes the importance of spatial conceptualizations during prelinguistic cognitive development, Smith emphasizes the importance of spatial conceptualizations and (shared) spatial focus when considering word learning (Samuelson, Smith, Perry, & Spencer, 2011). To be able to focus on particular spatial locations and spatial interactions, primitive spatial conceptualization need to be available. Our socially predisposed brain then apparently attempts to derive the current focus of the currently monitored caretaker, probably to be able to anticipate their intentions based on their currently unfolding actions. It has been observed that word learning is facilitated when objects are presented in the same spatial locations, and word learning is even possible when pointing at the location in space and naming the object, which was previously encountered at that location (Samuelson et al., 2011).

The conceptualization of the world thus starts mainly with the conceptualization of the body, the surrounding space, and relative spatiotemporal interactions. Once these features are sufficiently accurately conceptualized, they can be experienced in other contexts and with respect to other perceptions and sensorimotor interactions. As the available visual information progressively improves over the first year, progressively more fine-grained conceptualizations of objects, faces, humans and animals, and other entities in the environment are possible. Again, it can be expected that these conceptualizations focus on behavior-relevant or behavior-manipulatable properties. Animals and humans, for example, appear, move around, and disappear, whereby it is difficult for the baby to control their motions. On the other hand, inanimate objects do not move on their own and when they are in reach, they can be directly manipulated.

Thus, when starting to learn words, systematically conceptualized structures about space, spatial relations, entities, and entity properties are available. Due to the structure inherent in our world, a main distinction in such grounded conceptualizations is between space and entity properties. Any entity may generally be perceived anywhere in space – although individual objects may be perceived in some spatial areas and locations more often than in others – and this spatial location can typically change quickly, but systematically. Moreover, any entity has typical behaviorally and motivationally relevant properties, where these properties not only characterize particular entities, but they also allow one to generalize and abstract about the particular object, allowing the formation of object classes.

From a computational, grounded, developmental perspective, it is probable that words and grammar are not purely linguistic entities, but are interactively grounded in perception and behavior during ontogenetic development. The supporting evidence for this suspicion is still accumulating, although hypotheses of embodied, prelinguistic concept development, and its prerequisite for developing language competence have been proposed across disciplines (Barsalou, 1999; Barsalou, Breazeal, & Smith, 2007; Butz, 2013; Clark, 2013; Evans, 2015; Gallese, 2009; Gallistel, 2011; Grush, 2004; Howell, Jankowicz, & Becker, 2005; Mahon, 2015; Meltzoff, 1988; Roy, 2005b; Sugita, Tani, & Butz, 2011).

13.5.2 Phonological and lexical development

During language development, children face the challenge of mapping speech sounds produced by their caretakers and others onto the perceived, partially conceptualized environment. Infants need to figure out fundamental features of language on their own, or rather, they need to figure out how to infer the fundamental language features from the stream of auditory and other linguistic signals in combination with the perceived structure in the environment, and particularly the structure that is currently apparently being discussed. From a linguistic perspective, the child is directly confronted with (more or less complex) utterances, which are generally perceived as a continuous speech stream. For example, a mother may tell her child /lʊklɪtlkɪtiːzhʌŋgri/, which needs to be separated by the child's mind into its components, that is, "look, little kitty is hungry." However, when the child is observing the cat while the mother talks to her – most likely the mother will actually point at the cat when is utters /lʊk/ – the child probably thinks about what may be the matter with the cat, or rather, what may be the peculiar thing about the cat that is worth talking about. Thus, before actually understanding the uttered words, children have prelinguistic, conceptual structures available, which they will use in the attempt to map and thus comprehend uttered speech sounds.

Apart from the challenge to develop prelinguistic, conceptual structures, single sounds and sound sequences need to be extracted in the continuous speech stream (inferring Hockett's "discreteness" feature). For example, the mother's utterance /kɪti/ refers to the cat, that is, the animal that is moving around, purring, and meowing. This utterance needs to be eventually singled out from the surrounding speech sounds to understand about which entity (i.e. the cat) the mother is talking about. In effect, children face the symbol grounding problem, attempting to ground the perceived symbols in the concurrently perceived semantics. Adults do this all the time as well, but with much more elaborate prior knowledge. Nonetheless, the feeling of "what is he talking about" is probably well known to all of us, in which case we currently fail to match the perceived semantics in the world or in a conversation with the perceived utterances produced by the speaker. As a result, while children learn their mother tongue, and while adults learn new terms, there is a continuous interaction with currently perceived semantics, which is closely related to the conversational, semantic common ground, and with the concurrently perceived speech signals and the syntactic expectations (as explained, for example, by the parallel architecture; Jackendoff, 2002).

The good news when facing the challenge of mapping perceived language onto world semantics is that, as we have detailed previously, languages are not structured arbitrarily. When considering phonology, languages exhibit different types of regularities. Different languages comprise different sets of phonemes. For example, English and German have some vowels in common (like /a e i o u/), but German has also some vowels that English does not have, such as /y/ as in "Tür" (=door) or /ø/ as in "böse" (=bad). Possible sequences that can be formed with these phonemes follow language-specific patterns, which can generally be formalized by phonological rules. For example, particular consonant clusters, that is, sequences of consonants, can appear in a language, while others do not appear.

Such systematicities, as well as modifications in intonation and breaks in language utterances, help to identify word boundaries. In the utterance about the cat, for example, /tlk/ is not a valid consonant cluster in English, but /tl/ is. Thus, it is very likely that after the /tl/ sound, a new word begins with a /k/ sound. Similar regularities also exist on the morphological level. For example, in the English plural formation /-s/ is used after voiceless sounds (like /t/ or /k/, that is, sound that are produced without vocal cord vibration), /-z/ is used after voiced sounds (like /d/ or /g/), and /-ɪz/ is used after sibilants (like /s/ or /ʃ/).

It has been shown that even before children start to really understand words, they implicitly acquire the phonological rules of their mother tongue. While newborns are still able to hear subtle differences in speech sounds, even if the particular sounds do not belong to their mother tongue, this ability is lost after eight to ten months. Speech sound differentiation begins to focus on the sound differences that are relevant in the mother tongue, effectively narrowing the categorical perception and focusing it on the speech-relevant categories. For example, it was shown that one month old infants show a stronger separation of sounds that express different English phonemes than of sounds that differ acoustically by the same amount, but that nonetheless express the same English phoneme (Eimas, Siqueland, Jusczyk, & Vigorito, 1971). Thus, from birth on children may be characterized as little acoustic statisticians, who extract phonological rules, phonotactic constraints, morphological structures, and other regularities and constraints.

While analyzing sound perceptions, however, from birth infants also improve their own vocalization capabilities. During the first two months or so, children use their vocal apparatus to mainly articulate their own bodily and mental state, expressing what they think about the current environmental situation in rudimentary forms, uttering vegetative sounds, fussing, and crying. At this stage, the body itself also limits the producible sounds because of the still relatively small size of the oral cavity and the position of the larynx. Beginning after about two months, cooing and laughing commences. Moreover, due to the rapid growth of the head and neck, the variety of vowel-like sounds that can be produced increases significantly. Thus, evolutionarily determined embodied predispositions initially inhibit and then enable progressively more elaborate development (note also the relation to the retinal development and its likely effect on the development of the visual system, cf. Section 4.2.2).

Between four and six months of age, infants begin to actively test their vocal apparatus, attempting to yell, to whisper, to squeal, to growl, and so forth. Moreover, they tend to develop a sense of object permanence, enabling them to maintain an object in mind even when it temporarily disappears (Baillargeon, 1987) – a capability that seems essential to bind words to particular objects and other entities and to be able to refer to these objects even in their absence.

After about six months, *canonical babbling* commences, where infants attempt to replicate sound sequences, such as /dadada/, which are still to a large degree independent of the mother tongue. Very often the children have preferred sound sequences, which are produced more often. Interestingly, deaf children also show manual babbling around the same age – apparently in anticipation of sign language-based communication. During this period until about ten months, the first adult-like vocalizations can be noticed. At the same time, it appears that infants begin to notice similarities between perception and production, they begin to prefer native-like stress patterns, and show behavior that indicates the recognition of familiar words. After about ten months, infants begin to show clear signs of voluntary control over their own vocalizations. *Protowords*, which are word-like sounds that infants use in particular contexts or for referring to particular objects or persons, come into usage and eventually real words become recognizable.

During the first year of language development, the types of interactions that are encountered with parents and one's own abilities, such as phonological abilities, already influence learning progress. Akin to Vygotsky's Zone of Proximal Development (cf. Section 2.4.2), children learn from dyadic interactions, automatically focusing on faces, attempting to imitate, and learn social turn taking. That is, infants attempt to read signals from their caretakers, show initiatives, and wait for retrospective initiatives in turn. Moreover, they begin to follow the caretaker's gaze at around half a year, where this gaze behavior is strongly supported by other cues of the caretaker, which may elicit (joint) attention (Senju & Csibra, 2008). Clear evidence also exists that language acquisition does not work solely by a passive perception of language. For example, it has been shown that the television program as the sole spoken language input for a child of deaf parents resulted in speech capabilities much below age level (Sachs, Bard, & Johnson, 1981).

From about one year on, children begin to use their first words as names and labels for people and objects (starting most often with "mama"). Interestingly, soon words are not used simply as a reference for something, but rather they are used intentionally as a *holophrasis*, where, for example, the word "mama" can stand for "there is mama," "I want mama," or "mama help me." Later, *telegraphic speech* develops, where short multi-word utterances without function words or grammatical morphemes are used to express relations and more complex intentions (Brown & Fraser, 1964). Often, such word combinations allow multiple forms of interpretation, of which the child progressively becomes aware, such that developmental pressure toward grammatical sentence structures is encountered. For example, "more cookie" may mean that the child wants more cookies, that there are no more cookies, or that the brother or sister had more cookies, and so forth.

At about one and a half years, a *vocabulary explosion* commences, during which vocabulary increases from about 50 words at 18 months to about 200 words at two years and to more than 10,000 words at about six years when entering school. Words and names are learned individualizing plants and animals – even those that have only been seen in a book – characters in fictional stories, objects, tools, artifacts, and humans.

However, the vocabulary explosion not only includes concrete nouns and names. Adjectives come into play, which allow further characterizations of particular individuals, objects, and other entities. Verbs describe activities of individuals and adverbs allow the further characterization of these, as adjectives do for nouns. Nouns and verbs then call for the further differentiation of actor, recipient, and possibly the tool by means of which or through which the addressed interactive process unfolds. Thus, depending on the language, particular conjunctions, prepositions, and word declensions are used to differentiate the roles of the addressed entities in a sentence.

Moreover, when talking about space and time and spatiotemporal sequences, temporal and spatial relationships need to be made explicit, such as things happened before or after a time, or one object was located in front or behind another object. Furthermore, hypothetical speech needs to be possible to enable mental time traveling to a potential future, an hypothetical event, or a hypothetical story. *Counterfactuals*, such as, "had she done that, things would have developed differently," are a particularly important construct in the reasoning literature, which shows that language enables the explicit imagination of alternative events from those that have actually happened (Pearl, 2000). Finally, when starting to combine several sentences, pronouns help to refer to the same person or entity around which the story unfolds. Thus, clearly word learning goes hand-in-hand with grammatical learning.

13.5.3 Grounding and developing grammar

Seeing the word learning progression, it does not come as a surprise that grammatical learning accompanies word acquisition during language learning. To date, however, there is no artificial program that comes even close to developing an understanding of the grammar of a natural language akin to how humans do. This is mainly because these programs do not know or learn semantics, that is, they do not learn from interactions with the world, but only by passive data analysis. It should be noted that learning semantics is by no means easy, but, nonetheless, our brains are able to do so.

In fact, Chomsky's Universal Grammar argument rests on the *poverty of stimulus argument*, which has been put forward also in many other circumstances considering cognitive development. Chomsky argues under consideration of the complexity of grammar, postulating that the amount of data (that is, words and sentences) a child gathers during language learning is far too small to learn the grammar of a language. He stated that:

> [The] narrowly limited extent of the available data [...] leaves little hope that much of the structure of language can be learned by an organism initially uninformed as to its general character. (Chomsky, 1965, p. 58.)

Without getting into a long discussion about to what extent the poverty of stimulus argument applies to grammar learning, it should be noted that Chomsky's "innate" Universal Grammar addresses the predisposition for effective grammar learning even in light of a large, but not huge set of stimuli, that is, auditorily perceived language data. As it is now known that many prelinguistic conceptualizations are available that can support language learning (cf. Section 13.5.1), it seems very likely that the Universal Grammar develops ontogenetically very early in life and probably even before birth, grounded in the sensorimotor experiences that are gathered and explored during cognitive development.

Recent advances in artificial intelligence and cognitive robotics suggest that components of the Universal Grammar can be found in structures that foster versatile behavioral and intentional control. Furthermore, similar structures appear to enable the expansion of these capabilities for recognizing behaviors and intentions of other people, thus enabling more effective social cooperation and interaction. By analyzing particular interactions, the formation of action hierarchies and action grammars could be seen (Pastra & Aloimonos, 2012; Wörgötter et al., 2013). The action grammar in Pastra and Aloimonos (2012) was shown to enable the formation of action hierarchies, the combination of actions into temporal sequences, and the distinction of actor, recipient, tool, and the goal of the interaction. Moreover, the system was shown to permit the generation of action-grounded recursions, which is also present in natural languages. The analysis of Wörgötter et al. (2013) shows that object interactions can be structured into an action hierarchy within which conceptual similarities can be found that focus on the unfolding object manipulation independent of the actual object, actor, and potential tool identities. It essentially allows the conceptualization of, for example, "cutting", regardless of what is actually cut and how it is cut.

Additionally, insights from developmental linguistic studies show that children have a tendency to develop grammatical languages, even if they are not exposed to a proper grammar. Children who grow up with a *pidgin language* tend to develop a fully structured creole language out of it. That is, when confronted during cognitive development with a mix of languages that was converted into a pidgin with highly simplified and limited grammatical rules, a community of children will tend to unify this mix and generate a fully functional grammar out of it. Evidence from deaf children further supports the argument: deaf children that are not exposed to sign language tend to develop their own signs including grammatical structure (Goldin-Meadow, 2003). Thus grammar learning seems to be supported by prelinguistic compositional concept structures.

As we saw in the previous chapter, to enable flexible interactions with objects and other persons, entities need to be temporarily positioned in space relative to each other. Moreover, the properties of the represented entities need to be available and selectively activated when considering particular interactions and goals. Thus, working memory must be functional and must enable the temporal activation of multiple, possibly even identical, entities as well as their current relative spatial, conceptual, or fully abstracted state of interaction. When combined with language input and sharing a state of joint attention, the expectation must be that the conversational partner will talk about something that is informative. What is informative are possible actions and interactions that may be executed (possibly in a hypothetical, possibly in a future situation, but also at the moment of the conversation), and while learning from an adult, conceptually interesting aspects of the environment are expected to be pointed out.

Note that all indications for grammar learning come after basic word learning and even more so after the development of key behavioral competencies. With respect to *holophrasis* and *telegraphic speech* we have seen that basic grammatical abilities do not typically develop before the age of one year. However, action-grounded grammatical structures are learned from the beginning, abstracting over the actual interactions in a hierarchical schematic condition-action-effect based goal-oriented manner. When the vocabulary grows, the need for more differentiating speech becomes necessary. Interesting events want to be communicated, hypothetical scenarios want to be discussed, possible social interactions want to be debated. As a result, the need for a more complex grammar arises. Modern human language displays this complexity in that expected structures are found and action-grounded systematicities are mapped onto the grammar of the particular language, with which the developing mind is confronted.

While grammatical abilities of children progressively improve, interestingly, particular systematicities in syntax development can be detected. A common example is that of forming the past tense of irregular verbs in English. Initially, past tenses for individual verbs tend to be learned, such as walked and ran. However, at a certain point in time children tend to overgeneralize the past tense rule to simply add the morpheme "-ed" to any verb, thus falsely switching to "runned" for a while until learning the exception from the rule "ran". This indicates that language grammars are learned starting with examples, then attempting to generalize, and once a rule is recognized it is generally applied. Finally, differentiations of these generalizations seem to lead to the learning of exceptions. "The exception proves the rule" (cf. also Section 2.2.1) is valid not only for manual behavior, but also for linguistic behavior: while some manual interactions must be made, for example, by applying a rare, exceptional type of grasp, some verbs need to be applied in particular contexts (such as the past tense) in a rule-breaking manner.

13.6   Common and individual meaning

We have seen that linguistics, developmental psychology, and computational cognitive science suggest that both, embodied experiences of the world and the language encountered during ontogenetic development, influence our interpretation of words and sentences. As a consequence, interesting philosophical questions arise. How individualized is the perception of actual word meanings? Does a universal word meaning exist? Moreover, when considering conversations or when reading text corpora, how overlapping are our interpretations of the conversation or of a read text?

13.6.1   Word meaning

Aristotle considered the question of word meaning in *De Interpretatione* (cf. Section 2.2.3). In particular, he acknowledged the arbitrary link between words and actual entities, and emphasized that while we may use different words, speech expressions, and written forms for the same entity, it is still the same mental experience as the addressed entity is the same thing in the world. Plato went somewhat further, highlighting that an entity may have different names, but only the inherent meaning of the word can describe the true essence of an entity, which is the "idea" of the entity. Plato states that we even name individual letters, referring to the "idea" of the particular letter regardless in which form it is written. In naming it, we do not use the single sound of the letter itself (such as /s/), but we enhance it with the vowel "e" (that is /es/) (Schleiermacher, 1985). Nonetheless, the idea behind the letter, including its significance for language and writing, stay the same, even if we name it differently.

When reflecting on the qualia problem, how then do we individually perceive the meaning of a particular word? As meaning arises out of a particular context and we ground any idea behind a word into our own experiences, the activated associations are inevitably individual and subjective. However, even if they are individualized, they are probably – as already Socrates and Plato implied – entities in and ideas about the environment, reflecting actual environmental properties. For example, the idea of a "glass" is basically very similar in all humans, because glasses have particular physical properties, which are perceived by all our brains basically in the same manner. In particular, glasses have a containment property, can hold fluids, and have a rim from which drinking is suitably possible. The idea of a glass is thus equivalent in these property characteristics, which are individually encoded in our brain, but which nonetheless refer to the same general property. Thus, while aspects and details of language perception are individual – and these individual differences depend on the genetic predispositions and the experiences throughout one's life – the idea behind it may be considered to be *universal*. However, as meaning is grounded in experiences, this universality is not God-given or purely physical, but it is also influenced by the bodily morphology as well as the behavioral and cognitive abilities of humans and by the existence of other entities in the environment.

Meaning is represented by the words and structures that are present, commonly used, and inherent in natural languages. It has been shown that even the somewhat arbitrary masculine/feminine articles of nouns, for example, in German and Spanish, influence to a small, but significant degree how we tend to think about a certain word. One example is the word bridge, which has in German the feminine article "die" ("die Brücke") and in Spanish the masculine article "el" ("el puente"). Accordingly, when people are asked to name properties of a bridge, German native speakers tend to mention more often adjectives such as "elegant" or "beautiful", while Spanish native speaker tend to highlight somewhat more masculine-associated properties such as "big" or "sturdy" (Boroditsky, Schmidt, & Phillips, 2003). Moreover, speakers of different languages tend to remember particular properties or features of the environment more explicitly than others. For example, the preposition "on" in English can refer to several types of spatial relationships, including a picture that is hanging on the wall or a cup that is standing on the table. In German, on the other hand, two distinct prepositions would be used, which are "an" and "auf". As a consequence, studies indicate that native English and native German children tend to cluster relationships of "on" differently, where the German children tend to make a stronger distinction between the two types of "on". However, children and adults tend to only show such clustering and other effects when language is involved in the conducted experiments. Thus, many studies indicate that while language experience strongly influences the developing linguistic structures, non-linguistic structures are influenced not at all (Munnich & Landau, 2003).

The idea that language may influence or even dominantly structure the way we think is usually traced back to the American linguist Benjamin Lee Whorf (1897–1941) and is known as the *Whorfian hypothesis*. Because Whorf also related his hypothesis to the works of the ethnologist Edward Sapir (1884–1939), the hypothesis is also termed the *Sapir-Whorf hypothesis*. While Whorf's hypotheses strongly suggested that language has a causal and unpreventable influence on our thoughts, more recent research proposes a rather interactive process with mutual influences (Gentner & Goldin-Meadow, 2003)

The meaning of symbols is thus both, individual and universally shared, where the extent of the sharing depends on the accessibility of the meaning that a symbol refers to. The agreement of what the symbol refers to is determined by the society in which the language is actually used, including all forms of spoken and written forms of the language. In a sense, this agreement solves the problem of associating symbols with existing entities. In fact, the development of such agreements has been solved by simulating communicative interactions in artificial agents, claiming that the symbol-grounding problem has been solved (Steels, 2008). However, these simulations have provided existing entities that demanded to be symbolized. The deeper symbol grounding problem (Sugita & Butz, 2011), which we discussed in Section 3.4.1, rather asks the question where the entities come from that ask for symbolization and how these symbols can be learned to be combined grammatically. In large parts of this book, we have shown how such entities can emerge from the embodied, sensorimotor grounded experiences gathered while actively interacting with the environment.

13.6.2 Meaning in conversations

During a conversation then, individual differences yield individual interpretations of the conversation. Moreover, seeing that the current perspective of the individual partners inevitably differ and also to a degree each mind's focus differs from the others, the common ground in a conversation only exists theoretically. In reality, each conversational partner assumes a particular common ground, within which the conversation unfolds. Clearly, the individually assumed common grounds typically do not fully coincide. The extent to which the common grounds overlap depends on the quality of the conversation and the overlap in knowledge and understanding about the world and the topic on which the conversation focuses. During a good conversation, the individual mind sets grow together, forming a concert-like structure where each instrument converses with the other, accompanies the other, and complements it. In essence, language is the mediator that focuses on developing individual common grounds, typically making them progressively more similar.

The most coinciding common grounds may occur at large social events, including theater, music, opera, or sports events. At a sports event, for example, the emotionally shared common ground becomes particularly contagious and largely shared particularly when an amazing play unfolds (such as a successful Hail Mary throw in American football or a successful bicycle goal kick in soccer), which is significantly different from the normally observed events, but yet fully imaginable and very significant in terms of the implications for the game (win/loss). In such cases, linguistic communication is hardly necessary because the emotional experiences are mutually shared and bodily expressed.

The sports example highlights that it is not language alone that develops a common ground. Any type of interaction or cooperation does so, and various mechanisms have been revealed that lead to coordinated interactions (Sebanz et al., 2006). Additionally, gesture research suggests that gestures are used, among other things, for disambiguating utterances. Gestures also appear to be generative and are thus implemented in parallel with words to provide additional redundant or complementary information about the meaning, which is intended to be communicated (Wagner et al., 2014). In recent studies on text interpretation and reasoning, indicators were found that even very subtle metaphors, such as when comparing crime with a beast or a virus, may influence how people reason about possible solutions for the described situation (cf. Thibodeau & Boroditsky, 2013; but see also Steen, Reijnierse, & Burgers, 2014 for difficulties in reproducing the results).

Thus, conversations and texts are interpreted depending on many obvious and subtle cues in language and even in gestures and mimic. Speakers or writers want to convey a particular thought or idea, which is, often subconsciously, influenced by their own take on the matter. Similarly, readers or listeners interpret information by means of their interpretative capabilities, attempting to take all cues about the intended meanings (including subtle metaphors and gestures) into account while developing an individually constructed, assumed common ground. Individual experiences thus reflect individualized universalities, such as individualized interpretations of words, which are nonetheless grounded in our environmental reality. Sentences, stories, or linguistically transferred pieces of information in general are thus also inevitably perceived and interpreted very individually. However, because the conveyed information is grounded in our reality, usually there is a deeper truth behind it, which helps to find approximately correct interpretations, to establish a common ground, and thus to have productive communications.

13.7   Conclusions and implications

In the last chapter, we showed how conceptualizations can develop when systematically abstracting from behaviorally-grounded, sensorimotor encodings in an event-oriented manner. We have shown that spatial, property, entity, and action conceptualizations can develop. When interacting with the environment, conceptualizations are played out on various hierarchical levels, forming action grounded and environmentally grounded taxonomies. By these means, it appears that humans become generally language ready, having sufficiently conceptualized, symbolizable structures in their mind. Moreover, event-oriented, schematic encodings of interactions, including condition and effect encodings, are closely related to structures found in Chomsky's Universal Grammar. Thus, it has been proposed that the Universal Grammar is grounded in behavior-oriented conceptualizations, making children ready not only to symbolize conceptualizations, but also to grammatically combine them in a compositional manner.

During language processing, multiple, generative components are working together to comprehend or produce linguistic expressions. During language production, phonology, syntax, and the grounded semantics produce linguistic utterances in concert, mutually supporting, and aligning with each other. Similarly, during language comprehension all three components contribute to disambiguate the perceived linguistic expressions (including auditory utterances as well as mimicry and gestures) in a syntactically correct and semantically meaningful manner. In particular, assumed common grounds are developed by speaker and listener, which encode the currently unfolding conversation with the implied meaning. Individual common grounds particularly develop based on two influences: the perceived communicative acts, including the reactions of the listener to our utterances, and the own private ground, that is, our own semantic knowledge and current assumptions.

Various linguists and cognitive scientists have suggested that the imaginations that unfold during a conversation, or also while reading a text, can be described by mental simulations of the described situation (Barsalou, 1999, 2008, 2009; Clark, 2013, 2016; Hohwy, 2012; Lakoff & Johnson, 1980). These simulations are closely related to the developing individual common ground. Situated simulations support the interpretation of the linguistically perceived information, probing it for errors and inconsistencies, which can imply a misunderstanding, a speech error by the speaker, a misinterpretation by the listener, or even an intended prevarication by the speaker. Object words, for example, have been shown to directly prepare particular interactions with those objects (Bub et al., 2008). Sentences about object interactions have been shown to activate particular perspectives on the described scene dependent on the involvement of the actor, which was manipulated by the used pronoun ("I", "you", or "he/she") in the sentence (Brunyé, Ditman, Mahoney, Augustyn, & Taylor, 2009). Finally, described motions on a concrete and abstract level – such as "close the drawer" or "pass on an idea" – have been shown to prime corresponding directional motion (Glenberg & Kaschak, 2002).

Additionally, it appears that situated simulations also strongly support anticipations about implications, expected next pieces of information, expected responses, or expected conclusions (Barsalou, 2009; Barsalou et al., 2007). In fact, we tend to laugh when expectations are somewhat violated, leading to surprise, similar to the surprise we have described with respect to forward, anticipatory processing as well as with respect to the predictive brain principle. Furthermore, seeing that situated simulations are inevitably conceptual and compositional and form an event-oriented, hierarchical taxonomy, action-grounded grammatical structures are simulated. These structures facilitate the generation of sentences that convey further information about the current situated simulation, including interesting and disambiguating aspects of it, as well as possible further developments.

Thus, cognition and particularly abstract, symbol-oriented thought, is mediated by language, but it is fundamentally grounded in sensor, motor, motivational, and sensorimotor event-oriented taxonomies. Information processing mechanisms are flexibly applied (mediated by decision making and attention principles) within these interactive taxonomies, forming situated simulations about the current state of physical or mental affairs. As a result, metaphors across taxonomies can be made and have been characterized numerous times (Bergen, 2012; Lakoff & Johnson, 1980). The developmentally grounded conceptualizations particularly imply that spatial, property, behavioral, or motivationally-grounded metaphors are expectable. For example, when talking about one's "circle of friends", friends are viewed as entities in an enclosing, interactive, communicative circle. In a related manner, when "grasping" an idea, an idea is encoded as an entity that is grasped, thus manipulatable and understood. While it is still hotly debated how much abstract thoughts and complex concepts, such as "quantum mechanics" or a "democracy", are embodied (Arbib, Gasser, & Barrés, 2014), many indicators suggest that the brain tends to localize any thought in suitable spatial and other kinds of frames of reference. By means of such encodings, different thoughts are related to each other as long as proximal relations, that is, similarity estimates between the different thoughts are available. For example, it can be rather accurately stated that a democracy is closer to a republic than to a dictatorship or even to an anarchy.

Particularly because symbols become entities on their own, which are detachable from current space and time, they can serve as referents for a particular entity or thought in the absence of the actual entity or imagined event. Thus, mentalizing and perspective taking abilities are strongly facilitated when a sufficiently complex language has evolved. Our brains have essentially found a way to transfer action-grounded, conceptualized reality into a linguistic reality, thus enabling totally new levels of thought abstraction, reasoning, perspective taking, mental time traveling, and hypothetical thinking. Thus, language strongly supports thought abstraction. However, due to the behaviorally grounded nature of language, abstract thought can be expected to be typically related to spatial and other concepts, as was suggested, for example, with respect to spatial reasoning abilities (Knauff, 2013). Numbers are typically spatially localized, where the writing direction determines if small numbers are further to the left, which is the case when writing from left to right, or to the right, which is the case when writing from right to left (Wood, Willmes, Nuerk, & Fischer, 2008). Furthermore, even verbal working memory items seem to be co-localized sequentially in mental space as a function of ordinal position, once again most likely following the cultural, writing direction determined temporal direction (van Dijck & Fias, 2011). Indeed, as was emphasized throughout the book, spatial aspects reappear in cognitive science on many levels, starting with simple behavioral influences and cognitive development through to abstract forms of reasoning, working memory, and number cognition.

In the following final chapter of the book, we wrap up what we have learned and summarize how the brain comes into being from a behaviorally and developmentally grounded perspective on cognition.

13.8   Exercises

1. What is the difference between allomorphs and allophones. Give examples!

2. Relate Searl's speech acts with the 16 language characteristics of Hockett.

3. Consider the relation of Peirce's symbol characterization and the semiotic triangle. They are closely related. What are differences?

4. With reference to the semiotic triangle, argue that ideas behind words are partially individual, but partially also universal.

5. In Chapter 3 we have introduced morphological attractors. Give examples of such attractors and discuss to which extent these attractors may be individually experienced, but universally comprehensible.

6. Which skills does the brain need to accomplish when considering reading and writing? How may it be that our brain develops reading- and writing-specific areas, despite the fact that evolution did not have much time to evolve genes for enabling the development of such skills.

7. In the last chapter, we had seen that particular manual behavioral skills have developed in premotor areas in an arrangement that maps well to the motor homunculus. Consider the location of Broca's area and relate its location to the insights from the previous chapter.

8. Argue in which way the position and shape of the larynx in humans may be related to morphological intelligence.

9. Five aspects plus culture were mentioned that must have contributed to the evolution of human language capabilities. Give an example of a challenging, cooperative task and show how in order to master this task as a group, the optimization of all six aspects can greatly facilitate the mastery of the task.

10. First indicators of settlements of homo sapiens in Australia (about 45k years ago) and in America (about 16k years ago) go hand in hand with the extinction of the megafauna (for example, mammoths; cf. Harari, 2011). Considering the distinct capabilities of homo sapiens, what might have been reasons for the extinction of the megafauna. Sketch-out a hypothetical scenario that eventually might have led to the extinction of the megafauna.

11. Give examples of holophrasis and possible alternative interpretations of them. Enhance the holophrasis into two possible disambiguations by means of telegraphic speech and yet into two further possible disambiguations by means of full sentences. Argue in this respect why language production development in toddlers, which typically starts will holophrases, then expands to the utterance of telephrases, and finally ends up with the generation of fully grammatical sentences, encounters particular kinds of learning pressures during this developmental process. Which role do positive rewards (after successful communication) and frustrations (after miscommunication) play in this respect?

12. Chomsky's universal grammar posits that young children have an inborn universal grammar, which makes them ready to learn the grammar of a human language. Embodied cognitive science suggests that this universal grammar may actually be acquired from experience. Relate event schemata with conditions, actions, and effect encodings to simple grammatical trees.

13. Formulate a reasonably complex sentences about manipulating an object with a tool. Sketch-out the context-free grammatical tree and note the relation of the individual noun phrases to each other and the verb-specified interaction, characterizing the interaction event.

14. Consider the sentence "Lucy informed Mike about the weather forecast" and relate it to the similar sentences "Lucy gave Mike the fork" and "Lucy shared the fork with Mike." Detail the close similarities between the sentences, but also the involved abstractions in the former sentence when compared with the two latter ones.

15. Think of a simple conversation, for example, about the weather. Characterize the two personal grounds and the developing common ground while the one conversation partner informs the other one about the most recent weather forecast.

Chapter 14

Retrospection and future perspectives

14.1   Retrospection

This book has proposed a developmental, embodied perspective on how the mind comes into being. First, we have shown that traditional, symbolic artificial intelligence inevitably suffers from fundamental problems in cognitive science: the symbol grounding problem, the frame problem, and the binding problem. Symbols – even when embedded in a symbol network (that is, a symbolic ontology) or a complex logic – have no meaning on their own because they are not grounded in our environment, that is, in reality. Moreover, symbols do not provide a focus, that is, a determination of relevance beyond their location in the symbolic network or logic. Thus, the frame problem cannot be solved on the symbolic level, because relevancies would need to be encoded and determined explicitly all the time, which is very difficult and time consuming. Finally, symbols are singular entities without any binding abilities on their own. Although a symbolic ontology may provide set-based bindings, the natural manner in which humans bind bits of information about the environment into wholes seems to remain unachievable. These observations also entail the qualia problem; because symbols have no meaning on their own, the qualitative feel, which humans inherently associate with any kind of symbol, is inaccessible by the symbol alone.

We have also shown that embodied artificial intelligence approaches can solve many behavioral tasks by means of rather simple, but suitable sensorimotor couplings. Even important perceptual information can be shown to be extractable much easier when suitable sensor- and morphologically-grounded perceptual predispositions are available. Similarly, motor control can be issued much easier when the bodily morphology supports particular dynamics and inhibits others – as most illustratively shown by the passive walker. When coupling particular perceptual information with motor activities, seemingly goal-directed intelligent behavior can be generated, as shown by the Braitenberg vehicles. In biology, similar observations have been made and have motivated many developments in artificial intelligence. Subsumption architectures have been used to create somewhat intelligent robots, but the intelligence of such robotic architectures remains limited. Probably the most important aspect that is missing in these systems is their behavioral flexibility; they lack context-dependent decision making abilities.

Considering ontogentic and phylogenetic perspectives, flexible decision making and control abilities must have evolved to outperform less intelligent species. Moreover, the abilities must be learned because the world is full of different situations and different contexts so it would not be helpful to genetically encode such abilities (however, the predisposition to develop such abilities during a lifetime should be genetically encoded). During ontogenetic development, the competence of bodily control develops hand in hand with the development of conceptual knowledge. Even before birth, the fetus develops important behavioral abilities and bodily knowledge and is thus born with basic conceptual knowledge about its body and the fact that there appears to be an "outside" environment. After birth, this "outside" world is actively explored.

With respect to phylogenetic development, we have emphasized that genotypic building blocks determine phenotypic traits. Moreover, we have shown that the human niche of cooperation, deep social intelligence, and linguistic communication must have evolved from common ancestors of humans and apes. As these common ancestors did not live too long ago (about six million years ago), from an evolutionary perspective the steps toward human intelligence cannot have been particularly huge ones – albeit certainly very significant ones. Thus, we must acknowledge that human intelligence builds on and develops from bodily and mental abilities that we share with many other species. However, the level of thought abstraction that humans reach certainly goes beyond the abilities of any other species.

With the question how this human cognitive development can be accomplished in mind, we then addressed how much can be achieved by means of reward-oriented learning. Behaviorism in psychology, behavioral biology, and reinforcement learning (RL) in artificial intelligence have explored this question for decades and have shown that reward-based learning can be found in many species – even including worms – and that reward-based learning can lead to quite clever and adaptive behavioral abilities. However, the studies have also shown that the adaptive abilities are limited, because behavior cannot be flexibly switched and the learning of unconstrained behavioral policies takes a long time. Thus, reward-based learning can adapt behavior only to a limited extent to particular contexts. On the other hand, behavioral optimization, given suitable pre-wired sensorimotor control programs, can be optimized rather effectively.

To be able to develop even more versatile behavioral planning and decision making abilities, hierarchical models of the environment – including one's own body – need to be learned. Accordingly, we introduced the concept of anticipatory behavior, that is, behavior that is also controlled by the anticipated consequences of the behavior itself. To enable anticipatory behavior, predictive models about how the body and environment work need to be available; and these models can be learned following the ideomotor principle, starting with learning predictive, control-oriented models about one's own body.

The inversion of such predictive models enables goal-directed, versatile behavioral planning and control. Indeed, because such anticipatory behavior control is not only useful for versatile decision making, but also for speeding-up behavioral adaptations and control while executing a behavior, anticipatory behavioral abilities have not only evolved in more intelligent species, but seem to be present in rudimentary forms in nearly all species. Thus, while the principle once again may be rather old from an evolutionary perspective, in humans it has been recruited to enable planning and decision making on rather abstract levels.

Coupled with a bodily motivational system, which strives for bodily and mental homeostasis, anticipatory behavior can then yield very versatile behavioral capabilities. Indeed, when stretching the term "behavior" to include mental behavior in its definition, it appears that our abstract thought capabilities – including our ability to reason in an approximately logical manner – are grounded in such anticipatory behavioral principles. However, in relation to hierarchical RL (Section 5.3.3) and to anticipatory behavior (Chapter 6), we concluded that it remains a fundamental challenge how the necessary predictive models of the environment can be learned and structured effectively during cognitive development.

After having provided a basic introduction to neuroscience and the brain, we then addressed how the human brain develops the necessary predictive models. Focusing first on the visual modality, we have seen that redundant and complementary bits of information are available even when only considering the visual modality. Objects and other entities in the environment can be perceived in multiple ways. Visual motion signals provide depth and distance cues, which are well-suited to interact with the environment, avoiding unwanted bodily impacts with the rest of the environment and facilitating object identification and interaction. Moreover, dynamic motion patterns provide information about object identity and behavior, including the behavior of other humans. On the other hand, static visual edge signals provide information about object boundaries. Moreover, the deduction of shapes given edge signals enables object identification as well as the deduction of object orientations and sizes. Several other visual cues, such as texture and color, provide further information to disambiguate and thus to accurately identify objects and other entities. While vision thus provides complementary sources of information, in deeper cortical areas these are integrated to form location-independent object and other entity encodings as well as to form entity-independent, body-relative spatial encodings.

As the available bottom up information varies greatly in different contexts, such as under different lighting conditions or when being confronted with occlusions, bottom up sensory information is complemented by top-down expectations. The fusion of these two information aspects facilitates the identification of, as well as the interaction with, particular objects and other entities. However, given uncommon or unfavorable circumstances, overly confident top-down expectations may lead to illusions, such as the visual perception of illusionary contours or of incorrect object sizes. In general, it appears that bottom-up information interacts with top-down expectations in a Bayesian, that is, in a probabilistic manner, striving to create a maximally consistent interpretation of the relevant aspects of the external world. Due to the modularization into dominantly spatial and dominantly identity-oriented encodings, top-down expectations can easily focus on space, on identity, or on both.

The effort of creating and maintaining consistent interpretations about the incoming perceptions also takes place across sensory modalities. To relate different modalities, however, predictive models in the form of spatial transformations are necessary to map the modalities onto each other. These mappings will typically depend on the body-grounded orientation of sensors relative to each other. Due to relative encodings and spatial mappings, the perception of an object is possible with multiple modalities, such as by means of touch or vision, and is generally independent of where exactly the object is located in space, as long as it is perceivable. Nonetheless, object encodings associate and thus pre-activate those locations, where they are typically perceived, thus facilitating their localization at typical locations. As a result, the brain learns internal predictive models about its environment – and particularly about how particular objects, entities, and other environmental circumstances are perceived and how they typically behave over time.

When learning such predictive models, however, it is impossible to consider all available sensory information simultaneously, a fact that is also highlighted by the frame problem. Thus, it is necessary to focus the brain's processing, learning, planning, and reasoning resources on those aspects of the environment that seem relevant. Particularly when considering motor interactions, it soon becomes apparent that our body is only able to execute a few things at the same time – and the more active decision making and control is involved, the harder it becomes to execute several actions concurrently. Thus, relevancy is tied to behavior, where those aspects of the environment are relevant that ensure successful behavior. Again, behavior includes mental behavior, such as the mere perception or identification of an object without actively manipulating it – beyond scanning it by means of a sequence of eye fixations and saccades. Attentional mechanisms enable the brain to focus on relevant aspects. As in the sensory and multisensory cases, attention also has a bottom up and a top down component. Bottom up attention typically helps to identify uncommon or unexpected things, while top-down attention enables us to focus on particular things and to ignore others. Good predictive models (about what to expect) make it easier to focus attention.

Back to decision making and motor behavior, which actually blends into and is closely related to attention as mental behavior, we have seen that on the motor side a hierarchically structured cascade of motor control mechanisms develops in our body and mind. On the lowest level, simple muscle-based control loops unfold; on higher, cortical levels, control commands selectively activate and modulate these control loops. As a result, motor primitives and the resulting environmental interactions can be invoked and controlled with the help of an intricate system of control and self-stabilization mechanisms. Moreover, the motor primitives can be optimized by means of policy gradients – where a successful interaction results in the experience of high reward and thus behavioral reinforcement and optimization. Basic motor primitives and control routines can then be combined into motor complexes, which define the invocation of a blended sequence of motor primitives.

During decision making, then, it is necessary to choose and thus focus on the execution of one particular behavior, given many redundant interactions. Computationally speaking, the motor primitives need to be selectively activated and parameterized, given the current circumstances and motivation-dependent goals. The brain appears to manage this by pursuing a compromise between executing fully optimized behavioral routines and sticking to habitual behavioral interactions in anticipation of the current task.

To lift motor control to an even higher, more abstract level, the concept of interaction events and event boundaries is fundamental. Event-based segmentations seem to be not only relevant when perceiving environmental interactions, but also when executing them. When thus partitioning the continuous stream of sensorimotor information into events and event boundaries, hierarchical models can be learned. When further focusing on the determining factors that (i) enable the behavioral invocation of particular events, that is, the relevant contextual circumstances, and (ii) that bring particular events about by means of one's own motor behavior, the hierarchical models become factorized and event schemata can be developed. That is, the developing predictive models focus on those aspects of the environment that are critical for anticipating upcoming interaction consequences and the final consequences of actions can be anticipated, given the conditional circumstances are satisfied.

When furthermore acknowledging that own motor behavior invokes forces on the environment and that other entities in the environment and particularly other humans can also produce similar forces, behavior becomes an abstract force that manipulates aspects in the environment. Thought then can be viewed as the manipulation of abstract concepts and their relations in the brain by means of imagined forces.

As a result of the development of such embodied, predictive models, the human brain becomes language ready. In particular, behavior-oriented conceptualizations of the environment by means of event schemata, with their factorized components, reflect many aspects of a generative grammar. Event schemata thus offer themselves as the universal grammar, that is, the structure that strongly facilitates the learning of a human language. Indeed, developmental psycholinguists have shown that spatial concepts about the world are present in infants, which – as we have suggested – can be learned from sensorimotor experiences. Once object types become sufficiently differentiated given further sensorimotor experiences, these objects become progressively differentiated linguistically. Moreover, event schemata enable the flexible composition of terms. To avoid ambiguities in expressions, then, the grammatical tools available in the particular language are employed to assign subject, verb, and object, their particular roles in a sentence – such as actor, recipient, tool, location, or circumstance – and their relation to each other.

Sensorimotor-grounded, cognitive development thus offers a solution to the symbol grounding problem: symbols are objects, entities, actions, or relational encodings of particular, behaviorally suitably structured sensorimotor experiences. These structures include the encoding of spatial relationships, individual identities, behavioral properties of these identities, motivational relevancies, possible motion dynamics, and the capacity for invoking particular forces. Initially, such structures develop mainly about concrete, physical items and motor interactions. However, when learning about abstract systems, such as mathematics, physics, or biology, it appears that the same structures are recruited and adapted. Besides the symbol grounding problem, sensorimotor-grounded, cognitive development also offers a solution to the frame problem because event-oriented abstractions and factorizations conceptualize environmental interactions – thus directly identifying the behaviorally relevant aspects. Finally, the binding problem is also solved; because the brain continuously attempts to maintain a consistent, behavior-oriented interpretation of the outside world, it binds behaviorally relevant units, such as objects, together to maximize behavioral versatility.

Clearly, our brain solves these problems. With this book we hope to have given an idea about which fundamental mechanisms and developmental predispositions are necessary so that the brain can actually solve these problems. Moreover, we hope to have given an idea about which modularized, hierarchical, predictive neural structures need to develop to enable our cognitive capabilities. In short, we hope to have given an idea not only about how the mind comes into being, but also about the fundamental structures within which our minds exist and dynamically unfold.

14.2 Some underrepresented aspects

Obviously, we were not able in this book to cover everything that is part of our minds. Nonetheless, we hope to have laid the foundation and to have fostered an understanding of highly important concepts and components. Moreover, we hope that we have been able to at least give a feel, if not a basic conceptual understanding, of what embodied cognition and embodied cognitive development actually implies. In particular, we hope that the book has shed some light on how an understanding of concepts can develop – beginning with very bodily sensorimotor experiences – and how our ability to compositionally combine these concepts in an infinite, but meaningful manner with the help of language comes about. In the following paragraphs, we touch upon some considerations and aspects of cognition that have not been covered.

Beyond its introduction in Section 2.4.4, *working memory* was mentioned in several chapters, but was never addressed explicitly in further detail. Research on working memory is ongoing and so far an encompassing account of all aspects of working memory is still missing. From the perspective put forward in this book it can be stated that working memory must serve behavioral purposes. One purpose may be to enable the planning and execution of action sequences or to temporarily pursue a subgoal, such as when avoiding an obstacle while pursuing an object. Another purpose may be to be able to keep track of multiple objects and entities – even if they are temporarily out of sight. As we discussed in Chapter 11, working memory is closely related to attention, keeping an internal focus on particular aspects, such as words, objects, or behavioral interactions. It is probably due to this relationship that working memory often significantly differs in its exact nature, because different environmental aspects and abstract concepts are encoded in distinct manners in our brains.

Equally underrepresented and only touched upon in the language chapter is the importance of *social cognition*. Particularly for communication, but also for executing efficient social interactions, such as during particular behavioral cooperations, social cognition is mandatory. Without acknowledging that other humans have similar, but different behavioral and cognitive capabilities and skills, the effective coordination of interactions becomes very hard and inflexible. Without acknowledging that animals also have particular skills and behavioral – if not also cognitive – tendencies, interactions with animals would become similarly hard and inflexible. Humans appear to develop social skills from birth onwards, such that the perception of the experienced social reality strongly determines how individual minds structure themselves.

In a similar related vein, *tool usage* was not addressed in further detail, albeit it seems to strongly contribute to the development of our minds and our ability to manipulate the environment. When we become skilled tool users, the tool becomes part of our body and our behavioral repertoire expands. Other humans can indeed be viewed as social tools, "with a mind of their own". Similarly, we can view ourselves as a tool and can mimic tools with our hands. Thus, tool usage enhances our perspective on the environment and enables the subjectification of tools as well as the objectification of ourselves as tools – and also the perspective that others are "tools", which are mainly relevant during social interactions.

Reasoning – and particularly logical reasoning – was touched upon in the previous two chapters. The literature on reasoning in cognitive science it extensive, however, and cannot be addressed here in further detail. This book essentially addressed the cognitive foundations that can bring about abstract, logical reasoning capabilities. Recent work on reasoning has suggested that spatial encodings offer foundations for reasoning and human thought (Knauff, 2013). In this book, we have shown why space is so fundamental and why relative spatial encodings must develop in our brains: because our body interacts with space and other things in the environment are located in this space in varying positions and orientations – albeit (luckily) in a somewhat systematic manner. Thus, spatial representations are mandatory to be able to plan and execute goal-directed, context-dependent sequences of actions. Seeing that planning is about motor behavior, and reasoning is about abstract cognitive behavior, which abstracts motor forces to conceptual forces, the step to reasoning is not a large one. Moreover, motor behavior is boundedly optimal and was shown to often choose habitual behavior and context-appropriate behavioral primitives over fully task-specific optimal behavior. It comes as no surprise that cognition has been shown to be boundedly rational as well – applying useful heuristics and production rules for reasoning and decision making purposes, which may be suboptimal (Anderson, 1990; Gigerenzer & Todd, 1999; Gray, Sims, Fu, & Schoelles, 2006; Simon, 1969).

Finally, spatial cognition has hardly been addressed. We did, however, consider sequential, partially spatial planning of manual interactions (Chapter 12) and we mentioned the importance of the hippocampus for enabling navigation in the environment as well as for learning about the spatial outlines of a new environment (such as an unknown building or city, cf. Section 10.4). Research on spatial cognition has addressed the importance of landmarks for orientation, the challenge to integrate maps and map knowledge with actual episodic knowledge, and has identified various types of challenges when facing a navigation task (such as the presence or absence of knowledge about the general direction of a goal location, knowledge about landmarks, map knowledge etc., cf., for example, Wiener et al., 2009). In relation to embodied spatial cognition, it has recently been shown that mental travel can actually prime the orientation in which a particular place is recalled – especially when the mental travel can be easily imagined (Basten, Meilinger, & Mallot, 2012). We suggest that spatial cognition is thus strongly sensorimotorically and developmentally grounded. The relation to episodic memory and the integration of episodic experiences into a cognitive map, which abstracts over the temporal aspects of the episode, however, go beyond the scope of this book (but see our short considerations in Section 10.4).

14.3 Consciousness and cognitive modeling

A final important aspect that has not been addressed much is the matter of consciousness. Even more fundamental than consciousness, which remains hard to define precisely (beyond "the state of being awake" or similar), however, is the question of what keeps our mind going. What keeps us actually awake and makes us behave? How come we have goals, including our current, concrete ones and also more abstract and general goals in life – such as the pursuance of happiness or success? The concept we have introduced in this book in this respect is the concept of internal homeostasis (cf. Section 6.5), which can yield an autopoietic system that strives to self-regulate its own body and mind (Maturana & Varela, 1980). When a system strives for internal homeostasis, mechanistically speaking this system will strive to maintain the values of certain states of its body and mind within certain boundaries. As a result, the system becomes self-motivated – being continuously motivated to maintain internal homeostasis, and thus, simplistically speaking, to stay alive.

Coupled with the developing predictive models, the resulting system becomes conscious of its environment, actively interacting with it in a goal-directed, anticipatory manner and, meanwhile, simulating the state of and unfolding progressions in the environment. In consequence, because most living systems seem to have forward predictive models available (even if some of them are extremely simple), they may be admitted to have rudimentary forms of consciousness (Tononi & Koch, 2015). The degree of consciousness that is reached by a system depends on the complexity, accuracy, and abstractness of its predictive models. When the model's structures become progressively more modularized, hierarchically organized, and event-oriented – in the manner we have detailed – the resulting, self-motivated system becomes progressively more goal-directed, more flexible in its behavior, and thus progressively more capable to plan and think about the world on abstract cognitive levels (Butz, 2008).

At least two more fundamental aspect seem necessary, however, to reach the level of human consciousness, which could be characterized as the ability to think about the past and the future, and even about fully abstract or imaginative environments and systems (including mathematics, life on Alpha Centauri, Star Trek, a god, a political system, or how life might have been in the stone age). One is the social cooperative component: because we perceive others similar to ourselves by employing the same cognitive apparatus (cf., for example, mirror neurons) and because we can individualize many others, we apparently also develop the capacity to individualize ourselves as being distinct from others. As we socially cooperate and interact with others, we need to become proficient in mind reading and in developing theories about the minds of others (Frith & Frith, 2005). That is, we need to know what others know and what they want. For example, it is useful to inform others about relevant things, but annoying when stating the obvious (such as: "to walk, put one foot in front of the other"). Similarly, it is good to know about the expertises of others to partition work load and current tasks most effectively. Of course, intrigues, false information, or the concealment of information also play an important role in social interactions; and theories of mind are important to avoid being tricked by others – or to successfully trick others for that matter.

The second is language, that is, the ability to systematically symbolize and concatenate our thoughts, which enables us to fully detach our thoughts from the current situation with its sensory and motor impressions. Moreover, language gives names to individuals including the "I" for ourselves, which makes it even easier to think about the self and others in an explicitly, individualized manner. These two components – the social and the language component (where the latter would not be possible without the former) – enables us to project ourselves into the past and the future and to take different perspectives (Buckner & Carroll, 2007). With respect to consciousness, they essentially enable us to become explicitly self-conscious because we can imagine another person – or even an imaginary person for that matter – watching us interacting with the world. We can thus judge our own actions from this external perspective, enabling us to have feelings of regret, to feel lucky, or to be angry at ourselves.

With these predictive model components at hand, which are all highly interactive, but well-structured, it appears that our mind maintains and processes neural activity and thus its current "mind state" with the purpose of maintaining internal homeostasis. In most humans, some homeostatic needs appear to have a social flavor. Evolution must have managed to genetically encode in us the need to communicate and interact with others and to maintain a place in this society of relevant individuals – a trait that is present in all social animals. Indeed, recent neuroscientific indicators suggest that our brain acts in a default network, maintaining internal homeostasis by maintaining and pursuing consistent thoughts (Buckner, Andrews-Hanna, & Schacter, 2008). Coupled with the principles of a predictive, pro-active, anticipatory neural encoding and processing, the resulting system will be able to act and think goal-directedly (Bar, 2009; Barsalou, 1999; Butz, 2008; Friston, 2009; Friston, Rigoli, Ognibene, Mathys, FitzGerald, & Pezzulo, 2015; Rao & Ballard, 1999); and formulations of free energy-based inference even integrate learning into such predictive systems (Butz, 2016; Friston, 2009; Friston et al., 2015).

When in a state during which we behaviorally interact with the environment pursuing a particular task, our mindset (Bar, 2009) will be focused on mastering those aspects of the environment that are task-relevant. That is, our brain will have those predictive models activated and pre-activated that are task-suitable – at least to the best of its current knowledge. Similarly, when thinking about something, the default network focuses on the inner states, memories, possible futures, and perspectives with which the imagination is perceived and processed. Conscious experience, that is, qualia is most likely made of these currently active forward models, which are, however, not restricted to only visual consciousness as put forward elsewhere (O'Regan & Noë, 2001), but which include forward models on multiple levels of abstraction and with respect to diverse sensory and motor modalities.

While the matter of consciousness, and particularly of qualia, certainly remains to be debated on various conceptual levels, we like to close this book by emphasizing the need for more elaborate and complex neurocognitive models. Only by means of modeling will it be possible to investigate further the developmental, sensorimotor pathway to cognition further from a computational perspective and thus to foster a deeper, mechanistic understanding of the mind. In accordance with Marr's three levels of understanding (cf. Section 2.5), to foster the understanding of the mind actual algorithms of these computational principles and implementations of these algorithms are necessary. However, how can such algorithms be properly evaluated, verified, or falsified? The availability of progressively more realistic virtual reality simulations seems to offer a solution: in such environments developing, self-regulating neurocognitive agents can be simulated. That is, cognitive development can be simulated without the need for expensive robotics hardware and without the need for a morphologically intelligent system – because the morphological intelligence can be implemented in software within the simulation.

It remains an open question if it is possible to create self-regulating, self-structuring, large-scale artificial cognitive systems; and it will certainly be important to shape these systems in a way that they develop traits that are useful for us. However, it may be the case that the techniques and knowledge necessary to create such systems are already out there. Clearly, the knowledge is imprinted in our genes, and despite the huge advances in science over the last decades there is absolutely no indication that our minds come about by means of some supernatural or hyper-computational mechanisms. It rather seems to be the case that predictive, self-regularization mechanisms, coupled with emergent structuring principles – both of which are guided by genetic predispositions – do the trick. We hope that this prospect creates excitement and opens the potential for understanding the human mind in its complete form on all three levels of understanding, as proposed by David Marr, and for creating useful, intelligent artificial systems in the future.

References

Ach, N. (1905). *Über die Willenstätigkeit und das Denken: Eine experimentelle Untersuchung mit einem Anhang über das Hippsche Chronoskop.* Göttingen: Vandenhoeck & Ruprecht.

Ackley, D. H., Hinton, G. E., & Sejnowski, T. J. (1985). A learning algorithm for boltzmann machines. *Cognitive Science, 9*(1), 147–169. doi: 10.1207/s15516709cog0901_7

Adams, J. A. (1971). A closed-loop theory of motor learning. *Journal of Motor Behavior, 3*(2), 111-150.

Aflalo, T. N., & Graziano, M. S. A. (2006). Possible origins of the complex topographic organization of motor cortex: Reduction of a multidimensional space onto a two-dimensional array. *The Journal of Neuroscience, 26,* 6288-6297.

Anderson, B. L., & Winawer, J. (2005). Image segmentation and lightness perception. *Nature, 434*(7029), 79–83. doi: 10.1038/nature03271

Anderson, J. R. (1990). *The adaptive character of thought.* Hillsdale, NJ: Lawrence Erlbaum Associates.

Anderson, J. R., & Schooler, L. J. (1991). Reflections of the environment in memory. *Psychological Science, 2*(6), 396-408. doi: 10.1111/j.1467-9280.1991.tb00174.x

Arbib, M. A. (2005). From monkey-like action recognition to human language: An evolutionary framework for neurolinguistics. *Behavioral and Brain Sciences, 28,* 105–167.

Arbib, M. A. (2006). *Action to language via the mirror neuron system.* Cambridge, UK: Cambridge University Press.

Arbib, M. A., Gasser, B., & Barrés, V. (2014). Language is handy but is it embodied? *Neuropsychologia, 55,* 57 - 70. doi: 10.1016/j.neuropsychologia.2013.11.004

Aristotle. (2014). *The complete works of aristotle: The revised oxford translation* (One-Volume Digital Edition, Kindle Edition ed.). Princeton University Press.

Austin, J. L. (1962). *How to do things with words.* New York: Oxford University Press.

Axelrod, R. (1984). *The evolution of cooperation.* New York: Basic Books.

Baillargeon, R. (1987). Object permanence in 31/2-and 41/2-month-old infants. *Developmental psychology, 23*(5), 655.

Ballard, D. H., Hayhoe, M. M., Pook, P. K., & Rao, R. P. N. (1997). Deictic codes for the embodiment of cognition. *Behavioral and Brain Sciences, 20*(4), 723–767.

Bar, M. (2009). The proactive brain: Memory for predictions. *Philosophical Transactions of the Royal Society B: Biological Sciences, 364,* 1235-1243. doi: 10.1098/rstb.2008.0310

Barlow, J. S. (2002). *The cerebellum and adaptive control.* Cambridge University Press.

Barsalou, L. W. (1999). Perceptual symbol systems. *Behavioral and Brain Sciences, 22,* 577–600.

Barsalou, L. W. (2008). Grounded cognition. *Annual Review of Psychology, 59,* 617-645.

Barsalou, L. W. (2009). Simulation, situated conceptualization, and prediction. *Philosophical Transactions of the Royal Society B: Biological Sciences, 364*(1521), 1281-1289. doi: 10.1098/rstb.2008.0319

Barsalou, L. W., Breazeal, C., & Smith, L. B. (2007). Cognition as coordinated non-cognition. *Cognitive Processing, 8,* 79-91.

Bartlett, F. C. (1932). *Remembering.* Cambridge: Cambridge University Press.

Barto, A. G., & Mahadevan, S. (2003). Recent advances in hierarchical reinforcement learning. *Discrete Event Dynamic Systems*, *13*, 341-379.

Basten, K., Meilinger, T., & Mallot, H. A. (2012). Mental travel primes place orientation in spatial recall. *Spatial Cognition*, *LNAI 7463*, 378-385.

Bavelier, D. (1994). Repetition blindness between visually different items: the case of pictures and words. *Cognition*, *51*(3), 199–236. doi: 10.1016/0010-0277(94)90054-X

Bear, M. F., Connors, B. W., & Paradiso, M. A. (2007). *Neuroscience: Exploring the brain.* Baltimore, MA: Lippincott Williams & Wilkins.

Beilock, S. L. (2009). Grounding cognition in action: expertise, comprehension, and judgment. In M. Raab, J. G. Johnson, & H. R. Heekeren (Eds.), *Mind and motion: The bidirectional link between thought and action* (Vol. 174, p. 3 - 11). Elsevier. doi: 10.1016/S0079-6123(09)01301-6

Bekkering, H., Wohlschlager, A., & Gattis, M. (2000). Imitation of gestures in children is goal-directed. *The Quarterly Journal of Experimental Psychology Section A*, *53*(1), 153–164. doi: 10.1080/713755872

Belardinelli, A., Herbort, O., & Butz, M. V. (2015). Goal-oriented gaze strategies afforded by object interaction. *Vision Research*, *106*, 47–57. doi: 10.1016/j.visres.2014.11.003

Belardinelli, A., Stepper, M. Y., & Butz, M. V. (2016). It's in the eyes: Planning precise manual actions before execution. *Journal of Vision*, *16*(1), 18. doi: 10.1167/16.1.18

Bellman, R. (1957). *Dynamic programming* (First Princeton Landmarks in Mathematics edition, 2010 ed.). Princeton, NY: Princeton University Press.

Bergen, B. K. (2012). *Louder than words: The new science of how the mind makes meaning.* New York: Basic Books.

Bernstein, N. A. (1967). *The co-ordination and regulation of movements.* Oxford: Pergamon Press.

Beyer, H.-G., & Schwefel, H.-P. (2002). Evolution strategies - a comprehensive introduction. *Natural Computing*, *1*(1), 3-52. doi: 10.1023/A:1015059928466

Bidet-Ildei, C., Kitromilides, E., Orliaguet, J.-P., Pavlova, M., & Gentaz, E. (2014). Preference for point-light human biological motion in newborns: Contribution of translational displacement. *Developmental Psychology*, *50*, 113-120. doi: 10.1037/a0032956

Binet, A., & Simon, T. (1905). New methods for the diagnosis of the intellectual level of subnormals. *L'annee Psychologique*, *12*, 191–244.

Binet, A., & Simon, T. (1916). New methods for the diagnosis of the intellectual level of subnormals. In E. S. Kite (Ed.), *The development of intelligence in children* (pp. 39–90). The Training School at Vineland, New Jersey. (translaged by E.S. Kite; originally published in 1905)

Bishop, C. M. (2006). *Pattern recognition and machine learning.* Secaucus, NJ, USA: Springer-Verlag New York, Inc.

Blakemore, S. J., Wolpert, D., & Frith, C. (2000). Why can't you tickle yourself? *Neuroreport*, *11*(11), 11-16.

Bloomfield, L. (1933). *Language* ([Nachdr. d. Ausg. New York, 1965] ed.). London: Allen & Unwin.

Boroditsky, L., Schmidt, L. A., & Phillips, W. (2003). Sex, syntax, and semantics. In D. Gentner & S. Goldin-Meadow (Eds.), *Language in mind: Advances in the study of language and cognition.* (p. 61-79). A Bradford Book.

Botvinick, M., & Cohen, J. (1998). Rubber hands 'feel' touch that eyes see. *Nature*, *391*, 756. doi: 10.1038/35784

Botvinick, M., & Weinstein, A. (2014). Model-based hierarchical reinforcement learning and human action control. *Philosophical Transactions of the Royal Society of London B: Biological Sciences*, *369*(1655). doi: 10.1098/rstb.2013.0480

Braitenberg, V. (1984). *Vehicles: Experiments in synthetic psychology.* Cambridge, MA: MIT Press.

Broadbent, D. E. (1958). *Perception and communication.* Oxford, GB: Pergamon Press.

Broderick, P. C., & Blewitt, P. (2006). *The life span: human development for helping professionals* (2nd ed.). Upper Saddle River, NJ: Pearson Education Inc.

Brodmann, K. (1909). *Vergleichende lokalisationslehre der grosshirnrinde: in ihren prinzipien dargestellt auf grund des zellenbaues*. Leipzig: Barth.

Brooks, R. A. (1990). Elephants don't play chess. *Robotics and Autonomous Systems, 6*, 3-15.

Brown, R., & Fraser, C. (1964). The acquisition of syntax. In *Acquisition of language: Report of the fourth conference sponsored by the committee on intellective processes research of the social science research council* (Vol. 29, p. 43-79). Society for Research in Child Development.

Brown-Schmidt, S. (2009). The role of executive function in perspective taking during online language comprehension. *Psychonomic Bulletin & Review, 16*(5), 893-900. doi: 10.3758/PBR.16.5.893

Brozzoli, C., Ehrsson, H. H., & Farnè, A. (2014). Multisensory representation of the space near the hand: From perception to action and interindividual interactions. *The Neuroscientist, 20*(2), 122-135. doi: 10.1177/1073858413511153

Brunyé, T. T., Ditman, T., Mahoney, C. R., Augustyn, J. S., & Taylor, H. A. (2009). When you and i share perspectives: Pronouns modulate perspective taking during narrative comprehension. *Psychological Science, 20*(1), 27-32. doi: 10.1111/j.1467-9280.2008.02249.x

Bub, D. N., Masson, M. E. J., & Cree, G. S. (2008). Evocation of functional and volumetric gestural knowledge by objects and words. *Cognition, 106*(1), 27–58. doi: 10.1016/j.cognition.2006.12.010

Buckner, R. L., Andrews-Hanna, J. R., & Schacter, D. L. (2008). The brain's default network. *Annals of the New York Academy of Sciences, 1124*(1), 1–38. doi: 10.1196/annals.1440.011

Buckner, R. L., & Carroll, D. C. (2007). Self-projection and the brain. *Trends in Cognitive Sciences, 11*, 49-57.

Bundesen, C. (1990). A theory of visual attention. *Psychological Review, 97*(4), 523–547. doi: 10.1037/0033-295X.97.4.523

Bundesen, C., Habekost, T., & Kyllingsbaek, S. (2005). A neural theory of visual attention: Bridging cognition and neurophysiology. *Psychological Review, 112*, 291-328. doi: 10.1037/0033-295X.112.2.291

Butz, M. V. (2002a). *Anticipatory learning classifier systems*. Boston, MA: Kluwer Academic Publishers.

Butz, M. V. (2002b). Biasing exploration in an anticipatory learning classifier system. In P. L. Lanzi, W. Stolzmann, & S. W. Wilson (Eds.), *Advances in learning classifier systems: Fourth international workshop, IWLCS 2001 (lnai 2321)* (p. 3-22). Berlin Heidelberg: Springer-Verlag.

Butz, M. V. (2006). *Rule-based evolutionary online learning systems: A principled approach to LCS analysis and design*. Berlin Heidelberg: Springer-Verlag.

Butz, M. V. (2008). How and why the brain lays the foundations for a conscious self. *Constructivist Foundations, 4*(1), 1-42.

Butz, M. V. (2013). Separating goals from behavioral control: Implications from learning predictive modularizations. *New Ideas in Psychology, 31*(3), 302-312. doi: 10.1016/j.newideapsych.2013.04.001

Butz, M. V. (2016). Towards a unified sub-symbolic computational theory of cognition. *Frontiers in Psychology, 7*(925). doi: 10.3389/fpsyg.2016.00925

Butz, M. V., Herbort, O., & Hoffmann, J. (2007). Exploiting redundancy for flexible behavior: Unsupervised learning in a modular sensorimotor control architecture. *Psychological Review, 114*, 1015-1046.

Butz, M. V., & Hoffmann, J. (2002). Anticipations control behavior: Animal behavior in an anticipatory learning classifier system. *Adaptive Behavior, 10*, 75-96.

Butz, M. V., Kutter, E. F., & Lorenz, C. (2014). Rubber hand illusion affects joint angle perception. *PLoS ONE*, *9*(3), e92854. doi: 10.1371/journal.pone.0092854

Butz, M. V., Lanzi, P. L., Llorà, X., & Loiacono, D. (2008). An analysis of matching in learning classifier systems. *Genetic and Evolutionary Computation Conference, GECCO 2008*, 1349-1356.

Butz, M. V., Lanzi, P. L., & Wilson, S. W. (2008). Function approximation with XCS: Hyperellipsoidal conditions, recursive least squares, and compaction. *IEEE Transactions on Evolutionary Computation*, *12*, 355-376.

Butz, M. V., Linhardt, M. J., & Lönneker, T. D. (2011). Effective racing on partially observable tracks: Indirectly coupling anticipatory egocentric sensors with motor commands. *IEEE Transactions on Computational Intelligence and AI in Games*, *3*, 31-42.

Butz, M. V., Shirinov, E., & Reif, K. L. (2010). Self-organizing sensorimotor maps plus internal motivations yield animal-like behavior. *Adaptive Behavior*, *18*(3-4), 315–337.

Butz, M. V., Sigaud, O., & Gérard, P. (2003). Anticipatory behavior: Exploiting knowledge about the future to improve current behavior. In M. V. Butz, O. Sigaud, & P. Gérard (Eds.), *Anticipatory behavior in adaptive learning systems: Foundations, theories, and systems* (pp. 1–10). Berlin Heidelberg: Springer-Verlag.

Buzsaki, G., & Moser, E. I. (2013). Memory, navigation and theta rhythm in the hippocampal-entorhinal system. *Nat Neurosci*, *16*(2), 130-138. doi: 10.1038/nn.3304

Byrge, L., Sporns, O., & Smith, L. B. (2014). Developmental process emerges from extended brain-body-behavior networks. *Trends in Cognitive Sciences*, *18*(8), 395 - 403. doi: 10.1016/j.tics.2014.04.010

Caggiano, V., Fogassi, L., Rizzolatti, G., Pomper, J. K., Thier, P., Giese, M. A., & Casile, A. (2011). View-based encoding of actions in mirror neurons of area f5 in macaque premotor cortex. *Current Biology*, *21*(2), 144-148. doi: 10.1016/j.cub.2010.12.022

Caggiano, V., Fogassi, L., Rizzolatti, G., Thier, P., & Casile, A. (2009). Mirror neurons differentially encode the peripersonal and extrapersonal space of monkeys. *Science*, *324*, 403-406. doi: 10.1126/science.1166818

Castiello, U. (2005). The neuroscience of grasping. *Nat Rev Neurosci*, *6*(10), 726-736. doi: 10.1038/nrn1775

Cherry, E. C. (1953). Some experiments on the recognition of speech, with one and with two ears. *The Journal of the Acoustical Society of America*, *25*(5), 975-979. doi: http://dx.doi.org/10.1121/1.1907229

Chikkerur, S., Serre, T., Tan, C., & Poggio, T. (2010). What and where: A Bayesian inference theory of attention. *Vision Research*, *50*, 2233-2247. doi: 10.1016/j.visres .2010.05.013

Chomsky, N. (1965). *Aspects of the theory of syntax*. Cambridge, MA: MIT Press.

Chomsky, N. (2002). *Syntactic structures* (2nd edition ed.). Walter de Gruyter. (first published in 1957)

Christiansen, M. H., & Kirby, S. (2003a). Language evolution: consensus and controversies. *Trends in Cognitive Sciences*, *7*(7), 300–307. doi: 10.1016/S1364-6613(03)00136-0

Christiansen, M. H., & Kirby, S. (2003b). Language evolution: The hardest problem in science? In M. H. Christiansen & S. Kirby (Eds.), *Studies in the evolution of language* (p. 1-15). Oxford, GB: Oxford University Press.

Cisek, P. (2007). Cortical mechanisms of action selection: The affordance competition hypothesis. *Philosophical Transactions of the Royal Society B: Biological Sciences*, *362*(1485), 1585-1599. doi: 10.1098/rstb.2007.2054

Clark, A. (1999). An embodied cognitive science? *Trends in Cognitive Science*, *3*(9), 345-351.

Clark, A. (2013). Whatever next? predictive brains, situated agents, and the future of cognitive science. *Behavioral and Brain Science*, *36*, 181-253.

Clark, A. (2016). *Surfing uncertainty: Prediction, action and the embodied mind.* Oxford, UK: Oxford University Press.

Clark, H. H., & Marshall, C. K. (1981). Definite reference and mutual knowledge. In A. K. Koshi, B. Webber, & I. A. Sag (Eds.), *Elements o f discourse understanding* (p. 10-63). Cambridge: Cambridge University Press.

Clowes, M. B. (1971). On seeing things. *Artificial Intelligence*, *2*(1), 79–116. doi: 10.1016/0004-3702(71)90005-1

Colwill, R. M., & Rescorla, R. A. (1985). Postconditioning devaluation of a reinforcer affects instrumental learning. *Journal of Experimental Psychology: Animal Behavior Processes*, *11*(1), 120-132.

Colwill, R. M., & Rescorla, R. A. (1990). Evidence for the hierarchical structure of instrumental learning. *Animal Learning & Behavior*, *18*(1), 71-82.

Cook, R., Bird, G., Catmur, C., Press, C., & Heyes, C. (2014). Mirror neurons: From origin to function. *Behavioral and Brain Sciences*, *37*, 177-192. doi: 10.1017/S0140525X13000903

Corballis, M. C. (2009). The evolution of language. *Annals of the New York Academy of Sciences*, *1156*(1), 19–43. doi: 10.1111/j.1749-6632.2009.04423.x

Corina, D. P., McBurney, S. L., Dodrill, C., Hinshaw, K., Brinkley, J., & Ojemann, G. (1999). Functional roles of broca's area and smg: Evidence from cortical stimulation mapping in a deaf signer. *NeuroImage*, *10*(5), 570–581. doi: 10.1006/nimg.1999.0499

Coulom, R. (2007). Efficient selectivity and backup operators in monte-carlo tree search. In H. van den Herik, P. Ciancarini, & H. Donkers (Eds.), *Lecture notes in computer science* (Vol. 4630, p. 72-83). Springer Berlin Heidelberg. doi: 10.1007/978-3-540-75538-8_7

Craik, F. I. M., & Lockhart, R. S. (1972). Levels of processing: A framework for memory research. *Journal of Verbal Learning and Verbal Behavior*, *11*(6), 671–684. doi: 10.1016/S0022-5371(72)80001-X

Creem-Regehr, S. H., & Lee, J. N. (2005). Neural representations of graspable objects: are tools special? *Cognitive Brain Research*, *22*(3), 457–469. doi: 10.1016/j.cogbrainres.2004.10.006

Damasio, A. R., & Geschwind, N. (1984). The neural basis of language. *Annual Review of Neuroscience*, *7*(1), 127–147. doi: 10.1146/annurev.ne.07.030184.001015

Darwin, C. (1859). *The origin of species by means of natural selection.* Penguin Books. (1968 edition)

Dawkins, R. (1976). *The selfish gene.* Oxford, UK: Oxford University Press.

Dawkins, R. (1986). *The blind watchmaker: Why the evidence of evolution reveals a universe without design.* W. W. Norton. Paperback.

Dawkins, R. (1997). *Climbing mount improbable.* WW Norton & Company.

Deacon, T. W. (1997). *The symbolic species: The co-evolution of language and the brain.* W. W. Norton & Company.

de la Mettrie, J. O. (1748). *Man a machine.* Whitefish, MT: Kessinger Legacy Reprints. (reprint from 2010)

Dennet, D. (1984). Cognitive wheels: The frame problem in ai. In C. Hookway (Ed.), *Minds, machines, and evolution* (p. 128-151). Cambridge University Press.

Desimone, R., & Duncan, J. (1995). Neural mechanisms of selective visual attention. *Annual Review of Neuroscience*, *18*, 193–222. doi: 10.1146/annurev.ne.18.030195.001205

Desmurget, M., & Grafton, S. (2000). Forward modeling allows feedback control for fast reaching movements. *Trends in Cognitive Sciences*, *4*, 423-431.

Desmurget, M., Reilly, K. T., Richard, N., Szathmari, A., Mottolese, C., & Sirigu, A. (2009). Movement intention after parietal cortex stimulation in humans. *Science*, *324*(5928), 811-813.

Deutsch, J. A., & Deutsch, D. (1963). Attention: Some theoretical considerations. *Psychological Review*, *70*(1), 80–90. doi: 10.1037/h0039515

Diba, K., & Buzsaki, G. (2007). Forward and reverse hippocampal place-cell sequences during ripples. *Nat Neurosci*, *10*(10), 1241–1242. doi: 10.1038/nn1961

Dietterich, T. G. (2000). Hierarchical reinforcement learning with the MAXQ value function decomposition. *Journal of Artificial Intelligence Research*, *13*, 227-303.

Dijkerman, H. C., & de Haan, E. H. F. (2007). Somatosensory processes subserving perception and action. *Behavioral and Brain Sciences*, *30*(2), 189-201. doi: 10.1017/S0140525X07001392

Doya, K., Ishii, S., Pouget, A., & Rao, R. P. N. (2007). *Bayesian brain: Probabilistic approaches to neural coding*. The MIT Press.

Duncan, J. (1984). Selective attention and the organization of visual information. *Journal of Experimental Psychology: General*, *113*, 501-517. doi: 10.1037/0096-3445.113.4.501

Duncan, J., Humphreys, G., & Ward, R. (1997). Competitive brain activity in visual attention. *Current Opinion in Neurobiology*, *7*(2), 255 - 261. doi: 10.1016/S0959 -4388(97)80014-1

Ehrenfeld, S., & Butz, M. V. (2013). The modular modality frame model: Continuous body state estimation and plausibility-weighted information fusion. *Biological Cybernetics*, *107*, 61-82. doi: 10.1007/s00422-012-0526-2

Ehrenfeld, S., Herbort, O., & Butz, M. V. (2013a). Modular, multimodal arm control models. In G. Baldassarre & M. Mirolli (Eds.), *Computational and robotic models of the hierarchical organization of behavior* (p. 129-154). Springer Berlin Heidelberg. doi: 10.1007/978-3-642-39875-9_7

Ehrenfeld, S., Herbort, O., & Butz, M. V. (2013b). Modular neuron-based body estimation: Maintaining consistency over different limbs, modalities, and frames of reference. *Frontiers in Computational Neuroscience*, *7*(148). doi: 10.3389/fncom.2013.00148

Ehrsson, H. H., Holmes, N. P., & Passingham, R. E. (2005). Touching a rubber hand: Feeling of body ownership is associated with activity in multisensory brain areas. *Journal of Neuroscience*, *25*, 10564-10573.

Ehrsson, H. H., Spence, C., & Passingham, R. E. (2004). That's my hand! Activity in premotor cortex reflects feeling of ownership of a limb. *Science*, *305*(5685), 875-877. doi: DOI:10.1126/science.1097011

Eimas, P. D., Siqueland, E. R., Jusczyk, P., & Vigorito, J. (1971). Speech perception in infants. *Science*, *171*(3968), 303–306. doi: 10..1126/science.171.3968.303

Elsner, B., & Hommel, B. (2001). Effect anticipation and action control. *Journal of Experimental Psychology: Human Perception and Performance*, *27*, 229-240.

Engel, A. K., Maye, A., Kurthen, M., & König, P. (2013). Where's the action? the pragmatic turn in cognitive science. *Trends in Cognitive Sciences*, *17*(5), 202 - 209. doi: 10.1016/j.tics.2013.03.006

Ericsson, K. A., & Kintsch, W. (1995). Long-term working memory. *Psychological Review*, *102*, 211-245. doi: 10.1037/0033-295X.102.2.211

Erlhagen, W., & Schöner, G. (2002). Dynamic field theory of movement preparation. *Psychological Review*, *109*(3), 545–572. doi: DOI:10.1037//0033-295X.109.3.545

Ernst, M. O., & Banks, M. S. (2002). Humans integrate visual and haptic information in a statistically optimal fashion. *Nature*, *415*(6870), 429–433.

Evans, V. (2015). What's in a concept? analog versus parametric concepts in LCCM theory. In E. Margolis & S. Laurence (Eds.), *The conceptual mind: New directions in the study of concepts* (p. 251-290). Cambridge, MA: MIT Press.

Fadiga, L., Craighero, L., & D'Ausilio, A. (2009). Broca's area in language, action, and music. *Annals of the New York Academy of Sciences*, *1169*, 448–458. doi: 10.1111/j.1749-6632.2009.04582.x

Farnè, A. (2015). Seeing to feel: Anticipating contact in hand–objects interactions. *Cognitive Processing*, *16*(Suppl. 1), S15.

Fazio, P., Cantagallo, A., Craighero, L., D'Ausilio, A., Roy, A. C., Pozzo, T., … Fadiga, L. (2009). Encoding of human action in broca's area. *Brain*, *132*, 1980-1988. doi: 10.1093/brain/awp118

Fetsch, C. R., Pouget, A., DeAngelis, G. C., & Angelaki, D. E. (2012). Neural correlates of reliability-based cue weighting during multisensory integration. *Nat Neurosci*, *15*(1),

146–154. doi: 10.1038/nn.2983

Fisher, S. E., Lai, C. S., & Monaco, A. P. (2003). Deciphering the genetic basis of speech and language disorders. *Annu. Rev. Neurosci.*, *26*, 57–80. doi: 10.1146/annurev.neuro.26.041002.131144

Fisher, S. E., & Scharff, C. (2009). Foxp2 as a molecular window into speech and language. *Trends in Genetics*, *25*(4), 166–177. doi: 10.1016/j.tig.2009.03.002

Fleischer, F., Caggiano, V., Thier, P., & Giese, M. A. (2013). Physiologically inspired model for the visual recognition of transitive hand actions. *The Journal of Neuroscience*, *33*, 6563-6580. doi: 10.1523/JNEUROSCI.4129-12.2013

Fleischer, F., Christensen, A., Caggiano, V., Thier, P., & Giese, M. A. (2012). Neural theory for the perception of causal actions. *Psychological Research*, *76*(4), 476-493. doi: 10.1007/s00426-012-0437-9

Fleischer, J. G. (2007). Neural correlates of anticipation in cerebellum, basal ganglia, and hippocampus. In M. V. Butz, O. Sigaud, G. Pezzulo, & G. Baldassarre (Eds.), *Anticipatory behavior in adaptive learning systems: From brains to individual and social behavior* (Vol. 4520, p. 19-34). Springer.

Fogassi, L., Ferrari, P. F., Gesierich, B., Rozzi, S., Chersi, F., & Rizzolatti, G. (2005). Parietal lobe: From action organization to intention understanding. *Science*, *308*, 662 - 667.

Fogassi, L., Gallese, V., Fadiga, L., Luppino, G., Matelli, M., & Rizzolatti, G. (1996). Coding of peripersonal space in inferior premotor cortex (area f4). *Journal of Neurophysiology*, *76*, 141–157.

Fogel, L. J., Owens, A. J., & Walsh, M. J. (1966). *Artificial intelligence through simulated evolution*. John Wiley.

Foster, D. J., & Wilson, M. A. (2006). Reverse replay of behavioural sequences in hippocampal place cells during the awake state. *Nature*, *440*, 680-683.

Freeman, J., Dale, R., & Farmer, T. (2011). Hand in motion reveals mind in motion. *Frontiers in Psychology*, *2*(59). doi: 10.3389/fpsyg.2011.00059

Fries, P. (2005). A mechanism for cognitive dynamics: Neuronal communication through neuronal coherence. *Trends in Cognitive Sciences*, *9*, 474-480.

Fries, P. (2015). Rhythms for cognition: Communication through coherence. *Neuron*, *88*, 220–235. doi: 10.1016/j.neuron.2015.09.034

Fries, P., Nikolic, D., & Singer, W. (2007). The gamma cycle. *Trends in Neurosciences*, *30*, 309-316.

Friston, K. (2009). The free-energy principle: a rough guide to the brain? *Trends in Cognitive Sciences*, *13*(7), 293 - 301. doi: 10.1016/j.tics.2009.04.005

Friston, K. (2010). The free-energy principle: A unified brain theory? *Nature Reviews Neuroscience*, *11*, 127- 138. doi: 10.1038/nrn2787

Friston, K., Rigoli, F., Ognibene, D., Mathys, C., FitzGerald, T., & Pezzulo, G. (2015). Active inference and epistemic value. *Cognitive Neuroscience*, *6*, 187-214. doi: 10.1080/17588928.2015.1020053

Frith, C., & Frith, U. (2005). Theory of mind. *Current Biology*, *15*(17), R644–R645. doi: 10.1016/j.cub.2005.08.041

Frith, C. D., & Wolpert, D. M. (Eds.). (2004). *The neuroscience of social interaction: Decoding, imitating and influencing the actions of others*. Oxford: Oxford University Press.

Frith, U., & Frith, C. D. (2003). Development and neurophysiology of mentalizing. *Philosophical Transactions of the Royal Society of London. Series B: Biological Sciences*, *358*(1431), 459-473. doi: 10.1098/rstb.2002.1218

Fritzke, B. (1995). A growing neural gas network learns topologies. *Advances in Neural Information Processing Systems*, *7*, 625-632.

Fyhn, M., Molden, S., Hollup, S., Moser, M.-B., & Moser, E. I. (2002). Hippocampal neurons responding to first-time dislocation of a target object. *Neuron*, *35*(3), 555–566. doi: 10.1016/S0896-6273(02)00784-5

Gallagher, S. (2005). *How the body shapes the mind*. New York: Oxford University Press.

Gallese, V. (2001). The 'shared manifold' hypothesis: From mirror neurons to empathy. *Journal of Consciousness Studies: Between Ourselves - Second-Person Issues in the Study of Consciousness*, *8*(5-7), 33-50.

Gallese, V. (2009). Motor abstraction: a neuroscientific account of how action goals and intentions are mapped and understood. *Psychological Research*, *73*(4), 486 - 498.

Gallese, V., Fadiga, L., Fogassi, L., & Rizzolatti, G. (1996). Action recognition in the premotor cortex. *Brain*, *119*, 593–609.

Gallese, V., & Goldman, A. (1998). Mirror neurons and the simulation theory of mind-reading. *Trends in Cognitive Sciences*, *2*(12), 493–501.

Gallistel, C. R. (2011). Prelinguistic thought. *Language Learning and Development*, *7*(4), 253–262. doi: 10.1080/15475441.2011.578548

Garcia, J. O., & Grossman, E. D. (2008). Necessary but not sufficient: Motion perception is required for perceiving biological motion. *Vision research*, *48*(9), 1144–1149.

Gaussier, P., Revel, A., Banquet, J. P., & Babeau, V. (2002). From view cells and place cells to cognitive map learning: Processing stages of the hippocampal system. *Biological Cybernetics*, *86*(1), 15–28. doi: 10.1007/s004220100269

Gazzaniga, M. S., Ivry, R. B., & Mangun, G. R. (2002). *Cognitive neuroscience: The biology of the mind* (Second Edition ed.). W. W. Norton & Company. Hardcover.

Geisler, W. S. (2007). Visual perception and the statistical properties of natural scenes. *Annular Review of Psychology*, *59*(1), 167–192. doi: 10.1146/annurev.psych.58.110405 .085632

Gelly, S., & Silver, D. (2011). Monte-carlo tree search and rapid action value estimation in computer go. *Artificial Intelligence*, *175*(11), 1856–1875. doi: 10.1016/j.artint.2011 .03.007

Gentner, D., & Goldin-Meadow, S. (2003). Whither whorf. In D. Gentner & S. Goldin-Meadow (Eds.), *Language in mind: Advances in the study of language and cognition.* (p. 3-14). A Bradford Book.

Gentner, R., & Classen, J. (2006). Modular organization of finger movements by the human central nervous system. *Neuron*, *52*, 731-742.

Gergely, G., Bekkering, H., & Kiraly, I. (2002). Developmental psychology: Rational imitation in preverbal infants. *Nature*, *415*(6873), 755–755. doi: 10.1038/415755a

Gibson, J. J. (1979). *The ecological approach to visual perception*. Mahwah, NJ: Lawrence Erlbaum Associates.

Giese, M., & Rizzolatti, G. (2015). Neural and computational mechanisms of action processing: Interaction between visual and motor representations. *Neuron*, *88*(1), 167–180. doi: 10.1016/j.neuron.2015.09.040

Giese, M. A., & Poggio, T. (2003). Neural mechanisms for the recogniton of biological movements. *Nature Reviews Neuroscience*, *4*, 179-192.

Gigerenzer, G., & Todd, P. M. (1999). *Simple heuristics that make us smart*. New York: Oxford University Press.

Gil-da Costa, R., Martin, A., Lopes, M. A., Munoz, M., Fritz, J. B., & Braun, A. R. (2006). Species-specific calls activate homologs of broca's and wernicke's areas in the macaque. *Nat Neurosci*, *9*(8), 1064–1070. doi: 10.1038/nn1741

Gleitman, H., Gross, J., & Reisberg, D. (2011). *Psychology* (8th ed.). New York: W. W. Norton & Company, Inc.

Glenberg, A. M., & Kaschak, M. P. (2002). Grounding language in action. *Psychonomic Bulletin & Review*, *9*(3), 558-565. doi: 10.3758/BF03196313

Glover, S., Rosenbaum, D. A., Graham, J., & Dixon, P. (2004). Grasping the meaning of words. *Experimental Brain Research*, *154*, 103-108. doi: 10.1007/s00221-003-1659-2

Goldberg, D. E. (1999). The race, the hurdle and the sweet spot: Lessons from genetic algorithms for the automation of innovation and creativity. In P. Bentley (Ed.), *Evolutionary design by computers* (p. 105-118). San Francisco, CA: Morgan Kaufmann.

Goldin-Meadow, S. (2003). Thought before language: do we think ergative? In D. Gentner & S. Goldin-Meadow (Eds.), *Language in mind: Advances in the study of language and cognition*. (p. 493-522). A Bradford Book.

Goodale, M. A., & Milner, A. D. (1992). Separate visual pathways for perception and action. *Trends in Neurosciences, 15*(1), 20 - 25. doi: 10.1016/0166-2236(92)90344-8

Gray, H. (1918). *Anatomy of the human body*. Philadelphia: Lea & Febiger.

Gray, W. D., Sims, C. R., Fu, W.-T., & Schoelles, M. J. (2006). The soft constraints hypothesis: A rational analysis approach to resource allocation for interactive behavior. *Psychological Review, 113*, 461-482. doi: 10.1037/0033-295X.113.3.461

Graziano, M. S. A. (2006). The organization of behavioral repertoire in motor cortex. *Annual Review of Neuroscience, 29*, 105-134.

Graziano, M. S. A., & Aflalo, T. N. (2007). Mapping behavioral repertoire onto the cortex. *Neuron, 56*(2), 239 - 251. doi: http://dx.doi.org/10.1016/j.neuron.2007.09.013

Graziano, M. S. A., & Cooke, D. F. (2006). Parieto-frontal interactions, personal space, and defensive behavior. *Neuropsychologia, 44*, 845–859.

Graziano, M. S. A., Taylor, C. S., & Moore, T. (2002). Complex movements evoked by microstimulation of precentral cortex. *Neuron, 34*(5), 841 - 851. doi: 10.1016/S0896-6273(02)00698-0

Greenwald, A. G. (1970). Sensory feedback mechanisms in performance control: with special reference to the ideo-motor mechanism. *Psychological Review, 77*, 73-99.

Grush, R. (2004). The emulation theory of representation: Motor control, imagery, and perception. *Behavioral and Brain Sciences, 27*, 377–96.

Hansen, N., & Ostermeier, A. (2001). Completely derandomized self-adaptation in evolution strategies. *Evolutionary Computation, 9*, 159-195.

Harari, Y. N. (2011). *Sapiens: A brief history of humankind*. London: Vintage.

Hard, B. M., Tversky, B., & Lang, D. S. (2006). Making sense of abstract events: Building event schemas. *Memory & Cognition, 34*(6), 1221-1235. doi: 10.3758/BF03193267

Hartwigsen, G., Baumgaertner, A., Price, C. J., Koehnke, M., Ulmer, S., & Siebner, H. R. (2010). Phonological decisions require both the left and right supramarginal gyri. *Proceedings of the National Academy of Sciences, 107*(38), 16494–16499. doi: 10.1073/pnas.1008121107

Haruno, M., Wolpert, D. M., & Kawato, M. (2001). Mosaic model for sensorimotor learning and control. *Neural Computation, 13*, 2201-2220. doi: 10.1162/089976601750541778

Haruno, M., Wolpert, D. M., & Kawato, M. (2003). Hierarchical mosaic for movement generation. In T. Ono, G. Matsumoto, R. Llinas, A. Berthoz, R. Norgren, H. Nishijo, & R. Tamura (Eds.), *Excepta medica international coungress series* (Vol. 1250, p. 575-590). Amsterdam, The Netherlands: Elsevier Science B.V. doi: 10.1016/S0531-5131(03)00190-0

Harvey, B. M., Klein, B. P., Petridou, N., & Dumoulin, S. O. (2013). Topographic representation of numerosity in the human parietal cortex. *Science, 341*(6150), 1123-1126. doi: 10.1126/science.1239052

Hayhoe, M. M., Shrivastava, A., Mruczek, R., & Pelz, J. B. (2003). Visual memory and motor planning in a natural task. *Journal of Vision, 3*(1), 49–63. doi: 10:1167/3.1.6

Hebb, D. O. (2002). *The organization of behavior: A neuropsychological theory*. Psychology Press. (originally published: 1949)

Heider, F. (1958). *The psychology of interpersonal relations*. New York: John Wiley.

Heider, F., & Simmel, M. (1944). An experimental study of apparent behavior. *The American Journal of Psychology, 57*(2), 243–259. doi: 10.2307/1416950

Herbart, J. F. (1825). *Psychologie als Wissenschaft neu gegründet auf Erfahrung, Metaphysik und Mathematik. Zweiter, analytischer Teil [Psychology as a science newly grounded on experience, metaphysics, and mathematics. second part: Analytics]*. Königsberg, Germany: August Wilhem Unzer.

Herbort, O., & Butz, M. V. (2007). Encoding complete body models enables task dependent optimal behavior. *Proceedings of International Joint Conference on Neural Networks,*

Orlando, Florida, USA, August 12-17, 2007, 1424-1429.

Herbort, O., & Butz, M. V. (2011). Habitual and goal-directed factors in (everyday) object handling. *Experimental Brain Research, 213*, 371-382. doi: 10.1007/s00221-011-2787-8

Herbort, O., & Butz, M. V. (2012). The continuous end-state comfort effect: Weighted integration of multiple biases. *Psychological Research, 76*, 345-363. doi: 10.1007/s00426-011-0334-7

Herbort, O., Butz, M. V., & Hoffmann, J. (2005). Towards an adaptive hierarchical anticipatory behavioral control system. In C. Castelfranchi, C. Balkenius, M. V. Butz, & A. Ortony (Eds.), *From reactive to anticipatory cognitive embodied systems: Papers from the AAAI fall symposium* (p. 83-90). Menlo Park, CA: AAAI Press.

Herbort, O., Butz, M. V., & Pedersen, G. K. M. (2010). The SURE_REACH model for motor learning and control of a redundant arm: From modeling human behavior to applications in robotics. In O. Sigaud & J. Peters (Eds.), *From motor learning to interaction learning in robots* (pp. 85–106). Springer.

Herbort, O., Ognibene, D., Butz, M. V., & Baldassarre, G. (2007). Learning to select targets within targets in reaching tasks. *6th IEEE International Conference on Development and Learning, ICDL 2007*, 7 - 12.

Herculano-Houzel, S. (2009). The human brain in numbers: A linearly scaled-up primate brain. *Frontiers in Human Neuroscience, 3*(31). doi: 10.3389/neuro.09.031.2009

Hinton, G. E. (2002). Training products of experts by minimizing contrastive divergence. *Neural Computation, 14*(8), 1771-1800. doi: 10.1162/089976602760128018

Hinton, G. E., Dayan, P., Frey, B. J., & Neal, R. M. (1995). The wake-sleep algorithm for unsupervised neural networks. *Science, 268*, 1158-1161.

Hinton, G. E., Osindero, S., & Teh, Y.-W. (2006). A fast learning algorithm for deep belief nets. *Neural Computation, 18*(7), 1527–1554. doi: 10.1162/neco.2006.18.7.1527

Hirel, J., Gaussier, P., Quoy, M., Banquet, J. P., Save, E., & Poucet, B. (2013). The hippocampo-cortical loop: Spatio-temporal learning and goal-oriented planning in navigation. *Neural Networks, 43*, 8–21. doi: 10.1016/j.neunet.2013.01.023

Hockett, C. F. (1960). Logical considerations in the study of animal communication. In W. E. Lanyon & W. N. Tavolga (Eds.), *Animal sounds and communication* (p. 392-430). Washington, DC: American Institute of Biological Sciences.

Hockett, C. F., & Altmann, S. A. (1968). A note on design features. In T. A. Sebook (Ed.), *Animal communication: Techniques of study and results of research* (p. 61-72). Bloomington: Indiana University Press.

Hoffmann, J. (1986). *Die Welt der Begriffe*. Weinheim: Beltz-Verlag.

Hoffmann, J. (1993). *Vorhersage und Erkenntnis: Die Funktion von Antizipationen in der menschlichen Verhaltenssteuerung und Wahrnehmung. [Anticipation and cognition: The function of anticipations in human behavioral control and perception.]*. Göttingen, Germany: Hogrefe.

Hoffmann, J. (1996). Bedeutung · konzepte bedeutungskonzepte: Theorie und anwendung in linguistik und psychologie. In J. Grabowski, G. Harras, & T. Herrmann (Eds.), (p. 88-119). VS Verlag für Sozialwissenschaften.

Hoffmann, J. (2003). Anticipatory behavioral control. In M. V. Butz, O. Sigaud, & P. Gérard (Eds.), *Anticipatory behavior in adaptive learning systems: Foundations, theories, and systems* (p. 44-65). Berlin Heidelberg: Springer-Verlag.

Hohwy, J. (2012). Attention and conscious perception in the hypothesis testing brain. *Frontiers in Psychology, 3*, 96. doi: 10.3389/fpsyg.2012.00096

Holland, J. H. (1975). *Adaptation in natural and artificial systems*. Ann Arbor, MI: University of Michigan Press. (second edition, 1992)

Holland, J. H., & Reitman, J. S. (1978). Cognitive systems based on adaptive algorithms. In D. A. Waterman & F. Hayes-Roth (Eds.), *Pattern directed inference systems* (pp. 313–329). New York: Academic Press.

Holmes, N. P., & Spence, C. (2004). The body schema and multisensory representation(s) of peripersonal space. *Cognitive Processing, 5*, 94-105.

Hommel, B., Müsseler, J., Aschersleben, G., & Prinz, W. (2001). The theory of event coding (TEC): A framework for perception and action planning. *Behavioral and Brain Sciences, 24*, 849-878.

Horn, J. L., & Cattell, R. B. (1967). Age differences in fluid and crystallized intelligence. *Acta psychologica, 26*, 107–129.

Howell, S. R., Jankowicz, D., & Becker, S. (2005). A model of grounded language acquisition: Sensorimotor features improve lexical and grammatical learning. *Journal of Memory and Language, 53*(2), 258 - 276. doi: 10.1016/j.jml.2005.03.002

Hubbard, E., & Ramachandran, V. S. (2003). The phenomenology of synaesthesia. *Journal of Consciousness Studies, 10*(8), 49-57.

Hubel, D. H. (1993). Evolution of ideas on the primary visual cortex, 1955-1978: A biased historical account. In J. Lindsten (Ed.), *Nobel lectures in physiology or medicine, 1981-1990* (p. 24-56). Singapore: World Scientific Publishing Co.

Huffman, D. A. (1971). Impossible objects as nonsense sentences. In B. Meltzer & D. Michie (Eds.), *Machine intelligence* (Vol. 6, p. 295-324). Edinburgh University Press.

Hultborn, H., & Nielsen, J. B. (2007). Spinal control of locomotion – from cat to man. *Acta Physiologica, 189*(2), 111–121. doi: 10.1111/j.1748-1716.2006.01651.x

Hume, D. (1748). *Philosophical essays concerning human understanding.* London, Britain: Millar.

Hume, D. (1789). *A treatise of human nature* (reprinted from the original ed. in 3 volumes ed.; L. A. Selby-Bigge, Ed.). Oxford, UK: The Clarendon Press.

Ijspeert, A. J. (2008). Central pattern generators for locomotion control in animals and robots: A review. *Neural Networks, 21*(4), 642–653. doi: 10.1016/j.neunet.2008.03.014

Ijspeert, A. J., Nakanishi, J., Hoffmann, H., Pastor, P., & Schaal, S. (2013). Dynamical movement primitives: Learning attractor models for motor behaviors. *Neural Computation, 25*(2), 328–373. doi: 10.1162/NECO_a_00393

Ijspeert, A. J., Nakanishi, J., & Schaal, S. (2002). Movement imitation with nonlinear dynamical systems in humanoid robots. *Proceedings of IEEE international conference on robotics and automation*, 1398–1403.

Itti, L., & Koch, C. (2001). Computational modeling of visual attention. *Nature Reviews Neuroscience, 2*, 194-203.

Jackendoff, R. (2002). *Foundations of language. brain, meaning, grammar, evolution.* Oxford University Press.

James, W. (1890). *The principles of psychology.* New York: Dover Publications.

James, W. (1981). *The principles of psychology* (Vol. 2). Cambridge, MA: Harvard University Press. (originally published: 1890)

Jellema, T., & Perrett, D. I. (2006). Neural representations of perceived bodily actions using a categorical frame of reference. *Neuropsychologia, 44*(9), 1535 - 1546. doi: 10.1016/j.neuropsychologia.2006.01.020

Johansson, G. (1973). Visual perception of biological motion and a model for its analysis. *Perception & Psychophysics, 14*, 201-211. doi: 10.3758/BF03212378

Johnson, C. P., & Blasco, P. A. (1997). Infant growth and development. *Pediatrics in Review, 18*, 224-242. doi: 10.1542/pir.18-7-224

Johnson, M. (1987). *The body in the mind: The bodily basis of meaning, imagination and reason.* Chicago: University of Chicago Press.

Kahneman, D., Treisman, A., & Gibbs, B. J. (1992). The reviewing of object files: Object-specific integration of information. *Cognitive Psychology, 24*(2), 175 - 219. doi: 10.1016/0010-0285(92)90007-O

Kanwisher, N. G. (1987). Repetition blindness: Type recognition without token individuation. *Cognition, 27*(2), 117–143. doi: 10.1016/0010-0277(87)90016-3

Kerri, J., & Shiffrar, M. (Eds.). (2013). *People watching: Social, perceptual, and neurophysiological studies of body perception.* Oxford University Press.

Kilner, J. M., Friston, K. J., & Frith, C. D. (2007). Predictive coding: an account of the mirror neuron system. *Cognitive Processing, 8*(3), 159-166. doi: 10.1007/s10339-007-0170-2

Kilner, J. M., & Lemon, R. N. (2013). What we know currently about mirror neurons. *Current Biology, 23*(23), R1057 - R1062. doi: 10.1016/j.cub.2013.10.051

Knauff, M. (2013). *Space to reason. a spatial theory of human thought.* Cambridge, MA: MIT Press.

Kneissler, J., & Butz, M. V. (2014). Learning spatial transformations using structured gain-field networks. *Artificial Neural Networks and Machine Learning–ICANN 2014*, 683–690.

Kneissler, J., Drugowitsch, J., Friston, K., & Butz, M. V. (2015). Simultaneous learning and filtering without delusions: a bayes-optimal combination of predictive inference and adaptive filtering. *Frontiers in Computational Neuroscience, 9*(47). doi: 10.3389/fncom.2015.00047

Kneissler, J., Stalph, P. O., Drugowitsch, J., & Butz, M. V. (2014). Filtering sensory information with xcsf: Improving learning robustness and robot arm control performance. *Evolutionary Computation, 22*, 139-158. doi: 10.1162/EVCO_a_00108

Knott, A. (2012). *Sensorimotor cognition and natural language syntax.* Cambridge, MA: MIT Press.

Kober, J., & Peters, J. (2011). Policy search for motor primitives in robotics. *Machine Learning, 84*(1-2), 171-203. doi: 10.1007/s10994-010-5223-6

Kohonen, T. (2001). *Self-organizing maps* (3rd ed.). Berlin Heidelberg: Springer-Verlag.

Konczak, J., Borutta, M., Topka, H., & Dichgans, J. (1995). The development of goal-directed reaching in infants: Hand trajectory formation and joint torque control. *Experimental Brain Research, 106*, 156-168.

Konidaris, G., Kaelbling, L., & Lozano-Perez, T. (2014). Constructing symbolic representations for high-level planning. *Proceedings of the Twenty-Eighth AAAI Conference on Artificial Intelligence*, 1932-1940.

Konidaris, G., Kaelbling, L. P., & Lozano-Perez, T. (2015). Symbol acquisition for probabilistic high-level planning. In *Proceedings of the twenty fourth international joint conference on artificial intelligence* (p. 3619-3627).

Koza, J. R. (1992). *Genetic programming: on the programming of computers by means of natural selection* (Vol. 1). MIT press.

Kraft, D., Pugeault, N., Baseski, E., Popovic, M., Kragic, D., Kalkan, S., ... Krüger, N. (2008). Birth of the object: Detection of objectness and extraction of object shape through object action complexes. *International Journal of Humanoid Robotics, 5*(2), 247–265.

Krizhevsky, A., Sutskever, I., & Hinton, G. E. (2012). Imagenet classification with deep convolutional neural networks. In F. Pereira, C. J. C. Burges, L. Bottou, & K. Q. Weinberger (Eds.), *Advances in neural information processing systems 25* (pp. 1097–1105). Red Hook, NY: Curran Associates, Inc.

Kuhn, T. (1962). *The structure of scientific revolutions.* Chicago: Chicago University Press.

Kunde, W. (2001). Response-effect compatibility in manual choice reaction tasks. *Journal of Experimental Psychology: Human Perception and Performance, 27*(2), 387-394.

Lachmair, M., Dudschig, C., De Filippis, M., de la Vega, I., & Kaup, B. (2011). Root versus roof: automatic activation of location information during word processing. *Psychonomic Bulletin & Review, 18*, 1180-1188.

Lachman, R., Lachman, J. L., & Butterfield, E. (1979). *Cognitive psychology and information processing: An introduction.* Lawrence Erlbaum Associates Hillsdale, NJ.

Lakoff, G. (1987). *Women, fire, and dangerous things: What categories reveal about the mind.* Chicago: University of Chicago Press.

Lakoff, G., & Johnson, M. (1980). *Metaphors we live by* (Vol. 1980). Chicago, IL: The Universty of Chicago Press.

Lakoff, G., & Johnson, M. (1999). *Philosophy in the flesh: The embodied mind and its challenge to western thought.* New York, NY: Basic Books.

Land, M. F., & Tatler, B. W. (2009). *Looking and acting vision and eye movements in natural behaviour.* Oxford University Press.

Latash, M. L. (2008). *Synergy.* Oxford University Press.

Layher, G., Giese, M. A., & Neumann, H. (2014). Learning representations of animated motion sequences—a neural model. *Topics in Cognitive Science, 6*(1), 170–182. doi: 10.1111/tops.12075

LeCun, Y., Bottou, L., Bengio, Y., & Haffner, P. (1998). Gradient-based learning applied to document recognition. *Proceedings of the IEEE, 86*(11), 2278-2324. doi: 10.1109/5.726791

Libertus, K., & Needham, A. (2010). Teach to reach: The effects of active vs. passive reaching experiences on action and perception. *Vision Research, 50*(24), 2750 - 2757. (Perception and Action: Part I) doi: 10.1016/j.visres.2010.09.001

Littman, M. L. (2015). Reinforcement learning improves behaviour from evaluative feedback. *Nature, 521*(7553), 445–451. doi: 10.1038/nature14540

Locke, J. (1690). *An essay concerning human understanding* (2014th ed.). Hertfordshire, GB: Wordsworth Editions Limited.

Lonini, L., Forestier, S., Teuliere, C., Zhao, Y., Shi, B. E., & Triesch, J. (2013). Robust active binocular vision through intrinsically motivated learning. *Frontiers in Neurorobotics, 7*(20). doi: 10.3389/fnbot.2013.00020

Loos, H. S., & Fritzke, B. (1998). *DemoGNG (version 1.5).* online. Retrieved from `http://www.sund.de/netze/applets/gng/full/tex/DemoGNG/DemoGNG.html` (retrieved 05/2013, newest version at http://www.demogng.de)

Lovelace, C. o. (1842). Translator's notes to an article on Babbage's analytical engine. In R. Taylor (Ed.), *Scientific memoirs* (Vol. 3, p. 691-731).

Luksch, T., Gienger, M., Mühlig, M., & Yoshiike, T. (2012). Adaptive movement sequences and predictive decisions based on hierarchical dynamical systems. *25th IEEE/RSL International Conference on Intelligent Robots and Systems (IROS)*, 2082-2088.

Ma, W. J., & Pouget, A. (2008). Linking neurons to behavior in multisensory perception: A computational review. *Brain Research, 1242*, 4-12.

Mack, A., & Rock, I. (1998). *Inattentional blindness.* Cambridge, MA: MIT Press.

Mahon, B. Z. (2015). Missed connections: A connectivity constrained account of the representation and organization of object concepts. In E. Margolis & S. Laurence (Eds.), *The conceptual mind: New directions in the study of concepts* (p. 79-116). Cambridge, MA: MIT Press.

Mahon, B. Z., Kumar, N., & Almeida, J. (2013). Spatial frequency tuning reveals interactions between the dorsal and ventral visual systems. *Journal of Cognitive Neuroscience, 25*(6), 862–871. doi: 10.1162/jocn_a_00370

Mandler, J. M. (2004). Thought before language. *Trends in Cognitive Sciences, 8*(11), 508 - 513. doi: 10.1016/j.tics.2004.09.004

Mandler, J. M. (2012). On the spatial foundations of the conceptual system and its enrichment. *Cognitive Science, 36*(3), 421–451. doi: 10.1111/j.1551-6709.2012.01241.x

Maravita, A., Spence, C., & Driver, J. (2003). Multisensory integration and the body schema: Close to hand and within reach. *Current Biology, 13*, 531-539.

Marr, D. (1982). *Vision: A computational investigation into the human representation and processing of visual information.* Cambridge, MA: MIT Press.

Martin, A. (2007). The representation of object concepts in the brain. *Annual Review of Psychology, 58*(1), 25-45. (PMID: 16968210) doi: 10.1146/annurev.psych.57.102904.190143

Martinetz, T. M., Berkovitsch, S. G., & Schulten, K. J. (1993). "Neural-gas" network for vector quantization and its application to time-series prediction. *IEEE Transactions on Neural Networks, 4*, 558-569.

Masson, M. E. J., Bub, D. N., & Breuer, A. T. (2011). Priming of reach and grasp actions by handled objects. *Journal of Experimental Psychology: Human Perception and Performance, 37*, 1470-1484. doi: 10.1037/a0023509

Maturana, H., & Varela, F. (1980). *Autopoiesis and cognition: The realization of the living.* Boston, MA: Reidel.

McCarthy, J., & Hayes, P. (1968). *Some philosophical problems from the standpoint of artificial intelligence.* Stanford University USA.

McCarthy, J., Minsky, M. L., Rochester, N., & Shannon, C. E. (2006). A proposal for the dartmouth summer research project on artificial intelligence, august 31, 1955. *AI Magazine, 27*(4), 12. doi: 10.1609/aimag.v27i4.1904

McCulloch, W. S., & Pitts, W. (1943). A logical calculus of the ideas immanent in nervous activity. *The bulletin of mathematical biophysics, 5*(4), 115–133.

McGurk, H., & MacDonald, J. (1976). Hearing lips and seeing voices. *Nature, 264*(5588), 746–748. doi: 10.1038/264746a0

Meltzoff, A. N. (1988, Oct). Imitation of televised models by infants. *Child Development, 59*(5), 1221–1229.

Meltzoff, A. N. (1995). Understanding the intentions of others: Re-enactment of intended acts by 18-month-old children. *Developmental Psychology, 31*(5), 838–850. doi: 10 .1037/0012-1649.31.5.838

Michotte, A. (1946). *The perception of causality.* Louvain, Ed. de l'Institut Supérieur de Philosophie. (translated into English by C. A. Mace, 1962)

Milner, A. D., & Goodale, M. A. (1995). *The visual brain in action.* New York, NY: Oxford University Press.

Milner, A. D., & Goodale, M. A. (2008). Two visual systems re-viewed. *Neuropsychologia, 46*(3), 774 - 785. doi: 10.1016/j.neuropsychologia.2007.10.005

Minsky, M. (1988). *The society of mind.* New York, NY: Simon and Schuster.

Minsky, M., & Papert, S. (1969). *Perceptrons: An introduction to computational geometry.* Cambridge, Mass.: MIT Press.

Mishkin, M., Ungerleider, L. G., & Macko, K. A. (1983). Object vision and spatial vision: Two cortical pathways. *Trends in Neurosciences, 6*, 414 - 417. doi: 10.1016/0166 -2236(83)90190-X

Moll, H., & Meltzoff, A. N. (2011). Perspective-taking and its foundation in joint attention. In *Perception, causation, and objectivity* (p. 286-304).

Moser, E. I., Kropff, E., & Moser, M.-B. (2008). Place cells, grid cells, and the brain's spatial representation system. *Annual Review of Neuroscience, 31*(1), 69-89. doi: 10.1146/annurev.neuro.31.061307.090723

Muelling, K., Kober, J., & Peters, J. (2010). Learning table tennis with a mixture of motor primitives. In *Humanoid robots (humanoids), 2010 10th ieee-ras international conference on* (p. 411-416). doi: 10.1109/ICHR.2010.5686298

Munnich, E., & Landau, B. (2003). The effects of spatial language on spatial representation: setting some boundaries. In D. Gentner & S. Goldin-Meadow (Eds.), *Language in mind: Advances in the study of language and cognition.* (p. 113-155). Cambridge, MA: MIT Press.

Neisser, U., & Becklen, R. (1975). Selective looking: Attending to visually specified events. *Cognitive Psychology, 7*(4), 480–494. doi: 10.1016/0010-0285(75)90019-5

Newell, A., Shaw, J. C., & Simon, H. A. (1959). Report on a general problem-solving program. In *IFIP congress* (p. 256-264). Santa Monica, CA: The Rand Corporation.

Newell, A., & Simon, H. A. (1961). GPS, a program that simulates human thought. In H. Billing (Ed.), *Lernende automaten* (p. 109-124). München: Oldenbourg.

Newtson, D. (1973). Attribution and the unit of perception of ongoing behavior. *Journal of Personality and Social Psychology, 28*(1), 28–38. doi: 10.1037/h0035584

Nichols, M. J., & Newsome, W. T. (1999). The neurobiology of cognition. *Nature, 402*, C35-C38.

Niemi, P., & Näätänen, R. (1981). Foreperiod and simple reaction time. *Psychological Bulletin*, *89*(1), 133–162. doi: 10.1037/0033-2909.89.1.133

Noland, E., & Hirsch, S. A. (Eds.). (1902). *The greek grammar of Roger Bacon.* Cambridge: Cambridge University Press.

Norman, D. A. (1968). Toward a theory of memory and attention. *Psychological Review*, *75*(6), 522–536. doi: 10.1037/h0026699

Nowak, M., & Sigmund, K. (1993). A strategy of win-stay, lose-shift that outperforms tit-for-tat in the prisoner's dilemma game. *Nature*, *364*(6432), 56–58. doi: 10.1038/364056a0

Ogden, C. K., & Richards, I. A. (1923). *The meaning of meaning: a study of the influence of language upon thought and of the science of symbolism* (8. ed., 1956 ed.). London: Routledge and Kegan Paul LTD.

Ognibene, D., Rega, A., & Baldassarre, G. (2006). A model of reaching integrating continuous reinforcement learning, accumulator models, and direct inverse modeling. *From Animals to Animats*, *9*, 381-393.

O'Regan, J. K., & Noë, A. (2001). A sensorimotor account of vision and visual consciousness. *Behavioral and Brain Sciences*, *24*(5), 883–917.

Oudeyer, P.-Y., Kaplan, F., & Hafner, V. V. (2007). Intrinsic motivation systems for autonomous mental development. *IEEE Transactions on Evolutionary Computation*, *11*, 265-286. doi: 10.1109/TEVC.2006.890271

Pashler, H. E. (1998). *The psychology of attention.* Cambridge, MA: MIT Press.

Pastra, K., & Aloimonos, Y. (2012). The minimalist grammar of action. *Philosophical Transactions of the Royal Society B: Biological Sciences*, *367*, 103-117. doi: 10.1098/rstb.2011.0123

Patel, G. H., Kaplan, D. M., & Snyder, L. H. (2014). Topographic organization in the brain: searching for general principles. *Trends in Cognitive Sciences*, *18*(7), 351 - 363. doi: 10.1016/j.tics.2014.03.008

Pavlov, I. (1904). The nobel prize in physiology or medicine 1904. In *Nobel prizes and laureates.* Nobelprize.org. Retrieved from `http://www.nobelprize.org/nobel_prizes/medicine/laureates/1904/` (retrieved 2015.12.29)

Pavlova, M. A. (2012). Biological motion processing as a hallmark of social cognition. *Cerebral Cortex*, *22*(5), 981-995. doi: 10.1093/cercor/bhr156

Pearl, J. (2000). *Causality. models, reasoning, and inference.* New York: Cambridge University Press.

Peirce, C. S. (1934). *Collected papers of Charles Sanders Peirce: Volume V. pragmatism and pragmaticism* (C. Hartshorne & P. Weiss, Eds.). Cambridge, MA: Harvard University Press.

Peirce, C. S. (1960). *Collected papers of Charles Sanders Peirce: Principles of philosophy and elements of logic* (C. Hartshorne, P. Weiss, & A. W. Burks, Eds.). Cambridge, MA: Harvard University Press.

Pelikan, M. (2005). *Hierarchical Bayesian optimization algorithm: Toward a new generation of evolutionary algorithms.* Springer-Verlag.

Perani, D., Dehaene, S., Grassi, F., Cohen, L., Cappa, S. F., Dupoux, E., ... Mehler, J. (1996). Brain processing of native and foreign languages. *NeuroReport*, *7*(15-17), 2439-2444.

Perrett, D. I., Smith, P. A. J., Mistlin, A. J., Chitty, A. J., Head, A. S., Potter, D. D., ... Jeeves, M. A. (1985). Visual analysis of body movements by neurones in the temporal cortex of the macaque monkey: A preliminary report. *Behavioural Brain Research*, *16*, 153–170. doi: 10.1016/0166-4328(85)90089-0

Peters, J., & Schaal, S. (2008). Reinforcement learning of motor skills with policy gradients. *Neural Networks*, *21*, 682-697.

Petrides, M., Cadoret, G., & Mackey, S. (2005). Orofacial somatomotor responses in the macaque monkey homologue of broca's area. *Nature*, *435*(7046), 1235–1238. doi: 10.1038/nature03628

Pezzulo, G., Donnarumma, F., & Dindo, H. (2013). Human sensorimotor communication: A theory of signaling in online social interactions. *PLoS ONE*, *8*(11), e79876–. doi: 10.1371/journal.pone.0079876

Pfeifer, R., & Bongard, J. C. (2006). *How the body shapes the way we think: A new view of intelligence*. Cambridge, MA: MIT Press.

Plato, & Jowett, B. T. (1901). *Dialogues of plato: With analyses and introductions*. Charles Scribner's Sons. (Vol 1.)

Poggio, T., & Bizzi, E. (2004). Generalization in vision and motor control. *Nature*, *431*, 768-774.

Pouget, A., Dayan, P., & Zemel, R. S. (2003). Inference and computation with population codes. *Annual Review of Neuroscience*, *26*, 381-410.

Pouget, A., & Snyder, L. H. (2000). Computational approaches to sensorimotor transformations. *Nature Neuroscience*, *3*, 1192-1198.

Prinz, W. (1990). A common coding approach to perception and action. In O. Neumann & W. Prinz (Eds.), *Relationships between perception and action* (p. 167-201). Berlin Heidelberg: Springer-Verlag.

Prinz, W. (1997). Perception and action planning. *European Journal of Cognitive Psychology*, *9*, 129-154.

Pullum, G. K., & Gazdar, G. (1982). Natural languages and context-free languages. *Linguistics and Philosophy*, *4*(4), 471–504. doi: 10.1007/BF00360802

Purves, D., Augustine, G. J., Fitzpatrick, D., Hall, W. C., LaMantia, A. S., McNamara, J. O., & & Williams, S. M. (Eds.). (2004). *Neuroscience*. Sunderland, MA: Sinauer Associates, Inc.

Pylyshyn, Z. W. (2009). Perception, representation, and the world: The FINST that binds. In L. Dedrick DonTrick (Ed.), *Computation, cognition, and Pylyshyn* (Vol. xvii, p. 3-48). Cambridge, MA, US: MIT Press.

Quiroga, R. Q., Reddy, L., Kreiman, G., Koch, C., & Fried, I. (2005). Invariant visual representation by single neurons in the human brain. *Nature*, *435*(7045), 1102–1107. doi: 10.1038/nature03687

Rabiner, L. R. (1990). A tutorial on hidden Markov models and selected applications in speech recognition. In A. Waibel & K.-F. Lee (Eds.), *Readings in speech recognition* (p. 267 - 296). San Mateo, CA: Morgan Kaufmann Publishers Inc.

Ramachandran, V. S., & Blakeslee, S. (1998). *Phantoms in the brain: Probing the mysteries of the human mind*. New York, NY: HarperCollins Publishers Inc.

Rao, R. P. N., & Ballard, D. H. (1998). Development of localized oriented receptive fields by learning a translation-invariant code for natural images. *Computational Neural Syststems*, *9*, 219-234.

Rao, R. P. N., & Ballard, D. H. (1999). Predictive coding in the visual cortex: A functional interpretation of some extra-classical receptive-field effects. *Nature Neuroscience*, *2*(1), 79-87.

Raymond, J. E., Shapiro, K. L., & Arnell, K. M. (1992). Temporary suppression of visual processing in an rsvp task: An attentional blink? *Journal of Experimental Psychology: Human Perception and Performance*, *18*, 849–860. doi: 10.1037/0096-1523.18.3.849

Rechenberg, I. (1973). *Evolutionsstrategie Optimierung technischer Systeme nach Prinzipien der biologischen Evolution*. Stuttgart-Bad Cannstatt: Friedrich Frommann Verlag.

Reeve, C. D. C. (Ed.). (2004). *Plato republic*. Indianapolis, IN: Hackett Publishing Company.

Rensink, R. A. (2002). Change detection. *Annual Review of Psychology*, *53*(1), 245-277. (PMID: 11752486) doi: 10.1146/annurev.psych.53.100901.135125

Rensink, R. A., O'Regan, J. K., & Clark, J. (1997). To see or not to see: the need for attention to perceive changes in scenes. *Psychological Science*, *8(5)*, 368–373.

Rizzolatti, G., & Craighero, L. (2004). The mirror-neuron system. *Annual Review of Neuroscience*, *27*, 169-192.

Rizzolatti, G., Fadiga, L., Gallese, V., & Fogassi, L. (1996). Premotor cortex and the recognition of motor actions. *Cognitive Brain Research, 3*, 131-141.

Rizzolatti, G., Riggio, L., Dascola, I., & Umiltá, C. (1987). Reorienting attention across the horizontal and vertical meridians: Evidence in favor of a premotor theory of attention. *Neuropsychologia, 25*(1, Part 1), 31–40. doi: 10.1016/0028-3932(87)90041-8

Rochat, P. (2010). The innate sense of the body develops to become a public affair by 2-3 years. *Neuropsychologia, 48*, 738 - 745. doi: 10.1016/j.neuropsychologia.2009.11.021

Rochat, P., & Striano, T. (2000). Perceived self in infancy. *Infant Behavior and Development, 23*(3-4), 513 - 530. doi: 10.1016/S0163-6383(01)00055-8

Rohde, M., Di Luca, M., & Ernst, M. O. (2011). The rubber hand illusion: Feeling of ownership and proprioceptive drift do not go hand in hand. *PloS one, 6*(6), e21659.

Rolke, B., & Hofmann, P. (2007). Temporal uncertainty degrades perceptual processing. *Psychonomic Bulletin & Review, 14*(3), 522-526. doi: 10.3758/BF03194101

Rolls, E. T., Stringer, S. M., & Elliot, T. (2006). Entorhinal cortex grid cells can map to hippocampal place cells by competitive learning. *Network: Computation in Neural Systems, 17*(4), 447–465. doi: 10.1080/09548980601064846

Rosenbaum, D. A. (2010). *Human motor control* (2nd ed.). San Diego: Academic Press/Elsevier.

Rosenbaum, D. A., Slotta, J. D., Vaughan, J., & Plamondon, R. (1991). Optimal movement selection. *Psychological Science, 2*, 86–91.

Rovee-Collier, C. (1997). Dissociations in infant memory: Rethinking the development of implicit and explicit memory. *Psychological Review, 104*(3), 467.

Rowland, D., & Moser, M.-B. (2013). Time finds its place in the hippocampus. *Neuron, 78*(6), 953–954. doi: 10.1016/j.neuron.2013.05.039

Roy, D. (2005a). Grounding words in perception and action: computational insights. *Trends Cogn Sci, 9*(8), 389–396. doi: 10.1016/j.tics.2005.06.013

Roy, D. (2005b). Semiotic schemas: a framework for grounding language in action and perception. *Artificial Intelligence, 167*(1-2), 170–205. doi: 10.1016/j.artint.2005.04.007

Rubin, D. C. (2006). The basic-systems model of episodic memory. *Perspectives on Psychological Science, 1*(4), 277–311.

Rueschemeyer, S.-A., Lindemann, O., van Rooij, D., van Dam, W., & Bekkering, H. (2010). Effects of intentional motor actions on embodied language processing. *Experimental Psychology, 57*(4), 260 - 266. doi: 10.1027/1618-3169/a000031

Rumelhart, D. E., Hinton, G. E., & Williams, R. J. (1988). Learning representations by back-propagating errors. In T. A. Polk & C. M. Seifert (Eds.), *Cognitive modeling* (p. 213-220). Cambridge, MA: MIT Press.

Rumelhart, D. E., McClelland, J. L., & the PDP Research Group. (1986). *Parallel distributed processing: Explorations in the microstructure of cognition, volumes 1 and 2.* Cambridge: MIT Press.

Sachs, J., Bard, B., & Johnson, M. L. (1981). Language learning with restricted input: Case studies of two hearing children of deaf parents. *Applied Psycholinguistics, 2*(01), 33–54.

Salinas, E., & Sejnowski, T. J. (2001). Correlated neuronal activity and the flow of neural information. *Nature Reviews Neuroscience, 2*, 539-550.

Salzman, C. D., Britten, K. H., & Newsome, W. T. (1990). Cortical microstimulation influences perceptual judgements of motion direction. *Nature, 346*(6280), 174–177. doi: 10.1038/346174a0

Samuelson, L. K., Smith, L. B., Perry, L. K., & Spencer, J. P. (2011). Grounding word learning in space. *PLoS ONE, 6*(12), e28095. doi: 10.1371/journal.pone.0028095

Sandamirskaya, Y., Zibner, S. K., Schneegans, S., & Schöner, G. (2013). Using dynamic field theory to extend the embodiment stance toward higher cognition. *New Ideas in Psychology, 31*(3), 322 - 339. doi: 10.1016/j.newideapsych.2013.01.002

Schaal, S., Ijspeert, A., & Billard, A. (2003). Computational approaches to motor learning by imitation. *Philosophical Transaction of the Royal Society of London: Series B, Biological Sciences, 358*, 537–547.

Schleiermacher, F. (1985). *Platons werke* (Neuausgabe der zweiten verbesserten Auflage (Berlin 1817-26) bzw. der ersten Auflage des dritten Theils (Berlin 1828) ed.). Akademie Verlag Berlin.

Schmidhuber, J. (1991). Curious model-building control systems. *Proc. International Joint Conference on Neural Networks, 2*, 1458-1463.

Schmidt, R. A. (1975). A schema theory of discrete motor skill-learning. *Psychological Review, 82*(4), 225-260.

Schrödinger, E. (1944). *What is life?* (Canto edition, 1992 ed.). Cambridge, UK: Cambridge University Press.

Schrodt, F., & Butz, M. V. (2015). Learning conditional mappings between population-coded modalities. In *Machine learning reports 03/2015* (p. 141-148). (ISSN:1865-3960 http://www.techfak.uni-bielefeld.de/fschleif/mlr/mlr_03_2015.pdf)

Schrodt, F., Layher, G., Neumann, H., & Butz, M. V. (2015). Embodied learning of a generative neural model for biological motion perception and inference. *Frontiers in Computational Neuroscience, 9*(79). doi: 10.3389/fncom.2015.00079

Schwann, T. (1839). *Mikroskopische Untersuchungen über die Übereinstimmung in der Struktur und dem Wachsthum der Thiere und Pflanzen [Microscopic researches into the accordance in the structure and growth of animals and plants].* Berlin: Sander.

Searle, J. R. (1969). *Speech acts.* Cambridge: Cambridge University Press.

Searle, J. R. (1976). A classification of illocutionary acts. *Language in Society, 5*, 1–23. doi: 10.1017/S0047404500006837

Searle, J. R. (1980). Minds, brains, and programs. *Behavioral and brain sciences, 3*(03), 417–424.

Sebanz, N., Bekkering, H., & Knoblich, G. (2006). Joint action: Bodies and minds moving together. *Trends in cognitive sciences, 10*, 70-76.

Segal, S. J., & Fusella, V. (1970). Influence of imaged pictures and sounds on detection of visual and auditory signals. *Journal of Experimental Psychology, 83*(3, Pt.1), 458–464. doi: 10.1037/h0028840

Senju, A., & Csibra, G. (2008). Gaze following in human infants depends on communicative signals. *Current Biology, 18*(9), 668–671. doi: 10.1016/j.cub.2008.03.059

Serre, T., Wolf, L., Bileschi, S., Riesenhuber, M., & Poggio, T. (2007). Robust object recognition with cortex-like mechanisms. *IEEE Transactions on Pattern Analysis and Machine Intelligence, 29*, 411-426.

Shadmehr, R., & Krakauer, J. W. (2008). A computational neuroanatomy for motor control. *Experimental Brain Research, 185*(3), 359–381.

Shannon, C. E. (1950). Xxii. programming a computer for playing chess. *Philosophical magazine, 41*(314), 256–275.

Shapiro, K. L., Raymond, J. E., & Arnell, K. M. (1994). Attention to visual pattern information produces the attentional blink in rapid serial visual presentation. *Journal of Experimental Psychology: Human Perception and Performance, 20*(2), 357–371. doi: 10.1037/0096-1523.20.2.357

Shieber, S. M. (1987). Evidence against the context-freeness of natural language. In W. Savitch, E. Bach, W. Marsh, & G. Safran-Naveh (Eds.), *The formal complexity of natural language* (Vol. 33, p. 320-334). Dordrecht, The Netherlands: D. Reidel Publishing Comp. doi: 10.1007/978-94-009-3401-6_12

Sigaud, O., Butz, M. V., Kozlova, O., & Meyer, C. (2009). Anticipatory learning classifier systems and factored reinforcement learning. In G. Pezzulo, M. V. Butz, O. Sigaud, & G. Baldassarre (Eds.), *Anticipatory behavior in adaptive learning systems: From psychological theories to artificial cognitive systems* (p. 321-333). Berlin, Heidelberg: Springer-Verlag. doi: 10.1007/978-3-642-02565-5_18

Silver, D., Huang, A., Maddison, C. J., Guez, A., Sifre, L., van den Driessche, G., ...

Hassabis, D. (2016). Mastering the game of Go with deep neural networks and tree search. *Nature*, *529*(7587), 484–489. doi: 10.1038/nature16961

Simon, H. A. (1969). *Sciences of the artificial.* Cambridge, MA: MIT Press.

Simon, T. J., Hespos, S. J., & Rochat, P. (1995). Do infants understand simple arithmetic? a replication of wynn (1992). *Cognitive Development*, *10*(2), 253–269. doi: 10.1016/0885-2014(95)90011-X

Simons, D. J., & Chabris, C. F. (1999). Gorillas in our midst: Sustained inattentional blindness for dynamic events. *Perception*, *28*, 1059-1074.

Smith, L., & Gasser, M. (2005). The development of embodied cognition: Six lessons from babies. *Artificial Life*, *11*(1-2), 13–29. doi: 10.1162/1064546053278973

Smolensky, P. (1986). Information processing in dynamical systems: Foundations of harmony theory. In D. E. Rumelhart & J. L. McClelland (Eds.), *Parallel distributed processing* (Vol. 1, p. 194-281). Cambridge: MIT Press.

Sommerville, J. A., & Woodward, A. L. (2005). Pulling out the intentional structure of action: the relation between action processing and action production in infancy. *Cognition*, *95*, 1 - 30. doi: 10.1016/j.cognition.2003.12.004

Soto-Faraco, S., Ronald, A., & Spence, C. (2004). Tactile selective attention and body posture: assessing the multisensory contributions of vision and proprioception. *Percept Psychophys*, *66*(7), 1077-1094.

Spearman, C. (1904). "general intelligence," objectively determined and measured. *The American Journal of Psychology*, *15*(2), 201–292.

Stalnaker, R. C. (1978). Assertion. In P. Cole (Ed.), *Pragmatics* (p. 315-332). New York: Academic Press.

Stalph, P., Rubinsztajn, J., Sigaud, O., & Butz, M. V. (2012). Function approximation with lwpr and xcsf: A comparative study. *Evolutionary Intelligence*, *5*, 103-116. doi: 10.1007/s12065-012-0082-7

Steels, L. (2008). The symbol grounding problem has been solved. So what's next? In M. de Vega, A. M. Glenberg, & A. C. Graesser (Eds.), *Symbols and embodiment: Debates on meaning and cognition* (p. 223-244). New Haven: Academic Press.

Steen, G. J., Reijnierse, W. G., & Burgers, C. (2014). When do natural language metaphors influence reasoning? a follow-up study to thibodeau and boroditsky (2013). *PLoS ONE*, *9*(12), e113536. doi: 10.1371/journal.pone.0113536

Storn, R., & Price, K. (1997). Differential evolution–a simple and efficient heuristic for global optimization over continuous spaces. *Journal of global optimization*, *11*(4), 341–359.

Strack, F., Martin, L. L., & Stepper, S. (1988). Inhibiting and facilitating conditions of the human smile: A nonobtrusive test of the facial feedback hypothesis. *Journal of Personality and Social Psychology*, *54*, 768-777.

Stulp, F., & Sigaud, O. (2013). Robot skill learning: From reinforcement learning to evolution strategies. *Paladyn, Journal of Behavioral Robotics*, *4*, 49-61. doi: 10.2478/pjbr-2013-0003

Stulp, F., & Sigaud, O. (2015). Many regression algorithms, one unified model: A review. *Neural Networks*, *69*, 60 - 79. doi: 10.1016/j.neunet.2015.05.005

Suchan, J., & Karnath, H.-O. (2011). Spatial orienting by left hemisphere language areas: a relict from the past? *Brain*, *134*(10), 3059–3070. doi: 10.1093/brain/awr120

Sugita, Y., & Butz, M. V. (2011). Compositionality and embodiment in harmony. In P.-Y. Oudeyer (Ed.), *Amd newsletter* (Vol. 8, p. 8-9). IEEE CIS.

Sugita, Y., Tani, J., & Butz, M. V. (2011). Simultaneously emerging braitenberg codes and compositionality. *Adaptive Behavior*, *19*, 295-316. doi: 10.1177/1059712311416871

Sutton, R. S., & Barto, A. G. (1998). *Reinforcement learning: An introduction.* Cambridge, MA: MIT Press.

Sutton, R. S., Precup, D., & Singh, S. (1999). Between MDPs and semi-MDPs: A framework for temporal abstraction in reinforcement learning. *Artificial Intelligence*, *112*, 181-211.

Taube, J. S. (2007). The head direction signal: Origins and sensory-motor integration. *Annual Review of Neuroscience, 30,* 181-207.

Thibodeau, P. H., & Boroditsky, L. (2013). Natural language metaphors covertly influence reasoning. *PLoS ONE, 8*(1), e52961. doi: 10.1371/journal.pone.0052961

Thorpe, S. J., & Fabre-Thorpe, M. (2001). Seeking categories in the brain. *Science, 291*(5502), 260–263. doi: 10.1126/science.1058249

Thurman, S. M., & Grossman, E. D. (2008). Temporal "bubbles" reveal key features for point-light biological motion perception. *Journal of Vision, 8*(3), 28.

Todorov, E. (2004). Optimality principles in sensorimotor control. *Nature Reviews Neuroscience, 7*(9), 907-915.

Todorov, E., & Jordan, M. I. (2002). Optimal feedback control as a theory of motor coordination. *Nature Neuroscience, 5*(11), 1226-1235.

Tomasello, M. (2014). *A natural history of human thinking.* Harvard University Press.

Tomasello, M., Carpenter, M., Call, J., Behne, T., & Moll, H. (2005). Understanding and sharing intentions: The origins of cultural cognition. *Behavioral and Brain Sciences, 28,* 675–691. doi: 10.1017/S0140525X05000129

Tononi, G., & Koch, C. (2015). Consciousness: here, there and everywhere? *Philosophical Transactions of the Royal Society of London B: Biological Sciences, 370*(1668). doi: 10.1098/rstb.2014.0167

Treisman, A. M. (1964). Verbal cues, language, and meaning in selective attention. *The American Journal of Psychology, 77*(2), 206–219.

Treisman, A. M., & Gelade, G. (1980). A feature-integration theory of attention. *Cognitive Psychology, 12*(1), 97–136. doi: 10.1016/0010-0285(80)90005-5

Trommershäuser, J., Maloney, L. T., & Landy, M. S. (2003a). Statistical decision theory and the selection of rapid, goal-directed movements. *Journal of the Optical Society of America A, 20,* 1419-1433.

Trommershäuser, J., Maloney, L. T., & Landy, M. S. (2003b). Statistical decision theory and trade-offs in the control of motor response. *Spatial Vision, 16,* 255-275.

Turella, L., Wurm, M. F., Tucciarelli, R., & Lingnau, A. (2013). Expertise in action observation: recent neuroimaging findings and future perspectives. *Frontiers in Human Neuroscience, 7*(637). doi: 10.3389/fnhum.2013.00637

Turing, A. M. (1950). Computing machinery and intelligence. *Mind, 59,* 433-460.

Umiltà, M. A., Kohler, E., Gallese, V., Fogassi, L., Fadiga, L., Keysers, C., & Rizzolatti, G. (2001). I know what you are doing: A neurophysiological study. *Neuron, 31,* 155-165.

Ungerleider, L. G., & Haxby, J. V. (1994). "what" and "where" in the human brain. *Current Opinion in Neurobiology, 4(2),* 157–65.

van Dijck, J.-P., & Fias, W. (2011). A working memory account for spatial-numerical associations. *Cognition, 119*(1), 114–119. doi: 10.1016/j.cognition.2010.12.013

Vanrie, J., Dekeyser, M., & Verfaillie, K. (2004). Bistability and biasing effects in the perception of ambiguous point-light walkers. *Perception, 33,* 547–560.

Vigorito, C. M., & Barto, A. G. (2010). Intrinsically motivated hierarchical skill learning in structured environments. *Autonomous Mental Development, IEEE Transactions on, 2*(2), 132 -143. doi: 10.1109/TAMD.2010.2050205

von Hofsten, C. (2003). On the development of perception and action. In J. Valsiner & K. J. Connolly (Eds.), *Handbook of developmental psychology* (p. 114-140). London: Sage.

von Hofsten, C. (2004). An action perspective on motor development. *Trends in Cognitive Science, 8,* 266-272.

von Holst, E., & Mittelstaedt, H. (1950). Das Reafferenzprinzip (Wechselwirkungen zwischen Zentralnervensystem und Peripherie.). *Naturwissenschaften, 37,* 464-476.

von Humboldt, W. (1973). *Werke in 5 Bänden. Bd 3. Schriften zur Sprachphilosophie.* Darmstadt: JGCotta.

Vygotsky, L. (1978). Interaction between learning and development. In *Readings on the development of children* (p. 34-41). Cambridge, MA: Harvard University Press.

Wagner, P., Malisz, Z., & Kopp, S. (2014). Gesture and speech in interaction: An overview. *Speech Communication, 57*, 209–232. doi: 10.1016/j.specom.2013.09.008

Watkins, C. J. C. H. (1989). *Learning from delayed rewards* (Unpublished doctoral dissertation). King's College, Cambridge, UK.

Watson, J. B. (1930). *Behaviorism.* New York, NY: WW Norton & Co.

Welford, A. T. (1952). The 'psychological refractory period' and the timing of high-speed performance—a review and a theory. *British Journal of Psychology. General Section, 43*(1), 2–19. doi: 10.1111/j.2044-8295.1952.tb00322.x

Werbos, P. J. (1974). Beyond regression: New tools for prediction and analysis in the behavioral sciences. *Harvard University, Cambridge, USA.*

Whitney, D. E. (1969). Resolved motion rate control of manipulators and human prostheses. *IEEE Transactions on Man-Machine Systems, 10*, 47-53.

Wiener, J. M., Büchner, S. J., & Hölscher, C. (2009). Taxonomy of human wayfinding tasks: A knowledge-based approach. *Spatial Cognition & Computation, 9*(2), 152–165. doi: 10.1080/13875860902906496

Wierstra, D., Schaul, T., Glasmachers, T., Sun, Y., Peters, J., & Schmidhuber, J. (2014). Natural evolution strategies. *Journal of Machine Learning Research, 15*(1), 949–980.

Wilimzig, C., Schneider, S., & Schöner, G. (2006). The time course of saccadic decision making: Dynamic field theory. *Neural Networks, 19*(8), 1059–1074. doi: 10.1016/j.neunet.2006.03.003

Wilson, S. W. (1995). Classifier fitness based on accuracy. *Evolutionary Computation, 3*(2), 149-175.

Witt, J. K., Proffitt, D. R., & Epstein, W. (2005). Tool use affects perceived distance, but only when you intend to use it. *Journal of Experimental Psychology: Human Perception and Performance, 31*, 880–888.

Wolpert, D. M., & Kawato, M. (1998). Multiple paired forward and inverse models for motor control. *Neural Networks, 11*, 1317–1329. doi: 10.1016/S0893-6080(98)00066-5

Wolpert, D. M., Miall, R. C., & Kawato, M. (1998). Internal models in the cerebellum. *Trends in Cognitive Science, 2*, 338-347. doi: 10.1016/S1364-6613(98)01221-2

Wood, G., Willmes, K., Nuerk, H.-C., & Fischer, M. H. (2008). On the cognitive link between space and number: A meta-analysis of the SNARC effect. *Psychology Science, 50*(4), 489–525.

Woodward, A. L. (1998). Infants selectively encode the goal object of an actor's reach. *Cognition, 69*(1), 1 - 34. doi: 10.1016/S0010-0277(98)00058-4

Wörgötter, F., Aksoy, E. E., Krüger, N., Piater, J., Ude, A., & Tamosiunaite, M. (2013). A simple ontology of manipulation actions based on hand-object relations. *Autonomous Mental Development, IEEE Transactions on, 5*(2), 117-134. doi: 10.1109/TAMD.2012 .2232291

Wynn, K. (1992). Addition and subtraction by human infants. *Nature, 358*(6389), 749–750. doi: 10.1038/358749a0

Yi, W., & Ballard, D. (2009). Recognizing behavior in hand-eye coordination patterns. *International Journal of Humanoid Robotics, 6*(3), 337–359. doi: 10.1142/ S0219843609001863

Zacks, J. M., Kumar, S., Abrams, R. A., & Mehta, R. (2009). Using movement and intentions to understand human activity. *Cognition, 112*(2), 201–216. doi: 10.1016/ j.cognition.2009.03.007

Zacks, J. M., Speer, N. K., Swallow, K. M., Braver, T. S., & Reynolds, J. R. (2007). Event perception: A mind-brain perspective. *Psychological Bulletin, 133*(2), 273–293. doi: 10.1037/0033-2909.133.2.273

Zacks, J. M., & Tversky, B. (2001). Event structure in perception and conception. *Psychological Bulletin, 127*(1), 3–21. doi: 10.1037/0033-2909.127.1.3

Index

The manufacturer's authorised representative in the EU for product
safety is Oxford University Press España S.A. of El Parque Empresarial
San Fernando de Henares, Avenida de Castilla, 2 - 28830 Madrid
(www.oup.es/en or product.safety@oup.com). OUP España S.A. also acts
as importer into Spain of products made by the manufacturer.
Printed and bound by CPI Group (UK) Ltd, Croydon, CR0 4YY

06/07/2026
02157621-0001